# Textile Science and Clothing Technology

**Series Editor**

Subramanian Senthilkannan Muthu, Chief Sustainability Officer
Green Story, Canada
Tung Chung, Lantau, Hong Kong

This series aims to broadly cover all the aspects related to textiles science and technology and clothing science and technology. Below are the areas fall under the aims and scope of this series, but not limited to: Production and properties of various natural and synthetic fibres; Production and properties of different yarns, fabrics and apparels; Manufacturing aspects of textiles and clothing; Modelling and Simulation aspects related to textiles and clothing; Production and properties of Nonwovens; Evaluation/testing of various properties of textiles and clothing products; Supply chain management of textiles and clothing; Aspects related to Clothing Science such as comfort; Functional aspects and evaluation of textiles; Textile biomaterials and bioengineering; Nano, micro, smart, sport and intelligent textiles; Various aspects of industrial and technical applications of textiles and clothing; Apparel manufacturing and engineering; New developments and applications pertaining to textiles and clothing materials and their manufacturing methods; Textile design aspects; Sustainable fashion and textiles; Green Textiles and Eco-Fashion; Sustainability aspects of textiles and clothing; Environmental assessments of textiles and clothing supply chain; Green Composites; Sustainable Luxury and Sustainable Consumption; Waste Management in Textiles; Sustainability Standards and Green labels; Social and Economic Sustainability of Textiles and Clothing.

Dhiraj Kumar · Mohammad Shahid
Editors

# Science and Technology of Silk

## Volume I: Silk Production

Springer

*Editors*
Dhiraj Kumar
Department of Zoology
Centre for Skill Development
and Seri-Technology
Hansraj College
University of Delhi
New Delhi, Delhi, India

Mohammad Shahid
Department of Chemistry
Koneru Lakshmaiah Education Foundation
(Deemed to be University)
Guntur, Andhra Pradesh, India

ISSN 2197-9863 ISSN 2197-9871 (electronic)
Textile Science and Clothing Technology
ISBN 978-3-032-16229-8 ISBN 978-3-032-16230-4 (eBook)
https://doi.org/10.1007/978-3-032-16230-4



This Springer imprint is published by the registered company Springer Nature Switzerland AG
The registered company address is: Gewerbestrasse 11, 6330 Cham, Switzerland

# Preface

Silk, long celebrated as the "queen of fabrics" or the "thread of gold," has captivated humans for millennia with its elegance, strength, and versatility. Its lustrous appearance, smooth texture, and comfort have made it a symbol of luxury, cultural identity, and innovation. *Science and Technology of Silk—Volume I: Silk Production* provides a comprehensive overview of silk, covering its origin, production, properties, cultural significance, and the latest scientific and technological advancements in this field.

Chapter 1, "Thread of Gold—The Silk: An Introduction to Silk and Sericulture," introduces the fundamental aspects of silk, including its classification, history, and the sericulture process—the cultivation of silkworms for silk production. The chapter highlights the most sought-after silk varieties and outlines the steps involved in producing silk yarn and fabric. Chapter 2, "Indian Silk: Origin, Properties, and Notable Handicrafts", focuses on Indian silk, describing its unique physical properties, cultural significance, and role in traditional handicrafts, while also exploring the connection between science, technology, and the preservation of India's rich silk heritage.

Chapter 3, "Advancements in Mulberry Silk Improvements-Genetic Improvements and Beyond", discusses recent genetic and biotechnological advancements in mulberry silk, including the use of CRISPR-Cas9, marker-assisted selection, and genomic studies to enhance yield, quality, and stress resistance. It also explores non-genetic innovations such as climate control, probiotics, nanotechnology, and eco-friendly production methods that contribute to sustainable silk manufacturing. Chapter 4, "Mulberry Silk Sericulture: Historical Perspectives and Modern Practices", traces the history of sericulture from ancient China to India, Japan, and Europe, emphasizing the roles of trade, technology, and environmental considerations, and demonstrating how modern practices enhance silk quality while sustaining rural livelihoods. Chapter 5 examines "Non-mulberry Silks: Diversity, Origins, and Production Techniques", including tasar, muga, and eri, their unique physical properties, production techniques, and cultural significance. The chapter highlights the importance of conserving traditional sericulture practices while exploring economic and environmental implications. Chapter 6, "Silk Yarn and Its Quality Evaluation", addresses the preparation and grading of silk yarn, emphasizing the importance of maintaining quality at every stage to ensure high-quality fabric production and competitiveness in global markets.

Chapter 7, "Advances in Extraction and Characterization of Silk Proteins for Future Uses", presents sustainable methods for extracting and analyzing silk fibroin and sericin, along with their applications in biomedical, pharmaceutical, and cosmetic fields. Chapter 8, "Potential of Silkworm Pupae as a Nutritious Food Source for Humans", explores the nutritional value of silkworm pupae, including high-quality protein, essential amino acids, fatty acids, vitamins, minerals, and bioactive compounds, highlighting their potential to improve food security and combat malnutrition.

Chapters 9 and 10 focus on *Spider Silk*, explaining its exceptional combination of strength, elasticity, and biocompatibility, as well as the different types produced by spiders. They also discuss applications in high-performance textiles, biomedical devices, tissue engineering, and regenerative medicine. Chapter 10 also discusses artificial and recombinant production techniques to scale spider silk for industrial and medical use.

Chapter 11, "Sustainability Aspects of Silk Production: Environmental Concerns and Solutions," examines the ecological challenges of conventional silk production, including water use, pesticide dependence, and carbon emissions, and presents eco-friendly solutions and circular economy approaches. Chapter 12, "Beyond the Loom: A Sustainable Prospect for Silk Manufacturing," highlights organic sericulture, precision agriculture, composting, humane practices, and certification systems that promote environmentally conscious and socially responsible silk production. Finally, Chap. 13, "From Cocoon to Community: The Social Impact of Silk," highlights the economic and social significance of silk globally and in India, encompassing rural employment, income generation, women's empowerment, cultural preservation, and its contributions to trade and tourism. The chapter highlights the importance of sustainable and ethical silk production in promoting inclusive growth.

By presenting the historical, cultural, scientific, and technological aspects of silk, this volume aims to provide readers with a complete understanding of this remarkable fiber. It bridges traditional knowledge with modern research and innovations, offering a valuable resource for researchers, designers, policymakers, and anyone interested in the science, culture, and future of silk.

Guntur, India Mohammad Shahid
New Delhi, India Dhiraj Kumar

# Contents

# Editors and Contributors

## About the Editors

**Dr. Dhiraj Kumar** currently serves as an assistant professor in the Department of Zoology at Hansraj College, University of Delhi, India. He is also the coordinator of the Centre for Skill Development and Seri-technology at Hansraj College. Dr. Kumar is a post-doctorate from Soochow University, China, and later worked as Dr. D. S. Kothari, a Postdoctoral Fellow at the School of Studies in Zoology at Jiwaji University, India. He also served as an assistant professor in the Department of Zoology, G. G. Vishwavidyalaya (A Central University), Bilaspur, India. He has published extensively in renowned journals and is the author of over 35 peer-reviewed articles, four books, seven book chapters, three international proceedings, and granted a patent of his credit. He is a reviewer of several SCI journals and a member of many scientific bodies. He also received several prestigious awards for his contributions to science and education, including the Research Excellence Award. His research interests include molecular biology, metagenomics, gut microbiome relationships, biomedicine, and natural products.

**Dr. Mohammad Shahid** Ph.D., is an assistant professor in the Department of Chemistry, Koneru Lakshmaiah Education Foundation, India. Before joining KLEF, he worked at various esteemed institutions, including Jamia Millia Islamia, New Delhi; Soochow University, China; Henan University, China; University of Glasgow, UK; Institute of Chemical Technology, Mumbai; Dr. K. N. Modi University, Tonk; and Mewar University, Chittorgarh. He earned his Master's degree in Chemistry from Shibli National College, Azamgarh, in 2006 and his Ph.D. in Chemistry from Jamia Millia Islamia, New Delhi, in 2014. Dr. Shahid has received several prestigious fellowships to support his research pursuits, including grants from the UGC, India, the China Postdoctoral Fellowship, the EU Marie Curie Fellowship, and the Dr. D. S. Kothari Fellowship. He has contributed to over 50 scientific publications with ~4500 citations and an h-index of 27. In addition, has also edited and authored several books with renowned international publishing houses.

## Contributors

**Ritwik Acharya** Department of Sericulture, Raiganj University, Raiganj, Uttar Dinajpur, West Bengal, India

**Khushi Agarwal** Department of Biological Sciences, School of Basic and Applied Sciences, Dayananda Sagar University, City Innovation Campus, Bengaluru, India

**Mahmuda Akter** Department of Apparel Engineering, Bangladesh University of Textiles, Dhaka, Bangladesh

**Mst. Romana Akter** College of Textile and Clothing, Qingdao University, Qingdao, Shandong, China

**Monika Bhadauria** Toxicology and Pharmacology Lab, Department of Zoology, Guru Ghasidas Vishwavidyalaya, Bilaspur, Chhattisgarh, India

**Pronama Biswas** Department of Biological Sciences, School of Basic and Applied Sciences, Dayananda Sagar University, City Innovation Campus, Bengaluru, India

**Dhitri Borah** Department of Zoology, Biswanath College, Gauhati University, Biswanath, Assam, India

**Sohini Chakraborty** Department of Chemistry, Stella Maris College (Autonomous), University of Madras, Chennai, India

**Md. Kamrul Hassan Chowdhury** Department of Apparel Engineering, Bangladesh University of Textiles, Dhaka, Bangladesh

**P. Darshan** Department of Biotechnology, Sapthagiri College of Engineering, Bangalore, India

**Sayandeep Debnath** Textile Design Department, National Institute of Fashion Technology, Mumbai, India

**Jinal Dinesh** Department of Biological Sciences, School of Basic and Applied Sciences, Dayananda Sagar University, City Innovation Campus, Bengaluru, India

**Debnirmalya Gangopadhyay** Department of Sericulture, Raiganj University, Raiganj, Uttar Dinajpur, West Bengal, India

**Ravi Kumar Goswami** Department of Zoology, Hindu College, University of Delhi, Delhi, India

**Parul Goutam** Textile Design Department, National Institute of Fashion Technology, Mumbai, India

**Vijaykumar Guna** Department of Mechanical Engineering, RNS Institute of Technology, Bengaluru, Karnataka, India

**Divya Gupta** Amity Institute of Biotechnology, Amity University, Gwalior, Madhya Pradesh, India

**Nadim Ibn Sayed** Department of Textile Engineering, European University of Bangladesh, Mirpur, Dhaka, Bangladesh

**R. Janavi** Department of Biotechnology, Sapthagiri College of Engineering, Bangalore, India

**Mansi Kamble** Textile Design Department, National Institute of Fashion Technology, Mumbai, India

**Farhan Khan** Department of Textile Engineering, National Institute of Textile Engineering and Research, Nayarhat, Savar, Dhaka, Bangladesh

**Bronson Kumar Khangembam** Department of Zoology, Tripura University (A Central University), Suryamaninagar, Tripura, India

**B. Kiran Kumar** Department of Biotechnology, Sapthagiri College of Engineering, Bangalore, India

**Dhiraj Kumar** Centre for Skill Development and Seri-Technology, Department of Zoology, Hansraj College, University of Delhi, Delhi, India

**Aishwarya Lomte** Department of Biotechnology, Sapthagiri College of Engineering, Bangalore, India

**Saima Mahjabin** Department of Textile Engineering, Fareast International University, Banani, Dhaka, Bangladesh

**Amit Kumar Mandal** Department of Sericulture, Raiganj University, Raiganj, Uttar Dinajpur, West Bengal, India

**Pankaj Mandal** Department of Sericulture, Raiganj University, Raiganj, Uttar Dinajpur, West Bengal, India

**Dishali Mathur** Department of Biological Sciences, School of Basic and Applied Sciences, Dayananda Sagar University, City Innovation Campus, Bengaluru, India

**Shah Md. Maruf Hasan** Department of Apparel Engineering, Bangladesh University of Textiles, Dhaka, Bangladesh

**Rony Mia** Wilson College of Textiles, North Carolina State University, Raleigh, NC, USA

**Krishnakshi Misra** Molecular Endocrinology and Reproductive Biology Research Laboratory, Department of Zoology, Rajiv Gandhi University, Itanagar, Arunachal Pradesh, India

**Tithi Mitra** Department of Silpa-Sadana, Visva-Bharati (A Central University), Sriniketan, India

**Esmat Jahan Mitu** Department of Apparel Engineering, Bangladesh University of Textiles, Dhaka, Bangladesh

**A. Z. M. Mofasser** Department of Textile Engineering, European University of Bangladesh, Mirpur, Dhaka, Bangladesh

**Mohit** Centre for Skill Development and Seri-Technology, Department of Zoology, Hansraj College, University of Delhi, Delhi, India

**Sunil S. More** Department of Biological Sciences, School of Basic and Applied Sciences, Dayananda Sagar University, City Innovation Campus, Bengaluru, India

**Veena S. More** Department of Biotechnology, Sapthagiri College of Engineering, Bangalore, India

**Abhijit Mukherjee** Department of Fashion Technology, National Institute of Fashion Technology, Kolkata, West Bengal, India

**M. S. A. Muthukumar Nadar** Division of Biotechnology, Karunya Institute of Technology and Sciences, Coimbatore, India

**Satendra Kumar Nirala** Laboratory of Natural Products, Department of Rural Technology and Social Development, Guru Ghasidas Vishwavidyalaya, Bilaspur, Chhattisgarh, India

**O. S. K. Paavana** Division of Biotechnology, Karunya Institute of Technology and Sciences, Coimbatore, India

**Asmita Pramanik** Department of Fashion Technology, National Institute of Fashion Technology, Kolkata, West Bengal, India

**Rafaj Hossain Rakin** Department of Apparel Engineering, Bangladesh University of Textiles, Dhaka, Bangladesh

**Rupsa Roychowdhury** Department of Food Engineering and Technology, Institute of Chemical Technology, Mumbai, India

**Sankar Roy Maulik** Department of Silpa-Sadana, Visva-Bharati (A Central University), Sriniketan, India

**Sonal Saluja** Department of Zoology, Dyal Singh College, Karnal, Kurukshetra University, Haryana, India

**Siam Sarker** Department of Apparel Engineering, Bangladesh University of Textiles, Dhaka, Bangladesh

**Anupam V. Sharma** Department of Zoology, Hindu College, University of Delhi, Delhi, India

**Shubhajit Shaw** Department of Sericulture, Raiganj University, Raiganj, Uttar Dinajpur, West Bengal, India

**Shubham Singh** Department of Zoology, Government E. Raghvendra Rao P.G. Science College, Bilaspur, Chhattisgarh, India

**Bhoomika Sridhar** Department of Biological Sciences, School of Basic and Applied Sciences, Dayananda Sagar University, City Innovation Campus, Bengaluru, India

**R. Srinivasan** Department of Biotechnology, Sapthagiri College of Engineering, Bangalore, India

**S. P. Sruthi** Division of Biotechnology, Karunya Institute of Technology and Sciences, Coimbatore, India

**Trishnakhi Swargiari** Department of Zoology, Bhawanipur Anchalik College, Barpeta, Assam, India

# Chapter 1
# Thread of Gold—The Silk: An Introduction to Silk and Sericulture

**Vijaykumar Guna and Sohini Chakraborty**

## 1.1 Introduction to Silk History

Silk is considered one of the oldest known animal fibers, produced from the larvae of certain insects. Most other conventionally used fabrics, such as cotton, linen, and polyester, are either derived from plant sources or produced synthetically. Throughout history, silk has been regarded as the "queen of fabrics" due to its lustrous appearance, superior mechanical properties, comfort, compatibility with different dyes, and ability to adapt to various twisting forms (Padaki et al. 2015). Silk fibers are naturally spun from the fibrous protein secreted by hundreds of different silkworm species available throughout the world. Although several insects have been reported to produce silk, the filament produced by *Bombyx mori* is considered the highest quality for silk production. Despite the growing competition from synthetic fibers, silk has sustained a significant position in the development of both high-quality and luxury goods, due to its exceptional appearance and distinctive properties.

Silk is the oldest textile fiber discovered in China, dating back to around the twenty-seventh-century BCE. The production of silk fibers dates back to the Yang Shao period. According to Chinese history, Lady Hsi-Lin-Shih, the empress of Shi Huang Di, discovered silk when a cocoon fell into her tea under a mulberry tree. The discovery of this strong filamentous fiber suitable for weaving marked the beginning of silk's journey. However, the secret of silk and silk production remained in China for thousands of years, and silk became a part of the clothing of royal families, symbolizing royalty and wealth. During the Shang dynasty, as trade expanded

V. Guna (✉)
Department of Mechanical Engineering, RNS Institute of Technology, Channasandra, Bengaluru 560098, Karnataka, India
e-mail: vijayguna89@gmail.com

S. Chakraborty
Department of Chemistry, Stella Maris College (Autonomous), University of Madras, Chennai, India



D. Kumar and M. Shahid (eds.), *Science and Technology of Silk*, Textile Science and Clothing Technology, https://doi.org/10.1007/978-3-032-16230-4_1

westward, the production of silk was introduced to other cultures. Initially, during the second-century BCE, silk culture spread to Korea with the migration of Chinese migrants (Naskar et al. 2014; Herrera and Garcia-Bertrand 2018). The infamous Silk Road, also known as the Silk Routes, which connected various regions of Asia, including Japan, India, and Persia, with the Mediterranean world, North Africa, and Europe, emerged in the latter half of the first-century BC (Indian Silk Export Promotion Council, http://www.theindiansilkexportpromotioncouncil.com/history-of-indian-silk.php). Furthermore, silk specifically reached Sicily, Spain, Arabia, and Africa during the seventh century and finally reached Italy, where cities like Venice became centers for silk production (Naskar et al. 2014). From Italy, the industry spread to France and grew during the eighteenth century (International Sericultural Commission, https://inserco.org/en/evolution_of_silk). This ancient network of trade routes, stretching between the East and the West, allowed for the exchange of goods as well as cultural and traditional practices. Besides utilizing silk for the production of fabric, the fibers of silk were also used to fabricate bowstrings, fishing lines, and even strings for musical instruments, demonstrating the historical significance and diverse uses of silk throughout history (Naskar et al. 2014).

Globally, silk production is a significant agro-based industry, with several countries contributing to the annual supply of silk fibers. The major silk-producing countries include China, India, Uzbekistan, Vietnam, North Korea, Brazil, and Japan, among others. However, China remains the largest global producer and primary supplier of silk, contributing almost 50,000 metric tons, while India stands in the second place with 38,913 metric tons (International Sericultural Commission 2023). The legacy of silk as a luxury fabric remains of significant importance and continues to influence global trade through the New Silk Road (Li et al. 2015).

## 1.2 Silk Sources and Their Classification

Silk is produced by a variety of silkworm species, which are classified into two main types: mulberry and non-mulberry silks. These silks differ in characteristics such as texture, color, and quality. Figure 1.1 shows the cocoons produced by various mulberry and non-mulberry silkworm species (Malay et al. 2016).

### *1.2.1 Mulberry Silk*

Mulberry silk is one of the most widely cultivated commercial silk fibers, which is produced from the Lepidoptera insect belonging to the Bombycidae family. The mulberry silkworm, also known as the *Bombyx mori*, is crucial to the textile industry as it is instrumental in producing high-quality silk fibers. Mulberry silkworms are fully domesticated monophagous insects, which feed only on mulberry leaves throughout their larval stage and have been reared for more than 2000 years.

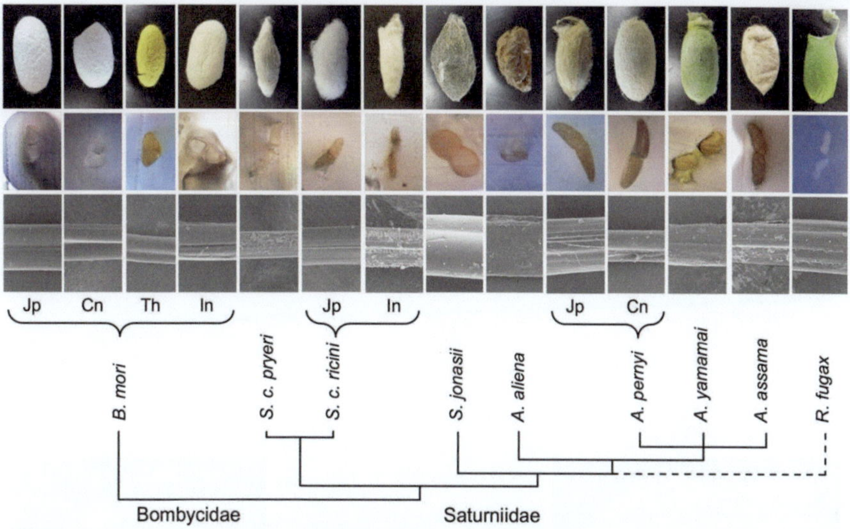


**Fig. 1.1** Pictures of cocoons (not to scale) from different silkworm species (first row); images represent the cross-sectional view of silk fibers in cyanoacrylate glue (second row); SEM micrographs of fiber surfaces (third row). The phylogenetic classification for both mulberry and non-mulberry silkworm species. Adapted and edited with permission from reference (Malay et al. 2016), copyright 2016, Springer Nature

These silkworms are classified based on various factors, including the number of generations per year (voltinism), place of origin, and molting characteristics (Ashraf and Qamar 2023). Figure 1.2 represents the geographical distribution of *Bombyx mori* and other silk-producing silkworm species, along with their characteristic cocoons produced (Lee et al. 2022).

#### 1.2.1.1 Classification Based on Origin

Classification of mulberry silkworms based on their place of origin is shown in Table 1.1 (Murugesh Babu and Babu 2013; Food and Agriculture Organization of the United Nations, https://www.fao.org/4/x2099e/x2099e03.htm).

#### 1.2.1.2 Classification by Voltinism

Silkworms can also be classified based on the number of life cycles per year (Table 1.2) (Takeda 2009; Babu and Kozłowski 2012).

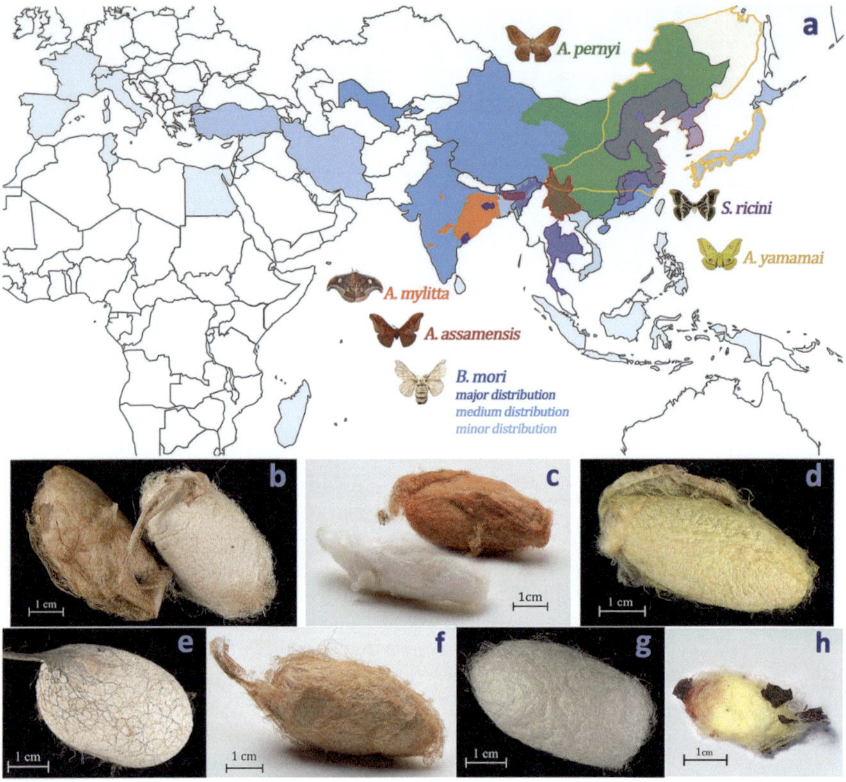


**Fig. 1.2** Geographical distribution of silk moths by their place of origin (**a**) and their characteristic cocoons (**b–h**). **a** Colored regions represent the indigenous habitat of the silk moth; **b** *A. pernyi*; **c** *S. ricini*; **d** *A. yamamai*; **e** *A. mylitta*; **f** *A. assamensis*; **g** *B. mori*; **h** *B. mandarina*. Adapted with permission from reference (Lee et al. 2022), copyright 2022, Springer Nature

#### 1.2.1.3 Classification Based on Molting

During their development, the silkworm larvae shed their skin and stop feeding on mulberry leaves. This process is known as molting. Based on the number of molts in the larval stage, silkworms are classified into trimolters, tetramolters, or pentamolters. Additionally, tetramolters are most widely reared for commercial purposes in comparison to bimolters and hexamolters (Dingle et al. 2005).

### *1.2.2 Non-mulberry Silk*

The silkworm species can also be classified into two categories: mulberry and non-mulberry. The non-mulberry silkworm is a wild, heterogeneous species that belongs

**Table 1.1** Classification of the mulberry silkworms based on their origin (Murugesh Babu and Babu 2013; Food and Agriculture Organization of the United Nations, https://www.fao.org/4/x2099e/x2099e03.htm)

| Races | Native | Characteristics of larvae | Color of cocoons | Shape of cocoon | Voltinism | Characteristics of silk fiber |
|---|---|---|---|---|---|---|
| Chinese | China | (a) Tolerance to high temperatures, while being sensitive to high humidity | White, golden yellow, green, red, or beige | Elliptical, spherical, or spindle | Univoltine, bivoltine, or multivoltine | Fine and good reliability |
| Japanese | Japan | Short larval stage | White, green, or yellow | Peanut | Univoltine or bivoltine | Thick, Short, and inferior quality |
| Indian | India and Southeast Asia | (a) Long larval stage (b) Tolerance to both high temperature and humidity | White, green, or yellow | Spindle | Multivoltines | Fine and good reliability |
| Thai | Northern and northeastern parts of Thailand | | Yellow | | | Coarse fiber |
| European | Europe and Central Asia | (a) Long larval stage (b) Sensitive for high temperature and humidity | | Longer elliptical | Multivoltines | Good reliability |

to the Saturniidae family of the order Lepidoptera (Fig. 1.3) (Peigler 2020). These silkworms are widely found across various regions, including India, China, Japan, Thailand, Indonesia, and Namibia. Non-mulberry silkworms, which belong to the Saturniidae family, are known to produce a wide variety of silk, including tasar, muga, eri, and fagaria. In contrast, the silkworms from the Lasiocampidae family produce a distinct silk variety known as shashe.

**Table 1.2** Classification of mulberry silkworms based on the voltinism (Takeda 2009; Babu and Kozłowski 2012)

| Voltinism | Generations/year | Characteristics of larvae | Characteristics of cocoons |
|---|---|---|---|
| Univoltines | One | (a) Larvae are large in size<br>(b) Longer larval duration | (a) Increased weight of the cocoon, shell, and cocoon filament, along with a higher shell ratio<br>(b) Superior quality of filament from cocoon |
| Bivoltines | Two | (a) Larvae are more robust<br>(b) Longer larval duration | The weight of the cocoon, shell, and filament, along with the shell ratio, is lower compared to univoltines |
| Multivoltines or polyvoltines | More than two | (a) Larvae are robust and can withstand high temperatures and humidity<br>(b) Shorter larval duration | (a) Smaller cocoon size and cocoon weight and the shell ratio and cocoon filament weight are lower than that observed in bivoltines<br>(b) Finer cocoon filament |

#### 1.2.2.1 Tasar or Tussah Silk

Tasar silk, a variety of wild silk commonly referred to as "Dasali Pattu," is obtained from the cocoons of the tasar silkworm, as represented in Fig. 1.4 (Zhao et al. 2022), which are characterized by their hardness and compact structure (Babu and Kozłowski 2012). Furthermore, tasar silk inherently exhibits distinct light golden to dark brown colors (Karthik et al. 2017) and consists of two main classifications: (a) temperate and (b) tropical. The temperate tasar silkworm species consists of *Antheraea pernyi*, which is located in various regions of China, and *A. yamamai*, native to Japan, as well as *A. roylei*, *A. frithi*, and *A. pernyi* (also referred to as *A. proylei*) in northeastern India (Kar et al. 2013). These silkworms are known for their finer and superior quality of silk (Jasmine et al. 2014). In contrast, *Antheraea mylitta* (tropical tasar silkworm) is predominantly restricted to regions of Central India. This silkworm primarily feeds on *Terminalia tomentosa* (Asan) and *Terminalia arjuna* (Arjuna) plants and is naturally reared on trees in open spaces. The Indian tropical tasar silkworm produces the highest amount of silk and can be either bivoltine or trivoltine, depending on the habitat and species variety (Babu and Kozłowski 2012; Jasmine et al. 2014; Kundu et al. 2012).

#### 1.2.2.2 Muga Silk

Muga silks, also known as mejankori silk, are produced by multivoltine *Antheraea assamensis* silkworms and are primarily found in the northeastern part of India, particularly in Assam. Similar to tasar, these silkworms are semi-domesticated on

**Fig. 1.3** Various silk cocoons of non-mulberry silkworms, particularly those belonging to the Saturniidae family. Adapted with permission from Reference (Peigler 2020), copyright 2020, Elsevier

the trees of open spaces, and they sustain themselves on the aromatic leaves of som and soalu plants. The silk fibers produced are inherently golden-yellow in color (Fig. 1.5) (Gupta et al. 2015), which cannot be easily bleached or damaged by sunlight or common detergents. However, the color of the cocoons varies marginally according to the region and variety of *A. assamensis* reared. Muga silk fibers have better mechanical properties compared to other commercial silk fibers. Additionally, the fabric produced from muga silk is highly durable and has a high demand in both national and international markets (Babu and Kozłowski 2012; Devi et al. 2011; Dutta et al. 2013).

**Fig. 1.4** Tussah silkworm cocoon. Adapted with permission from Reference (Zhao et al. 2022), copyright 2022, Multidisciplinary Digital Publishing Institute

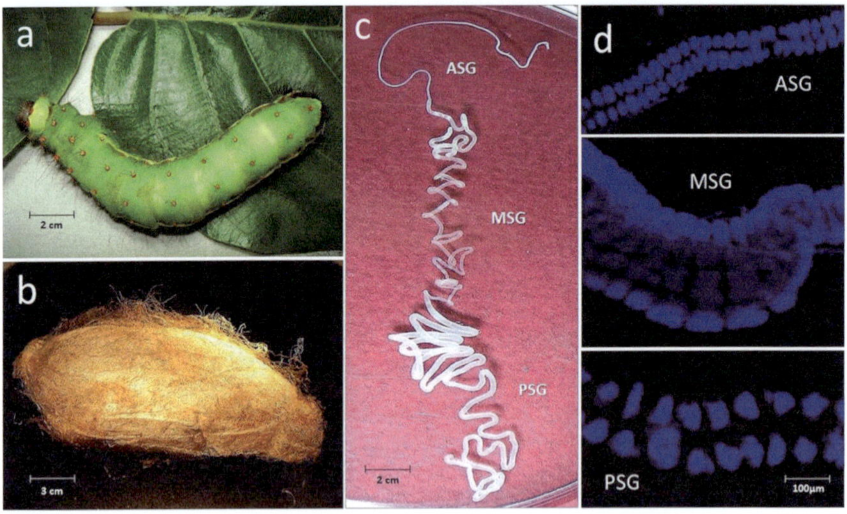


**Fig. 1.5** *Antheraea assama* silkworm species **a** larva, **b** cocoon, **c** silk gland, and **d** fluorescent micrographs of different portions of silk gland, including anterior silk gland (ASG), middle silk gland (MSG), and posterior silk gland (PSG). Adapted and edited with permission from Reference (Gupta et al. 2015), copyright 2015, Springer Nature

#### 1.2.2.3 Eri Silk

Eri silk is widely distributed across the globe, particularly in regions of South Asia, East Asia, and parts of Europe and South America. Similar to the mulberry silkworm, the *Philosamia ricini (Samia ricini/Cynthia ricini)* silkworm produces Eri silk. It can also be domesticated, and they mostly feed on the leaves of castor (*Ricinus communis*) and kesseru (*Heteropanax fragrans*). Eri cocoons are fibrous and soft, which makes them unsuitable for reeling into a continuous thread (Chattopadhyay et al. 2017). However, eri silks are known as "peace silk" due to their ethical production process, wherein the caterpillars are allowed to transform into moths and complete their natural life cycle, rather than being killed to harvest the silk filaments as in traditional silk production (Karthik et al. 2017). The size, shape, and color of the cocoons obtained from the wild eri silkworm *Samia canningi* are shown in Fig. 1.6 (Chattopadhyay et al. 2023). Furthermore, eri silk possesses softness and warmth, or insulating characteristics similar to those of wool, making it a promising material for commercial applications (Murugesh Babu and Babu 2013).

#### 1.2.2.4 Fagaria Silk

Fagaria silk is obtained from silkworms of the *Attacus* genus, especially *Attacus atlas*, the largest silk-producing moth, which is native to Southeast Asia. These silkworms are characterized as bivoltine and polyphagous. Moreover, Fagaria silk yarns exhibit superior tensile strength compared to both muga and tasar spun silks (Croft et al. 2022).

#### 1.2.2.5 Cricula Silk

The silkworm *Cricula trifenestrata* produces Cricula silk, which is distributed across South Asian countries, including Indonesia, Northeast India, the Philippines, and Java. Among the different occurring non-mulberry silkworms, *Cricula trifenestrata* is the smallest moth and produces a distinct golden perforated cocoon (Kundu et al. 2012; Croft et al. 2022).

#### 1.2.2.6 Shashe Silk

Shashe silk fibers are produced by *Gonometa postica* of the Lasiocampidae family. The spined larvae feed on various East African podo trees and *Acacia* species, forming cocoons with superior toughness and durability (Kundu et al. 2012; Croft et al. 2022).

#### 1.2.2.7 Anaphe Silk

The anaphe silk fibers are obtained from the silkworm belonging to *the Anaphe* genus. Various species, including *A. moloneyi Druce*, *A. panda Boisduval*, *A. reticulata Walker*, *A. carteri Walsingham*, *A. vesta Butler*, and *A. infracta Walsingham,* are responsible for producing anaphe silk, and they are primarily found across various regions of southern and central Africa. These silkworms form communal cocoons,

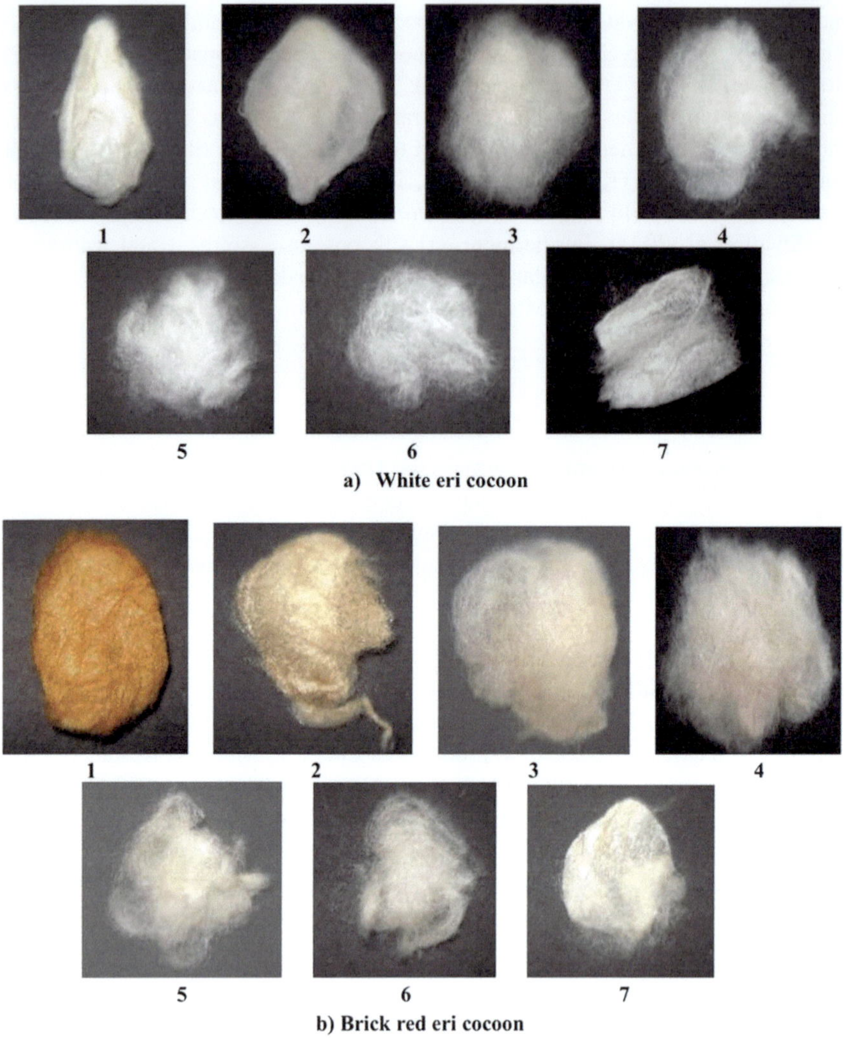


**Fig. 1.6** Photographs of eri silk and fibers extracted from various layers of the cocoon, including the floss, upper, middle, lower, and pelade layer. Adapted with permission from Reference (Chattopadhyay et al. 2023), copyright 2023, Taylor & Francis

encompassing thin layers of silk fibers. These cocoons are then harvested and processed into raw silk, which is both soft and moderately lustrous. The silk derived from *A. infracta* is commonly called "book," while the silk from *A. moloneyi* is referred to as "tissnian-tsamia" and "koko." Additionally, anaphe silk fabric exhibits excellent elasticity and superior strength compared to mulberry silk. Consequently, it is employed in the fabrication of velvet and plush fabrics.

#### 1.2.2.8 Coan Silk

The white cocoons, measuring approximately 8.9 × 7.6 cm, are produced by the *Pachypasa otus D. species*, which is found in various parts of the Mediterranean region. These silkworms are polyphagous in nature, feeding on a wide range of plants, including pine, ash, cypress, juniper, and oak trees. During ancient times, Coan silk was specifically employed in the production of crimson-dyed garments for Roman dignitaries. However, commercial production eventually declined due to limited yields and the emergence of superior-quality silk (Babu and Kozłowski 2012).

### *1.2.3 Spider Silk*

Spiders are members of the order Araneae, which consists of ~ 37,000 species. These species produce silk with varying structures, each adapted for specific purposes, including shelter, breeding offspring, and capturing prey. Spiders utilize silk throughout their lifetime to construct webs and are noted for their superior properties. In comparison to synthetic fibers, the silk spun by spiders is known for its superior mechanical strength and flexibility. Furthermore, spider silk fibers are less dense (1.3 g/cm$^3$) than steel (7.8 g/cm$^3$). Despite possessing significant mechanical characteristics, spider silk is seldom utilized for large-scale commercial production due to the challenges associated with domesticating these spider species (Jung et al. 2021). Additionally, the fiber yield from a single silkworm cocoon ranges from 600 to 900 m. In contrast, the ampullate gland of a spider produces only about 137 m of silk, and a fully constructed spider web contains only 12 m of silk (Kluge et al. 2008). However, the limited commercial spider silk production is primarily associated with Madagascan species such as *Nephila madagascariensis*, *Miranda aurantia*, and *Eperia*. Moreover, the durability and resistance of this variety of silk to various environmental conditions, such as high temperature and humidity, make it a suitable material for use as crosshairs in optical instruments, wear-resistant clothing, and bulletproof vests (Karthik et al. 2017). *Nephila clavipes*, a golden orb-weaving species, produces a variety of silks for different functions, as shown in Table 1.3 (Hakimi et al. 2007). Additionally, Table 1.4 and Fig. 1.7 provide an overview of the natural functions of different spider silks (Hu et al. 2006).

### *1.2.4 Marine Silk*

Several marine organisms, including mussels, barnacles, and oysters, produce a silk-like byssus to withstand the various challenges posed by harsh underwater environments. Among the various sources of silk, marine silk, which produces a unique and distinct silk, is the least studied. Around 12,000 species of the insect order Trichoptera

**Table 1.3** Several types of silks produced by *Nephila clavipes* (Hakimi et al. 2007)

| Type of silk | Gland | Spinneret | Function of the silk |
|---|---|---|---|
| Dragline | Major ampullate | Anterior | Orb web frame, radii, safety line |
| Viscid | Flagelliform | Posterior | Prey capture, core fibers of adhesive spiral |
| Glue-like | Aggregate | Anterior and posterior | Prey capture, adhesive silk of spiral |
| Minor ampullate | Minor ampullate | Medial | Orb web frame reinforcement |
| Cocoon | Tubuliform | Posterior | Reproduction |
| Wrapping | Aciniform | Anterior | Wrapping captured pray, inner egg sac |
| Attachment | Pyriform | Anterior | Attachment disk and joining fibers |

**Table 1.4** Natural functions of different spider silks (Hu et al. 2006)

| Spider silks and their natural functions | | |
|---|---|---|
| Silk | Location | Core fiber proteins |
| Major ampullate dragline | Web frame, safety line | MaSpl, MaSp2 |
| Minor ampullate | Web reinforcement and temporary capture silk | MiSpl, MiSp2 |
| Flagelliform | Capture spiral | Flag |
| Aciniform | Wrapping silk, small diameter egg case fiber | AcSpl |
| Tubuliform | Large diameter egg case silk fiber | TuSpl, ECP-1, ECP-2 |
| Aggregate | Glue coating for capture spiral | Unknown |
| Pyriform | Attachment disk and joining fiber | Unknown |

have been identified and further grouped into the suborders Spicipalpia, Annulipalpia, and Integripalpia. Caddisflies are commonly referred to as "underwater architects," which belong to the order Trichoptera, and are capable of creating a wide range of aquatic structures from silk, which primarily serve to provide protection against predators, food storage, and facilitate oxygenated water across abdominal gills. Some species of Integripalpia spin cocoons are semipermeable, leading to the formation of an internal osmotic environment (Standring et al. 2025).

*Crassicorophium bonellii*, which belongs to the Amphipoda crustacean, is another species of marine animals that produces sticky, fibrous secretions through their specialized secretory legs. The adhesive material produced exhibits excellent properties in terms of strength, durability, extensibility, and resistance to marine water. Furthermore, extrusion-spun silk is used to construct open-end tubes, which can be

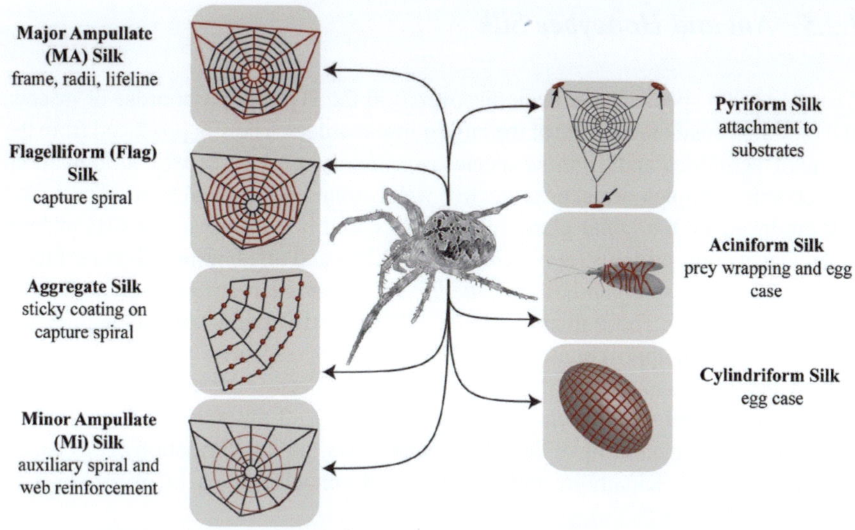


**Fig. 1.7** Schematic representation of the different spider silk types utilized for specific applications. Adapted with permission from reference (Eisoldt et al. 2011), copyright 2011, Elsevier

secured horizontally to various substrates and utilized for shelter. The produced silk differs in properties and composition from regular silk (Kronenberger et al. 2012).

Mussels typically produce a bundle of threads called byssus threads, which are soft and flexible on one end, while the other end is stiff and strong, remaining attached to the substrate. In comparison to tendons, which have a stiffness of $6 \times 10^8$ Pa, byssal threads are less rigid ($8.5 \times 10^7$ Pa). Additionally, their extensibility (0.44) is five times higher than that of tendons (< 0.10). However, the composition and characteristics of the bundle of threads differ among species. For instance, *Mytilus* sp. generates a bundle of 50 to 100 threads and is 4–5 cm long. *Pinna nobilis* produces a significant number of fine filaments and/or threads popularly referred to as "sea silk" (Reddy 2019). *Pinna squamosa* produces a strong brown filament, which is carded and spun into silk called "fish wool." However, the production of fish wool is mainly restricted to the Taranto region of Italy (Murugesh Babu and Babu 2013).

The nematocytes (also referred to as cnidoblasts or cnidocytes) found in the tentacles of hydra, jellyfish, corals, and sea anemones are considered specialized stinging cells that play an important role in prey capture and defense. *Hydra magnipapillata* contains silk-like cnidoin protein that provides high elasticity and rapid coiling, leading to exceptionally fast nematocyst discharge (Jung et al. 2021).

### 1.2.5 Ant and Honeybee Silk

About 144,000 species have been discovered in the Hymenoptera order of insects, and they are considered as one of the largest insect orders. The silk produced from the larvae of *Vespoidea* and *Apoidea* species is primarily used as the structural material for cocoons, complex nests, or protective webs around parasitized hosts. The coiled silk produced by the labial gland of bees, ants, and hornet larvae consists of low-molecular-weight proteins (approximately 30 kDa) and an alanine-rich core. Fibers extracted from the silk glands of *Apis mellifera adansonii* larvae have superior toughness and flexibility. These fibers have a toughness of 167 MPa, with an engineering stress and strain of approximately 132 MPa and 200%, respectively (Sutherland et al. 2012; Campbell et al. 2014; Maitip et al. 2015). Insects of the order Hymenoptera produce coiled coil silk, which is utilized for a variety of purposes, including pupation protection, the building of large communal nests, and the formation of egg cases, as seen in Fig. 1.8 (Kirshboim and Ishay 2000; Brannoch et al. 2017; Weisman et al. 2008).

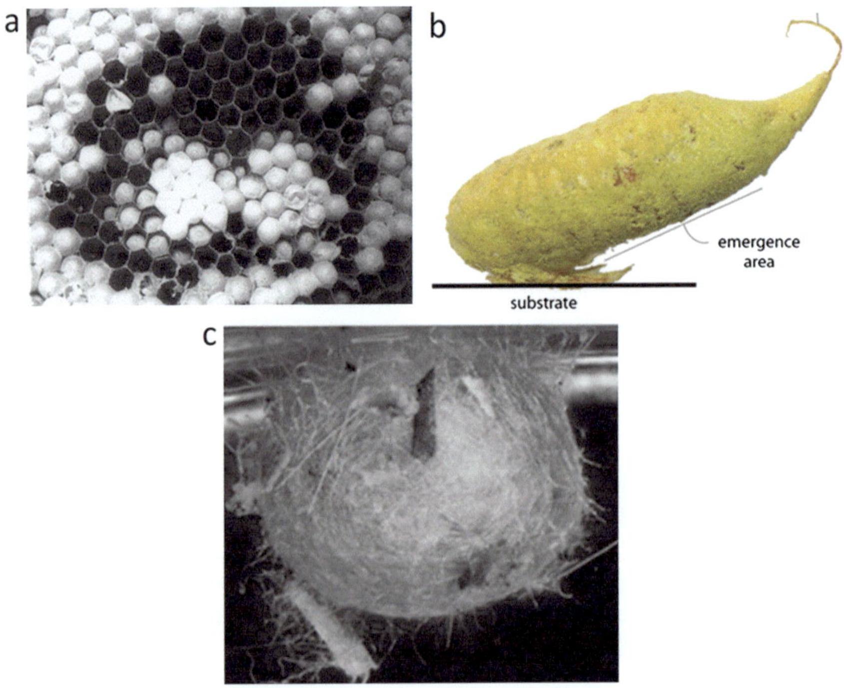


**Fig. 1.8** Coiled coil silk produced by stinging insects **a** hornet nest silk. Adapted with permission from reference (Kirshboim and Ishay 2000), copyright 2000, Elsevier; **b** mantis egg case. Adapted and edited with permission from reference (Brannoch et al. 2017), copyright 2017, ZooKeys; **c** lacewing cocoon. Adapted and edited with permission from reference (Weisman et al. 2008), copyright 2008, American Chemical Society

**Fig. 1.9** Silk and/or cocoons produced from different ant species, particularly Oecophylla (upper-left position), *Brachymyrmex patagonicus* (upper-right position), *Leptogenys crustosa* (lower-left position), and *Rhytidoponera victoriae* (lower-right position). Adapted with permission from reference (Campbell et al. 2014), copyright 2014, Elsevier

The silk produced by the final-instar larva of *Oecophylla smaragdina* (weaver ant) is utilized to form a non-woven mesh with an average fiber diameter of 766 nm. Compared to webs made from collagen, regenerated silk, and other biopolymers, these nanofibers exhibit superior mechanical properties (Reddy 2019). Figure 1.9 shows the silk and/or cocoons obtained from different ant species (Campbell et al. 2014).

## 1.3 Production of Silk in the Silk Industry

Several moth species produce silk that varies in different properties, size, texture, and color. However, the textile industry relies largely on silk obtained from fully domesticated *Bombyx mori* silkworms. Sericulture is an agro-based industry that operates on a two-step process: The first step involves cultivating mulberry trees, followed by rearing silkworms on the leaves of mulberry for the production of silk cocoons. Silkworms are monophagous insects that feed exclusively on fresh mulberry leaves (*Morus*, family Moraceae). However, both silkworm rearing and mulberry cultivation are often considered interdependent processes due to the difficulties in transporting and storing mulberry leaves for extended periods (Takeda 2009; Chauhan and Tayal

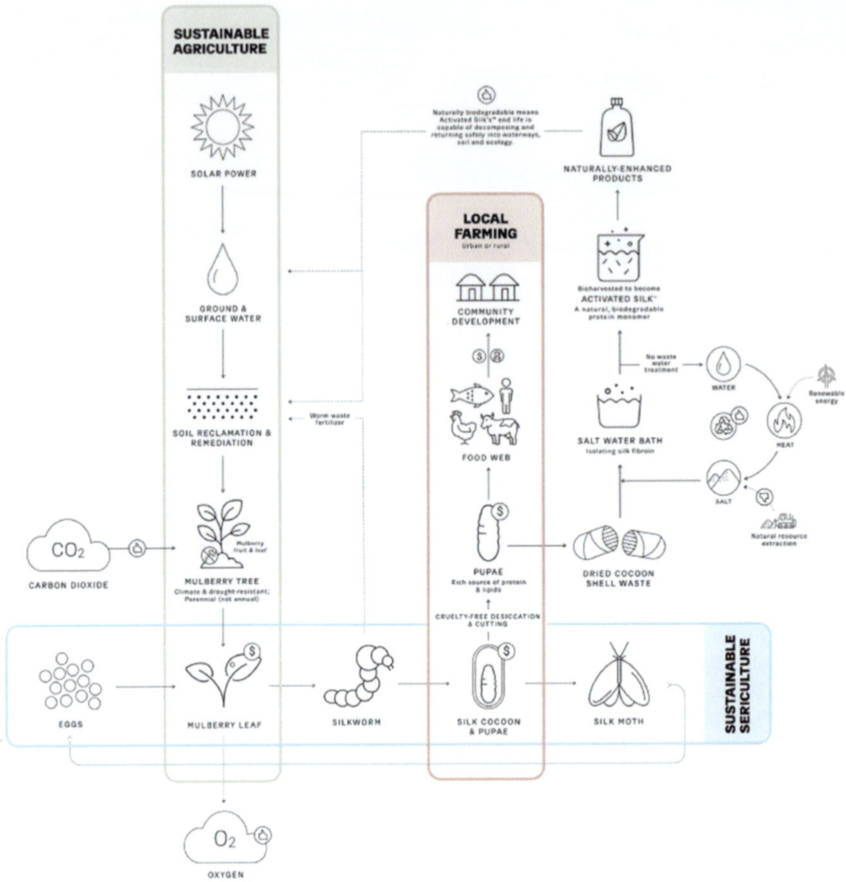


**Fig. 1.10** Schematic outline of the various stages of the sericulture process and their interdependencies, including associated supporting industries, key inputs and outputs, and the resulting silk-derivative products. Adapted with permission from reference (Altman and Farrell 2022), copyright 2022, Elsevier

2017). Figure 1.10 provides a schematic outline of the various stages of the sericulture process and their interdependencies, including associated supporting industries, key inputs and outputs, and the resulting silk-derivative products (Altman and Farrell 2022).

### 1.3.1 *Rearing of Silkworm*

The process of silkworm rearing involves several steps: obtaining and transporting mulberry leaves, providing fresh leaves to the larvae, mounting the larvae for cocoon

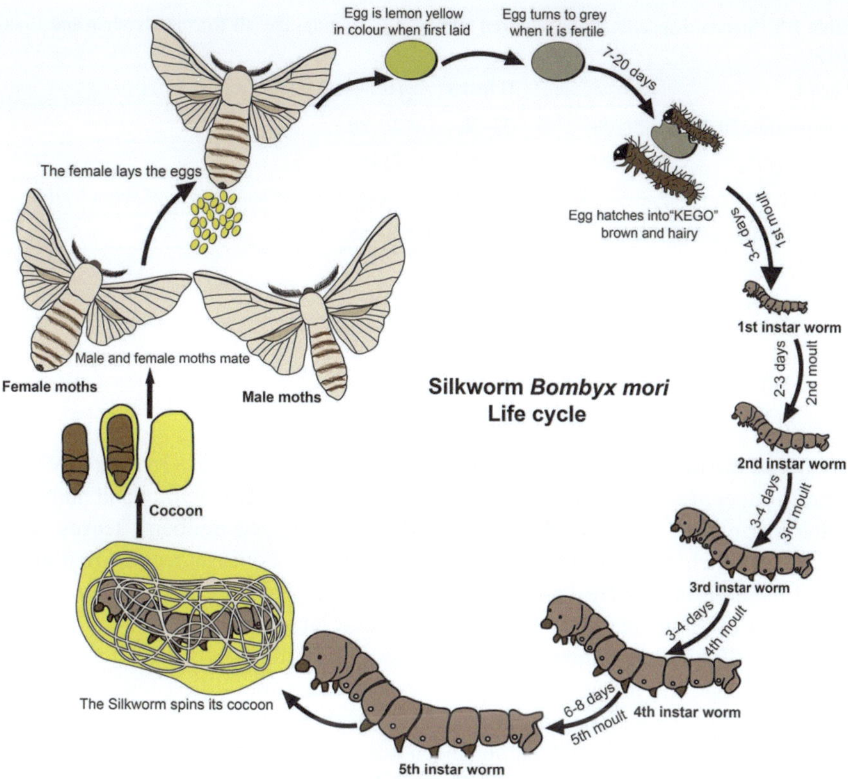


**Fig. 1.11** Life cycle of a silk worm. Adapted with permission from reference (Hassan et al. 2024), copyright 2024, Elsevier

formation, and collecting and shipping the cocoons. The vast majority of global silk producers aim to achieve the highest quality silk threads. The life cycle of a silkworm (Fig. 1.11) (Hassan et al. 2024) ranges between 6 and 8 weeks, and each stage of silkworm rearing directly depends on temperature. At a temperature ranging from 20 to 30 °C, eggs hatch within 10 days after laying, and the larvae immediately begin feeding on finely chopped mulberry leaves upon hatching. Silkworm rearing during the larval stage is categorized into two phases: the early/young silkworm period (I to III instar) and the late/grown silkworm period (IV to V instar), with differences in rearing techniques at each stage (Hassan et al. 2024).

#### 1.3.1.1 Young Instar Silkworm Rearing

The primary principle of rearing young instar silkworm involves placing hatched egg larvae in the rearing facilities at comparatively high temperatures (26–28 °C) and humidity (75–90%). However, both the temperature and humidity are lowered

**Table 1.5** Standard specifications required for rearing of young-age silkworm (Chauhan and Tayal 2017)

| Factor | I instar | II instar | III instar |
|---|---|---|---|
| Temperature (0 °C) | 27–28 | 27–28 | 25–26 |
| Humidity (%) | 85–90 | 80–85 | 75–80 |
| Leaf size (sq. cm) | 0.5–1.5 | 1.5–4.0 | 4.0–6.0 (entire leaf in case of shoot feeding) |
| Quantity of leaf (kg) | 5–6 | 14.0–15.0 | 50–55 |
| Bed area (sq. m) | | | |
| (a) In the beginning | 0.36 | 1.35 | 4.05 |
| (b) At the end | 1.35 | 4.05 | 9 |
| Bed cleaning | No cleaning | Twice | Daily |

steadily as the instars progress. The hatched larvae are placed on a rearing tray with paraffin paper at the bottom, which maintains the optimal temperature and humidity in the rearing bed. Afterward, they are fed tender, minced mulberry leaves. This larval stage is characterized by a very high growth rate. For instance, the first instar larvae increase in size by 15-fold within 50 h of feeding, the II instars grow five times their original size in 45 h, and the III instar larvae triple in size within 60 h of feeding (Table 1.5) (Chauhan and Tayal 2017). Furthermore, during the young silkworm stage, the larvae exhibit a high rate of digestion with a low quantity of food ingestion. The various physiological activities of larvae are determined by the quality of the feed and depend mainly on temperature, as silkworms are cold-blooded animals (Takeda 2009; Chauhan and Tayal 2017).

#### 1.3.1.2 Grown Silkworm Rearing

In the late-age rearing, silkworms grow a thousandfold by consuming a large portion of mulberry leaves. During the V instar stage, silkworms consume ~ 88% of the mulberry leaves provided and grow about seven to eight times in size compared to the fourth instar. In comparison to young-age rearing, late-age rearing involves relatively lower temperatures of approximately 26 °C for the fourth instar and 24 °C for the fifth instar. Similarly, the ideal humidity range is between 65 and 70% (Table 1.6) (Chauhan and Tayal 2017). Silk glands develop rapidly during the III instar stage and reach their maximum for the biosynthesis activity of silk proteins by the completion of the fifth instar stage (Murugesh Babu and Babu 2013; Takeda 2009; Chauhan and Tayal 2017).

**Table 1.6** Standard specifications required for rearing of late-age silkworms (Chauhan and Tayal 2017)

| Factor | III age | IV age | V age |
|---|---|---|---|
| Temperature (0 °C) | 25–26 | 24–25 | 23–24 |
| Relative humidity (%) | 75–80 | 70–75 | 65–70 |
| Quantity of leaf (kg) | 50–55 | 175–200 | 1000–1200 |
| Bed area (sq. m) | 4.05–9.00 | 9.00–20.7 | 20.7–43.2 |

### *1.3.2 Spinning of Cocoons*

The larvae undergo four molting phases and become 9 cm long caterpillars. When the mature silkworm reaches the cocoon-spinning stage, appropriate structures, such as support frames, shelving units, and bundles of twigs or straw, are provided to facilitate the process. The silkworm forms cocoons by moving its head in a pattern and produces silk protein fibroin through two salivary glands known as sericteries. They then combine it with other liquid gummy proteinaceous substances called sericin, produced through specialized openings called spinnerets present beneath their jaws. The liquid secreted solidifies when exposed to air and forms a continuous 700–1500 m long filament in 72 h (Chen et al. 2019). The silkworm becomes completely enclosed within a cocoon, where fibers are held together by sericin or silk gum (Murugesh Babu and Babu 2013; Takeda 2009; Chauhan and Tayal 2017; Chen et al. 2019).

### *1.3.3 Reeling of Cocoons*

The process of harvesting silk fibers from the cocoon is referred to as reeling. However, to emerge naturally from their cocoons, silk moths release enzymes that degrade the cocoon material, facilitating their escape. As a result, the cocoons are collected before the damage occurs and before the silk moth emerges. The extraction of silk filament from the cocoons is carried out by immersing the cocoons in hot boiling water for the removal of the proteinaceous silk gum (sericin) that binds the filaments together. The composition of silk filament consists of two proteins: fibroin (76%) and sericin (22%). The reeled silk, known as raw silk, is collected in skeins during the filature process in factories.

### *1.3.4 Production of Silk Yarns*

The process of "throwing" converts reeled silk into a yarn or thread, which is similar to the spinning of other natural fibers. The raw silk skeins are washed using warm water with soap or oil to further remove sericin and are then classified based on their color, size, and quantity. After drying, the skeins are placed on reels and wound onto bobbins, where the silk strands are twisted to the required degree. Strands may be doubled and twisted in the same or opposite directions. The yarn is then passed between rollers to ensure a uniform width along its entire length. However, the number of twists can be adjusted to obtain different forms of silk yarns. Moreover, any residual sericin can be eliminated through "degumming," a process that involves washing the yarn with soap and water, without altering the natural shine and soft texture of silk (Karthik et al. 2017; Altman and Farrell 2022).

## 1.4 Main Stages of Silk Fabric Production

Silk fabric manufacturing involves various stages, such as soaking, winding, doubling, twisting, warping, and pirn winding. Figure 1.12 represents the flowchart of the several key steps in the manufacturing of silk fibers (Babu 2020).

### *1.4.1 Soaking*

Prior to weaving, the raw silk fibers are subjected to soaking, mainly for the removal of sericin, a natural silk gum that binds fibers together. Consequently, improving their flexibility facilitates the easy handling of fibers. A solution containing 3% soap as a softening agent, 4% oil as a lubricant, neutral soft water (1:9 ratio), and a small amount of borax as a solvent is utilized to treat fibers. The process of soaking is generally carried out at 40 °C for 12–14 h, making the silk fibers suitable for weaving.

### *1.4.2 Winding*

The process of transferring the softened silk fibers onto bobbins is referred to as "winding." The operation is carried out by using winding machines equipped with 50–100 double-flanged bobbin units, during which irregularities in the fibers, including hard gum spots, loose ends, and knots, are eliminated. This stage is specifically conducted under controlled conditions with a relative humidity of $65 \pm 2\%$ RH to maintain the quality of silk fibers. Typically, 100–150 g/day of silk is produced from each drum.

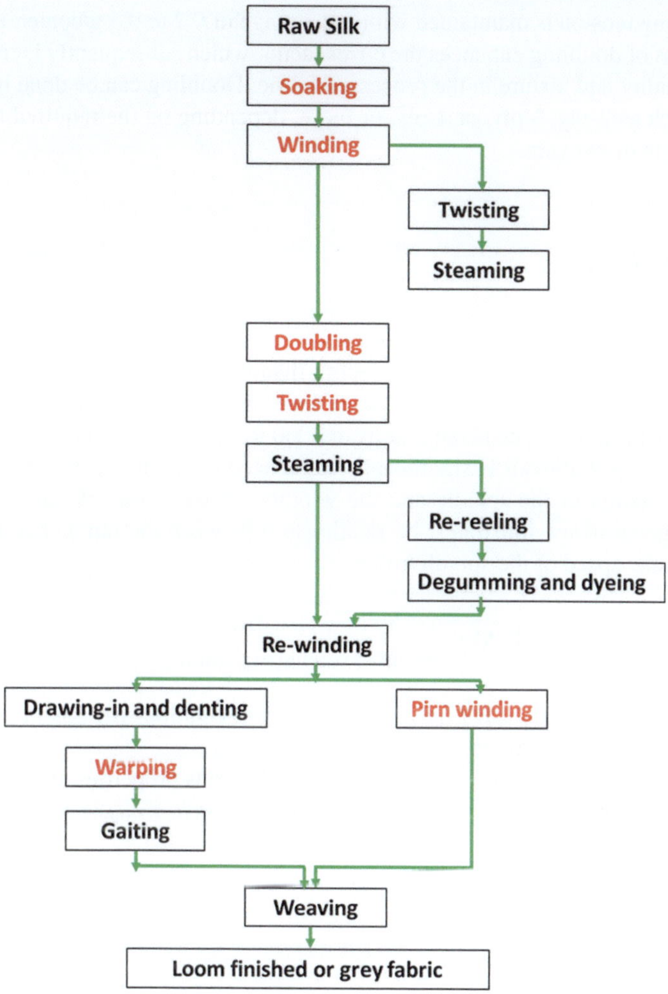


**Fig. 1.12** Flowchart on the manufacturing process of silk fabric. Redrawn with permission from reference (Babu 2020), copyright 2020, Elsevier

### *1.4.3 Doubling*

In this stage, multiple threads from the winding bobbins are placed in parallel and wound to enable the twisting of the threads into yarn. Similar to winding, doubling increases the strength of the yarn by reducing the irregularities and weak spots. The winding bobbins are positioned on the creel, and the yarn passes through individual thread stop-motion devices, tensioners, and thread guides before being wound onto a horizontally mounted double-flanged bobbin. A doubling machine is comparable to a winding machine, exhibiting a processing capacity two to four times greater. Further, the doubling tension is maintained within the range of 0.2 to 0.3 g/denier. However, the process of doubling enhances the cover factor, which subsequently increases the overall quality and texture in the processed fabric. Doubling can be done in various forms, such as 2-ply, 3-ply, or 4-ply, or more, depending on the required thickness and strength of the yarn.

### 1.4.4 Twisting

Twisted silk yarns are commonly used for weaving, and the number of twists mainly depends on the type of fabric being manufactured. Uptwisters are specialized machines used for twisting silk, with doubling bobbins mounted vertically on spindles that rotate at speeds ranging from 8000 to 10,000 rpm. The degree of twist applied to the silk threads is significantly influenced by the interaction between the speed of rotation of the spindle and the winding speed of the take-up roller. The equation given below illustrates the relationship between the number of turns per meter and the speed of the spindle/roller.

$$\text{TPM} = \frac{\text{Spindle speed(rpm)}}{\text{Front - roller speed(mpm)}},$$

where TPM refers to turns per meter, rpm denotes revolutions per minute, and mpm represents meters per minute.

Additionally, the twisting process affects the lustrous properties of silk. With an increase in the number of twists, the thread becomes less *lustrous*, resulting in a more opaque appearance.

### 1.4.5 Warping

In the warping process, individual threads from the creels are collected and wound onto a large roller referred to as a beam. In addition, warping results in the formation of uniformly spread threads to facilitate the weaving into a fabric on a loom. The number of ends required for the fabric depends on the density of the warp ends and the width of the fabric, which in turn helps determine the number of sections required. Sectional, beam, and ball warping are the primary warping methods employed in the textile industry, with sectional warping preferred for fine denier and greater numbers of ends, and ball warping being widely implemented in handloom weaving.

### *1.4.6 Pirn Winding*

The preparation of weft yarn requires pirn winding, which can be carried out using both automatic and non-automatic machines. Additionally, pirns used for power looms are larger in size and contain a higher yarn capacity, while handloom pirns are comparatively smaller and prepared using hand-operated chakras (Murugesh Babu and Babu 2013; Takeda 2009; Babu 2020).

## 1.5 Production of Spun Silk

Spun silk is produced by utilizing various waste materials generated in all sectors during silk manufacturing. During the mulberry silk reeling process, various by-products are formed, which can be classified as cocoon waste, reeling waste, and thread waste. Cocoon waste includes floss, spelaia (Italian), discarded, pierced, double, and stained cocoons, which are produced due to defects and damage in the cocoons. Reeling waste consists of cookers, reeler, and boiling-off waste, generated during the silk extraction from the cocoons. Lastly, thread waste is produced during re-reeling, winding, and weaving processes, including throwers' and weaving waste (hard waste).

Spun silk yarn production can be categorized into three major steps: degumming, dressing, and spinning as seen from Fig. 1.13 (Yin et al. 2021).

The main purpose of degumming is the removal of sericin and facilitation of the mechanical processing of fibers in later stages. The methods available for the degumming of waste silk include the soap and soda boil method and the enzymatic

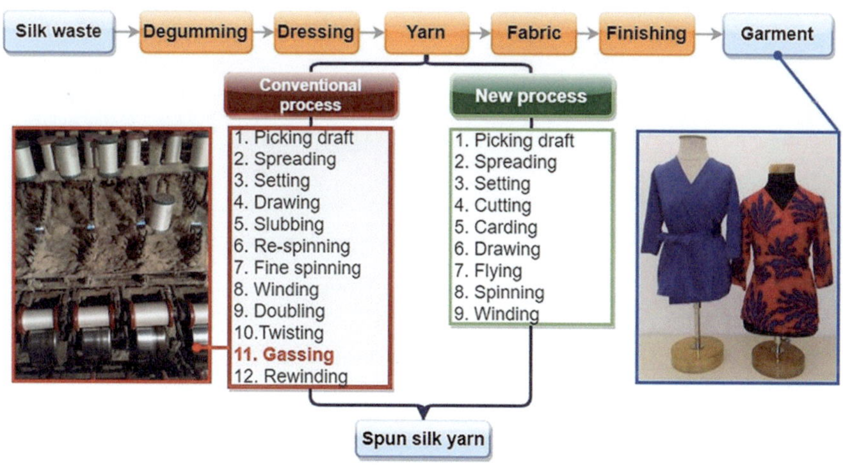


**Fig. 1.13** New production process of spun silk fiber. Adapted with permission from reference (Yin et al. 2021), copyright 2021, Elsevier

degumming process. The degummed raw material is processed further during the dressing phase using an opener machine. This machine processes the entangled mass of fibers into parallel, uniformly arranged filaments. Subsequently, the fibers are carefully combed to form combed drafts. Through the circular dressing machine, three grades of drafts are extracted. Fibers with lengths varying from 65 to 78 mm are the result of the first dressing process and are classified as the 1st-grade draft. The 2nd-grade draft, measuring between 50 and 65 mm in length, is obtained by re-dressing the remaining silk. The final dressing produces the 3rd-grade draft, ranging from 40 to 50 mm in length. The spinning process consists of two main stages: pre-spinning and post-spinning. During the pre-spinning stage, the combed drafts are processed into a consistent and continuous sliver, which is subsequently drawn and spun into roving. Post-spinning involves several processes, including doubling, gassing, and reeling. The post-spinning process draws and twists the roving into a yarn. The twisted yarns are then passed through the flame with a deal of friction, which is called "gassing." The loose, protruding fibers, and small neps are burned away by the flame, which also contributes to the brilliance of the fiber. The burnt particles are removed by the process of "cleaning," which involves passing the singed thread between the rotating steel rollers. After gassing and cleaning, the yarn is subjected to reeling frames, resulting in the formation of hanks (Takeda 2009; Yin et al. 2021). Subsequently, Fig. 1.14 (Yin et al. 2021) represents the production of spun silk yarns through the modified spinning method.

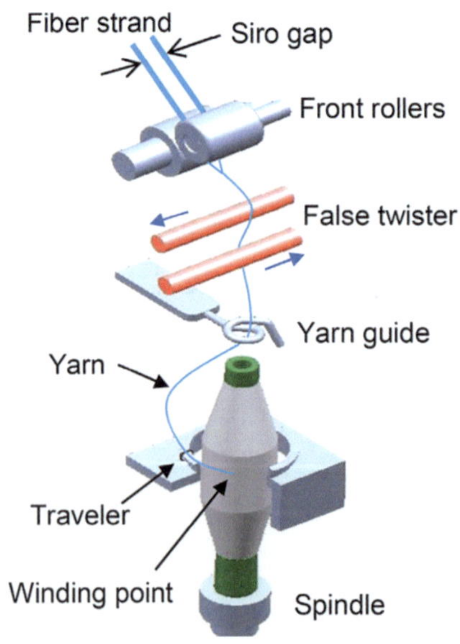


**Fig. 1.14** Schematic representation of the modified spinning method for the production of spun silk yarns. Adapted with permission (Yin et al. 2021). Copyright 2021, Elsevier

## 1.6 Sustainability Issues in Silk Production

Silk is a highly valued textile fiber due to its superior properties, including strength, elasticity, absorbency, and excellent dyeing capabilities. However, silk production involves several sustainability issues, from mulberry cultivation to the dyeing and finishing stages of the fabric. The Waste and Resources Action Program reports that silk yarn production generates a carbon footprint of 6964 kg $CO_2$ equivalent ($CO_2e$) per metric ton, which exceeds the carbon footprint of polyester yarn by 2.58 times. Furthermore, the increased carbon footprint associated with silk yarn manufacturing can be attributed to the increased consumption of water and energy. Some previous studies estimate that the energy required to produce silk fiber is approximately 1800 MJ per kilogram, which is considerably higher compared to fibers like nylon or cotton. The cocoon cooking process alone utilizes 51% of the total energy required for raw silk production. Besides, it has been noted that globally, 35% of the total silk production results in heterogeneous fibrous waste and/or silk waste, with around 11 million tons of silk waste generated annually. Table 1.7 represents the Environmental Impact Assessment of raw silk production in comparison to other natural and synthetic fibers (Astudillo et al. 2015). Additionally, one of the most considerable sources of emissions in sericulture is the release of chemicals and silk sericin into both drainage systems and groundwater during the chemical degumming stage. Hence, it is crucial to develop sustainable yarn technologies that reduce the carbon footprint, recycle silk waste, and explore alternatives for more environmentally friendly degumming processes. In an effort to promote cleaner production, researchers have investigated eco-friendly alternatives to the traditional degumming process, including hydrated lime water, steam treatment, and ultrasound-assisted enzymatic treatments. Similarly, some mills have utilized worsted gilling machines to comb silk fibers as an alternative to the traditional dressing process, thereby enhancing processing efficiency. Furthermore, some attempts have been made to report a gassing-free production route for manufacturing mulberry-spun silk yarns. The elimination of gassing substantially saved production costs while being more environmentally friendly and energy-saving (Yin et al. 2021; Gaviria et al. 2023).

Conventional mulberry silk production involves boiling cocoons to extract silk fibers by killing the larvae inside. This raises a concern about animal welfare and is criticized by many animal welfare and rights activists. The substitution of conventional silk fiber with Ahimsa silk (peace silk), wild silk, and organic silks has been widely promoted to mitigate ethical concerns, allowing silkworms to emerge naturally before the harvesting of cocoons. Moreover, the rough handloom silks commonly known as “Matka silk” in India are produced from the damaged cocoons in the weft, which can also be deemed as a promising alternative to mulberry silk. The implementation of closed-loop water management systems, utilizing natural dyes and others, could further enhance the sustainability of silk production (Karthik et al. 2017).

**Table 1.7** Environmental Impact Assessment of raw silk production in comparison to other natural and synthetic fibers (Astudillo et al. 2015)

| Category | $GWP_{100}$ (kg $CO_{2eq}$/kg) | CED R. (MJ/kg) | CED NR. (MJ/kg) | Ecotoxicity (CTUe /kg) | ALO ($m^2a$/ kg) | BWF ($m^3$/ kg) | FE (g $P_{eq}$/ kg) |
|---|---|---|---|---|---|---|---|
| Raw silk Karnataka | 51.5 | 1349.9 | 110.1 | 522.8 | 19.7 | 24.6 | 4.8 |
| Cotton fibers (China) | 3.4 | 19.7 | 0.1 | 71.2 | 7.8 | 7 | 0.8 |
| Nylon 66 | 8 | 1.3 | $7.0*10^{-4}$ | $6.0*10^{-4}$ | $2.0*10^{-4}$ | 0.2 | 0.3 |
| Sheep wool at farm (USA) | 18.5 | 81.7 | 0.1 | 3.4 | 53.5 | 0.2 | 0.5 |

*Note* GWP100—global warming potential over 100 years; CED R. and CED NR.—renewable and non-renewable cumulative energy demand, respectively; ALO—agricultural land occupation; FE—freshwater eutrophication; BWF—blue water footprint

## 1.7 Recent Trends in Silk Production

### *1.7.1 Fluorescent Silk*

Silk is a natural protein fiber produced by arthropods, mainly silkworms and spiders. Among the various types of silk, the most commonly used one is obtained from the domesticated *B. mori* silkworm, which is used to make fabrics and has been utilized for over 5000 years. A new trend in silk production involves the development of fluorescent silk. Several methods, such as modifying silkworm genes or altering the diet, as well as post-processing techniques applied to natural silk, are considered for producing colored silk (Das and David 2022). Among the various methods, the introduction of dyes into silk fibroin through dietary modification of silkworms has emerged as a promising approach. Consequently, it reduces the environmental impact of conventional dyeing methods and results in improved color fastness (Tansil et al. 2012). For instance, a large amount of recombinant silk with three different fluorescent colors (green, red, and orange) was produced using transgenic silkworms. The cocoons were subjected to low-temperature drying and cooking, which allowed the reeling of colored, fluorescent silk through an automatic reeling machine without any color loss. Although the extracted silk exhibited decreased mechanical performance compared to conventional silk fiber, it is still suitable for fabric production (Iizuka et al. 2013).

### *1.7.2 Automation in Silk Production*

Recently, several attempts have been made to apply artificial intelligence (AI) to the sericulture chain, ranging from silkworm rearing to the production of raw silk fibers. For instance, to obtain raw silk of superior quality, cocoons must be evaluated for defects prior to the reeling process. However, cocoon sorting is a labor-intensive process, and studies have been conducted to improve both the quality of silk and its production efficiency by utilizing an automated prototype for cocoon sorting through a multi-step approach and machine learning algorithms. The developed prototype demonstrated an efficiency of sorting 80 cocoons per minute. Overall, the efficiency of the selection process can be improved by adding more machines in parallel, all coordinated with one another, further leading to a higher rate of cocoons processed per minute (Vasta et al. 2023). Furthermore, the silk textile processing sector is an agro-based industry that requires standardization of raw materials, similar to other industrial chains, to produce high-quality products. However, environmental factors (temperature, humidity, light intensity, and gases) significantly influence the production of high-quality silk. Therefore, efforts are being made to implement smart automation techniques such as the Internet of Things (IoT) to maintain the rearing environment and track real-time data through a network system (Jyothi et al. 2021). Additionally, silk farming faces several other challenges, including egg counting, optimum environmental control, disease, and defective disease detection, which also influence the quantity and quality of silk obtained. The traditional approaches to addressing these challenges incur huge labor costs. Consequently, AI, machine learning (ML), and cloud computing are considered to provide a revolutionized solution to address the challenges related to the silk industry (Raju et al. 2022).

The production of silk filaments generates waste or by-products termed as "silk waste," which is often under-utilized or discarded. The growing interest in a circular economy has led to the evaluation and study of methods of silk waste recycling. Traditionally, low-grade yarns or fabrics are fabricated from recycled silk waste. However, with properties similar to raw silk, waste silk is also considered as a potential resource for the development of sustainable materials (Wang et al. 2023). Recently, recycling and valorization of waste silk have been studied widely in various fields, including reinforcement for composites, energy conversion, tissue engineering, multi-functional electrodes, and others. Based on these advances, the future of recycling silk provides a range of opportunities. For instance, the development of three-dimensional (3D) reinforcements for composites from waste silk, which includes the fabrication of 3D waste silk fabric involving 3D textile structures (e.g., 3D braiding, weaving, knitting, and needling) are significantly increasing to enhance the strength and performance of the composite. Another potential alternative involves the development and utilization of low-cost and green technologies for the recovery of silk protein from waste silk. In addition, the production of colored silk fibroin filaments obtained by dissolving colored waste silk textiles helps in preventing the environmental pollution from dyeing processes. Moreover, the development of multi-functional electrodes can also be considered in utilizing the waste silk as a material for fabricating electrodes after carbonization, with promising applications in the field of batteries and wearable electronics (Lu et al. 2022; Sashina and Yakovleva 2023).

### *1.7.3 Advancements in Non-mulberry Silk Production*

The rise in awareness of non-mulberry silks (Ahimsa, Peace, or Vanya silk) in recent years can be attributed to growing concerns for ethical and sustainable products. Previously, no significant attention was given to the development and diversification of products derived from these silks. However, with the growing awareness, industries are focusing more on developing new designs, patterns, and products within the Vanya silk industry, thereby expanding the market demand for these silks (Babu 2020). In the future, research and development are considered to bring changes in the silk industry, particularly through advancements in the finishing of silk fabrics, such as stain-repellent/release, anti-static, crease-recovery, flame-retardant, antimicrobial, and other finishes (Choudhury and Basu 2015). Traditionally, silk was used primarily for textiles. However, currently, silk has found potential application in a variety of fields such as nutrition, cosmetics, pharmaceuticals, biomaterials, biomedical, and bioengineering.

## 1.8 Summary and Conclusion

Silk, often referred to as the "queen of fabrics," holds a significant place in history and forms an essential part of the economic and social landscape of several nations. The legacy of silk as a luxury fabric remains of significant importance and continues to influence global trade through the New Silk Road. Its production is rooted in ancient Chinese practices, and it has evolved since then into a global agro-industrial enterprise. In this chapter, we explore the multifaceted world of silk, from its historical evolution to the technological and sustainable aspects of its production. Various types of silk, produced from different organisms and characterized by their unique properties, including mechanical properties, aesthetic value, and ecological adaptation, have been highlighted. Sericulture, the process of rearing silkworms, has also been thoroughly explored. This chapter further discusses the industrial processes involved in silk production, including cocoon spinning, reeling, yarn production, and weaving, which transform raw cocoons into silk fabric. Additionally, unconventional sources of silk, such as spider silk and marine silk, have also been detailed. Although their commercial production is a challenge, they can be used for bioengineering applications owing to their high mechanical stability and biochemical properties. Furthermore, in the context of sustainability, several innovative approaches have been discussed, highlighting the importance of waste valorization. Renewed interest in indigenous silks has also been advocated to address the growing ethical concerns over silk production. The integration of recent technological advancements, such as AI and IoT, has been discussed to promote the development of smart silk-based materials for various applications. Therefore, in conclusion, the enduring legacy of silk, along with its contemporary challenges and innovation, reiterates its continued relevance. The future of silk lies in sustainable practices, interdisciplinary research, and the embrace of diverse silk sources to meet the global demands.

**Declarations**

**Conflict of Interests** The authors do not have any conflicts of interest to declare.

## List of Abbreviations

| | |
|---|---|
| MaSp2 and MaSp2 | Major ampullate spidroin 1 and 2 |
| MiSp1 and MiSp2 | Minor ampullate spidroin 1 and 2 |
| Flag | Flagelliform Spidroin |
| AcSpl | Aciniform Spidroin 1 |
| TuSpl | Tubuliform Spidroin 1 |
| ECP-1 | Egg Case Protein 1 |
| ECP-2 | Egg Case Protein 2 |

## References


Altman GH, Farrell BD (2022) Sericulture as a sustainable agroindustry. Clean Circ Bioeconomy 2:100011

Ashraf H, Qamar A (2023) Silkworm Bombyx mori as a model organism: a review. Physiol Entomol 48(4):107–121

Astudillo MF, Thalwitz G, Vollrath F (2015) Life cycle assessment of silk production–a case study from India. In: Muthu SS (ed) Handbook of life cycle assessment (LCA) of textiles and clothing. Woodhead Publishing, pp 255–274

Babu KM (2012) Silk production and the future of natural silk manufacture. In: Kozłowski RM (ed) Handbook of natural fibres. Woodhead Publishing, pp 3–29

Babu KM (2020) Silk–production and future trends. In: Kozłowski RM, Mackiewicz-Talarczyk M (eds) Handbook of natural fibres. Woodhead Publishing, pp 121–145

Brannoch SK, Wieland F, Rivera J, Klass KD, Béthoux O, Svenson GJ (2017) Manual of praying mantis morphology, nomenclature, and practices (Insecta, Mantodea). ZooKeys 13(696):1

Campbell PM, Trueman HE, Zhang Q, Kojima K, Kameda T, Sutherland TD (2014) Cross-linking in the silks of bees, ants and hornets. Insect Biochem Mol Biol 48:40–50

Chattopadhyay D, Chakraborty A, Chatterjee SM (2017) Studies on degumming of eri silk cocoons. J Text Inst 108(8):1327–1339

Chattopadhyay D, Chakraborty A, Chatterjee SM (2023) Microstructural characteristics of eri silk fibre in different layers of cocoon. J Nat Fibers 20(1):2146828

Chauhan TP, Tayal MK (2017) Mulberry sericulture. In: Omkar (ed) Industrial entomology. Springer Nature, Singapore, pp 197–263
Chen S, Liu M, Huang H, Cheng L, Zhao HP (2019) Mechanical properties of Bombyx mori silkworm silk fibre and its corresponding silk fibroin filament: a comparative study. Mater Des 181:108077
Choudhury AK (2015) Advances in the finishing of silk fabrics. In: Basu A (ed) Advances in silk science and technology. Woodhead Publishing, pp 81–110
Croft AS, Spessot E, Bhattacharjee P, Yang Y, Motta A, Wöltje M (2022) Gantenbein B. biomedical applications of silk and its role for intervertebral disc repair. JOR Spine 5(4):e1225
Das S, David J (2022) Colored and luminescent silk. J Nat Fibers 19(14):9420–9433
Devi D, Sarma NS, Talukdar B, Chetri P, Baruah KC, Dass NN (2011) Study of the structure of degummed Antheraea assamensis (muga) silk fibre. J Text Inst 102(6):527–533
Dingle JG, Hassan E, Gupta M, George D, Anota L, Begum H (2005) Silk production in Australia: a report for the rural industries research and development corporation.
Dutta S, Talukdar B, Bharali R, Rajkhowa R, Devi D (2013) Fabrication and characterization of biomaterial film from gland silk of muga and eri silkworms. Biopolymers 99(5):326–333
Eisoldt L, Smith A, Scheibel T (2011) Decoding the secrets of spider silk. Mater Today 14(3):80–86
Food and Agriculture Organization of the United Nations. Silk reeling and testing manual. https://www.fao.org/4/x2099e/x2099e03.htm. Accessed 20 May 2025
Gaviria A, Jaramillo-Quiceno N, Motta A (2023) Restrepo-Osorio A. Silk wastes and autoclaved degumming as an alternative for a sustainable silk process. Sci Rep 13(1):15296
Gupta KA, Mita K, Arunkumar KP, Nagaraju J (2015) Molecular architecture of silk fibroin of Indian golden silkmoth. Antheraea Assama Sci Rep 5(1):12706
Hakimi O, Knight DP, Vollrath F, Vadgama P (2007) Spider and mulberry silkworm silks as compatible biomaterials. Composites B Eng 38(3):324–337
Hassan MA, Basha AA, Eraky M, Abbas E, El-Samad LM (2024) Advancements in silk fibroin and silk sericin-based biomaterial applications for cancer therapy and wound dressing formulation: a comprehensive review. Int J Pharm, 124494
Herrera RJ, Garcia-Bertrand R (2018) Ancestral DNA, human origins, and migrations. Academic Press
Hu X, Vasanthavada K, Kohler K, McNary S, Moore AM, Vierra CA (2006) Molecular mechanisms of spider silk. Cell Mol Life Sci 63:1986–1999
Iizuka T, Sezutsu H, Tatematsu KI, Kobayashi I, Yonemura N, Uchino K, Nakajima K, Kojima K, Takabayashi C, Machii H, Yamada K (2013) Colored fluorescent silk made by transgenic silkworms. Adv Funct Mater 23(42):5232–5239
Indian Silk Export Promotion Council. History of Indian silk. http://www.theindiansilkexportpromotioncouncil.com/history-of-indian-silk.php. Accessed 20 May 2025
International Sericultural Commission. Statistics (2023). https://inserco.org/en/statistics. Accessed 20 May 2025
International Sericultural Commission. Evolution of silk. https://inserco.org/en/evolution_of_silk. Accessed 20 May 2025
Jasmine S, Mandal BB (2014) Types and properties of non-mulberry silk biomaterials for tissue engineering applications. In: Kundu SC (ed) Silk biomaterials for tissue engineering and regenerative medicine. Woodhead Publishing, pp 275–298
Jung D, Lee J, Park TY, Yang YJ, Cha HJ (2021) Diverse silk and silk-like proteins derived from terrestrial and marine organisms and their applications. Acta Biomater 136:56–71
Jyothi S, Mamatha DM, Raju PJ, Sultana S, Seetharamulu J (2021) A study on the sericulture farm automation system using the internet of things (IoT). In: Proceedings of the 2nd international conference on computational and bio engineering. Springer Singapore
Kar S, Talukdar S, Pal S, Nayak S, Paranjape P, Kundu SC (2013) Silk gland fibroin from Indian muga silkworm Antheraea assama as potential biomaterial. Tissue Eng Regen Med 10:200–210
Karthik T, Rathinamoorthy R (2017) Sustainable silk production. In: Muthu SS (ed) Sustainable fibres and textiles. Woodhead Publishing, pp 135–170

Kirshboim S, Ishay JS (2000) Silk produced by hornets: thermophotovoltaic properties—a review. Comp Biochem Physiol Part A Mol Integr Physiol 127(1):1–20

Kluge JA, Rabotyagova O, Leisk GG, Kaplan DL (2008) Spider silks and their applications. Trends Biotechnol 26(5):244–251

Kronenberger K, Dicko C, Vollrath F (2012) A novel marine silk. Sci Nat 99:3–10

Kundu SC, Kundu B, Talukdar S, Bano S, Nayak S, Kundu J, Mandal BB, Bhardwaj N, Botlagunta M, Dash BC, Acharya C (2012) Nonmulberry silk biopolymers. Biopolymers 97(6):455–467

Lee B, Pires E, Pollard AM, McCullagh JS (2022) Species identification of silks by protein mass spectrometry reveals evidence of wild silk use in antiquity. Sci Rep 12(1):4579

Li P, Qian H, Howard KW, Wu J (2015) Building a new and sustainable "Silk Road economic belt." Environ Earth Sci 74:7267–7270

Lu L, Fan W, Ge S, Liew RK, Shi Y, Dou H, Wang S, Lam SS (2022) Progress in recycling and valorization of waste silk. Sci Total Environ 830:154812

Maitip J, Trueman HE, Kaehler BD, Huttley GA, Chantawannakul P, Sutherland TD (2015) Folding behavior of four silks of giant honey bee reflects the evolutionary conservation of aculeate silk proteins. Insect Biochem Mol Biol 59:72–79

Malay AD, Sato R, Yazawa K, Watanabe H, Ifuku N, Masunaga H, Hikima T, Guan J, Mandal BB, Damrongsakkul S, Numata K (2016) Relationships between physical properties and sequence in silkworm silks. Sci Rep 6(1):27573

Murugesh Babu K (2013) Silk reeling and silk fabric manufacture. In: Babu KM (ed) Silk processing, properties and applications. Woodhead Publishing, Delhi, pp 33–55

Naskar D, Barua RR, Ghosh AK, Kundu SC (2014) Introduction to silk biomaterials. In: Kundu SC (ed) Silk biomaterials for tissue engineering and regenerative medicine. Woodhead Publishing, pp 3–40

Padaki NV, Das B, Basu A (2015) Advances in understanding the properties of silk. In: Basu A (ed) Advances in silk science and technology. Woodhead Publishing, pp 3–16

Peigler RS (2020) Wild silks: their entomological aspects and their textile applications. In: Kozłowski RM, Mackiewicz-Talarczyk M (eds) Handbook of natural fibres. Woodhead Publishing, pp 715–745

Raju CG, Sarkar S, Canamedi V, Parameshwaranaik J, Sarkar S (2022) A review of silk farming automation using artificial intelligence, machine learning, and cloud-based solutions. ICDAL, Springer Nature Singapore, Singapore

Reddy N (2019) Silk: materials, processes, and applications. Woodhead Publishing

Sashina ES, Yakovleva OI (2023) The current state and prospects of recycling silk industry waste into nonwoven materials. Fibers 11(6):56

Standring S, Heckenhauer J, Stewart RJ, Frandsen PB (2025) Unraveling the genetics of underwater caddisfly silk. Trends Genet 41(6):537–546

Sutherland TD, Weisman S, Walker AA, Mudie ST (2012) The coiled coil silk of bees, ants, and hornets. Biopolymers 97(6):446–454

Takeda S (2009) Sericulture. In: Resh VH, Cardé RT (eds) Encyclopedia of insects. Academic Press, pp 912–914

Tansil NC, Koh LD, Han MY (2012) Functional silk: colored and luminescent. Adv Mater 24(11):1388–1397

Vasta S, Figorilli S, Ortenzi L, Violino S, Costa C, Moscovini L, Tocci F, Pallottino F, Assirelli A, Saviane A, Cappellozza S (2023) Automated prototype for Bombyx mori cocoon sorting attempts to improve silk quality and production efficiency through multi-step approach and machine learning algorithms. Sensors 23(2):868

Wang H, Cao L, Li Y, Wen R, Xu G (2023) Silk nonwoven: recycling and upcycling of waste silk into bio-based materials with high value-added properties. Fibers Polym 24(10):3489–3500

Weisman S, Trueman HE, Mudie ST, Church JS, Sutherland TD, Haritos VS (2008) An unlikely silk: the composite material of green lacewing cocoons. Biomacromol 9(11):3065–3069

Yin R, Xiang YF, Zhang ZH, Tao XM, Gluck JM, Chiu K, Lam W (2021) Cleaner production of mulberry spun silk yarns via a shortened and gassing-free production route. J Clean Prod 278:123690
Zhao Y, Liu Y, Li F, Li X, Du Z, Guo T, Yan Y, Lu Y, Song Z (2022) Experimental research on the physical and mechanical properties of the Tussah Cocoon for leaf-stripping machine development. Agriculture 12(11):1792

# Chapter 2
# Indian Silk: Origin, Properties, and Notable Handicrafts

**Sayandeep Debnath, Mansi Kamble, Parul Goutam, and Rupsa Roychowdhury**

## 2.1 Introduction

Usually spun by silkworms, silks are spontaneously spun fibrous protein filaments. The silk fiber of the Bombyx mori silkworm has been utilized by humans for over 5000 years as the "queen of textiles" and is of exceptional quality. Because silk fibers are strong, extensible, biocompatible, and biodegradable, they have exceptional mechanical qualities (Asakura et al. 2002). In addition to the *Bombyx mori* silkworm, there are additional wild silk moths belonging to the *Saturniidae* family, also referred to as "Vanya" or non-mulberry silk, including Eri, Tasar, and Muga, which produce silk fiber popularly used in textile fabric manufacturing. These moths are found across Southeast Asia (Ahmed et al. 2015). The use of silk for textile applications was a significant part of ancient Chinese civilization, dating back approximately 5000 years. To be precise, the practice of sericulture (the cultivation of silkworms to produce silk) and silk weaving was attributed to the ingenious Hsi Ling-Shih, the wife of the Yellow Emperor. The earliest domesticated silk usage in China dates back to the 3rd millennium BCE, with remnants found at the site of Qianshanyang. The earliest documented evidence of silk in China dates back to the Shang dynasty period. The mastery of silk production techniques was guarded within China until the Western Han Period around the second-century BCE. The techniques of degumming and reeling, which are pivotal in silk production, have long been regarded as enigmatic secrets of the Chinese silk industry. These processes involved extracting the outer sericin gum from silk fibers, leaving behind the inner fibroin, and then cleaning, inspecting, and sending the material for further processing. Silk has historically been

S. Debnath (✉) · M. Kamble · P. Goutam
Textile Design Department, National Institute of Fashion Technology, Mumbai, India
e-mail: sayandeep.debnath@nift.ac.in; sayandeepdebnath@yahoo.in

R. Roychowdhury
Department of Food Engineering and Technology, Institute of Chemical Technology, Mumbai, India



D. Kumar and M. Shahid (eds.), *Science and Technology of Silk*, Textile Science and Clothing Technology, https://doi.org/10.1007/978-3-032-16230-4_2

traded globally as a prized luxury fabric and was also valued for its therapeutic use in treating war wounds. The first sterile silk suture was introduced into clinical practice in 1869 by Joseph Lister (Vazifdar and Debnath 2025). It is reported that the silk was introduced to India from China in the fourth-century BC (Alam 1999). Over time, the tradition of silk cultivation in India has flourished, culminating in the creation of some of the world's most renowned crafts. Today, India is the second-largest producer of silk in the world, reported to have produced 38,913 metric tons (MT) of silk in the FY24, including all four types of natural silks, viz. mulberry, Eri, Tasar, and Muga. The activities of sericulture alone in India are spread across 52,360 villages (India Brand Equity Foundation 2025). Raw silk serves as a material for a wide range of garments, such as shirts, suits, ties, blouses, lingerie, pajamas, and jackets. Hand-spun mulberry silk is commonly employed in the production of comforters and sleeping bags, while processed mulberry silk is used to create a wide variety of textured fabrics such as dupion, plain silk, deluxe, satin, chiffon, chignons, crepe, and brocade. Apart from Muga silk, India produces various silk varieties, including Eri silk, Tasar silk. Eri silk is soft, moderately sheen, and used for shawls, sarees, garments, and hosiery. Tasar silk is coarse, lower sheen, ideal for sarees and dresses. Muga silk, indigenous to Assam and celebrated as the "golden silk," is the most expensive variety of silk in India. It is traditionally woven into Mekhela Chador and sarees, while also being crafted into neckties, umbrellas, shoes, lampshades, and other fashion products. Notably, Muga silk has also been experimentally explored as a substitute for Zari in sarees. Major production is in Manipur, Himachal Pradesh, Uttar Pradesh, Chhattisgarh, Assam, Meghalaya, and Jammu & Kashmir. Due to the availability of various silk-based crafts in India and the presence of skilled artisans, India's textile heritage is diverse across regions. Northern regions, such as Jammu & Kashmir, are known for the production of silk and Sozani embroidery, while western regions, including Rajasthan, Gujarat, and Maharashtra, showcase Kota Doria and Lehariya textiles. Central Regions like Madhya Pradesh and Chhattisgarh offer Chanderi, Kosa silk, and Maheshwar sarees. Eastern regions, such as Bihar, Jharkhand, Odisha, and West Bengal, offer Tasar and Kuchai silks. Northeastern Regions like Assam and Meghalaya offer Eri silk loin loom weaving. Southern Regions like Andhra Pradesh, Telangana, Karnataka, Kerala, and Tamil Nadu offer various silk-based fabrics (Silk Mark Organisation of India, n.d.). Silk thus stands as one of the most significant fibers in India. This chapter explores the classification, structure, and properties of silk fibers, as well as the distribution of silk-based crafts across India. It also highlights Geographical Indication (GI)-tagged silk products as an important aspect of the discussion. The recognition of silk-based crafts through Geographical Indications (GI) not only safeguards the authenticity and cultural heritage of these traditions but also enhances their market value and global visibility. By ensuring fair recognition and economic returns for artisans, GI tags serve as a vital tool for sustaining livelihoods while preserving the unique identity of India's silk heritage. This chapter also discusses the notable silk-based recognized crafts (GI and Non-GI) available in India.

## 2.2 Classification of Silk Fiber

The following classification of Indian silks could be formed based on the silk's origin. In essence, silk is a naturally occurring polymer made up of amino acids joined by a peptide group. The silkworm *Bombyx mori* (caterpillar stage) secretes silk, which is mostly a protein-based filament that solidifies into a fiber when exposed to air. Other species of silk worms that extrude protein fiber are likewise classified as silk, although they have different genus names, like Tasar, Muga, Eri, and Anaphe. They are called, as is, "wild silk" or "Vanya Silk" or "non-mulberry" silk (Bureau of Indian Standards 2020). Fibroin and sericin are the two main compounds that make up silk. Fibrin remains at the center of the fiber extruded from the spinneret of a silkworm, covered and bonded together by sericin, which also fills in the spaces to increase the cocoon fiber's tenacity (Zhang et al. 2011). An illustration of the arrangement of Fibroin and Sericin in the cross-section of silk fiber is presented in Fig. 2.1, depicting also the differences between wild silk and cultivated silk, such as Bombyx mori.

As seen in Fig. 2.1, each filament extruding from each gland of the silkworm is referred to as "Brin." Two brins (one from each gland on either side of the body of the silkworm) are cemented together by an adhesive-like compound or sericin at the spinneret of the silkworm to emerge as a silk strand or "bave" (Bureau of Indian Standards 2020).

The classification of silk and source species available in India is tabulated in Table 2.1.

To have a better understanding of the different fibers, such as mulberry, Tasar/Tussah, Muga, and Eri/Endi, understanding of the fine structure of the fibers in terms

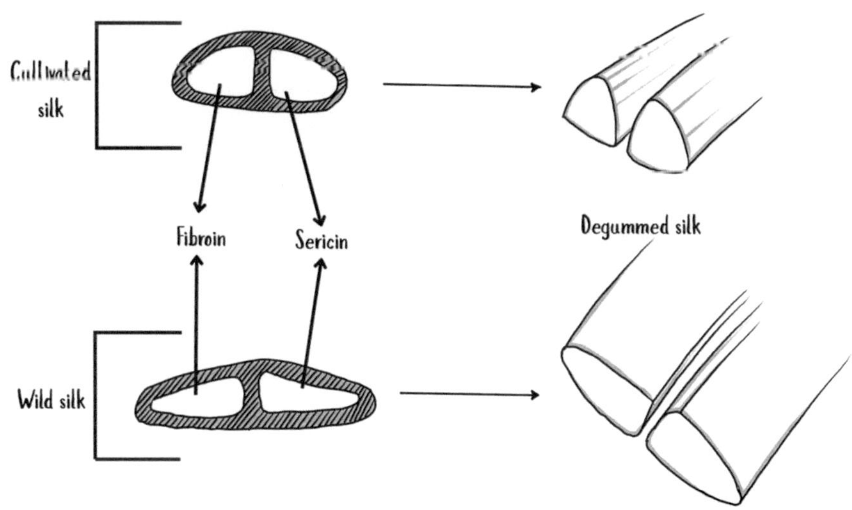


**Fig. 2.1** Difference in arrangement of fibroin and sericin in the cross-section of cultivated silk and wild silk [redrawn from Chand et al. (2023)]

**Table 2.1** Classification of silk available in India

| Silk type | Source species | Key features | References |
|---|---|---|---|
| Mulberry | Bombyx mori | Yellow, white or greenish yellow color, fine, lustrous, uniform; highest tensile strength | Bureau of Indian Standards (2020), Bajwa et al. (2019) |
| Tasar/ Tussah | Antheraea mylitta | Natural copperish color, textured appearance | Bureau of Indian Standards (2020), Cheung et al. (2009) |
| Muga | Antheraea assamensis | Silk of golden yellow color, Natural golden shine, fine, strong | Bureau of Indian Standards (2020), Sangma and Mahale (2020) |
| Eri/Endi | Samia ricini | White or brick red color obtained, Ahimsa Silk, soft, airy | Bureau of Indian Standards (2020), Singh et al. (2023), Laheru et al. (2025) |

of fiber constituents, physical properties, mechanical properties, and crystallization properties is important. The following section not only discusses these, but it also discusses the mechanism of silk formation in the glands of silk-spinning worms with a specific reference to *Bombyx Mori.*

## 2.3 Fine Structure of Silk Fiber

### *2.3.1 Fiber Constituents and Formation of Silk Fiber*

Silk fiber comprises mostly fibroin and sericin, with precise ratios and amino acid distributions. Fibroin remains in the core of the fiber as depicted in Fig. 2.1, and sericin surrounds the fibroin, binding two or more fibroin filaments in the raw silk, making it a bi-component entity secreted from the spinneret of the moth (Chand et al. 2023). To derive the usable fiber from raw silk, the outer coating of sericin is removed through a process called *degumming*. This treatment exposes the inner fibroin component, which is finer, lustrous, and primarily employed as silk filaments for producing yarns and fabrics in the textile industry (Rheinberg 1991). Silk fiber is composed predominantly of fibroin (70–80%) and sericin (20–30%), along with minor constituents such as carbohydrates (1.2–1.6%), inorganic matter (0.6–0.7%), waxes (0.4–0.8%), and pigments (0.2–0.3%) (Silva et al. 2022). Structurally, fibroin forms the inner core, imparting mechanical strength to the fiber, whereas sericin serves as the outer glue-like layer binding the filaments together. Each silk fiber consists of two fibroin filaments encased in a sericin coating. Both proteins are characterized by repetitive amino acid sequences that facilitate β-sheet structure formation. Fibroin is typically represented by the repeating sequence [GAGAGS]n, while a portion of the sericin sequence is denoted as GSVSSTGSSSNTDSST, where the residues G, A, S, T, V, N, and D correspond to glycine, alanine, serine, threonine, valine, asparagine, and aspartic acid, respectively (Akai 1984).

Though the major components of silk remain the same for all the varieties, the percentage of fibroin, sericin, and other constituents present in the cocoon varies for each species of silk-spinning worm. Table 2.2 presents the classification of cocoon shells based on the different percentages of components present.

Silk sericin is produced in the labial glands of B. mori, commonly referred to as silk glands (Silva et al. 2022). The silk glands are long and paired structures originating from the labial region. They are anatomically and physiologically divided into three major compartments: the anterior silk gland (which forms the excretory duct), the middle silk gland (which secretes three types of sericin), and the posterior silk gland (which secretes fibroin) (Fig. 2.2) (Kunz et al. 2016; Song et al. 2024).

**Table 2.2** Classification of cocoon shell components expressed in percentage

| Component | Mulberry | Tasar | Muga | Eri |
|---|---|---|---|---|
| Fibroin | 66–72 | 78–85 | 80–86 | 82–88 |
| Sericin | 25–32 | 14–17 | 12–16 | 11–13 |
| Wax | 0.3–0.4 | 1–2 | 0.5–1 | 1.5–2.2 |
| Minerals, ash, and others | 0.7–0.8 | 3–4 | 2–3 | 2–3 |

Adapted with permission from Padaki et al. (2015)

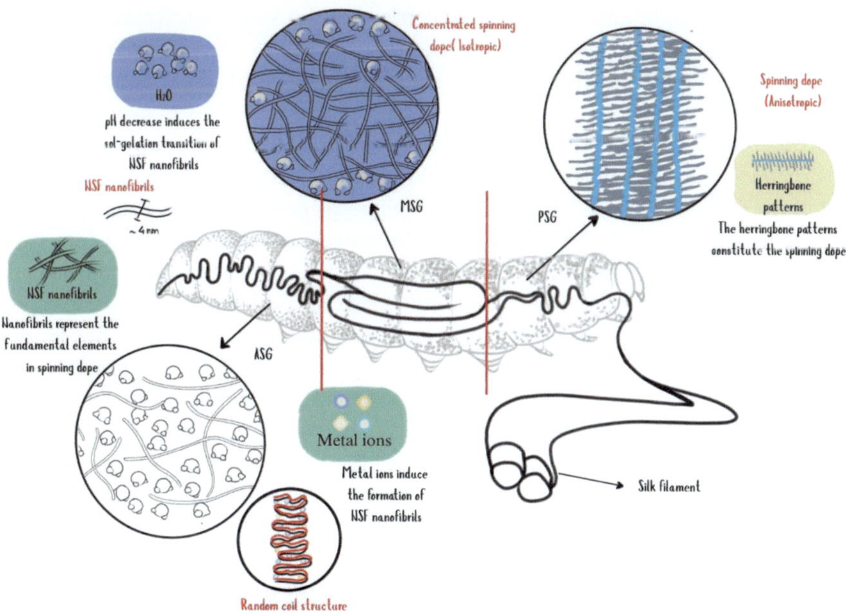


**Fig. 2.2** Anterior gland, middle silk gland, and posterior gland in B. mori [redrawn from Song et al. (2024)]

- Fibroin is stored here as a soluble protein complex, preventing premature aggregation (Inoue et al. 2000).
- Fibroin is stored in high concentration in a metastable state; sericin is secreted here and coats fibroin (Inoue et al. 2000; Wang et al. 2012).
- Along the gland, there is a gradual change from a random coil/α-helix to a β-sheet, accompanied by changes in mechanical (viscous/elastic) behavior, increasing aggregate size, etc. As a result, natural flow/residence in the anterior parts of the gland prepares the fibroin by modifying its conformation/aggregation before spinning (Jin et al. 2013).
- The narrowing duct applies elongational flow and shear, aligning fibroin molecules (Jin and Kaplan 2003; Holland et al. 2007).
- Inside the anterior silk gland, fibroin stored as a dense fluid begins to organize into nanofibrils. At this stage, fibroin does not yet form uniform fibrils, but instead assembles into discontinuous globular domains connected by thin linkers. This morphology represents an intermediate self-assembly state, facilitating the later formation of continuous, aligned nanofibrils (Hu et al. 2020).
- The discontinuous protofibrillar assemblies evolve into uniform nanofibrils. Structural transition from random coil/α-helix to β-sheet nanocrystals stabilizes the fibrils. This stage creates the basic nanoscale building blocks of silk fiber (Wang et al. 2012).
- Mature nanofibrils laterally assemble into parallel nano-bundles, oriented along the gland axis. These bundles form the mesoscopic core architecture of the silk thread (Koh et al. 2015).
- The left and right anterior silk glands merge at the spinneret, forming a single extrusion path. The spinneret's morphology, especially the silk press and duct geometry, facilitates fiber shaping and determines final silk dimensions by applying shear and compressive forces (Guo et al. 2019).
- As the fibroin dope exits the spinneret, shear forces and water removal induce a structural transition from soluble random coil/α-helix (silk I) to β-sheet crystallites (silk II). These β-sheets act as physical crosslinks, locking the structure into a mechanically stable fiber. Moreover, environmental triggers (pH, ion concentration, dehydration, shear) collectively stabilize silk fibroin into β-sheet domains. The bundled nanofibrils, once aligned and compressed in the spinneret, become densely packed into a continuous microfiber. This hierarchical packing ensures multi-scale organization: nanocrystals within fibrils, fibrils within bundles, bundles forming the fiber (Jin and Kaplan 2003). The previously aligned nano-bundles become closely packed and consolidated into a continuous microfiber within the spinneret. This assembly results in a multi-scale hierarchy: β-sheet nanocrystals within nanofibrils, which are bundled into microfibers, are critical for silk's strength and extensibility (Qi et al. 2017). As the silkworm pulls on the extruded fiber, tensile forces further align the β-sheet-rich nanofibrils along the fiber axis, reducing structural defects and improving mechanical properties such as toughness and stiffness (Jin and Kaplan 2003).

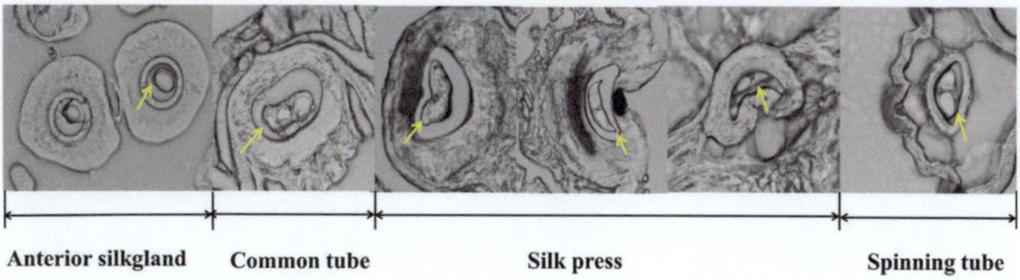


**Fig. 2.3** Shape-shifting of silk (yellow arrows) in the spinneret to form spinning tube [adapted from Guo et al. (2019) with permission]

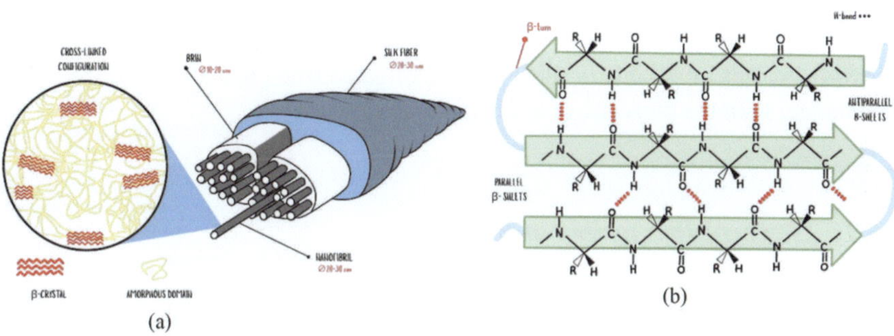


**Fig. 2.4** **a** Illustration showing crystalline and amorphous regions in silk fiber. **b** Illustration of β-sheet of protein in silk fiber [adapted from Reizabal et al. (2023)]

One important phenomenon referred to as "shape-shifting" takes place as presented in Fig. 2.3. As the silk feedstock reaches the spinneret, the two anterior silk gland ducts converge into a single, narrowing channel. This tapered geometry acts as a shape-shifting conduit, applying both shear and compressive forces that serve to align and compact the protein assemblies within the dope. The spinneret thereby functions as a precise molecular filter and form-shaper, dictating both the alignment and the final cross-sectional morphology of the emerging silk filament (Fig. 2.4a) (Guo et al. 2019). At the final outlet, the geometry of the spinneret orifice dictates the cross-sectional morphology of the emerging fiber. Simultaneously, the highly aligned fibroin undergoes β-sheet crystallization, locking the molecular order into a solid state that confers mechanical robustness (Fig. 2.4b) (Jin and Kaplan 2003).

## 2.3.2 *Surface and Cross-sectional Morphology of Silk Fiber*

Mulberry silk fibers have a smooth, uniform longitudinal appearance and triangular cross-section, as represented by the SEM images in Fig. 2.5. However, wild silk fibers are reportedly to have a significantly rough surface (Padaki et al. 2015). Moreover, all

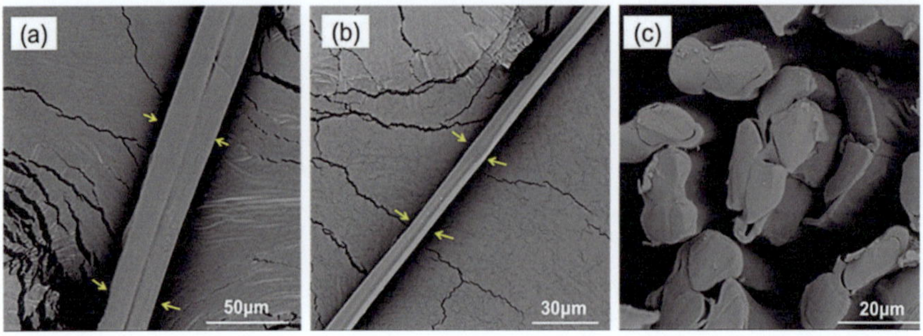


**Fig. 2.5** Cross-sectional/surface micrograph of silk fiber [adapted from Guo et al. (2019)]

the wild varieties of silk fibers show a flat, elongated triangular-like cross-sectional appearance, with the presence of voids in the fibers.

### 2.3.3 *General Properties of Silk Fiber*

The fineness of silk fibers is commonly measured in denier, which represents the mass in grams of 9 km of fiber. Another key parameter in silk characterization is the Nonbreaking Filament Length (NBFL), referring to the continuous length of silk filament present within a cocoon. Silk filaments are obtained through a process called reeling, in which 8–10 individual filaments drawn from separate cocoons are combined to produce a yarn of the desired count. For efficient reeling, it is essential to determine the available filament length in different types of cocoons (Padaki et al. 2015). Table 2.3 summarizes the fiber density, fineness, NBFL, and total filament length for various silk varieties.

Among all the varieties, mulberry silks are the finest, ranging from 2–3 deniers. In terms of NBFL, mulberry varieties also show a range of 400–600 m, whereas the wild varieties show a range of 0.05–250 m, as presented in Table 2.3.

**Table 2.3** Physical properties of silk fiber [adapted from Padaki et al. (2015)]

| Type of silk cocoon | Fiber density (g/cm$^3$) | Fiber fineness (den) | NBFL (m) | Total filament length (m) |
|---|---|---|---|---|
| Mulberry (bivoltine) | 1.34 | 2–3 | 700–800 | 1200–1600 |
| Mulberry (multivoltine) | 1.34 | 2–3 | 400–600 | 900–1200 |
| Tasar | 1.31 | 8–12 | 100–250 | 750–900 |
| Muga | 1.30 | 4–7 | 150–250 | 600–800 |
| Eri | 1.30 | 3–4 | 0.05–2.0 | 400–500 |

Silk, being a protein fiber, is composed of amino acids. Each amino acid contains at least two functional groups, viz. an amino group (–$NH_2$), which can accept a proton to form –$NH_3^+$ (basic group), and a carboxyl group (–COOH), which can lose a proton to form –$COO^-$ (acidic group) (Dash et al. 2008). Due to the presence of both basic and acidic groups, an interesting phenomenon occurs in silk. At acidic pH, the amino groups are protonated (–$NH_3^+$ dominates), and at alkaline pH, the carboxyl groups are deprotonated (–$COO^-$ dominates). Studies suggest, therefore, that the silk fiber exhibits an isoelectric point ranging from 3.6 to 5.2, at which silk carries equal positive and negative charges, resulting in a net charge of zero (Pham and Tiyaboonchai 2020). Since in both acidic and basic pH, both groups are ionized, the fibroin molecule carries both positive and negative charges simultaneously. Therefore, silk may be described as an amphoteric/polyampholyte protein with a defined isoelectric point (Sookne and Harris 1939).

The amphoteric nature of silk (due to amino and carboxyl groups in fibroin amino acids) strongly influences its behavior in textile applications. Consequently, the fibre exhibits versatility in coloration, sensitivity to pH during dyeing and finishing processes, and suitability for surface modification, contributing to the moisture management and comfort properties of silk textiles (Koh et al. 2015).

The densities of different silk fibers are reported to vary significantly, as presented in Fig. 2.6. This leads to substantial differences in microstructural parameters, including crystallinity, density, birefringence, dichroic ratio, and sonic modulus, across the different varieties. Studies (Sen et al. 2004; Javali et al. 2018) confirm that crystallinity in silk is the central structural factor influencing many of its measured physical properties.

As it is clear from Fig. 2.7, the crystallinity % of the fibers ranges from 25 to 50%. Fibers from different layers of the cocoon wall differ in their molecular orientation and crystallinity. Typically, the outer layers of the cocoon have higher crystallinity and orientation, making them stiffer and stronger. In contrast, the inner layers are often less crystalline, softer, and more extensible. The probable reason may be as the silkworm spins, the first layers are more uniform and aligned, while later layers

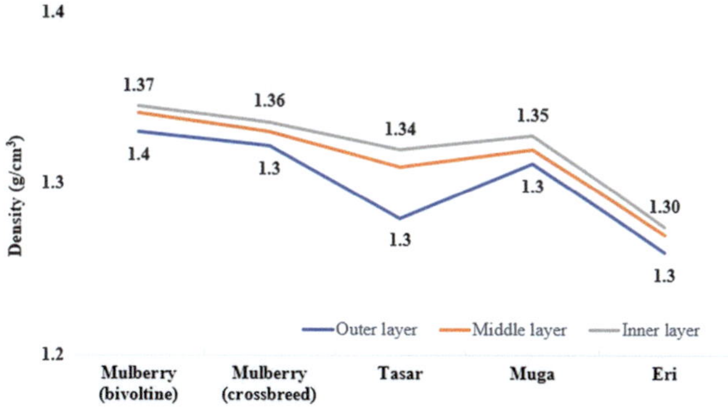


**Fig. 2.6** Densities (g/m$^3$) of different varieties of silk fibers (Babu 2020)

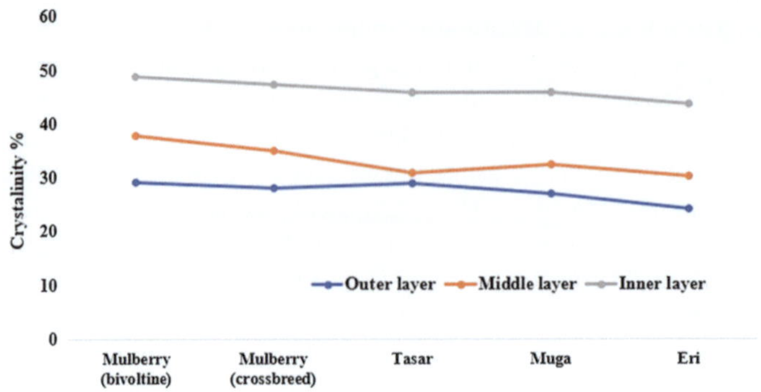


**Fig. 2.7** Crystallinity (%) of different varieties of silk fiber extracted from different layers of cocoon (Sen et al. 2004)

are laid down with variations, leading to differences in orientation and packing of fibroin molecules resulting such variation.

## 2.4 Processing of Silk Fiber

### *2.4.1 Fiber to Yarn*

Depending on the source of the cocoon, India's silk fiber spinning is often divided into two systems: mulberry and non-mulberry. Non-mulberry silks (Eri, Tasar, and Muga) are equally important for their distinctive qualities, cultural significance, and regional economic benefits. Nevertheless, mulberry silk spinning continues to serve as the cornerstone of India's organized silk industry. Depending upon the type of silk-spinning worm and the type of cocoon that they make, the spinning process of steps differs significantly. Since mulberry silks are finer, continuous in nature, and have the highest NBFL value as referred in Table 2.3, the processes for spinning may consist of steps such as hot water reeling, automatic reeling machines, and re-reeling to improve yarn uniformity. Non-mulberry silks differ in their sericin composition, filament length, and denier (as referred in Tables 2.2 and 2.3), involving different approaches to cooking, degumming, and spinning. For example, Tasar silk often requires permeation-based cooking due to its dense shell (Javali et al. 2018), while Eri silk, being spun rather than reeled, is processed more like cotton in carding and spinning systems, as after the moth is out of the cocoon, only shorter lengths of fibers are available for the spinning process (Andiappan et al. 2016). Similarly, Muga silk, prized for its natural golden luster, is manually reeled and reserved for luxury textiles (Jigyasu et al. 2023). Table 2.4 gives a comparison of the processing steps of different types of silks.

**Table 2.4** Comparative steps of processing different types of silks

| Factors | Mulberry | Tasar/Muga | Eri |
|---|---|---|---|
| Types of filaments/ cocoons | Closed; continuous filament suitable for reeling (Sonwalkar 1993a) | Tougher cocoon shells with reelable but sturdier filaments (Javali et al. 2018) | Discontinuous filament leaving staple like fibers (Zhou et al. 2020) |
| Route | Filament Reeling (Lee 1999) | Filament reeling (wet/ dry; mechanized) (Central Silk Board, n.d.-a) | Spun-silk (staple) spinning only (Sonwalkar 1993a) |
| Degumming | Standard cocoon cooking (hot water/ steam) to soften sericin (Lee 1999) | Stronger/optimized cooking (permeation/ alkaline assists) needed due to tough shell (Javali et al. 2018) | Pre-degumming before staple processing (Babu 2018) |
| Machineries | Charkha, cottage basin, filature, multi-end automatic (Sonwalkar 1993a) | Motorized reeling-cum-twisting machine, two-in-one reeling-cum-twisting machine (CSTRI) ( Central Silk Board, n.d.-a) | Opening, carding/ combing followed by ring/rotor spinning (Babu 2018) |
| Fiber/yarn characteristics | Fine, uniform filaments, high luster (Babu 2018) | Coarser filaments; higher denier; distinct yellow/golden shades (Jigyasu et al. 2023) | Short staple; wool-like hand feel; cannot form continuous filament (Zhou et al. 2020) |

One notable point to mention is that there are few popular names of yarns made from mulberry and Tasar, tabulated in Table 2.5.

In any discussion of silk yarn manufacturing, the process of reeling holds particular importance. The fibers present in the cocoon must be extracted in filament form before they can be twisted together to form the yarn. When a silkworm spins its cocoon, it produces a continuous filament that can be several meters in length, as mentioned

**Table 2.5** Variety of handspun yarns made from different silk fibers (Bureau of Indian Standards 2020)

| Type | Popular name | Description |
|---|---|---|
| Mulberry | Matka | The yarn spun by hand appliances cut of mulberry pierced and other waste cocoons |
| Tasar | Ghicha | The yarn drawn by hand out of Tasar cocoons without any twisting |
| | Jhuri | Yarn spun out of Tasar waste without subjecting it to opening processes |
| | Katia | The yarn spun out of Tasar waste after opening and cleaning |
| | Kete | A term used for handspun Tasar thread |

in Table 2.3. These filaments are adhered together with a substance called sericin, and they need to be softened before being removed from the cocoon. For mulberry silk, boiling the cocoons with hot water is effective enough. However, researchers have reported that a pressurized, multi-pan cooking method with controlled reeling speed (80–120 m/min) has produced significantly improved reliability for Indian bivoltine hybrid silk cocoons (Naik and TH 2005). The Central Silk Board (CSB) reports a two-pan cooking protocol for multivoltine mulberry cocoons, which is extendable to Tasar varieties (Central Silk Board, n.d.-b). Since Eri cocoons are open-mouthed and discontinuous, the aim for reeling is not applicable in this case. Eri cocoons are treated for degumming mainly to enable the subsequent spinning process (Chattopadhyay et al. 2017). However, depending on the types of cocoons and the post-cooking process, namely reeling, the process also varies. For mulberry silk, reeling techniques such as Charkha, cottage basin, filature, and multi-end automatic are mainly used (Sonwalkar 1993a). For Indian Tasar silk yarns, reeling could be in both dry and wet conditions (Central Silk Board, n.d.-a). For Muga silk, CSTRI has developed a reeling-cum-twisting machine (Central Tasar Research & Training Institute 2023). However, Eri silk doesn't require any reeling, as it is open-mouthed and has a discontinuous filament (Kumar et al. 2021a). In addition to this, the reeling process leaves certain unreliable parts of the cocoon for both cultivated and mulberry silk, which is regarded as the waste silk or silk noil. The same parts are often used as raw materials for spun silk. The yarns spun from the noils are often called noil yarn. Apart from the different terms mentioned in Table 2.5, there are other terms related to silk yarns based on techniques, application, and yarns spun from the waste. Table 2.6 presents a list of such most popular silk yarn names (Bureau of Indian Standards 2020, 1991).

### *2.4.2 Typical Constructions of Silk Fabrics*

Table 2.7 stresses that silk fabrics have area densities ranging from 20 to 45 g/m$^2$. The yarns used in the weft direction are either coarser or have a higher ply or almost equivalent linear density. Although the linear densities may be the same, the weft direction's strength is typically greater than the warp direction because of a higher number of picks per decimeter (dm) or ply.

## 2.5 Notable Silk-Based Handicrafts of India

India's silk heritage has a long history that dates back centuries. Each silk variety, such as mulberry, Tasar, Muga, and Eri, is intrinsically tied to specific regions and communities in India, giving rise to distinctive weaving and craft techniques. In the past, there have been examples of silk fabrics being patronized by the royals of the

**Table 2.6** Different terms related to silk fiber and yarn (Bureau of Indian Standards 2020)

| Terms (related to silk fiber and yarn) | Description |
|---|---|
| Galettame | Waste silk from the cocoon's inner part |
| Northrow | When two or more untwisted singles are doubled together, they only twist enough to hold the silk strands together. The process eliminates the creation of loops and produces an extremely soft yarn |
| Frison | Before reeling, the silk waste scraped from the cocoons' outer layers. Such garbage is also referred to as Jooth (Hindi), Knubs (English), and Strusa (Italian) |
| Floss | The cocoon's exterior layer contains low-quality silk that is scarcely accessible. Another name for it is Blaze (French). The word is also used to describe delicate silk yarn or untwisted singles that are used for wire insulation and needlework |
| Cordonnet | Two or more strands of soft, thick silk yarn twisted together in a single direction |
| Crepe-de-chine yarn | Hard twisted yarn, usually with 1 600 to 2 800 turns per meter generally made from 3 to 5 raw silk threads. It is used as weft in crepe-de-chine |
| Crepe-georgette Yarn | Twisted yarn usually with 2 000 to 3 600 turns per meter, generally made of two threads of raw silk |
| Grenadine twist | Hard-twisted plied silk yarn with 800 to 2400 turns per meter, composed of singles, which have been first twisted in the same direction and then have been twisted together in the reverse direction. It is used in the warp of grenadines which have to withstand severe friction |
| Croisure | A technique of twisting raw silk over itself in the process of reeling with the object of consolidating the different filaments constituting raw silk, rounding off the angularities of the thread and squeezing off the extra moisture and gum |
| Etching silk | A hard-twisted silk sewing thread used for outlining embroidery |
| Dupion | Silk reeled from double cocoon, also known as "Duppione" |
| Pure zari | A yarn having a silk core, wrapped with silver wire and may be electroplated with gold. The silk core is a two ply 16/18 denier soft twisted yarn dyed in a red or yellow color |

country from time to time. For example, in South India, the Chola and Vijayanagara dynasties extended patronage to silk weavers, particularly in Kanchipuram, establishing the foundation of the world-renowned Kanjivaram silk sarees (Murthy 1985). Similarly, in Assam, the Ahom royalty promoted the cultivation of Muga and Eri silks, elevating them into symbols of status and cultural identity (Goswami and Neog 2013). During the Mughal period, emperors such as Akbar actively encouraged the weaving of fine silks and brocades, leading to the rise of celebrated traditions like Banarasi silk with its rich zari work (Kumar 2006). In the same timelines, the town of Murshidabad in undivided Bengal emerged as a major silk-producing center

**Table 2.7** Few typical fabric constructions for handwoven silk fabric suitable for use in sarees or dhotis (Bureau of Indian Standards 1991)

| Warp count (tex) | Weft count (tex) | Ends/dm | Picks/dm | Mass (g/m$^2$) | Breaking load in strip method (Kg) |
|---|---|---|---|---|---|
| 3.1/3.6 (single) | 2.2/2.4 (3 Ply) | 390 | 410 | 40 | Warp—16, weft—44 |
| 2.2/2.4 (2 Ply) | 2.2/2.4 (3 Ply) | 390 | 350 | 45 | Warp—25, weft—34 |
| 3.6/3.8 | 2.4/2.6 (2 Ply) | 418 | 418 | 35 | Warp—18, weft—20 |
| 3.2/3.6 | 3.2/3.6 | 394 | 456 | 25 | Warp—16, weft—18 |
| 2.4/2.6 | 2.8/3.2 | 410 | 410 | 20 | Warp—14, weft—16 |
| 2.4/2.6 (2 Ply) | 2.2/2.4 (3 Ply) | 394 | 394 | 35 | Warp—18, weft—20 |
| 1.4/11.7 (2 Ply) | 2.4/2.6 | 434 | 434 | 25 | Warp—16, weft—16 |

in the seventeenth and eighteenth centuries, often referred to as the "Silk Mecca of Bengal." The Nawabs of Bengal, especially under Murshid Quli Khan, encouraged the growth of silk sericulture and weaving, integrating Murshidabad silk into courtly attire and diplomatic gifts (Chattopadhyay 2014). Bengal silk, renowned for its fineness and luster, also drew patronage from European trading companies, such as the British East India Company, which exported large quantities to Europe. The support of the Mughal court and regional Nawabs not only sustained Bengal silk weaving but also gave rise to a distinctive heritage of Baluchari sarees, woven with intricate figurative motifs narrating mythological and royal themes (Mukherjee 2005). Royal households often commissioned silk textiles for rituals, court ceremonies, and diplomatic exchanges, reinforcing silk's association with luxury, power, and spirituality. Till today, the legacy goes on with the people of India. To protect the intellectual property of silk products that possess a specific geographical origin and unique qualities linked to that origin, they are granted a Geographical Indication (GI) tag by the Government of India. Figure 2.8 shows all the major GI-tagged handicrafts of India. Some non-GI products are also mentioned in the map. India's significant silk-based crafts are depicted on the map, which is separated into zones for convenient comprehension, including the Eastern, Northeastern, Central, Western, Northern, and Southern regions.

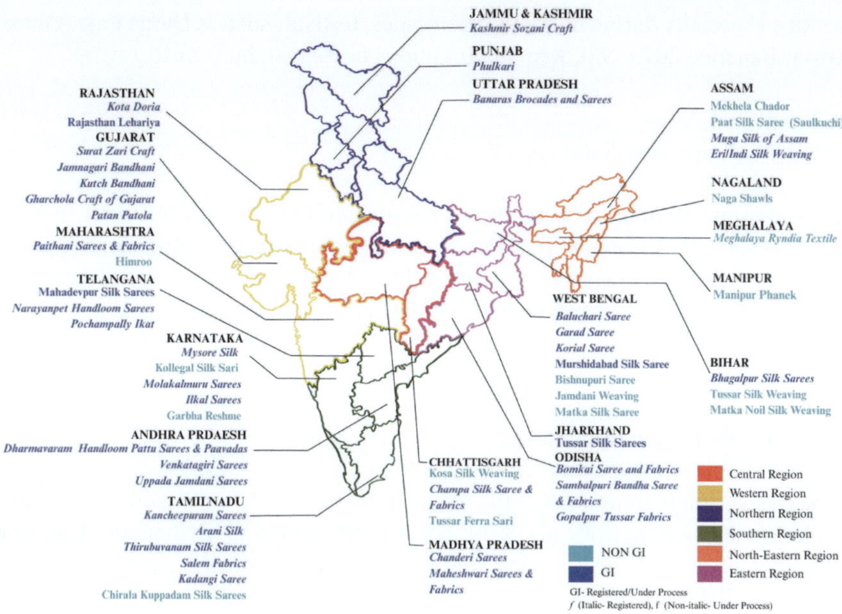


**Fig. 2.8** Region-wise notable silk-based handicrafts in India

## *2.5.1 Eastern Region*

The major silk crafts in Eastern India come from states such as West Bengal, Odisha, and Bihar. The crafts famous in this state are mentioned as follows.

### 2.5.1.1 Silk Crafts from West Bengal

#### 2.5.1.1.1 Baluchari Silk Saree

Baluchari is one of the iconic silk sarees, woven from mulberry silk, and is famous for its elaborate pallus and borders featuring brocaded depictions of mythological stories from the Ramayana and Mahabharata, as well as motifs inspired by courtly life, flora, and fauna, executed using extra-weft techniques (Mukherjee 2016; Sonwalkar 1993b).

#### 2.5.1.1.2 Garad Silk Saree

Garad saree weaving is linked with the sericulture heritage of Murshidabad and Malda district, with mulberry silk. These sarees are featured in minimalistic bright red borders and temple-style designs making them ideal for representing Bengali

Identity, especially during religious ceremonies, festivals such as Durga Puja (Ghosh 2014a; Banerjee 2013; Silk weaving traditions of Bengal, https://csb.gov.in).

##### 2.5.1.1.3 Korial Silk Saree

Sarees woven with mulberry silk have a contrasting look with a white body and red border, commonly termed as "Laal Paar." As red signifies fertility and auspiciousness, while white signifies purity and sanctity in Indian culture, this saree is another attire of choice for auspicious events such as Durga Puja rituals, weddings, and festive occasions (Sen 2015).

##### 2.5.1.1.4 Bishnupuri Silk Saree

Primarily woven with mulberry silk, Bishnupuri silk sarees are notable for designs that draw inspiration from the terracotta temple panels of Bishnupur, featuring depictions of mythological stories, floral patterns, and geometric shapes (Das and Mukherjee 2018).

##### 2.5.1.1.5 Murshidabad Silk

Murshidabad silk fabrics are primarily woven with mulberry silk, notable for its vibrant colors, minimalistic motifs, and elegant plain or striped borders. However, hand-painted and block-printed designs have also become popular, often incorporating floral, paisley, and geometric motifs, and hand-painted sarees using the kalamkari technique (Das and Mukherjee 2018).

##### 2.5.1.1.6 Matka Silk

Matka silk is produced primarily in West Bengal in districts such as Murshidabad and Birbhum. Unlike filament silk, Matka yarn retains a coarse, irregular texture, which imparts a matte appearance to the fabric (Chanu et al. 2024).

##### 2.5.1.1.7 Tasar Silk

Apart from these sarees mentioned above, there are cultures of weaving sarees with Tasar silk, particularly in the districts of Bankura, Purulia, and Birbhum, that are notable for their natural golden-beige sheen, coarse yet lustrous texture, and lightweight drape (Roy 2015).

#### 2.5.1.2 Silk Crafts from Bihar

The following major crafts are discussed, which are available in Bihar.

##### 2.5.1.2.1 Bhagalpur Silk Saree

Once known as the "Silk City of India," the city of Bhagalpur in Bihar has a long history of weaving Tasar silk. These sarees are notable for their golden sheen, durability, and coarse yet rich texture (Ghosh 2014b).

##### 2.5.1.2.2 Kuchai Silk

Woven with Tasar silk in Jharkhand, earlier a part of Bihar (New Indian Express 2011).

#### 2.5.1.3 Silk Crafts from Odisha

##### 2.5.1.3.1 Bomkai Saree

Originated in the village called Bomkai in the Ganjam area. Both cotton and silk sarees are available, which are famous (Behera 2024). The unique aspect of these sarees is the intermixing of the two techniques of tie-and-dye and embroidery. Bomkai, meaning bandhakala in the Odia language, employs ikat (binding) as well as detailed needlework on the pallu (one part of the saree worn over the shoulder), especially. Such intricate threadwork is made on bright, contrasting backgrounds (Bomkai: Draped in traditions of Odisha. 2021).

##### 2.5.1.3.2 Sambalpuri Ikat Silk Saree

Sambalpuri (Odisha) ikat sarees, often locally called "Bandha," are handwoven textiles famed for their intricate ikat (tie-dye) technique in which yarns are resist-dyed before weaving so that the design emerges directly on the loom. Practiced mainly in western Odisha in districts, namely Sambalpur, Bargarh, Sonepur, Bolangir, and neighboring districts, Sambalpuri ikat is produced in both cotton and silk; the silk versions are prized for their fine luster, sharp motifs, and traditional designs (conches, wheels, florals, and animals) that carry symbolic and ritual meanings. The craft is community-based (notably the Meher weaving community) (Das 2021).

2.5.1.3.3 Gopalpur Tasar Silk

Gopalpur Tussar (from Gopalpur village, Jajpur district, Odisha) is a traditional handwoven tussar (Tasar) silk fabric used for sarees, stoles, and related textiles. The cluster is noted for its extra-weft/tapestry techniques and characteristic motifs (Kumar et al. 2021b).

2.5.1.3.4 Habaspuri Silk

Habaspuri sarees are made from both silk and cotton. The saree features an amazing Anchal that is beautifully embellished with traditional patterns, such as fish, flowers, and tortoises, and is bordered by ribs (Danti). The body is adorned with little butties. The same classic designs in vertical ripples are used to embellish the border, which looks really lovely. Stoles (Dupatta): Dupattas are made from cotton that is 2/100 and 21,120 in size, and they are embellished with the same traditional motifs on the Anchal or pallu and the borders (HEPC India n.d.).

## *2.5.2 Northeastern Region*

### 2.5.2.1 Silk Crafts from Assam

2.5.2.1.1 Muga Silk

Assam is known for its traditional Muga silk, a naturally golden, non-mulberry (vanya) silk produced by the semi-domesticated silkworm Antheraea assamensis. It is prized for its natural sheen, durability, and cultural importance in Assamese ceremonial textiles (Directorate of Sericulture, Government of Assam n.d.).

2.5.2.1.2 Pat Silk

"Pat" (often written paat) refers to mulberry silk historically associated with the Sualkuchi weaving cluster of Assam; Pat silk is characterized by a fine white luster, which is widely used for mekhela-chador and conventional silk garments (Government of Assam, Handloom, Textiles & Sericulture Department n.d.).

2.5.2.1.3 Eri Silk

Eri (endii/eri) silk is produced by Samia ricini (and related species) and is a spun (non-reeled) silk known for thermal comfort (warmth), softness, and its use in shawls and traditional garments; Eri culture is important for rural livelihoods in Assam and is the subject of applied sericulture research on breeds, host plants, and rearing practices (Directorate of Sericulture, Government of Assam n.d.).

##### 2.5.2.1.4 Mekhela Chador

Native to Assam's Brahmaputra Valley, the Mekhela Chadar is a luxurious traditional garment made of silk. The hub of the Assam silk industry is "Sualkuchi," which is located a few kilometers from Guwahati, the state capital. The area can rightfully be referred to as "The Silk Heaven of India" and known for making Mekhla Chador using silk (Government of Assam 2026).

### 2.5.2.2 Silk Crafts from Manipur

#### 2.5.2.2.1 Manipur Phanek

The Phanek is a culturally embedded, handwoven wraparound skirt worn by Meitei women of Manipur, traditionally woven on loin-looms (back-strap looms) in cotton or silk, featuring stripes or block colors and ornate borders with mythological motifs like the khoijao (hook) or snake designs. It symbolizes femininity, social identity, and resistance, historically marking clan affiliation and notably worn by women during the Nupi Lan (women's uprisings) as an emblem of valor and community solidarity (Angom and Srivastava 2019; Ngangbam et al. 2024).

### 2.5.2.3 Silk Crafts from Meghalaya

#### 2.5.2.3.1 Ryndia Silk

Ryndia Silk, also known as "Peace Silk" or Ahimsa Silk, is a traditional Eri silk handspun and handwoven by women artisans of Meghalaya, especially in the Ri-Bhoi district's Umden-Diwon Eri Silk Village. Named after the Ryndia (castor) plant, which feeds the silkworms, this silk is produced through an eco-friendly, cruelty-free process: The silkworm emerges naturally before its cocoon is harvested. The fabric is naturally dyed using plant-based ingredients and woven on traditional looms (North Eastern Handicrafts & Handloom Development Corporation Ltd., n.d.; The Times of India 2025).

##### 2.5.2.4 Silk Crafts from Nagaland

###### 2.5.2.4.1 Naga Shawls

Assam produces three types of silk, viz. Muga, Pat (mulberry), and endi (also known as Eri). These silks are used to weave traditional fabrics in Nagaland on Loin loom, such as Naga Shawls which are prized for their longevity, vibrant color, and excellent quality (Chakravorty et al. 2010; Karolia and Prakash 2014).

##### 2.5.2.5 Silk Crafts from Mizoram

###### 2.5.2.5.1 Paun

Assam produces three types of silk, viz. Muga, Pat (mulberry), and endi (also known as Eri). These silks are used to weave traditional fabrics in Mizoram on Loin loom, such as Paun from Mizoram (Ramswamy and Kumar 2013).

### *2.5.3 Northern Region*

## 2.6 Silk Crafts from Jammu & Kashmir

### *2.6.1 Kashmir Sozani*

"Kashmir Sozani" is a delicate needlework technique used on a variety of fabrics to show designs inspired by the Kashmir Valley's flora and fauna done using silk and viscose yarn (Singh 2015).

#### 2.6.1.1 Silk Crafts from Punjab

##### 2.6.1.1.1 Phulkari

Phulkari, or flower embroidery, reflects the artistic sensibilities of Punjabi women and their ability to blend practicality with aesthetic appeal. Pat, a soft, untwisted silk floss thread used in the embroidery, is regionally dyed and sourced from areas such as Afghanistan, Turkistan, Kashmir, and Bengal (Gupta and Mehta 2014).

#### 2.6.1.2 Silk Crafts from Uttar Pradesh

##### 2.6.1.2.1 Banaras Brocade Sarees

One of India's richest centers for weaving crafts, Banaras is well-known for its brocade sarees and other clothing items. Banaras is renowned worldwide for its exquisite and priceless silk sarees and brocades. Jangla, Tanchoi, Vaskat, Cutwork, Tissue, and Butidar are some of the exclusive saree varieties made in the centers. They are composed of silk warp and silk weft, on a plain or satin ground base, and brocaded with additional weft patterns in various layouts that introduce Buties, Bells, Creepers, and Buttas in the ground, border, and Anchal to give them a lovely appearance (Verma and Mishra 2018).

### *2.6.2 Central Region*

#### 2.6.2.1 Silk Crafts from Madhya Pradesh

##### 2.6.2.1.1 Chanderi Sarees

Chanderi, a small town in Madhya Pradesh's Guna district, is well known for its exquisite silk work. The town's talented craftspeople weave silk sarees, which are renowned for their exquisite fineness, elaborate motifs, and delicate texture. These sarees are a prime example of the historic Chanderi weaving heritage, in which silk is expertly handcrafted into a product with cultural and aesthetic value (Mohanty 2022; Dhamija 2014).

##### 2.6.2.1.2 Maheshwari Sarees

Maheshwari fabric, woven from cotton or pure silk, is a popular saree in the Khargone district of Madhya Pradesh. Initially made from pure silk, it has evolved to include silk-based, cotton-based, and blended cotton-silk fabric to meet market demands. The sarees feature symmetrical stripes, checks, and floral patterns and were initially made from pure silk (Dhamija 2014).

#### 2.6.2.2 Silk-Based Crafts from Chhattisgarh

##### 2.6.2.2.1 Champa Silk Saree

The Chhattisgarh region is famous for its Champa silk sarees woven by the Devangan community. The production of silk, spinning, and other auxiliary activities is practiced by numerous different indigenous communities. Silk cocoons, gheecha (coarser yarn), katiya (waste material yarn), nassi (stem-based yarn), and other threads like golden or silver zari, fancy yarns, and extra warp or embroidery threads are among the raw materials needed to produce Champa silk (Indian Textiles & Crafts Repository n.d.).

##### 2.6.2.2.2 Kosa Silk Saree

The Devangan community in Chhattisgarh, particularly in the towns of Bilaspur, Raigarh, Korba, and Champa, is the primary practitioner of the Kosa silk-weaving craft. The cocoons of Indian silkworms, which are found on certain tree species like Arjun, Sal, and Saja, are used to make Kosa silk (Indian Textiles & Crafts Repository n.d.).

##### 2.6.2.2.3 Tussar Ferra Saree

The Devangan community in India is known for weaving tussar silk saris and fabrics, with other tribal communities also rearing and collecting the silk. The Devangan community has a deep-rooted connection to the art of weaving these textiles, and several tribal communities are involved in silk rearing, collection, and other supporting services. Tussar ferra saris are made using local tussar or kosa, mulberry, and artificial silk. A three-shuttle loom is used for weaving these saris (Indian Textiles & Crafts Repository n.d.).

### *2.6.3 Western Region*

#### 2.6.3.1 Silk-Based Crafts from Rajasthan

##### 2.6.3.1.1 Kota Doria

Kota Doria, also known as Masuria Malmal, is a light, airy, and comfortable fabric ideal for summer. Its softness and transparency make it graceful and part of fashion. The warp and weft use a 5:1 cotton-silk thread ratio, creating a fine checquered pattern with "khats," which are made so proficiently that the fabric becomes transparent (Chaudhary and Pant 2016).

##### 2.6.3.1.2 Rajasthan Lehariya

Rajasthan's popular dyeing technique, known as Leheriya, produces striped textiles in a variety of bright colors. This technique uses cotton or silk cloth and resists dyeing. The Leheriya pattern waveform consists of alternate colored and white cloth, with artisan colors used in single or multicolor schemes. The technique is named after the waveform, which is called Leheriya in Rajasthan (Chaudhary and Pant 2016).

#### 2.6.3.2 Silk-Based Crafts of Gujarat

##### 2.6.3.2.1 Surat Zari Craft

Surat is a small-scale cottage industry that produces intricate designs using silk and zari threads on handlooms. Zari is an intermediate product for textiles and forms a subset of the industry. The uniqueness of Surat zari thread lies in its luster, greater shine due to silver electroplating, and artisans' ability to draw wire up to 2500 yards per ounce using independently produced machinery (Garg and Sidhpuria 2015).

##### 2.6.3.2.2 Jamnagari Bandhani

Tie-dye, also known as Jamnagari Bandhani, is a very old method of cloth decoration.

Jamnagari Bandhani is used by practically all Hindu and Muslim communities in the Kachchh region for both daily wear and ceremonial purposes. In order to stop the color from leaking into the tied regions during the dyeing process, the cloth is tied in elaborate designs using nails, beads, or grain in the age-old Jamnagari Bandhani dyeing technique. Gajji silk, Georgette, Chiffon, Chinon, Tussar, mulberry silk, cotton silk, and fabric woven with cotton warp and silk weft are among the materials used as seeds for silk Jamnagari Bandhani (Bandhani 2016).

##### 2.6.3.2.3 Kutch Bandhani

One of India's oldest tie-dye traditions, Kutch Bandhani (also called Bandhej or Bandhni) is thought to have started with the Muslim Khatri minority in Gujarat's Kutch district. This technique, which comes from the Sanskrit word "bandhana" (to tie), entails choosing a light fabric, usually silk, washing and dyeing it a base color, then painstakingly tying tiny knots that serve as resistance points against dye penetration. The outcome of multiple dyeing and untying processes is a richly colored cloth with intricate patterns, such as dots, waves, squares, or abstractions, which gives Bandhani sarees their distinctively striking appearance (TOI Lifestyle Desk 2023).

##### 2.6.3.2.4 Gharchola

Gharcholas, originating in Gujarat's Khambhat district, are a traditional part of Gujarati weddings. They symbolize blessings and well-wishes to newlyweds. The craft involves pinching, tying, and dyeing, and is traditionally woven by hand using Gajji silk. Gharcholas have a distinctive checquered design, with 9, 12, and 52 square patterns being the most popular. Each check or box in the Gharchola is tied and colored to create Bandhani patterns (Bharatiya Vastra Shilpa Kosh, NIFT DEL-420 CFP—Gharchola).

##### 2.6.3.2.5 Patan Patola

The twin ikat fabric known as patola has no distinction between the right and incorrect sides. The Sanskrit word patta, which means silk, is the root of the word patolu. A silk bridal sari with vivid, striking colors is called a patola. One sari takes four to six months to weave, making patola one of the most intricately woven fabrics (Bharatiya Vastra Shilpa Kosh, NIFT DEL-54 CFP—Patan Patola).

#### 2.6.3.3 Silk-Based Crafts of Maharashtra

##### 2.6.3.3.1 Paithani Saree

Paithani, a traditional silk sari in Maharashtra, is a significant part of the socio-cultural fabric of the state. It is traditionally part of the trousseau of every Maharashtrian bride and is centered in Pune, Nashik, and Aurangabad. Yeola, a town and municipal council in Nashik District, is currently thriving with Paithani weaving activity. This silk sari is part of a culture that values thrift over flamboyance, showcasing the lavish spending of indigenous weavers to clothe their womenfolk in traditional silk and spun gold, especially on festive occasions (Bhatia and Desai 2014).

##### 2.6.3.3.2 Himroo Shawls

Indian artisans modified Persian themes to meet local needs, elevating them to the more earthy and colorful characteristics of Indian art overall. Glamorous Indian motifs overshadowed and surpassed motifs and texture. This handwoven fabric, which is an ethereal synthesis of cotton and silk, features exquisite combinations of green, blue, black, gold, and other romantic color combinations that were originally custom-made by the royal household, nobility, and courtiers (Bhatia et al. 2016).

## 2.6.4 *Southern Region*

### 2.6.4.1 Silk-Based Crafts of Andhra Pradesh

#### 2.6.4.1.1 Dharmavaram Handloom Pattu Sarees and Paavadais

The Dharmavaram silk sarees are well-known among silk weaves. Located in the Rayalaseema region of Andhra Pradesh's district of Anantapur, Dharmavaram is a small town with a thriving handloom cluster. Tussar silk is used in the weaving of these sarees. In India, particularly in the South, this silk saree is much sought after. The most sought-after bridal sarees have broad borders of solid colors with contrast pallu woven with brocaded gold designs (Savithri et al. 2013).

#### 2.6.4.1.2 Venkatagiri Sarees

Venkatagiri town continues to take pride in producing sarees that are almost as good as silk. Zari is a very fine silk thread woven on silver thread that has been polished with gold. The core of the tested zari is composed of pure silk, coiled around a small piece of copper foil, and then electroplated with pure gold (Bala 1999).

#### 2.6.4.1.3 Uppada Jamdani Sarees

Uppada Jamdani sarees are diaphanous silk sarees from Andhra Pradesh, India. Jamdani is a handwoven fabric with Persian and Bengali roots. The fabric is created by two weavers working in tandem on a pit loom, passing weft and warp threads for the background. The designs are then added, similar to embroidery. The name "Jamdani" translates to "flower vase" and has Persian origins. The weaving technique is unique and intricate (Government of Andhra Pradesh n.d.).

### 2.6.4.2 Silk-Based Crafts of Telangana

#### 2.6.4.2.1 Mahadevpur Tussar Silk Products

The saree, originating from the Mahadevpur forests, is a symbol of the region's biodiversity and cultural legacy. The raw Tussar silk, also known as "wild silk," is harvested from silkworms feeding on local tree leaves, contributing to its quality and texture. The saree, also known as "Kosa Silk," has a long history in India, symbolizing status and grace in various cultural ceremonies. It is worn during religious festivals and traditional ceremonies, showcasing the wearer's heritage. The saree is not just a garment but a family heirloom passed down through generations (Madhushree. 2024).

#### 2.6.4.2.2 Narayanpet Silk Sarees

Narayanpet handloom sarees are characterized by their contrasted look and small zari designs. They are best modeled in cotton-silk blends and light pure silk sarees. The process begins with dyeing the silk/cotton yarn. The borders and pallus are traditional, featuring a rich pallu and alternating red and white bands. Double-shade sarees are also available, featuring blends like pink-purple and maroon-mustard (Cotton Handloom and Silk Sarees 2026)

#### 2.6.4.2.3 Pochampally Ikat

Pochampally, located in Nalgonda district, is renowned for its Ikat sarees, which are woven in various towns in Nalgonda, Hyderabad, and Guntur districts. The manufacturing history of Pochampally Ikat sarees dates back to 1970, when village headmen decided to weave silk and cotton for a better living. The Ikat weaving is known as "Chit-ku" in these areas. The 8-m saree called Zjanani features Ikat bands in the end piece. The characteristic features of Pochampally Ikat sarees are their bold, bright patterning, inspired by modernization and choice. The modern Pochampally Ikat sarees are vibrantly colored and feature a wide range of modern designs. The weavers follow a specific weaving method, using silk warp and weft (16/18D to 20/22D and 22/24 D warp and weft 45). The "D" stands for denier (Savithri et al. 2013).

### 2.6.4.3 Silk-Based Crafts of Karnataka

#### 2.6.4.3.1 Mysore Silk

The Mysore silk cluster is a prominent silk-weaving industry in Mysore, Karnataka, India, known for its high-quality silk production and preservation of traditional art. Skilled artisans use pure mulberry silk to create luxurious saris and fabrics with intricate designs and vibrant colors (Bharatiya Vastra Shilpa Kosh, NIFT DEL-358 CFP—Mysore Silk).

#### 2.6.4.3.2 Kollegal Silk Saree

Kollegal, formerly known as Kaulava, got its name from two hermits named Kauhala and Gala, who had a significant role in the area's growth. Kollegal, sometimes referred to as the Silk City, is home to weavers who specialize in creating sarees and other garments out of pure silk. Kollegal's traditional and indigenous occupation is silk weaving. Kollegal silk sarees are incredibly distinctive in terms of their style and uniqueness. Kollegal manufactures its own raw silk and is well-known for reeling and twisting silk. Pit looms are used to create the majority of silk sarees in the district of Hanur (Digital Empowerment Foundation 2017)

2.6.4.3.3 Molakalmuru Sarees

Molakalmuru weaving, influenced by Kancheepuram and Dharmavaram weavers, resembles Kancheevaram sarees in motif and texture. Mostly done by pit looms, people in Molakalmuru are primarily engaged in weaving and agriculture, with the recent production of contemporary and traditional silk sarees (Government of Karnataka, Industries & Commerce Department n.d.).

2.6.4.3.4 Ilkal Sarees

The Ilkal saree is a traditional Indian feminine wear, named after the town of Ilkal in Karnataka state. It is woven using cotton warp for the body and art silk warp for the border and pallu portion, with pure silk sometimes used. The saree has been given a Geographical Indication (GI) tag (Karthiga 2017).

2.6.4.3.5 Garbha Reshme

Bhagyanagar, a village in the Kopal district of Karnataka, is renowned for producing beautiful silk sarees, known as garbha reshme or half silk sarees. These sarees, made of fine quality silk and cotton, are woven on a shuttle pit loom and have a width of 42 inches. Borders, such as Gomi, gadi dhadi, flower, diamond, and peacock, enhance the saree's beauty. Other varieties include Paras, Satin, Patti, Pineapple, Chrysanthemum, Mango, and Rudraksha (Badanayak 2017).

#### 2.6.4.4 Silk-Based Crafts from Tamil Nadu

2.6.4.4.1 Kancheepuram Silk Sarees

Kancheepuram, the golden city of a thousand temples, is renowned for its traditional silk sarees, which are a living example of the legendary craftsmanship of the weavers of the country. These sarees are known for their royal look, contrasting colors, Korvai work, marvelous motifs, and intricate designs. The Kanjeevaram sari is a popular choice for weddings in South India and other auspicious and religious occasions. The Kanjeevaram sari is made of heavy Kanjivaram silk, which is so durable that it can be washed in water at home. The heavier the silk, the better the quality. The creation of a Kanjeevaram sari is a challenging task, starting with the twisting of the silk thread, dyeing and drying in the sun, and then being interlocked together in an impossible-to-detach joint (Savithri et al. 2013).

#### 2.6.4.4.2 Arani Silk Sarees

Arani silk saris are renowned for their luxurious texture, intricate designs, and vibrant colors, crafted using ancient weaving techniques. Originating from mulberry silk, these saris have religious and legendary significance in Hindu mythology. Supported by South Indian kings, they have evolved to meet changing consumer preferences and market demands. The art of Arani silk weaving has evolved over time, with a diverse clientele and enduring popularity for weddings, festivals, and special occasions. Efforts to empower weavers include improving working conditions, fair wages, and resource access. Despite challenges like competition from machine-made textiles and changing customer preferences, the art continues to thrive (Bharatiya Vastra Shilpa Kosh, NIFT DEL-218 CFP—Arani Silk Sarees).

#### 2.6.4.4.3 Thirubuvanam Silk

Thirubuvanam silk sarees are handwoven textiles from Thirubuvanam, a historic silk-weaving town in Tamil Nadu. Made from high-quality filature silk for both warp and weft, these sarees are known for their luster, durability, and smooth texture. They feature intricate borders, pallus, and motifs inspired by temple architecture, peacock eyes, Rudraksha beads, and vibrant contrast colors (Arockiaraj 2019).

#### 2.6.4.4.4 Salem Silk Known as Salem Venpattu

Salem, 340 km from Chennai, is renowned for its silk and cotton dhotis and silk saris, including unique white silk dhotis called Salem venpattu. These dhotis are known for their luster and color, created by flattening the surface with a small brass or steel plate. They are often combined with angavastram, stole, and silk sattai for special occasions. Venpattu dhotis are woven on pit looms or raised pit looms, with dobbies used for mayilkan borders. The dhotis have gained "Geographical Indication" for their uniqueness and were branded under "India Handloom Brand" by the Government of India (Bharatiya Vastra Shilpa Kosh, NIFT DEL-15 CFP—Salem Silk).

#### 2.6.4.4.5 Kadangi Silk Sarees

Kandangi sarees, originally silk-woven in Karaikudi, were later adapted into cotton versions for convenience. These handloom sarees have checked patterns, earthen tones, and contrasting borders and are traditionally worn in a specific drape style. Historically, they were crafted with a coarse 40s yarn for durability, but contemporary versions may use a finer, 60s yarn (Raman 2019).

#### 2.6.4.4.6 Chirala Kuppadam Silk Sarees

Handcrafted on ancient pit looms in the Bapatla area of Andhra Pradesh, Chirala's Kuppadam silk sarees are renowned for their softness, resilience, and climate-adaptive comfort. Rich silk (and silk-zari) borders with temple patterns are what set the sarees apart, and weavers are now able to create wider borders of 12 to 15 inches thanks to a recent invention that involves inserting a spring in the shuttle (Times of India 2025).

## 2.7 The Importance of Indian Silk-Based Textile Crafts

A few notable silk-based textile crafts are presented in Table 8. As seen in the Kosha silk saree Fig. 2.9a, it has borders woven in contrast colors (maroon, white, and beige) with traditional tribal or temple motifs, geometric designs with zari embellishment. The contrasting colors may be maroon, black, red, or deep green, and the patterns may feature temple motifs, geometric designs, or sometimes floral patterns. Woven with Tasar silk gives a natural, authentic look to the fabric. Figure 2.9b shows a fabric of Himroo, featuring a cotton base (warp) for strength and silk (weft), which gives the fabric its lustrous designs. The union blending made the fabric softer, lighter, and more breathable than pure silk, which suited the Deccan's warm climate. Although the craft employs geometric shapes, florals, creepers, and miniature-like ornamental patterns as motifs, due to Indian influence, indigenized motifs such as paisleys, peacocks, and traditional floral borders may also be woven into these types of fabrics, making them more contemporary.

Figure 2.9c shows a Baluchari saree, made with pure silk in both warp and weft. Contrast-colored silk yarns create the motifs, along with the use of zari yarn. Earlier fabrics did not feature zari; however, the current sample exhibits a more contemporary version of the saree. Figure 2.9d shows a Maheswari saree, unique for its union blend, where the warp is made of silk yarn and the weft is made of cotton yarn. The extra warp technique, which creates intricate patterns in the border, is a hallmark of this craft. The craft draws its motifs from the Narmada River and the architecture of the Maheswar fort. Rhe border features a wave-like motif, locally known as "leheriyan," inspired by the waves of the Narmada River. Figure 2.9e shows a unique combination of work where the famous "Kantha" stitch is popularly made with running stitch motifs over a "Tasar" silk saree. With the touch of this embellishment, the fabric transforms into a value-added product. Figure 2.9f represents the famous Paithani saree from Maharashtra, characterized by being woven in pure silk. The motifs include peacock woven, not stitched, using the Traditional handloom tapestry technique. The unique aspect of these crafts showcases the extent of contemporizing, the method of blending the yarns, and the century-old silk craft tradition that has been passed down through generations of craftsmen. Over time, there have been changes to compete with modern fast fashion brands, but these few, among many, remain alive today.

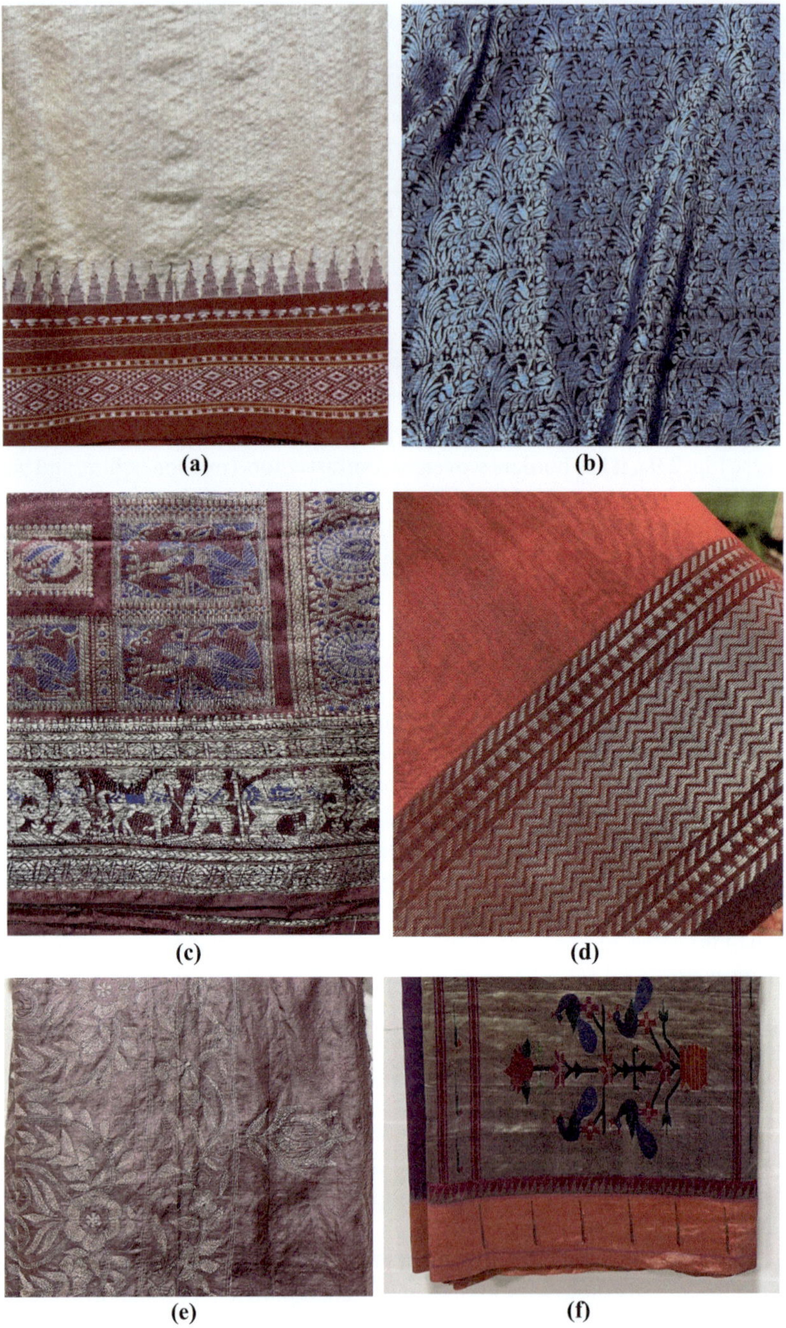


**Fig. 2.9** **a** Kosa silk saree from Jharkhand. **b** Himroo silk fabric from Aurangabad, Maharashtra, India. **c** Baluchari saree from West Bengal, India. **d** Maheshwari saree from Madhya Pradesh, India. **e** Tasar saree with Kantha stitch from West Bengal, India. **f** Paithani saree from Maharashtra, India. *Image Source* Author

## 2.8 Present Scenario and the Position of India

In 2023–24, the country produced 38,913 metric tons of raw silk and exported silk products worth Rs. 2027.56 crore, solidifying its position as the world's second-largest silk producer and top consumer (DD News 2025). The trend of silk production in India is presented in Fig. 2.10. There is a consistent upward trend in production, except for a few years between 2019 and 2021.

Figure 2.11 presents the volume of export, import, and the difference (export–import) in terms of crores (INR) from 2012–2013 to 2023–2024. It's notable that though the production of silk has increased throughout the years, there is a considerable import of silk.

Apparently, this piece of information may cause some confusion among readers. How come the second-highest producer of silk could import from other countries? If the data is compared, there is an increase in production from 31,906 metric tons in 2017–18 to 38,913 metric tons in 2023–24 (DD News 2025). India weaves growth:

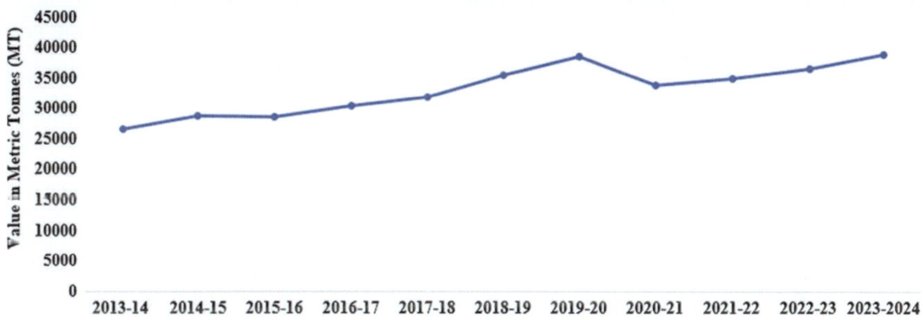


**Fig. 2.10** Production of silk for the past 11 years (MT) (Central Silk Board 2013, 2014, 2015, 2016, 2017, 2018, 2019, 2020, 2021, 2022, 2023, 2024)

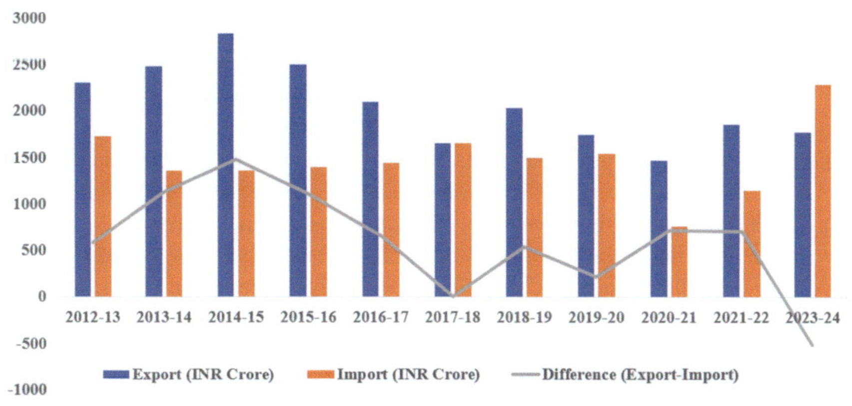


**Fig. 2.11** Import–export of silk for the past 12 years (Central Silk Board 2013, 2014, 2015, 2016, 2017, 2018, 2019, 2020, 2021, 2022, 2023, 2024)

Silk production rises to 38,913 MT, exports cross Rs. 2,000 crores, benefiting over 78,000 under government schemes (DD News 2025). There may be two major reasons for this. The first one is being the existence of a gap between supply and demand. Though there is an increase in silk production, it's yet to meet the demand. Secondly, the availability of cheaper silk from neighbouring country has resulted in continued imports (Saluja 2020). However, the credit for raw silk production in the country may be due to government schemes such as Catalytic Development Programme (CDP), North East Region Textile Promotion Scheme (NERTPS), Integrated Scheme for Development of Silk Industry (ISDSI), and Silk Samagra and Silk Samagra-2. The projected employment generation estimated in the silk sector, based on raw silk production as of January 2025, is 80.90 lakh persons (DD News 2025). The good sign is that India's imports of raw silk, followed by fabrics and made-ups, decreased by 32% from 3874 MT in 2022–23 to 2,650 MT in 2023–24 (DD News 2025).

The data in Figs. 2.10 and 2.11 clearly demonstrate the significance of silk in India's socio-economic structure. This is a sector where tradition and modern business melt together. Silk weaving has been a medium of storytelling for generations, sharing a common connection with people. It can be inferred that, apart from storytelling, silk plays a wider role in Indian society, extending from the production and export of the produced goods to importing some to meet market needs.

## 2.9 Conclusion

Silk stands as one of the most treasured natural fibers, renowned for its distinctive luster and its extensive significance in cultural, artistic, and practical applications. For centuries, silk has held a special place in global textile history, particularly in India's culture and economy. As discussed in the chapter, India is a major producer and exporter of silk worldwide. India has a rich history of producing both fine heritage textiles and modern ones. This chapter discusses the different silk fibers available in India, the different types of yarns, and their popular names. This chapter explains how silk is formed in the glands of silkworms and describes the fiber properties that make silk unique, including its triangular cross-section, which contributes to its shine. The importance of silk extends far beyond science and technology, as it is also a part of India's rich living history of handicrafts, where it is deeply linked to artistic identity, regional pride, and centuries-old workmanship. The conversation surrounding GI-tagged silk products, such as Banarasi, Kanchipuram, Muga, and others, highlights how silk manufacturing has become a symbol of regional uniqueness and authenticity, protecting both the artisans and their traditions. The chapter highlights silk's dual role as both a luxury material and a cultural medium that preserves livelihoods and traditional craftsmanship knowledge across generations, examining the wide range of silk-based handicrafts in India, including woven sarees, embroidered textiles, painted fabrics, and resist-dyed fabrics. Silk will always be an integral part of India's story of craft excellence and textile innovation, as it is both a scientific marvel and a symbol of history.

**AI Usage Declaration**:

AI tools were used solely for language and grammar assistance. After using these tools, authors have reviewed the content and take full responsibility for the originality and integrity of the final work.

## References


Ahmed SA, Singh NI, Sarkar CR (2015) Role of forest biodiversity in conservation of non-mulberry (vanya) silk in India. Munis Entomol Zool 10(1):342–357

Akai H (1984) The ultrastructure and functions of the silk gland cells of Bombyx mori. In: Insect ultrastructure. Springer, Berlin, Heidelberg, Germany, pp 323–364

Alam I (1999) History of sericulture in India to the 17th century. Proc Indian Hist Congr 60:339–352. http://www.jstor.org/stable/44144100

Andiappan M, Sundararajan V, Natarajan TS (2016) Comparison of eri and tasar silk fibroin scaffolds with mulberry silk fibroin scaffolds for tissue engineering applications. Int J Biol Macromol 86:662–667. https://doi.org/10.1016/j.ijbiomac.2016.01.100

Angom I, Srivastava M (2019) Manipuri textiles—a great traditional heritage of Northeast region. Asian Acad Res J Soc Sci HumIties 1(20):83–87

Arockiaraj DV (2019) Tamil Nadu's Thirubuvanam silk sari gets GI tag. The Times of India. https://timesofindia.indiatimes.com/city/chennai/tamil-nadus-thirubuvanam-silk-sari-gets-gi-tag/articleshow/68376518.cms

Asakura T, Kametani S, Suzuki Y (2002) Silk. In: Encyclopedia of polymer science and technology, pp 1–19

Babu KM (2018) Silk: processing, properties and applications. Woodhead Publishing/Elsevier, 2nd ed. https://doi.org/10.1016/C2016-0-04193-7. ISBN: 9780081025406

Babu KM (2020) Silk fibres–structure, properties and applications. In: Handbook of natural fibres. Woodhead Publishing, pp 385–416

Badanayak P (2017) Development of designs from ethnic Khana motifs on dress materials (Master's thesis). University of Agricultural Sciences, Dharwad. Retrieved February 8, 2026. https://krishikosh.egranth.ac.in/server/api/core/bitstreams/f1949b4c-a0d5-4c0b-bc6a-d6c1a3d9ed4d/content

Bajwa GA, Nawab Y, Umair M, Rizwan Z (2019) Techno-mechanical properties of cocoon, raw silk and filament of two mulberry silk worm (Bombyx mori strains). Materialwiss Werkstofftech 50(10):1287–1294

Bala KU (1999) Adopting Venkatagiri designs on Kameez sets. Doctoral dissertation, ANGRAU HSCH: textiles and clothing

Bandhani J (2016) GI Journal No. 83. Origin. Retrieved Sept 14, 2025, from https://www.origin-gi.com/wp-content/uploads/2017/06/2-jamnagari-bandhani.pdf

Banerjee S (2013) Textiles and traditions of Bengal. Seagull Books. Central Silk Board, Kolkata, p 2020

Behera S (2024) Harnessing tradition and preserving Bomkai sarees of Odisha. Home Sci J 10(1, Part C):922–931. Retrieved Aug 31, 2025, from https://www.homesciencejournal.com/archives/2024/vol10issue1/PartC/10-1-30-922.pdf

Bharatiya Vastra Shilpa Kosh, NIFT DEL-15 CFP—Salem silk. https://www.vastrashilpakosh.in/search/recordPreview/nift_del-15-cfp?searchTerm=salem%20silk&t=Salem%20venpattu. Accessed 30 Aug 2025

Bharatiya Vastra Shilpa Kosh, NIFT DEL-218 CFP—Arani silk sarees. https://www.vastrashilpakosh.in/search/recordPreview/nift_del-218-cfp?searchTerm=arani%20silk%20weaving&t=Arani%20silk. Accessed 30 Aug 2025

Bharatiya Vastra Shilpa Kosh, NIFT DEL-358 CFP—Mysore silk. https://www.vastrashilpakosh.in/search/recordPreview/nift_del-358-cfp?searchTerm=mysore%20silk&t=Mysore%20silk. Accessed 30 Aug 2025

Bharatiya Vastra Shilpa Kosh, NIFT DEL-420 CFP—Gharchola. https://www.vastrashilpakosh.in/search/recordPreview/nift_del-420-cfp?searchTerm=bandhani&t=Gharchola%20sari. Accessed 30 Aug 2025

Bharatiya Vastra Shilpa Kosh, NIFT DEL-54 CFP—Patan Patola. hhttps://www.vastrashilpakosh.in/search/recordPreview/nift_del-54-cfp?searchTerm=patan%20patola&t=Patan%20patola. Accessed 30 Aug 2025

Bhatia A, Bhagat S, Mathur R (2016) Himroo: earliest industry of Aurangabad. Int J Appl Home Sci 3(11):480–484

Bhatia R, Desai D (2014) Sustaining the traditional Paithani saree weaving of Maharashtra, India. In: International conference on innovative trends in multidisciplinary academic research, pp 332–338

Bomkai: Draped in traditions of Odisha (2021) Media India Group. https://mediaindia.eu/culture/bomkai-draped-in-traditions-of-odisha/

Bureau of Indian Standards (1991) IS 1583: Handloom Silk Dhoties and Printed Saris (2nd rev.). Bureau of Indian Standards, New Delhi

Bureau of Indian Standards (2020) Glossary of Textile Terms — Natural Fibres (Third Revision) (IS 232:2020). Bureau of Indian Standards. https://www.services.bis.gov.in

Central Silk Board (2013) Annual report 2012–13. Ministry of Textiles, Government of India. Retrieved Sep 14, 2025, from http://csb.gov.in

Central Silk Board (2014) Annual report 2013–14. Ministry of Textiles, Government of India. Retrieved Sep 14, 2025, from https://csb.gov.in/wp-content/uploads/2018/09/CSBAR-1314English.pdf

Central Silk Board (2015) Annual report 2014–15. Ministry of Textiles, Government of India. Retrieved Sep 14, 2025, from https://csb.gov.in/wp-content/uploads/2018/04/CSBAR-1415English.pdf

Central Silk Board (2016) Annual report 2015–16. Ministry of Textiles, Government of India. Retrieved Sep 14, 2025, from https://csb.gov.in/wp-content/uploads/2018/04/AAR-ENG-2015-16-part1.pdf

Central Silk Board (2017) Annual report 2016–17. Ministry of Textiles, Government of India. Retrieved Sep 14, 2025, from https://csb.gov.in/wp-content/uploads/2018/04/CSBAR1617English.pdf

Central Silk Board (2018) Annual report 2017–18. Ministry of Textiles, Government of India. Retrieved Sep 14, 2025, from https://csb.gov.in/wp-content/uploads/2019/01/annual-report-english-2017-18.pdf

Central Silk Board (2019) Annual report 2018–19. Ministry of Textiles, Government of India. Retrieved Sep 14, 2025, from http://csb.gov.in/wp-content/uploads/2020/03/csb-aar-18-19-english-low-res.pdf

Central Silk Board (2020) Annual report 2019–20. Ministry of Textiles, Government of India. Retrieved Sep 14, 2025, from https://csb.gov.in/wp-content/uploads/2021/03/CSB-ANNUAL-REPORT-2019-20-compressed-97-196-eng.pdf

Central Silk Board (2021) Annual report 2020–21. Ministry of Textiles, Government of India. Retrieved Sep 14, 2025, from https://csb.gov.in/wp-content/uploads/2024/05/CSB-Annual-Report-2020-21-Outer-cover-pages-dt-24-12-2021-merged_compressed.pdf

Central Silk Board (2022) Annual report 2021–22. Ministry of Textiles, Government of India. Retrieved Sep 14, 2025, from https://csb.gov.in/wp-content/uploads/2024/05/CSB-Annual-Report-2021-22-compressed.pdf

Central Silk Board (2023) Annual report 2022–23. Ministry of Textiles, Government of India. Retrieved Sep 14, 2025, from https://csb.gov.in/wp-content/uploads/2024/05/1-AR-2022-23-Final-Low.pdf

Central Silk Board (2024) Annual report 2023–24 [PDF]. Ministry of Textiles, Government of India. Retrieved Sep 14, 2025, from https://csb.gov.in/wp-content/uploads/2025/05/AAR-2023-24-Final-Low-resolution.pdf

Central Silk Board (n.d.-a) Processing of tasar cocoons. Retrieved Sep 12, 2025, from Silks—CSB https://silks.csb.gov.in/jalna/processing-of-tasar-cocoons

Central Silk Board (n.d.-b) Processing of mulberry cocoons. Sikkim South. Ministry of Textiles, Government of India. Retrieved Aug 31, 2025, from https://silks.csb.gov.in/sikkim/processing-of-mulberry-cocoons/

Central Silk Board (n.d.-c) Rearing of muga silkworm. Retrieved Aug 31, 2025, from https://silks.csb.gov.in/cachar/rearing-of-muga-silkworm/

Central Tasar Research & Training Institute (2023) Souvenir-cum-book of abstracts. Central Silk Board, Ministry of Textiles, Government of India. Retrieved Sep 12, 2025, from https://ctrti.res.in/wp-content/uploads/2023/03/Soveniar-Cum-Book-of-Abstracts.pdf

Chakravorty R, Dutta P, Ghose J (2010) Sericulture and traditional craft of silk weaving in Assam. Indian J Tradit Knowl 9(2):378–385

Chand S, Chand S, Raula B (2023) Usage of silkworm materials in various ground of science and research. J Nat Fibers 20(1):2139328

Chanu NV, Gogoi N, Baruah RR, Saikia M (2024) Effect of natural mordants on eri matka silk yarn dyed with the bark of mangifera indica dye. Int J Res Agron 7(7, Part J):779–784. https://doi.org/10.33545/2618060X.2024.v7.i7j.1131

Chattopadhyay S (2014) Murshidabad silk industry: its past glory and present crisis. Soc Chang Dev 11(2):65–77

Chattopadhyay D, Chakraborty A, Chatterjee SM (2017) Studies on degumming of eri silk cocoons. J Text Inst 108(8):1327–1339

Chaudhary K, Pant S (2016) Value addition of the Kota Doria through designing techniques. Int J Text Fash Technol 6(2):2250–2378

Cheung HY, Lau KT, Ho MP, Mosallam A (2009) Study on the mechanical properties of different silkworm silk fibers. J Compos Mater 43(22):2521–2531

Cotton Handloom and Silk Sarees (2026) Narayanpet District – Government of Telangana. Retrieved February 8, 2026, from https://narayanpet.telangana.gov.in/?s=Narayanpet+Silk+Sarees

Das TK (2021) Sambalpuri handloom weavers' livelihood: an analysis of Western Odisha clusters, India. Int J Community Soc Dev 3(3):215–235. https://doi.org/10.1177/25166026211038348

Das P, Mukherjee R (2018) Bishnupuri silk: tradition and adaptation in Bengal's textile industry. J Text Assoc 79(3):245–252

Dash R, Mandal M, Ghosh SK, Kundu SC (2008) Silk sericin protein of tropical tasar silkworm inhibits UVB-induced apoptosis in human skin keratinocytes. Mol Cell Biochem 311(1):111–119

DD News (2025) India weaves growth: Silk production rises to 38,913 MT, exports cross Rs 2,000 crore benefiting over 78,000 under government schemes. DD News. Retrieved Sep 14, 2025, from https://ddnews.gov.in/en/india-weaves-growth-silk-production-rises-to-38913-mt-exports-cross-rs-2000-crore-benefiting-over-78000-under-government-schemes/?utm

Dhamija J (2014) Sacred textiles of India. Abhinav Publications. ISBN: 978-81-7017-902-9

Digital Empowerment Foundation (n.d.) Kaulava. digital cluster development programme. Retrieved Sep 5, 2025, from https://dcdpindia.org/kaulava/

Digital Empowerment Foundation (2026) Kollegal, Karnataka: Baseline Survey Report. Retrieved Feb 8, 2026, from https://defindia.org/wp-content/uploads/2017/08/Kolegal-Baseline-Report_2017.pdf

Directorate of Sericulture, Government of Assam (n.d.) Know about silk and its production in Assam. Retrieved Aug 31, 2025, from https://sericulture.assam.gov.in/how-to/know-about-silk-and-its-production-in-assam

District Administration, Narayanpet District, Government of Telangana (n.d.) Handicraft|Narayanpet District, Government of Telangana. Narayanpet District. Retrieved Sep 2, 2025, from https://narayanpet.telangana.gov.in/handicraft/

Garg R, Sidhpuria M (2015) Strategic interventions to enhance competitiveness: a case of Surat zari industry in India. AI Soc 30(2):235–249. https://doi.org/10.1007/s00146-013-0506-7
Ghosh D (2014a) Tradition and transformation: a study on Bengal silk sarees. Indian J Tradit Knowl 13(3):606–612
Ghosh S (2014b) Traditional handloom weaving in Bhagalpur: a study of tussar silk industry. Indian J Tradit Knowl 13(3):606–612
Goswami P, Neog R (2013) Traditional muga and eri silk industry of Assam: an overview. Indian J Tradit Knowl 12(2):221–229
Government of Andhra Pradesh (n.d.) One district one product—Uppada Jamdhani sarees. In Kakinada District. Retrieved Sep 5, 2025, from Government of Andhra Pradesh https://kakinada.ap.gov.in/about-district/one-district-one-product/
Government of Andhra Pradesh (n.d.) One district one product – Uppada Jamdani Sarees. Retrieved February 8, 2026. https://kakinada.ap.gov.in/?s=Uppada+Jamdani+Sarees
Government of Assam (n.d.-a) Assam silk. Retrieved Aug 31, 2025, from https://assam.gov.in/citizen/425.assam.gov.in
Government of Assam (n.d.-b) Assam silk. Retrieved Sep 12, 2025 from https://assam.gov.in/citizen/425
Government of Assam (2026) Assam silk. Assam State Portal. Retrieved February 8, 2026, from https://assam.gov.in/citizen/425?utm
Government of Assam, Handloom, Textiles & Sericulture Department (n.d.) Rearing of silkworm. https://hts.assam.gov.in/portlet-innerpage/rearing-of-silkworm
Government of Karnataka, Industries & Commerce Department (n.d.) List of One District One Product (ODOP) products & their descriptions (Molakalmuru Sarees, GI No. 77). Retrieved February 8, 2026, from https://industries.karnataka.gov.in/uploads/media_to_upload1743831776.pdf
Guo N, Lu K, Cheng L, Li Z, Wu C, Liu Z et al (2019) Structure analysis of the spinneret from Bombyx mori and its influence on silk qualities? Int J Biol Macromol 126:1282–1287
Gupta AH, Mehta S (2014) Patterns of Phulkari: then and now. Bonfring Int J Ind Eng Manag Sci 4(4):179
Handloom Export Promotion Council (n.d.) Habaspuri saree – Fabrics (Odisha). Retrieved February 8, 2026, from https://hepcindia.com/gihandloom_products/odisha.html
HEPC India (n.d.) GI handloom products: Odisha. Retrieved Sep 12, 2025 from https://hepcindia.com/gihandloom_products/odisha.html
Holland C, Terry AE, Porter D, Vollrath F (2007) Natural and unnatural silks. Polymer 48(12):3388–3392
Hu L, Han Y, Ling S, Huang Y, Yao J, Shao Z, Chen X (2020) Direct observation of native silk fibroin conformation in silk gland of bombyx mori silkworm. ACS Biomater Sci Eng 6(4):1874–1879
India Brand Equity Foundation (2025) Silk textile industry in India, silk manufacturers in India. IBEF. Retrieved Aug 30, 2025, from https://www.ibef.org/exports/indian-silk-industry
Indian Textiles & Crafts Repository (n.d.) Indian textiles & crafts repository (Bharatiya Vastra Shilpa Kosh). Retrieved February 8, 2026, from https://www.vastrashilpakosh.in/
Inoue S, Tanaka K, Arisaka F, Kimura S, Ohtomo K, Mizuno S (2000) Silk fibroin of Bombyx mori is secreted, assembling a high molecular mass elementary unit consisting of H-chain, L-chain, and P25, with a 6: 6: 1 molar ratio. J Biol Chem 275(51):40517–40528
Intellectual Property India (2025) State wise registered GI of India (PDF). Intellectual Property India, Government of India. Available at: https://ipindia.gov.in/writereaddata/Portal/Images/pdf/State_wise_Registered_GI_of_India.pdf. Accessed 14th Sep 2025
Javali UC, Talukdar P, Rathod R (2018) Studies on tasar cocoon cooking using permeation method. J Inst Eng (India) Ser E 99(2):151–156. https://doi.org/10.1007/s40034-018-0111-6
Jigyasu DK, Patidar OP, Kumar A, Shabnam AA (2023) Muga silk insect (Antheraea assamensis). In: Commercial insects. CRC Press, pp 126–145
Jin HJ, Kaplan DL (2003) Mechanism of silk processing in insects and spiders. Nature 424(6952):1057–1061

Jin Y, Hang Y, Luo J, Zhang Y, Shao H, Hu X (2013) In vitro studies on the structure and properties of silk fibroin aqueous solutions in silkworm. Int J Biol Macromol 62:162–166
Karolia A, Prakash R (2014) Design and development of fashion accessories inspired from the hand-woven shawls of Nagaland. Indian J Tradit Knowl 13(2):416–426
Karthiga S (2017) Production and marketing management of Ilkal and Kanchipuram pure silk saree—a comparative analysis
Koh LD, Cheng Y, Teng CP, Khin YW, Loh XJ, Tee SY et al (2015) Structures, mechanical properties and applications of silk fibroin materials? Prog Polym Sci 46:86–110
Kumar R (2006) Costumes and textiles of royal India. Christie's Book, New Delhi. ISBN: 9781851495094
Kumar TS, Kumar MR, Kumar BS (2021a) Investigation of dimensional and physical properties of eri silk bi-layer knit fabrics for active wear applications. AATCC J Res 8(6):22–30
Kumar V, Kumari P, Yadav P, Kumar M (2021b) Ancient to contemporary—the saga of Indian handloom sector. Indian J Fibre Text Res 46:411–431
Kunz RI, Brancalhão RMC, Ribeiro LDFC, Natali MRM (2016) Silkworm sericin: properties and biomedical applications. Biomed Res Int 2016(1):8175701
Laheru G, Sultana K, Mukhopadhyay S (2025) Weaving sustainability, lessons from Khamir desi wool and Eri silk of Assam. In: Traditional textiles of the Indian subcontinent: sustainable engineering, manufacturing, and design, pp 83–105
Lee Y-W (1999) Silk reeling and testing manual. Food and Agriculture Organization of the United Nations. ISBN: 9251042934
Madhushree. (2024) Mahadevpur tassar silk saree. BongChong. Retrieved Sep 14, 2025, from https://bongchong.in/mahadevpur-tassar-silk-saree/
Mohanty T (2022) Craft, culture and entrepreneurship: weaving tradition in Chanderi. Arts Acad 2(2):144–172
Mukherjee S (2005) The textiles of Bengal. Seagull Books, Kolkata ISBN: 9781905422014
Mukherjee A (2016) Handloom weaving in Bengal: historical perspectives. Jadavpur University Press, Kolkata
Murthy KK (1985) Textiles in ancient India. Concept Publishing Company, New Delhi ISBN: 9788170220152
Naik SV, TH S (2005) Studies on bivoltine cocoon reeling part I: effect of cocoon cooking condition and reeling speed on reeling performance of Indian bivoltine hybrid cocoons. J Silk Sci Technol Jpn 14:61–65
National Institute of Fashion Technology (n.d.-a) Champa silk saree fabrics (NIFT DEL-268 CFP). Bharatiya Vastra Shilpa Kosh. Retrieved Aug 30, 2025, from https://www.vastrashilpakosh.in/search/recordPreview/nift_del-268-cfp?searchTerm=champa%20silk%20sari%20and%20fabric%20weaving&t=Champa%20silk%20saris%20and%20fabrics
National Institute of Fashion Technology (n.d.-b) Kosa silk saree (NIFT DEL-85 CFP). Bharatiya Vastra Shilpa Kosh. Retrieved Aug 30, 2025, from https://www.vastrashilpakosh.in/search/recordPreview/nift_del-85-cfp?searchTerm=kosa%20silk%20saree&t=Kosa%20silk%20weaving
National Institute of Fashion Technology (n.d.-c) Tussar ferra saree (NIFT DEL-402 CFP). Bharatiya Vastra Shilpa Kosh. Retrieved Aug 30, 2025, from https://www.vastrashilpakosh.in/search/recordPreview/nift_del-402-cfp?searchTerm=tussar%20ferra%20sari%20weaving&t=Tussar%20ferra%20sari
New Indian Express (2011) Weaving culture with fashion: Kuchai silk showcased in Bangalore. The New Indian Express. Retrieved Aug 31, 2025, from https://www.newindianexpress.com/cities/bengaluru/2011/nov/15/weaving-culture-with-fashion-310375.html
Ngangbam VC, Boruah RR, Konwar M, Borah P (2024) An analytical research on traditional attires of Meitei community of Manipur. Int J Adv Biochem Res 8(7):582–585. https://doi.org/10.33545/26174693.2024.v8.i7g.1555
North Eastern Handicrafts & Handloom Development Corporation Ltd. (n.d.) Ryndia—eri silk brand. Retrieved Aug 31, 2025, from https://nehhdc.com/ryndia

Padaki NV, Das B, Basu A (2015) Advances in understanding the properties of silk. Adv Silk Sci Technol, 3–16
Pham DT, Tiyaboonchai W (2020) Fibroin nanoparticles: a promising drug delivery system. Drug Deliv 27(1):431–448
Press Information Bureau (2025) Raw silk production in 2024–25. Ministry of Textiles, Government of India. Retrieved Sep 14, 2025, from https://www.pib.gov.in/PressReleasePage.aspx?PRID=2117111
Qi Y, Wang H, Wei K, Yang Y, Zheng RY, Kim IS, Zhang KQ (2017) A review of structure construction of silk fibroin biomaterials from single structures to multi-level structures? Int J Mol Sci 18(3):237
Raman S (2019) TN's Kandangi sari gets GI tag. What does this mean for the state's handloom market? The Times of India. Retrieved Sep 14, 2025, from https://timesofindia.indiatimes.com/city/chennai/tns-kandangi-sari-gets-gi-tag-what-does-this-mean-for-the-states-handloom-market/articleshow/71092224.cms
Ramswamy R, Kumar NJ (2013) Handloom production in tribal clusters: a case study of Thenzawl in Mizoram. IUP J Oper Manag 12(2):25
Reizabal A, Costa CM, Pérez-Álvarez L, Vilas-Vilela JL, Lanceros-Méndez S (2023) Silk fibroin as sustainable advanced material: material properties and characteristics, processing, and applications. Adv Funct Mater 33(3):2210764
Rheinberg L (1991) The romance of silk: a review of sericulture and the silk industry. Text Prog 21:1–43
Roy C (2015) Issues of productivity, employment and exploitation in the artisanal silk industry of West Bengal. Indian J Soc Nat Sci 4(2):49–68. https://doi.org/10.2139/ssrn.2826492
Saluja N (2020) KVIC seeks ban on import of silk from China. The Economic Times. Retrieved Sep 14, 2025, from https://economictimes.indiatimes.com/industry/cons-products/garments-/-textiles/kvic-seeks-ban-on-import-of-silk-from-china/articleshow/74708931.cms?utm
Sangma NA, Mahale G (2020) Physical properties of Muga silk and Tencel yarn. J Farm Sci 33(02):290–291
Savithri G, Sujathamma P, Ramanamma CH (2013) Glory of Indian traditional silk sarees. Int J Text Fash Technol 3(2):61–68
Sen A (2015) Symbolism in Bengal's festive attire: the red-and-white saree. Indian J Cult Stud 22(2):145–158
Sen K, Babu KM (2004) Studies on Indian silk. II. structure–property correlations. J Appl Polym Sci 92(2):1098–1115
Silk weaving traditions of Bengal. Government of India. https://csb.gov.in. Accessed 31 Aug 2025
Silva AS, Costa EC, Reis S, Spencer C, Calhelha RC, Miguel SP, Ribeiro MP, Barros L, Vaz JA, Coutinho P (2022) Silk sericin: a promising sustainable biomaterial for biomedical and pharmaceutical applications. Polymers 14:4931
Singh, S (2015) Geographical indication; a case study of 'Kashmir Sozani' and 'Kani Shawl'. New Man, 153
Singh S, Jigyasu DK, Patidar OP, Kumar A, Shabnam AA (2023) Eri silk insect (Samia ricini, Donovan). In: Commercial insects. CRC Press, pp 103–125
Silk Mark Organisation of India (nd) Types of silk. Silk Mark India. Retrieved Aug 30, 2025, from https://silkmarkindia.com/index.php/types-of-silk/
Song K, Wang Y, Dong W, Li Z, Xia Q, Zhu P, He H (2024) Decoding silkworm spinning programmed by pH and metal ions. Sci Bull 69(6):792–802
Sonwalkar TN (1993a) Hand book of silk technology. New Age International. ISBN: 8122404952
Sonwalkar TN (1993b) Handbook of silk technology. New Age International, New Delhi. ISBN: 8122404952
Sookne AM, Harris M (1939) Electrophoretic studies of silk. Text Res J 9(10):374–383. https://doi.org/10.1177/004051753900901002

The Times of India (2025) Meghalaya celebrates GI tag for Ryndia, Khasi handloom. The Times of India. Retrieved Aug 31, 2025, from https://timesofindia.indiatimes.com/city/guwahati/meghalaya-celebrates-gi-tag-for-ryndia-khasi-handloom/articleshow/123172281.cms

Times of India (2025) Chirala's Kuppadam silk sarees get nat'l honour under ODOP initiative. The Times of India. Retrieved Sep 14, 2025, from https://timesofindia.indiatimes.com/city/vijayawada/chiralas-kuppadam-silk-sarees-get-natl-honour-under-odop-initiative/articleshow/121834781.cms

TOI Lifestyle Desk (2023) Bandhani: the oldest tie-dye technique in India. Times of India. Retrieved Sep 5, 2025, from Times of India website

Vazifdar A, Debnath S (2025) Natural fibers: a look from a historical perspective. In: Advances in renewable natural materials for textile sustainability?. CRC Press, pp 1–36

Verma S, Mishra N (2018) Recognition and marketing opportunities of a "GI" tag in handloom product: a study of Banaras brocades and sarees. J Intellect Prop Rights 23(2–3):101–110

Wang Q, Yang Y, Chen X, Shao Z (2012) Investigation of rheological properties and conformation of silk fibroin in the solution of AmimCl. Biomacromol 13(6):1875–1881

Zhang X, Tsukada M, Morikawa H, Aojima K, Zhang G, Miura M (2011) Production of silk sericin/silk fibroin blend nanofibers. Nanoscale Res Lett 6(1):510

Zhou B, Chen P, Li X, Shao Z (2020) Structure and functions of cocoons constructed by eri silkworms (Samia ricini). J Mech Behav Biomed Mater 103:103560. https://doi.org/10.1016/j.jmbbm.2019.103560

# Chapter 3
# Advancements in Mulberry Silk Improvements-Genetic Improvements and Beyond

**Pronama Biswas, Jinal Dinesh, Bhoomika Sridhar, Khushi Agarwal, Dishali Mathur, and Sunil S. More**

## 3.1 Introduction

Mulberry silk, derived from the domesticated *Bombyx mori* (*B. mori*), has been a cornerstone of sericulture for centuries. The silk industry has long relied on selective breeding to enhance yield, fibre quality, and disease resistance. However, with progress in genetics and biotechnology, significant improvements have been achieved through a deeper understanding of the *B. mori* genome. Genetic modifications, mapping tools, and molecular breeding techniques have enhanced silk production, enabling the precise selection of traits and the development of superior *B. mori* strains. These improvements not only enhance silk yield but also improve the strength, elasticity, and adaptability of silk fibers, ensuring sustainability and commercial feasibility (Fang et al. 2015). The sequencing of the *B. mori* genome has provided key insights into the genetic basis of silk production. With approximately 432 million base pairs and over 16,000 protein-coding genes, this genomic data has enabled researchers to identify crucial genes involved in the synthesis of fibroin and sericin, cocoon formation, and environmental stress resistance. Understanding genetic variations among different *B. mori* strains have further facilitated targeted breeding programs, optimizing silk characteristics based on specific commercial and industrial demands. Recent breakthroughs in molecular genetics, such as QTL mapping and genome-wide association studies (GWAS), have strengthened the ability to link genetic markers with desirable silk traits, making selective breeding more efficient.

P. Biswas · J. Dinesh · B. Sridhar · K. Agarwal · D. Mathur · S. S. More (✉)
Department of Biological Sciences, School of Basic and Applied Sciences, Dayananda Sagar University, City Innovation Campus, Bengaluru, India
e-mail: sunilsmorey@gmail.com



D. Kumar and M. Shahid (eds.), *Science and Technology of Silk*, Textile Science and Clothing Technology, https://doi.org/10.1007/978-3-032-16230-4_3

Traditional breeding approaches have been instrumental in improving silk production; however, they are often time-consuming and dependent on phenotypic selection. Modern genetic tools, such as marker-assisted selection (MAS), have significantly accelerated the breeding process by allowing early identification of high-performing strains based on their genetic markers. This method enhances the precision of breeding programs, ensuring the selection of *B. mori* with optimized silk yield, robustness, and disease resistance. Additionally, linkage mapping and next-generation sequencing have facilitated the identification of genes responsible for economically valuable traits, further improving breeding outcomes (Mondal et al. 2024).

Beyond conventional breeding, genetic engineering has transformed silk production by enabling direct modifications of the *B. mori* genome. The advent of clustered regularly interspaced palindromic sequences (CRISPR-Cas9) and other gene-editing technologies has allowed precise alterations in silk-related genes, leading to enhanced fibre strength, elasticity, and resistance to external stressors. Transgenic *B. mori* expressing modified fibroin genes or spider silk proteins have demonstrated remarkable improvements in silk tensile properties, expanding their potential applications in high-performance textiles and biomedical materials. These genetic methods have opened new avenues for engineering silk with functional properties tailored to specific industrial needs (Zhao et al. 2022; Kant et al. 2024).

Another crucial aspect of genetic improvements in sericulture is the enhancement of *B. mori* disease resistance. Pathogen susceptibility remains a significant challenge in silk production, often resulting in substantial economic losses. By identifying and incorporating genes associated with resistance to viral, bacterial, and fungal infections, researchers have developed *B. mori* strains with improved immunity and survival rates. Genomic selection techniques further optimize these traits, ensuring that *B. mori* populations maintain high productivity even under adverse conditions. While much of the focus has been on *B. mori* genetics, improvements in mulberry plant genetics also play a vital role in supporting silk production. Mulberry leaves serve as the primary food source for *B. mori*, and genetic modifications aimed at enhancing leaf quality, nutritional content, and resistance to pests and environmental stresses have indirectly contributed to better silk yields. Advances in molecular breeding and gene editing in mulberry plants have led to the development of high-yield, disease-resistant mulberry varieties, ensuring a stable and nutritious food supply for *B. mori*.

As genetic research continues to advance, the future of silk production is poised to be transformed by emerging technologies, including synthetic biology, transcriptomics, and epigenetic modifications. These cutting-edge approaches hold the potential to fine-tune silk production at the molecular level, optimizing fibre properties with unprecedented precision. This chapter examines the genetic foundations of mulberry silk production, highlighting key advancements in genome mapping, selective breeding, genetic engineering, and disease resistance. By integrating accepted breeding techniques with modern genetic methods, the silk industry is poised for a new era of enhanced productivity, quality, and sustainability.

## 3.2 Genetic Foundations of Mulberry Silk

### *3.2.1 Overview of the* B. mori *Genome*

The *B. mori* is a lepidopteran insect that has been domesticated for over 5000 years for silk production. The genetic makeup of the *B. mori* is characterized by a well-mapped genome that has been extensively studied, particularly in relation to traits influencing silk production. The complete genome sequencing of *B. mori* has enabled researchers to identify key genes associated with cocoon yield and silk quality. For instance, Fang et al. conducted a genetic and genomic analysis revealing that specific QTLs are linked to cocoon yield traits, highlighting the impact of domestication on genetic diversity and trait expression in *B. mori* (Fang et al. 2020). Additionally, comparative transcriptomic analyses between domestic and wild strains have identified differentially expressed genes (DEGs) that play crucial roles in silk gland function, further elucidating the genetic basis of silk production (Fang et al. 2015). The *B. mori* genome has been fully sequenced, revealing approximately 14,500 protein-coding genes. Its genetic structure has been significantly shaped by artificial selection, resulting in a highly specialized organism for silk production, unlike its wild counterpart, *Bombyx mandarina* (*B. mandarina*).

### *3.2.2 Key Genes Involved in Silk Production*

The genetics of silk production in *B. mori* is intricately linked to several key genes that influence the structural properties and quality of silk fibres. The primary structural proteins involved in silk formation are the Fibroin Heavy Chain (Fib-H) and Fibroin Light Chain (Fib-L). Fib-H, which has a molecular weight of approximately 350 kDa, is responsible for the tensile strength of silk fibres, while Fib-L, at around 25 kDa, stabilizes the Fib-H through disulfide bonds, thus ensuring the integrity of the silk structure (Hu et al. 2012; Teramoto and Kojima 2015; Fungmongkonsatean et al. 2022). The interaction between these two chains is crucial for the formation of the silk fibre's characteristic strength and elasticity. Additionally, the protein P25, also known as fibro hexamerin, plays a significant role in stabilizing the Fib-H and Fib-L complex, contributing to the assembly of silk fibres (Tsubota et al. 2021; Qin et al. 2022). The presence of sericin proteins, encoded by genes such as sericin-1 (Ser1), sericin-2 (Ser2), and sericin-3 (Ser3), further influences silk quality by coating the fibroin fibres and enhancing fibres adhesion. Sericin typically accounts for 20–30% of the weight of the silk fibres and is essential for the overall structural integrity of the silk fibres, as it provides resistance and protection (Silva Pugina and Almeida Caiut 2021; Qin et al. 2022). Another important gene is BmKr-h1 (Krüppel-homolog 1), which regulates moulting and the development of silk glands. This gene is positively regulated by juvenile hormones, which are critical for the growth and development of *B. mori* larvae (Kayukawa et al. 2012; Zhu et al. 2021a). The expression of BmKr-h1 is crucial during the pupal stage, where it peaks in correlation with the levels of ecdysteroids, indicating its role in the transition from larval to adult forms (Zhu et al. 2021a, b). Moreover, the ATP-binding cassette (ABC) transporters BmABCG1 and BmABCG4 are involved in the transport of pigments, which can affect the coloration of silk fibres (Tsubota et al. 2021). The regulation of nutrient signalling through the Target of Rapamycin (TOR) pathway, involving genes such as BmTOR, BmRheb, and BmS6K, is also significant as it impacts silk gland size and protein synthesis, thereby influencing silk production (Mandal et al. 2012; Cao et al. 2023). In summary, the genetic framework governing silk production in *B. mori* encompasses a variety of genes that collectively influence the structural integrity, quality, and strength of silk fibre. The interplay between fibroin proteins, sericin, and regulatory genes such as BmKr-h1 and those involved in nutrient signalling underscores the complexity of silk biosynthesis and its evolutionary adaptations.

### *3.2.3 Genetic Variations Among* B. mori *Strains*

Genetic variations among *B. mori* strains, particularly between domesticated *B. mori* and its wild counterpart *B. mandarina*, have been extensively studied to understand the implications of domestication on genetic diversity, phenotypic traits, and silk production capabilities. The domestication process has led to significant genetic divergence, with domesticated strains exhibiting reduced genetic diversity compared to wild strains. For instance, Luan et al. conducted a study using RNA sequencing to identify genes associated with silk yield, revealing that, domesticated strains cluster distinctly from wild strains, indicating selective breeding for specific traits (Luan et al. 2018). This clustering is further supported by findings from Zhou et al., who noted that domesticated *B. mori* produce silk yields that are 3–5 times higher than those of wild *B. mori*, a result of long-term artificial selection (Zhou et al. 2020).

The genetic mapping of domesticated and wild *B. mori* has revealed specific genes under positive selection in domesticated strains. Wang et al. identified four genes with significantly higher fixation indices and lower nucleotide diversity in domesticated *B. mori*, suggesting that these genes may contribute to the behavioural traits that have been lost during domestication, such as climbing ability and mimicry (Wang et al. 2020a). Additionally, the comparative methyl omics study by Xiang et al. highlighted the role of DNA methylation in influencing gene expression differences between domesticated and wild *B. mori*, suggesting epigenetic factors may also play a role in the observed phenotypic variations (Xiang et al. 2013). Moreover, the analysis of olfactory gene expression has shown that domesticated *B. mori* exhibit reduced sensitivity in olfactory perception compared to their wild counterparts, with significant differences in the expression of olfactory-related genes (Qiu et al. 2018). This reduction in olfactory sensitivity is indicative of the broader behavioural changes that have occurred due to domestication. Kim et al. further emphasized the importance of gene expression profiles, noting significant differences in the expression patterns of gloverin genes between domesticated and wild strains, which may reflect adaptations to their respective environments (Kim et al. 2016).

In terms of genetic diversity, Guo et al. reported that domesticated *B. mori* have lost approximately 17% of nucleotide diversity compared to wild *B. mori*, which is indicative of the genetic bottleneck effect associated with domestication (Guo et al. 2011). This loss of diversity can have implications for the adaptability of domesticated strains to changing environmental conditions. Conversely, wild *B. mori* maintain a higher level of genetic diversity, which may confer greater resilience to environmental stresses (Meng et al. 2020). Further genomic studies have identified specific mutations associated with silk quality and fibres strength. Variations in Fib-H, Fib-L, and Sericin genes influence tensile strength, elasticity, and adhesion of silk fibres. Domesticated strains have undergone selection for increased fibroin gene expression, leading to thicker and more durable silk fibres (Qiu et al. 2018). Additionally, variations in the BmTOR signalling pathway genes regulate silk gland development and overall silk yield, with domesticated strains exhibiting upregulation of nutrient-sensing pathways to enhance silk production efficiency (Zhang et al. 2021a).

In addition to genetic divergence from wild strains, domesticated B. mori populations also exhibit regional variations. A comparative study of silkworm germplasm from China and Uzbekistan revealed differences in egg colour, shape, attachment, cocoon morphology, and silk filament properties (Fig. 3.1). Chinese strains predominantly produced elliptic cocoons, whereas Uzbekistan strains showed a higher proportion of peanut- and cylindrical-shaped cocoons. Silk filament length and thickness also varied significantly, with Uzbekistan strains producing longer and thicker filaments. These variations are likely influenced by regional breeding practices and environmental factors (Xu et al. 2024).

Interestingly, metabolic adaptations in domesticated *B. mori* also reflect their reliance on a specialized diet of mulberry leaves. Genomic analysis has identified differences in digestive enzyme gene expression, with domesticated strains exhibiting enhanced expression of amylase and protease genes to optimize nutrient absorption from mulberry leaves. This dietary specialization contrasts with the broader feeding preferences of wild *B. mori*, which have retained genetic flexibility in their digestive systems. Overall, these genetic variations highlight the profound effects of domestication on *B. mori* with selection pressures shaping both its physiological traits and silk production capabilities. While domesticated strains excel in high-yield silk production, their reduced genetic diversity and specialized adaptations make them more dependent on controlled environments, whereas wild strains retain greater ecological resilience and adaptability.

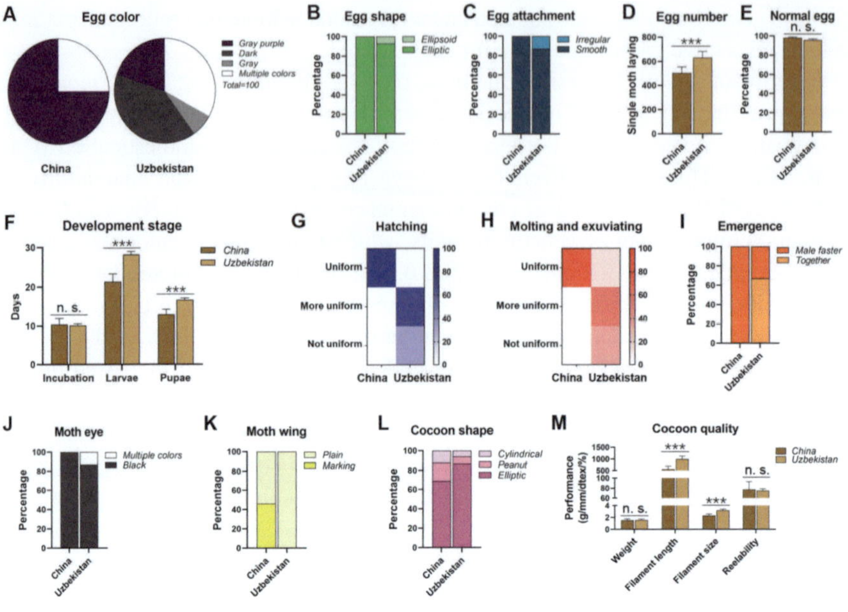


**Fig. 3.1 Morphological and performance characteristics of *B. mori*strains originating from China and Uzbekistan**: **a** Distribution of egg pigmentation. **b** Variability in egg morphology. **c** Characteristics related to egg adhesion. **d** Mean egg count per individual female. **e** Proportion of eggs exhibiting normal morphology. **f** Duration of different developmental phases. **g** Consistency in hatching time. **h** Patterns of larval molting and exuviation. **i** Variations in adult emergence timing. **j** Pigmentation differences in moth eyes. **k** Wing pattern diversity in moths. **l** Range of cocoon shapes. **m** Parameters assessing cocoon quality. Reprinted with permission from Xu et al. (2024)

## 3.3 Advancements in Genetic Mapping and Tools

Recent advancements in genetic mapping and the development of innovative tools have improved the study of *B. mori* genetics, providing deeper insights into gene function, regulation, and the genetic basis of important traits such as silk production, disease resistance, and environmental adaptation. Several breakthrough technologies and methodologies have significantly enhanced the resolution and efficiency of genetic studies in *B. mori*, enabling more precise identification of beneficial genes for breeding and improving *B. mori* strains.

One of the most significant developments in genetic mapping is the complete sequencing of the *B. mori* genome, which has paved the way for detailed studies of its genetic structure and functional genomics. The genome sequencing project, completed in 2004, provided a reference genome that serves as a foundation for mapping genes involved in critical traits. Developments in genetic mapping and tools for *B. mori* have significantly enhanced our understanding of its genetics, particularly in the context of silk production and other economically important traits. This section will discuss the development of molecular markers, linkage mapping and QTL analysis, and the genome sequencing and annotation of worm.

### 3.3.1 Development of Molecular Markers

Molecular markers are integral to genetic studies, breeding programs, and conservation strategies in *B. mori*, the domesticated *B. mori*. The utility of molecular markers, such as isozyme markers, has been well-documented in genetic diversity assessments, particularly when comparing *B. mori* strains from different geographical regions. For instance, isoenzymes offer a reliable method for evaluating genetic variations within and between populations, which plays a crucial role in MAS in breeding programs. By identifying specific genetic variations that affect traits like silk yield and fibres quality, these markers allow for more efficient selection of desirable traits in *B. mori* breeding. Additionally, growth in molecular diagnostics, especially through the integration of omics technologies, have provided enhanced tools for diagnosing diseases in *B. mori* and understanding pathogen interactions, thus helping mitigate health risks and improving overall *B. mori* production (Deepika et al. 2024).

One of the most widely used classes of molecular markers in *B. mori* genetics are Simple Sequence Repeats (SSRs), also known as microsatellites. These markers are particularly useful due to their high polymorphism and stability, making them an excellent choice for genetic diversity analysis, trait mapping, and breeding programs. The ability to create molecular maps with SSR markers has enabled researchers to identify key genetic relationships and improve *B. mori* strains over time. For example, SSR markers have been utilized to study and map traits related to silk production, allowing for better control of economically important characteristics in *B. mori* breeding. The development of these markers has significantly advanced the genetic mapping of *B. mori* and has facilitated the identification of traits that influence silk yield and other important traits, such as disease resistance and growth rate (Qian et al. 2019; Trochez-Solarte et al. 2019).

One of the most recent breakthroughs in *B. mori* genetics is the high-resolution genome assembly, which is instrumental in understanding the genetic basis of silk production. Tong et al. conducted a comprehensive study where they assembled 545 high-resolution *B. mori* genomes, including strains from wild, local, and improved *B. mori*. They extracted 264 complete Fib-H sequences from these genomes, which are crucial for understanding how silk production is regulated at the molecular level. This research underscores the utility of molecular markers in tracing genetic lineages and understanding evolutionary adaptations, shedding light on the genetic basis of traits such as silk yield and quality (Lu et al. 2023). This genome assembly also serves as a foundation for more precise MAS, allowing researchers to fine-tune breeding programs aimed at improving silk production.

Another important advancement in *B. mori* genetics is the application of SSR markers for assessing genetic diversity. In a study by Wei et al., a linkage map was constructed using 28 SSR markers to analyse the genetic diversity of 13 *B. mori* strains. The authors demonstrated the effectiveness of SSR markers in enhancing breeding programs by using them to estimate the relationships among the *B. mori* strains, providing insight into the potential for improving silk yield and disease resistance (Wei et al. 2013). This research illustrates how molecular markers can assist in breeding programs by identifying desirable genetic traits and improving the overall quality of *B. mori*.

Copy Number Variations (CNVs) are another type of genetic variation that has gained attention in *B. mori* research. Zhao et al. investigated CNVs among different *B. mori* strains and found that domesticated strains contained a significantly higher number of CNVs compared to wild strains. These CNVs may be linked to phenotypic differences such as silk production capabilities, suggesting that CNVs could play a role in the adaptability and resilience of *B. mori* to environmental changes (Zhao et al. 2014). This finding is important for understanding the genetic mechanisms that contribute to the higher silk yields seen in domesticated *B. mori*, as well as the increased resilience of wild strains to environmental stressors.

The development and application of molecular markers also have practical implications in enhancing disease resistance in *B. mori*. Hou et al. utilized SSR markers for marker-assisted selection to breed *B. mori* strains with improved resistance to the *Bombyx mori* Densonucleosis virus (BmDNV), a common pathogen that affects *B. mori* populations. In their study, they constructed a linkage map and identified specific SSR markers associated with resistance traits. These markers could be used to breed *B. mori* with enhanced disease resistance, thus improving the health and productivity of *B. mori* populations and reducing the reliance on chemical treatments for disease control (Hou et al. 2013). This case highlights the real-world applications of molecular markers in *B. mori* breeding, demonstrating how they can be used not only for improving silk production but also for addressing health-related challenges in *B. mori* populations. Furthermore, the use of transcriptomic data has become an essential tool in the development of molecular markers for *B. mori* genetics. Li et al. performed a high-throughput RNA sequencing (RNA-seq) analysis of *B. mori*, leading to the identification of novel genes and transcripts that could be used as markers for breeding programs. The integration of gene expression profiles with molecular marker development represents a significant advancement in the genetic mapping of *B. mori*, providing valuable information about the regulation of traits such as silk production, growth rate, and disease resistance (Hu et al. 2012). By incorporating transcriptomic data into the development of molecular markers, researchers can now select *B. mori* with more refined genetic profiles, leading to better breeding outcomes.

The development of molecular markers has been instrumental in advancing *B. mori* genetics. The case studies highlighted above demonstrate the diverse applications of these markers in improving silk production, enhancing disease resistance, and understanding the genetic basis of important traits in *B. mori* populations. As genomic technologies continue to evolve, the potential for molecular markers in *B. mori* genetics will only expand, providing new opportunities for research and breeding programs aimed at increasing the efficiency and sustainability of sericulture. The integration of genomic tools, including transcriptomics, SSR markers, and high-resolution genome assemblies, will continue to play a crucial role in shaping the future of *B. mori* breeding and silk production. With these changes, the industry can expect more robust, high-yielding, and disease-resistant *B. mori* strains, contributing to the growth and sustainability of the silk industry.

### *3.3.2 Linkage Mapping and QTL Analysis*

Linkage mapping and QTL analysis are fundamental to understanding the genetic mechanisms underlying economically significant traits in the *B. mori*, particularly in relation to its primary food source, mulberry. These genetic tools allow researchers to identify loci associated with key traits such as growth rate, cocoon yield, and disease resistance, ultimately optimizing breeding strategies to enhance sericulture productivity. Extensive genetic mapping efforts in *B. mori* have led to the development of high-resolution linkage maps facilitating QTL analysis. Yamamoto et al. (2008) constructed a bacterial artificial chromosome (BAC)-based linkage map encompassing 28 linkage groups, corresponding to the *B. mori*'s chromosomes. This foundational work has significantly contributed to identifying QTLs associated with mulberry-related traits (Yamamoto et al. 2008). Similarly, Mirhoseini et al. (2010) employed amplified fragment length polymorphism (AFLP) markers to map QTLs for economically significant traits, demonstrating the efficacy of molecular markers in *B. mori* genetics (Mirhoseini et al. 2010).

The interplay between *B. mori* and mulberry extends beyond nutrition, influencing genetic traits through dietary preferences. Paudel et al. (2020) demonstrated that *B. mori* larvae exhibits a preference for mulberry-based diets, attributed to specific olfactory receptors that detect mulberry-derived compounds. This dietary inclination directly impacts growth rates and other commercially valuable traits, making it a key factor in QTL studies (Paudel et al. 2020). Developments in QTL mapping have enabled the identification of genetic loci influencing cocoon weight and silk quality. Lü identified multiple QTLs associated with cocoon weight, providing valuable insights for breeding programs aimed at improving silk production. Likewise, Zhan et al. (2009) developed an integrated genetic linkage map that facilitated the precise mapping of genes and QTLs, underscoring the significance of genetic mapping in sericulture. The integration of high-density molecular markers, such as SSRs, has further refined linkage analysis (Zhan et al. 2009). Zou et al. (2021) highlighted the role of SSR markers in enhancing the accuracy of recombination analyses, which are critical for fine-mapping traits influenced by mulberry. This refinement allows for precise MAS, accelerating the development of high-yielding *B. mori* strains (Zou et al. 2021).

Beyond *B. mori* genetics, improving mulberry genotypes plays a crucial role in enhancing silk productivity. Manjunath et al. (2023) emphasized the need to develop pest- and disease-resistant mulberry cultivars to sustain healthy *B. mori* populations and maximize yield. This holistic approach to sericulture integrates host plant improvement with targeted *B. mori* breeding to enhance overall productivity and sustainability (Manjunath et al. 2023). Recent breakthroughs in QTL analysis have leveraged whole-genome resequencing to uncover genetic factors underlying silk production. Fang et al. (2020) analysed domesticated and wild *B. mori* strains, identifying selective sweep regions linked to cocoon yield traits. These findings offer evolutionary insights and potential markers for breeding superior *B. mori* varieties (Fang et al. 2020). In addition to traditional QTL mapping, statistical growth such as composite interval mapping have improved the precision of genetic analyses. Li et al. utilized this method to identify multiple QTLs associated with cocoon traits, demonstrating the impact of refined statistical techniques in genetic mapping (Fang et al. 2015). Furthermore, Li et al. identified QTLs governing silk protein synthesis pathways, providing critical genetic markers for enhancing fibroin and sericin production in breeding programs.

Overall, the integration of linkage mapping and QTL analysis has restructured *B. mori* genetics, enabling the discovery of key genetic loci responsible for silk yield, cocoon quality, and disease resistance. As genomic technologies advance, combining QTL analysis with transcriptomics and functional genomics will further unravel the genetic regulation of silk production, paving the way for developing superior *B. mori* strains with optimized genetic potential.

### *3.3.3 Genome Sequencing and Annotation of* B. mori

The genome sequencing and annotation of the *B. mori*, represents a significant advancement in Lepidopteran genetics, with profound implications for sericulture and biotechnology. The initial sequencing of the *B. mori* genome was completed in 2004, serving as a reference for subsequent studies and refined annotations (Xia et al. 2014). The genome, approximately 450 megabase (Mb) in size, provides a comprehensive resource for exploring biological processes such as development, metabolism, and disease resistance.

One of the most crucial outcomes of *B. mori* genome sequencing is the identification of genes involved in key biological pathways. For instance, Zhang et al. (2010) characterized apoptotic pathways in *B. mori*, identifying homologs of mammalian genes that regulate cell death and survival. Their findings demonstrated evolutionary conservation in apoptosis-related pathways, which play a role in *B. mori*'s responses to environmental stresses and pathogens (Zhang et al. 2010). In addition to functional gene annotation, large-scale transcriptomic studies have provided deeper insights into *B. mori* gene expression. Suetsugu et al. (2013) conducted extensive full-length complementary DNA (cDNA) sequencing, revealing diverse gene variants and alternative splicing events. These transcriptomic datasets have significantly improved our understanding of gene regulation, particularly in relation to silk protein synthesis and other economically relevant traits (Suetsugu et al. 2013). The integration of proteomics with genomic data has further enhanced annotation efforts. Levin et al. (2020) employed proteo transcriptomics in the *B. mori* BmN4 cell line, demonstrating how the combination of transcriptomic and proteomic data refines gene function predictions. This approach has been instrumental in identifying novel proteins and understanding their specific roles in *B. mori* biology (Levin et al. 2020). A significant portion of the *B. mori* genome consists of repetitive elements. Xu et al. (2013) compiled a database of transposable elements (TEs), revealing that they constitute approximately 40% of the genome. These elements influence gene expression and contribute to genetic diversity, which is essential for adaptation and evolutionary processes (Xu et al. 2013). Moreover, Zhao et al. (2013) identified segmental duplications in the genome, which complicate gene annotation but also provide insights into functional redundancy and evolutionary adaptations. Understanding these duplications is particularly relevant to traits such as silk quality and yield, key factors in sericulture (Zhao et al. 2013).

Furthermore, the development of databases such as KAIKObase has facilitated access to genomic data and annotations for researchers worldwide, promoting further studies on *B. mori* and related species (Yang et al. 2021). This resource is invaluable for both basic research and applied sciences, including breeding programs aimed at improving silk production. In summary, the genome sequencing and annotation of *B. mori* have laid the groundwork for a multitude of research avenues, enhancing our understanding of its biology and genetics. The integration of genomic, transcriptomic, and proteomic data continues to provide insights into the complex biological processes that underpin the life cycle and economic importance of this species.

## 3.4 Selective Breeding and Genetic Hybridization

Selective breeding and genetic hybridization have played a critical role in improving the quality and productivity of mulberry silk from *B. mori*. While traditional methods focused on phenotypic selection, modern approaches now use marker-assisted selection (MAS) and genome editing for greater precision. Hybridisation combines traits from different strains to produce superior hybrids with longer filaments, heavier cocoons, and better resistance to disease and environmental stress (Attri et al. 2024). As seen from Table 3.1, In Japan, automated feeding systems and climate-controlled environments have optimized production efficiency and silk quality. In India, high-yielding mulberry varieties and scientific breeding methods have significantly boosted silk production and economic growth. Meanwhile, in China's Zhejiang Province, a combination of organic farming and modern biotechnological interventions has resulted in higher silk yields and improved soil health. These approaches highlight how hybridization, along with technological methods, has strengthened the sericulture industry worldwide.

One of the key advancements in sericulture genetics is the use of molecular markers and genomic tools to improve *B. mori* breeding. Marker-assisted selection (MAS) and genome sequencing have enabled precise identification of traits such as silk protein composition, cocoon shell weight, and filament denier, accelerating the development of superior silkworm strains (Long et al. 2024). Hybridization between bivoltine and polyvoltine strains has led to high-yielding, climate-resilient hybrids.

**Table 3.1** Technological advancements in silk production across leading sericulture regions

| Region | Technological practices | Results |
|---|---|---|
| Japan | Implementation of automated feeding systems, environmental control units for temperature and humidity, and disease monitoring tools | Significant improvement in production efficiency, uniform cocoon quality, reduced labor dependency, and enhanced resilience against environmental stressors |
| India | Development and adoption of high-yielding and drought-resistant mulberry cultivars, use of scientific cross-breeding programs, and integration of silkworm probiotics and pest management strategies | Substantial rise in raw silk output, rural employment generation, improved silkworm health, and overall economic upliftment of sericulture-dependent communities |
| China (Zhejiang Province) | Synergistic use of organic farming (manual harvesting, pesticide-free cultivation) with cutting-edge biotechnology, including gene expression analysis and microbial biofertilizers | Elevated cocoon yield, improved silk fiber strength and fineness, enhanced soil fertility, and alignment with global sustainability standards |

To avoid inbreeding-related issues like reduced vitality and disease susceptibility, breeding programs now combine controlled hybridization with genome analysis to maintain diversity and enhance hybrid vigor (Giora et al. 2022; Rajawat et al. 2024). Gene-editing technologies, especially CRISPR/Cas9, have transformed silkworm breeding by enabling rapid, targeted changes in silk fibroin genes, resulting in stronger and more elastic fibers. Transgenic *B. mori* expressing foreign proteins have expanded silk's applications into biomedicine and advanced materials (Pavithra et al. 2024; Xu et al. 2024). A study by Tan et al. (2013) introduced a transgene-based female-specific lethality system in *B. mori*, enabling efficient genetic sexing for selective breeding. This system uses a female-specific promoter to activate a lethal gene, selectively eliminating female larvae while allowing male silkworms, preferred for their higher silk yield and quality, to survive. By ensuring the propagation of only superior male lineages, this approach enhances hybridization efficiency and supports key goals in sericulture, such as improving silk yield, disease resistance, and overall productivity. Figure 3.2 illustrates this genetic construct and its role in controlled hybridization strategies. (Tan et al. 2013). Since mulberry foliage is the exclusive food source for *B. mori*, mulberry breeding programs have been critical to improving silk yield and quality. Traditional and modern hybridization techniques have developed high-yielding, nutrient-rich, and drought-resistant cultivars to ensure a consistent leaf supply for commercial silkworm rearing (Saini et al. 2023). With climate change posing new threats, both silkworm and mulberry breeding now focus on developing heat- and disease-resistant strains to ensure sustainable silk production (Gundi et al. 2023). The convergence of molecular breeding, genetic engineering, and hybridization strategies has significantly advanced the efficiency, productivity, and sustainability of sericulture. The application of MAS, CRISPR/Cas9 gene editing,

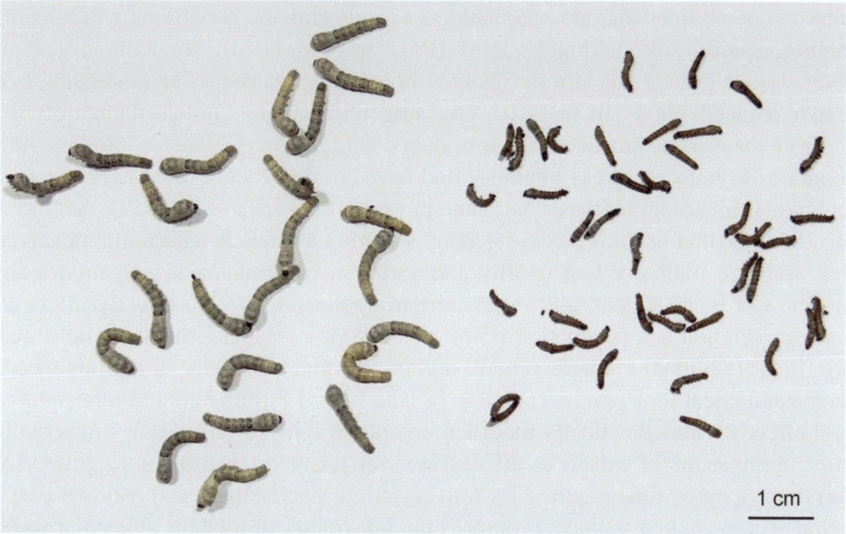


**Fig. 3.2** Female-specific lethality in OX4319 transgenic *B. mori* lines in the absence of dietary tetracycline. Male larvae (left) successfully progressed to the third instar stage, while larvae (right) presumed to be females were unable to molt beyond the first instar and died early in development. Scale bar: 1 cm. Reprinted with permission from Tan et al. (2013)

RNAi technology, and genetic sexing systems has streamlined *B. mori* breeding, enhancing both silk yield and disease resistance (Xu et al. 2011; Mondal et al. 2024; Rajawat et al. 2024).

### *3.4.1 Traditional Breeding Approaches*

Traditional breeding approaches have long been the foundation of genetic improvement in mulberry and *B. mori*, focusing primarily on selection and hybridization to enhance key traits such as leaf yield, pest resistance, climatic adaptability, and silk quality. The systematic breeding of mulberry, which began prominently in the mid-twentieth century, involved the controlled crossing of selected genotypes to develop F1 hybrids exhibiting hybrid vigor, a phenomenon that enhances the qualities of offspring compared to their parents (Mondal et al. 2024). Over time, these methods have led to significant improvements in mulberry growth rates, nutritional content, and resilience to adverse environmental conditions, directly impacting the productivity of the silk industry (Gundarapu et al. 2024). In traditional *B. mori* breeding, the selection process relies heavily on morphological and physiological evaluations, where breeders assess cocoon weight, shell percentage, filament length, and disease resistance to ensure the production of high-quality silk. Hybridization between bivoltine and polyvoltine strains has been a crucial strategy for developing robust *B. mori* lines that are adaptable to various climatic conditions while maintaining superior silk yield and quality (Rajawat et al. 2024). Maintaining genetic diversity is a critical concern in this type of breeding, as excessive inbreeding can lead to reduced vitality, increased disease susceptibility, and limited adaptability. To mitigate these risks, breeders have introduced wild mulberry relatives into breeding programs to improve stress tolerance and have employed controlled hybridization techniques to sustain heterosis and genetic variability (Saini et al. 2023). Additionally, the breeding extends beyond genetic selection to include agronomic practices that optimize mulberry leaf quality and yield, such as enhanced irrigation techniques, soil management, and micronutrient application, which directly affect *B. mori* growth and silk production (Giora et al. 2022). Despite growth in molecular breeding, traditional methods remain highly relevant, especially in regions where biotechnological tools are inaccessible or where local farming communities prefer cost-effective, non-genetically modified organisms (GMOs) approaches to sericulture improvement (Mondal et al. 2024). However, it has certain limitations, primarily due to the lengthy time required for trait selection, stabilization, and regional adaptation. These challenges have prompted the integration of modern molecular techniques with conventional breeding to accelerate genetic improvements in mulberry and *B. mori* strains (Gundarapu et al. 2024). Looking forward, bridging traditional breeding with molecular tools such as MAS and genome sequencing is expected to enhance breeding efficiency while maintaining the sustainability and economic feasibility of the silk industry (Rajawat et al. 2024). The integration of climate-resilient breeding strategies has become increasingly necessary as global temperature fluctuations and environmental stresses threaten silk production. Recent efforts in breeding have focused on developing heat-tolerant and disease-resistant *B. mori* hybrids, which exhibit higher survival rates and stable silk yield even in extreme environmental conditions (Gundi et al. 2023). Furthermore, crossbreeding between drought-resistant and high-yielding mulberry cultivars has resulted in varieties that maintain high leaf biomass and nutrient content despite water scarcity, offering a sustainable solution for sericulture in arid regions (Rajawat et al. 2024).

The significance of traditional breeding in sericulture remains undeniable, as it has been instrumental in enhancing silk yield, resilience, and quality through selective propagation and hybridization techniques. While these methods have established a strong foundation for genetic improvements in mulberry and *B. mori* breeding, challenges such as extended breeding cycles and environmental adaptability constraints have necessitated the integration of molecular tools to optimize efficiency. The incorporation of MAS, genome sequencing, and stress-tolerant hybridization has facilitated the development of high-yielding, climate-resilient mulberry and *B. mori* strains, ensuring the long-term sustainability of silk production. As climatic and ecological pressures intensify, the future trajectory of sericulture will rely on a synergistic approach that combines conventional breeding expertise with biotechnology, fostering greater genetic diversity, adaptability, and economic credibility in silk-producing regions (Mondal et al. 2024).

### 3.4.2 Development of High-Yield and Disease-Resistant B. mori Hybrids

The development of high-yield and disease-resistant *B. mori* hybrids has become a central focus in modern sericulture, aiming to enhance silk production efficiency while ensuring resilience against prevalent pathogens and environmental stressors. Hybridization strategies have leveraged natural genetic diversity within *B. mori*, enabling researchers to crossbreed superior strains to enhance silk yield, cocoon weight, and resistance to diseases such as *Bombyx mori* nucleopolyhedrovirus (BmNPV) (Mondal et al. 2024; Gundarapu et al. 2024). The evaluation index (E.I) method, first developed and widely applied in contemporary breeding programs, provides a quantitative framework for assessing multiple genetic traits simultaneously, allowing breeders to select the most robust hybrids based on cumulative performance rather than isolated traits (Kant et al. 2024). A significant breakthrough in recent years has been the integration of CRISPR/Cas9 technology to introduce genetic modifications that improve both silk yield and disease resistance. By precisely targeting genes associated with BmNPV susceptibility, researchers have successfully engineered *B. mori* strains with enhanced antiviral defences (Fig. 3.3), significantly improving survival rates and reducing production losses (Chen et al. 2017; Dong et al. 2019b). The development of multiplex CRISPR systems has further enabled the simultaneous editing of multiple genes, a strategy that has demonstrated enhanced resilience against viral replication and potentially reduces the dependency on chemical disease management methods, making sericulture more environmentally sustainable (Dong et al. 2020). Beyond gene editing, metabolomic analyses have provided valuable insights into the physiological mechanisms underlying silk production efficiency and disease resistance. Studies indicate that *B. mori* fed on fresh mulberry leaves exhibit superior growth rates and cocoon yields compared to those reared on artificial diets, reinforcing the importance of nutritional optimization in hybrid breeding programs (Li et al. 2023b). Additionally, recent research has explored immune response pathways in *B. mori*, revealing key genes, such as those encoding serine proteases, that play a pivotal role in pathogen resistance (Qin et al. 2012). These findings offer a new dimension for selective breeding, integrating both phenotypic selection and molecular screening to create disease-resistant hybrids with robust immune responses. Environmental adaptability is another critical factor influencing hybrid performance, particularly in regions experiencing climate change-induced fluctuations in temperature and humidity. Studies have demonstrated that selecting for heat-resistant traits in *B. mori* hybrids significantly enhances survival rates and silk productivity, ensuring year-round silk production despite environmental challenges (Umarov and Nasirillaev 2023). Alongside genetic modifications, crossbreeding with wild *B. mori* relatives has been explored to introduce natural resilience traits into domesticated strains, a strategy that has yielded hybrids with greater vigor and resistance to both biotic and abiotic stressors [NO_PRINTED_FORM]. Moreover, MAS has been increasingly employed to accelerate the breeding process, allowing for the precise identification of genetic markers linked to high-yield and disease-resistant traits (Giora et al. 2022). This approach enables breeders to streamline hybrid selection and improve genetic stability while maintaining high heterotic effects to maximize silk output. As the sericulture industry continues to evolve, the future of high-yield and disease-resistant *B. mori* hybrid development will rely on a multifaceted approach that integrates normal breeding expertise with advanced genetic engineering techniques. The combined use of CRISPR/Cas9, MAS, immune pathway analysis, and climate-resilient hybridization will be pivotal in creating next-generation *B. mori* hybrids that meet the growing demands of sustainable and economically viable silk production (Mondal et al. 2024; Rajawat et al. 2024; Gundarapu et al. 2024).

### *3.4.3 Success Stories in Hybrid* B. Mori *Strains*

The advancement of hybrid *B. mori* strains has significantly transformed the sericulture industry, demonstrating substantial improvements in silk yield, disease resistance, and adaptability to environmental challenges. By leveraging hybridization techniques, breeders have successfully developed superior *B. mori* strains that capitalize on heterosis, enabling higher productivity while ensuring resilience against pathogens such as BmNPV and fungal infections. One of the most notable success stories is the Bivoltine-2 hybrid strain, developed through the crossbreeding of Chinese native strains, which has exhibited a 20% increase in silk yield compared to wild strains, while demonstrating enhanced resistance to viral diseases (Xiao et al. 2022). Similarly, the Ubon Ratchathani 60–35 hybrid, developed in Thailand, has shown improved metabolic capacity in resisting fungal infections such as *Beauveria bassiana*, which has long posed a challenge to the sericulture industry (Veenstra 2019). Hybridization strategies that integrate genetic material from different ecological regions have further enhanced the adaptability of *B. mori* strains, as seen in the Xiangxian hybrid, which combines genetic traits from *B. mori* bred in the northern and southern regions of China, allowing for year-round sericulture in diverse climates (Furdui et al. 2014). The incorporation of MAS in breeding programs has further streamlined the development of hybrids, allowing breeders to identify genetic markers linked to higher silk yields and disease resistance, thus accelerating the selection process for improved strains (Giora et al. 2022). The Jingsong × Haoyue hybrid has exemplified the success of heterosis, producing significantly higher silk filament lengths and cocoon weights than its parent strains, reinforcing the importance of crossbreeding strategies for maximizing productivity (Ruiz and Almanza 2018). Beyond traditional hybridization, biotechnological methods have transformed *B. mori* breeding, with CRISPR/Cas9 genome editing enabling precise modifications in genes associated with silk production and pathogen resistance (Long et al. 2024). Research has successfully edited genes responsible for silk gland development, leading to hybrid strains capable of producing stronger and more

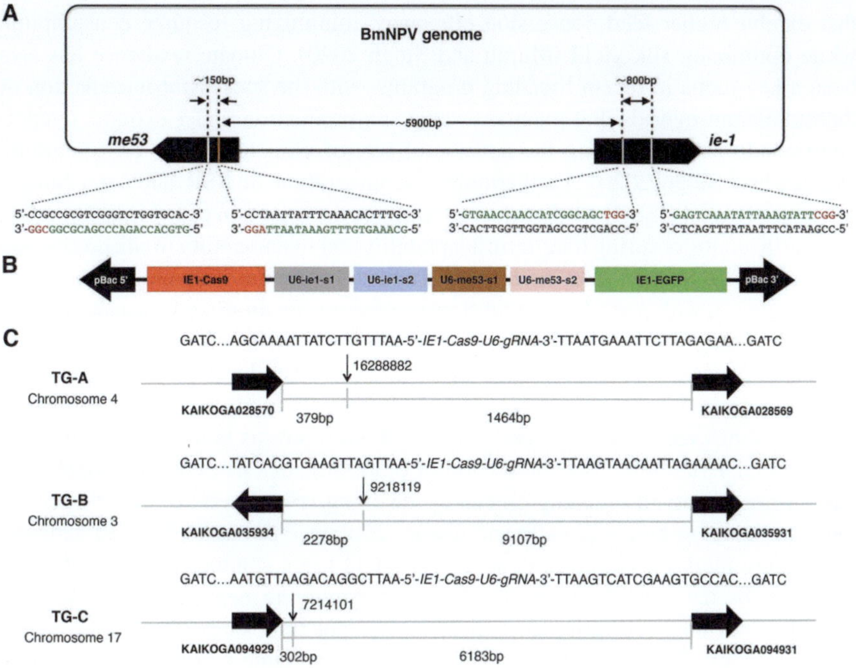


**Fig. 3.3** Development of a CRISPR/Cas9 system targeting *Bombyx mori* nucleopolyhedrovirus (BmNPV) in silkworms. **a** Diagram showing the positions of gRNA target sequences within the BmNPV genome. The genome is illustrated by a black rounded rectangle, with two major genes, me53 and ie-1, indicated by large black arrows. gRNA target regions are marked in green, while corresponding PAM sites are shown in red. Distances between target site pairs are labeled in base pairs. **b** Schematic layout of the pBac-EGFP-Cas9/4 × gRNAs construct. The Cas9 expression unit is highlighted in red. gRNA expression cassettes targeting ie1-s1, ie1-s2, me53-s1, and me53-s2 are represented in grey, blue, brown, and pink, respectively. The EGFP marker cassette is shown in green. **c** Insertion sites of the construct in three transgenic silkworm lines: TG-A, TG-B, and TG-C. Vertical arrows denote genomic insertion points. Chromosomal locations and partial sequences flanking the BfuCI restriction site at the 3′ and 5′ boundaries of the inserted construct are illustrated. Nearby genes are marked with horizontal black arrows, with the base pair distances between the insertion points and adjacent genes indicated. Reprinted with permission from Chen et al. (2017)

elastic silk fibres (Bi et al. 2019). Additionally, transcriptomic studies have identified key genetic pathways associated with cocoon development, offering insights into potential genetic modifications for future hybrid improvement (Li et al. 2024). Hybrid strains optimized for nutritional efficiency have also emerged, as seen in the Liangguang II hybrid, which demonstrated superior larval quality and increased cocoon yield when compared to conventional hybrids (Dewi et al. 2021). In India, field evaluations of newly developed bivoltine hybrids have shown enhanced adaptability to different agricultural conditions, confirming their potential for sustained cocoon production across various farming systems. The application of nutritional genomics in hybrid breeding has allowed for the development of *B. mori* strains that exhibit higher feed conversion efficiency, minimizing resource consumption while optimizing silk yield (Murali and Singh 2020). Climate resilience has also been a key focus in recent breeding programs, with the successful introduction of thermotolerant hybrids that maintain stable silk production under extreme temperature conditions, mitigating the adverse effects of climate change on sericulture (Murali and Singh 2020). Furthermore, the integration of SSR and Inter-Simple Sequence Repeat (ISSR) markers has allowed for the selection of hybrids that retain genetic diversity, ensuring long-term adaptability and sustainability in silk production (Vijayan 2010; Srivastava et al. 2011). Advances in cloning technologies have also been explored, with parthenogenetic cloning methods facilitating the acceleration of hybrid breeding cycles while preserving genetic traits associated with high productivity and disease resistance (Larina et al. 2021). These developments underscore the convergence of older hybridization techniques with modern genomic strategies, leading to the creation of high-performance *B. mori* strains tailored for large-scale commercial sericulture. The integration of bioinformatics and computational tools has further improved breeding efficiency, enabling researchers to identify genetic markers linked to high silk yield and disease resilience through advanced genomic sequencing methods (Rajawat et al. 2024). Field trials conducted in various sericulture hubs have validated the feasibility of deploying next-generation *B. mori* hybrids, with data indicating that these strains not only outperform wild breeds in terms of productivity but also exhibit stronger immune responses to prevalent pathogens, reducing mortality rates and boosting economic returns for sericulture farmers (Mondal et al. 2024; Gundarapu et al. 2024). The global success stories of hybrid *B. mori* strains affirm the growing importance of integrating traditional breeding methods with molecular genetics, ensuring that future hybrids continue to push the boundaries of silk yield and quality. As the sericulture industry continues to evolve, sustained research efforts in hybrid vigor optimization, climate adaptation, and disease resistance will be critical in securing the long-term potentiality of silk production, reinforcing its role as a cornerstone of economic sustainability and agricultural practices.

## 3.5 Genetic Engineering in Silk Production

Genetic engineering in silk production has transformed *B. mori* from a traditional breeding model into a bio-factory for advanced fibres and high-value proteins. Moving beyond phenotypic selection, modern techniques like CRISPR/Cas9, TALENs, and transgenesis have enabled precise modifications to improve silk yield, mechanical properties, and introduce novel functionalities (Yanagisawa et al. 2007). Transgenic *B. mori* expressing recombinant proteins such as human erythropoietin (EPO) and lactoferrin demonstrate the feasibility of using silk glands as bioreactors, offering scalable alternatives to conventional protein production (Seong et al. 2011; Teulé et al. 2012). Integration of spider silk genes has resulted in hybrid fibres with enhanced strength and elasticity, expanding silk's potential in biomedicine, tissue engineering, and high-performance materials (Teulé et al. 2012; Tang et al. 2021). Genetic modifications have also enabled silk with special properties: improved cell adhesion (via collagen/fibronectin peptides), biodegradability, fluorescence (through GFP), and antimicrobial resistance (via AMPs) (Mieszawska et al. 2010; Iizuka et al. 2013).

These advances align silk with applications in regenerative medicine, drug delivery, and sustainable materials. Genomic tools and RNA-seq studies have identified genes related to silk yield, stress tolerance, and growth, supporting targeted breeding and improved productivity under variable conditions (Luan et al. 2018; Xu et al. 2024). Combining traditional breeding with MAS and gene editing enhances silk quality while preserving genetic diversity (Maimoona Kanwal et al. 2019). Despite its promise, widespread use of genetically engineered *B. mori* faces challenges. Biosafety, ecological risks, and consumer acceptance remain key hurdles, and minimizing off-target effects in genome editing is an ongoing research priority.

### *3.5.1 Gene Editing with CRISPR/Cas9 in* **B. mori**

Gene editing with CRISPR/Cas9 technology presents a transformative approach in the genetic manipulation of the *B. mori*, significantly enhancing sericulture practices by developing high-yield, disease-resistant hybrids. The CRISPR/Cas9 system allows for precise alterations in the genome, enabling the activation, disruption, or modification of specific genes that play crucial roles in the biological processes of *B. mori*. This capability is particularly pivotal for advancing traits such as silk yield and resistance against pathogens like the BmNPV (Dong et al. 2019a). By enabling targeted modifications at desired genomic locations, CRISPR/Cas9 facilitates improvements in silk production through enhancements in yield, disease resistance, and overall resilience to environmental stressors. The application of this technology in *B. mori* genetics not only accelerates conventional breeding processes but also expands the potential for novel genetic improvements.

One primary application of the CRISPR/Cas9 system in *B. mori* genetic research is the enhancement of silk yield. Studies have demonstrated that targeting specific genes associated with the silk production process can result in higher silk yields. For example, the use of CRISPR/Cas9 to knock out or edit the *B. mori* vacuolar protein sorting 13D (BmVps13d) gene has been shown to lead to significant reductions in larval weight and silk yield, indicating the gene's critical role in silk production (Zhao et al. 2022). In this context, CRISPR/Cas9 not only serves as a means to modify undesirable traits but can also facilitate the identification and validation of essential genes for silk production. The Cas9 protein and guide RNAs work synergistically to create double-strand breaks at specific genomic locations. Following this, the cellular repair machinery attempts to fix the break, often leading to insertions or deletions that disrupt the target gene. This process can either knock out the gene or, with the assistance of donor DNA, introduce specific mutations or insertions that confer desirable traits (Baci et al. 2021; Rao et al. 2022). By using this method, researchers have been able to generate mutant lines in *B. mori* carrying specific gene mutations, which can then be used to study gene function and trait expression under various conditions (Brady et al. 2020).

Another notable application of CRISPR/Cas9 in *B. mori* is the targeted disruption of genes that confer susceptibility to viral infections, such as BmNPV. Research suggests that modifying genes linked to immune responses can enhance resistance against this virus, a significant threat to *B. mori* populations (Hu et al. 2023; Usama Hanif et al. 2025). By leveraging CRISPR/Cas9, scientists can create genetically modified *B. mori* that exhibit increased resistance to BmNPV, thereby reducing reliance on chemical interventions and fostering more sustainable sericulture practices. For instance, CRISPR/Cas9-mediated disruption of immediate early genes in BmNPV has significantly reduced viral replication and improved *B. mori* survival rates (Dong et al. 2019a). The ongoing refinement of CRISPR systems, including the development of multiplex editing capabilities, allows for the simultaneous targeting of multiple genes, further improving the efficiency of breeding approaches for disease resistance (Dong et al. 2019b).

Beyond improving silk yield and disease resistance, CRISPR/Cas9 has facilitated the production of therapeutic biomolecules within *B. mori*. The ability to express complex proteins, such as spider silk proteins, clearly illustrates this potential (Aramwit et al. 2023). Editing fibroin genes in the *B. mori* enables the enhancement of silk fibres properties, generating silk with unique mechanical or biodegradable properties suited for various biomedical applications. Additionally, the combination of CRISPR/Cas9 with transgenic approaches has been instrumental in developing novel *B. mori* lines capable of producing recombinant proteins for pharmaceutical and industrial applications (Aramwit et al. 2023). Furthermore, CRISPR/Cas9 technology aids in advancing genetic studies concerning *B. mori* domestication. Researchers have used gene-editing frameworks to explore the functional significance of genes influencing domestication-related traits, such as pigmentation and behavioural attributes (Tomihara et al. 2022). By generating targeted knockouts, researchers have elucidated the roles these genes play in distinguishing domesticated *B. mori* from their wild progenitors, *B. mandarina*, thereby informing future breeding programs aimed at preserving genetic diversity.

Methodological advancements coupled with the CRISPR/Cas9 system, have streamlined the development of diverse mutant lines in *B. mori*. Systems have been established to screen for stable CRISPR/Cas9-mediated mutations (Brady et al. 2020), allowing for efficient characterization and stabilization of mutant phenotypes. This capacity for rapid generation and assessment of mutant lines accelerates the pace of functional genomics in *B. mori* research, paving the way for new discoveries and applications in the field. Additionally, researchers have explored the potential of CRISPR/dCas9 for molecular imaging in living *B. mori* cells, enhancing the understanding of genomic organization and dynamics (Xing et al. 2020). This approach not only adds depth to genetic studies but also provides a framework for visualizing gene expression patterns in real-time, contributing to the understanding of gene regulation during development and response to environmental stimuli.

Despite the numerous advantages of CRISPR/Cas9 technology, challenges remain in its implementation within *B. mori* research. The efficiency of genome editing varies, and rigorous screening of edited lines is necessary to confirm successful modifications. Success rates for CRISPR/Cas9 editing events in *B. mori* can be inconsistent, often resulting in disruptions such as decreased hatching and survival rates, as noted in studies utilizing different guide RNA constructs (Hong et al. 2020). Additionally, ethical and environmental concerns surrounding the use of CRISPR technology in breeding programs necessitate careful assessments of long-term sustainability and genetic stability. Maintaining genetic diversity, while ensuring the stability of edited traits across generations, remains a challenge that must be addressed to prevent risks associated with monoculture and reduced genotypic variation (Gai et al. 2020).

### 3.5.2 Introduction of Foreign Genes for Functionalized Silk

The introduction of foreign genes into *B. mori* for the functionalization of silk represents a groundbreaking advancement in sericulture and biomaterial engineering. By leveraging genetic engineering, researchers can manipulate the *B. mori* genome to produce silk with enhanced mechanical, biological, and chemical properties, expanding its applications beyond textiles into biomedical, industrial, and environmental domains. This transformative approach has facilitated the synthesis of transgenic silk fibres that incorporate specialized proteins, yielding novel materials with improved performance and versatility. One of the most significant developments in this field is the incorporation of functional proteins into silk, enhancing its bioactive properties. For example, Wang et al. demonstrated that genetic modification of *B. mori* to express human fibroblast growth factor 1 (hFGF1) within the silk glands resulted in silk capable of promoting cell proliferation, underscoring its potential in tissue engineering and regenerative medicine (Wang et al. 2015). The simultaneous expression of GFP in these transgenic *B. mori* provided a visual confirmation of successful gene integration, while also demonstrating that silk can be engineered as a functional biomaterial. Similarly, research by Zhang et al. revealed that modifying silk protein expression can confer antibacterial properties, highlighting its potential for medical applications such as wound dressings and surgical sutures (Zhang et al. 2023).

The use of gene-editing technologies, particularly CRISPR/Cas9, has been instrumental in refining the incorporation of foreign genes into silk proteins. For instance, targeted genetic modifications of the *B. mori* ecdysone kinase-like 1 (*BmEcKL1*) gene have been shown to significantly enhance silk gland development, leading to increased silk production (Li et al. 2024). The introduction of exogenous genes related to growth regulation, such as *Ras* genes, has further optimized silk yield by improving glandular protein synthesis, demonstrating how metabolic pathways can be manipulated to augment both the quality and quantity of silk (Li et al. 2024). These genetic modifications not only enhance silk production but also create avenues for scaling up the commercial viability of engineered *B. mori*. Beyond augmenting silk properties, the introduction of spider silk genes into *B. mori* has generated significant interest due to the superior mechanical attributes of spider silk. Xu et al. successfully expressed spider silk genes in transgenic *B. mori*, yielding fibres with increased tensile strength and elasticity compared to conventional *B. mori* silk (Xu et al. 2018). This integration has immense implications for textile engineering and high-performance fibres development, as spider silk's lightweight, yet highly durable nature makes it ideal for applications ranging from protective gear to aerospace materials. Similarly, the production of fluorescent and coloured silk has been achieved through the genetic insertion of fluorescence genes, as demonstrated by Iizuka et al., who engineered *B. mori* to express various fluorescent protein, expanding the potential for silk in bio-imaging and biosensing applications (Iizuka et al. 2013).

In addition to mechanical and aesthetic enhancements, genetic modifications have facilitated the development of silk-based materials with inherent antimicrobial properties. The introduction of AMPs into silk fibroin has been explored as a means of preventing microbial contamination in medical and environmental applications (Zhang et al. 2023). Moreover, metabolic engineering approaches, such as upregulating *Myc* expression in silk glands, have been investigated to enhance DNA replication and, consequently, silk protein synthesis (Qian et al. 2021). These modifications demonstrate how targeted genetic interventions can optimize silk's biochemical composition for specialized applications. The integration of silk with inorganic materials has also been explored to develop multifunctional composite materials. For instance, the fusion of silk with silica-binding peptides has yielded hybrid biomaterials with enhanced structural stability, making them suitable for bone regeneration and implant coatings (Mieszawska et al. 2010). Such developments underscore silk's adaptability as a scaffold for biomedical applications, further reinforced by genetic modifications that enable the controlled expression of bioactive molecules within silk fibres.

Despite these promising developments, challenges remain in optimizing transgene expression levels and ensuring the stability of genetic modifications across generations. The efficiency of foreign gene integration into the *B. mori* genome often varies, requiring the refinement of expression vectors and regulatory sequences to achieve consistent and high-yield protein production. Moreover, regulatory and ethical considerations surrounding genetically modified *B. mori* necessitate comprehensive assessments of environmental impact and biosafety before large-scale commercial adoption can be realized (Gupta and Rastogi 2023).

In conclusion, the introduction of foreign genes into *B. mori* has revolutionized silk biotechnology, enabling the production of functionalized silk with enhanced mechanical, biological, and chemical properties. This advancement extends silk's applications into medicine, environmental science, and high-performance materials, transforming older perceptions of sericulture. As genetic engineering techniques continue to evolve, the future of silk production will likely see further refinements in transgene expression, expanded functional capabilities, and increased commercial viability, cementing silk's status as a versatile and valuable biomaterial in modern science and industry.

### 3.5.3 Achievements in Improving Silk Yield, Strength, and Quality

The advancements in enhancing silk yield, strength, and quality in *B. mori* have been driven by a combination of genetic engineering, molecular biology, breeding strategies, and innovative processing techniques. These developments not only improve the economic credibility of sericulture but also align with the growing demand for high-quality silk fibres across various industries, including textiles, biomedicine, and advanced materials. By integrating these breeding methods with cutting-edge biotechnology, researchers have achieved remarkable improvements in both the quantity and mechanical properties of silk fibres, ensuring their competitive edge in modern applications.

One of the most significant breakthroughs in silk production has been the strategic use of genetic modification techniques, particularly CRISPR/Cas9, to enhance silk gland development and increase silk protein synthesis. Targeted editing of genes such as *BmEcKL1* has led to larger silk glands, directly resulting in higher silk yields (Li et al. 2024). Similarly, overexpression of key regulatory genes like, *B. mori* glycosyltransferase 1-like (*BmGT1-L)* has been demonstrated to promote exocrine gland function, further optimizing silk output by improving fibroin secretion efficiency (Tang et al. 2020). These genetic methods have played a crucial role in creating *B. mori* strains capable of producing significantly larger quantities of silk without compromising quality.

Beyond increasing yield, researchers have focused on enhancing the mechanical properties of silk fibres by introducing spider silk genes into *B. mori*. Spider silk is renowned for its exceptional tensile strength and elasticity, and genetic modifications incorporating spider silk proteins, particularly flagelliform and dragline motifs, have resulted in hybrid silks that exhibit up to a 59.6% increase in strength and a 121.1% improvement in toughness compared to conventional *B. mori* silk (Kuwana et al. 2014; Wang et al. 2015). These innovations provide high-performance fibres suitable for applications requiring durability, such as medical sutures, protective textiles, and biodegradable composites. Additionally, the processing conditions of silk have been shown to play a crucial role in determining its mechanical properties. Research indicates that controlled silking speeds influence fibres crystallinity, with optimized processing yielding artificial *B. mori* silk that achieves tensile strengths of up to 0.9 GPa (Zhang et al. 2012). Such findings highlight the importance of integrating environmental controls into silk production to enhance fibres robustness.

The chemical composition and structural organization of silk fibres have also been manipulated to improve their properties. Mineral additives, such as calcium carbonate nanoparticles, have been successfully incorporated into silk fibroin, inducing high-density mesoscopic networks that enhance both strength and elongation properties (Zhang et al. 2023). Similarly, reinforcing silk with carbon nanotubes has resulted in composite fibres with significantly superior tensile strength and elasticity, broadening their potential for use in high-performance engineering and biomedical applications (Lu et al. 2024). These composite-based approaches demonstrate how integrating material science innovations with biological silk production can produce advanced functional fibres with tailored properties.

Apart from genetic and material innovations, improvements in silk quality have been achieved through refinements in silk reeling and extraction processes. Advanced silk reeling technologies, which allow for greater control over filament consistency, have contributed to higher-quality silk threads with improved neatness and durability (Gowrisankar et al. 2023). Additionally, dietary interventions have proven beneficial in enhancing silk production. Studies have shown that supplementing *B. mori* diets with nanoparticles, such as titanium dioxide ($TiO_2$), can significantly improve silk yield by enhancing amino acid absorption and protein synthesis efficiency (Rao et al. 2019). Similarly, hormonal treatments, particularly the application of juvenile hormones, have been used to extend larval feeding times, leading to increased silk production through prolonged protein accumulation (Li et al. 2016). These biochemical and physiological interventions provide alternative strategies for maximizing silk yield without relying solely on genetic modification.

The integration of transcriptomic and proteomic analyses has further refined our understanding of the molecular mechanisms governing silk production. Comparative studies between domesticated and wild *B. mori* strains have identified key genetic factors responsible for higher silk yields, offering valuable insights for selective breeding programs (Fang et al. 2015). The identification of QTLs linked to silk fibres properties has facilitated MAS, allowing breeders to develop superior hybrids with enhanced yield and strength characteristics (Pavithra et al. 2024; Xu et al. 2024). Such developments illustrate how genomic data can be leveraged to optimize sericulture at a fundamental level.

Moreover, the exploration of sustainable silk production methods has gained momentum, with researchers investigating alternative feed sources and ecological rearing practices to improve both yield and fibres quality. Organic additives, including plant-derived bioactive compounds, have been incorporated into *B. mori* diets to enhance growth rates and cocoon formation efficiency (Bhatti et al. 2019). These sustainable approaches align with modern agricultural practices aimed at reducing chemical inputs while maintaining high production standards. Furthermore, environmental modifications, such as selective exposure to external stimuli, have been explored to induce beneficial stress responses that enhance silk's mechanical properties, demonstrating how controlled rearing conditions can be strategically utilized to optimize silk output (Zhou et al. 2020).

In summary, the achievements in improving silk yield, strength, and quality have been driven by a multifaceted approach, combining genetic engineering, advanced breeding techniques, material science innovations, and optimized production methodologies. The integration of CRISPR-based modifications, spider silk gene incorporation, composite fibres engineering, and transcriptomic insights has revolutionized silk production, ensuring that modern sericulture can meet the evolving demands of high-performance industries. As research continues to push the boundaries of silk biotechnology, the future of silk production lies in further refining these improvements to create commercially viable, sustainable, and high-quality silk materials for diverse applications.

## 3.6 Functional Genomics and Transcriptomics

Functional genomics explores gene and protein functions on a genome-wide scale, integrating DNA sequences, gene expression, and protein interactions. It uncovers regulatory networks that control gene expression, development, and cellular processes, including both coding and non-coding RNAs, protein translation, and molecular interactions. These insights help model dynamic systems governing cell cycles and metabolic adaptations (Bunnik and Le Roch 2013). The emergence of high-throughput sequencing and molecular analysis techniques has enabled researchers to systematically investigate gene expression patterns, identify functional elements within the genome, and determine how genetic variations contribute to phenotypic diversity (Kumar 2016).

In the context of silk production, functional genomics plays a pivotal role in understanding the genetic basis of economically important silk traits. These traits are typically quantitative, regulated by multiple genes, and governed by intricate genetic mechanisms. Recent growth in QTL mapping have facilitated the identification of genetic regions associated with cocoon quality and silk yield (Luan et al. 2018). Molecular marker techniques, such as AFLPs and SSRs, have been widely employed in QTL analysis for *B. mori*. However, due to the limited resolution of conventional markers, narrowing down QTL regions to pinpoint specific genes remains challenging. The advent of next-generation sequencing (NGS) technologies has revolutionized genetic mapping by enabling high-density single nucleotide polymorphism (SNP) marker development through methods such as restriction-site-associated DNA sequencing (RAD-Seq). These advancements have significantly improved the resolution of genetic maps, facilitating the identification of candidate genes associated with silk yield and quality (Fang et al. 2020).

Complementary to functional genomics, transcriptomics provides a dynamic snapshot of gene expression by analysing the complete set of RNA molecules transcribed from the genome at any given time. Transcriptome profiling enables researchers to unravel molecular mechanisms underlying silk production by identifying DEGs, alternative splicing events, and novel transcripts (Li et al. 2012). RNA-Seq has emerged as a powerful and cost-effective tool for transcriptome analysis, offering higher sensitivity and accuracy than expressed sequence tag (EST) or microarray-based methods. In *B. mori* studies, RNA-Seq has led to the discovery of thousands of previously unannotated transcripts and novel exonic regions in silk-associated genes, significantly improving existing gene annotations. By integrating transcriptomic data with functional genomic approaches, researchers can gain deeper insights into gene regulatory networks, fibroin biosynthesis pathways, and genetic modifications that enhance silk production efficiency (Li et al. 2012).

The improvements in functional genomics and transcriptomics have not only expanded our understanding of silk-producing organisms but have also paved the way for genetic improvements in sericulture. Through high-throughput sequencing technologies, genome-wide association studies, and RNA-Seq-based transcriptome profiling, researchers can identify key regulatory genes, optimize genetic breeding strategies, and develop genetically superior *B. mori* strains with enhanced silk yield and quality. As these technologies continue to evolve, functional genomics and transcriptomics will remain at the forefront of sericulture research, driving innovations in silk biotechnology and genetic engineering.

### *3.6.1 RNA Sequencing to Understand Gene Expression During Silk Formation*

Understanding transcriptomic changes is essential for unravelling the molecular mechanisms that regulate silk production in *B. mori*. Among the most advanced tools for transcriptome analysis is RNA-seq that provides high-throughput and high-resolution data to analyse gene expression patterns. Unlike transcriptomic methods, RNA-seq offers the advantage of detecting both highly and lowly expressed genes with exceptional accuracy, facilitating the identification of novel transcripts, alternative splicing events, and SNPs (Li et al. 2012; De La Peña et al. 2022). This technique is especially useful in sericulture research, enabling a comprehensive analysis of differentially expressed genes during silk gland development, fibroin biosynthesis, metabolic adaptations. The standard workflow of RNA-seq in *B. mori* studies begins with RNA extraction from silk gland tissues, followed by messenger RNA (mRNA) enrichment or ribosomal RNA depletion to selectively isolate protein-coding transcripts. The extracted RNA is then reverse transcribed into cDNA, which is processed to generate a sequencing library by ligating adaptors. The prepared library is then subjected to high-throughput sequencing using platforms such as Illumina, producing millions of sequencings reads per sample. The final phase of RNA-seq involves bioinformatic analysis, where sequencing reads are aligned to a reference genome or de novo assembled into transcripts. This is followed by quantification of transcript abundance, normalization of gene expression across samples, and statistical modelling to identify differentially expressed genes. By comparing transcriptomic profiles across different silk gland developmental stages, environmental conditions, or genetic backgrounds, RNA-seq enables researchers to uncover the molecular mechanisms that drive fibroin synthesis, silk gland differentiation, and metabolic adaptations for silk production biology (Stark et al. 2019).

A notable application of RNA-seq in *B. mori* research involved examining the effects of selenium (Se), a micronutrient utilized in sericulture, on gene expression and silk yield. In a study, Li Jiang et al., conducted an RNA-seq analysis on the fat body of *B. mori* treated with 50 mM and 200 mM of Se, revealing significant changes in gene expression. At a lower Se concentration (50 mM), DEGs were primarily enriched in the peroxisome and fatty acid metabolism pathways, suggesting that Se played a role in nutritional regulation and antioxidant defence mechanisms. Conversely, at a higher Se concentration (200 mM), gene expression shifted towards glycerolipid metabolism and arachidonic acid metabolism, highlighting the activation of detoxification-related genes such as cytochrome P450s

(CYPs) and glutathione-S-transferases (GSTs). This suggests that selenium supplementation must be optimized to enhance silk production while avoiding stress-induced metabolic shifts that could hinder fibroin synthesis and cocoon formation (Jiang et al. 2020). Similarly, RNA-seq has been used to investigate the physiological effects of hydrogen sulphide ($H_2S$) exposure on *B. mori* gene expression, particularly in the silk gland and fat body. Transcriptomic data indicated that genes related to inflammation were significantly downregulated, supporting the hypothesis that $H_2S$ may have an anti-inflammatory role in *B. mori* growth. Moreover, genes involved in energy metabolism and nutrient transport pathways were upregulated, suggesting that $H_2S$ exposure enhances energy production during metamorphosis. Notably, fibroin synthesis-related genes, including, Fib-L, Fib-H, and P25, were also significantly upregulated, indicating that $H_2S$ may promote silk fibre production by increasing fibroin gene expression and silk gland activity. (Zhang et al. 2021a).

Comparative transcriptomics has further shed light on the genetic differences between domesticated and wild *B. mori* strains, providing insights into the molecular mechanisms underlying increased silk yield through selective breeding. Studies comparing the transcriptomes of wild and domesticated *B. mori* have revealed that during the early and intermediate stages of silk gland development, domesticated *B. mori* exhibit an upregulation of genes associated with DNA integration, nucleic acid binding, and transporter activity. These genetic modifications contribute to enhanced cell proliferation, leading to a larger posterior silk gland (PSG) in domesticated strains, which houses a greater number of fibroin-producing cells than its wild counterpart. During the late stage of silk gland development, domesticated *B. mori* demonstrate increased expression of genes involved in protein processing within the endoplasmic reticulum and ribosome pathways, highlighting an optimized fibroin synthesis system (Zhou et al. 2020). A more extensive transcriptomic analysis of the silk gland across different evolutionary stages, comparing wild, early domesticated, and improved *B. mori* strains, revealed that modern breeding has driven drastic gene expression changes favouring silk production. Specifically, genes associated with nitrogen metabolism were significantly upregulated in improved strains, indicating that selective breeding has facilitated a more efficient allocation of nitrogen resources to the silk gland. Additionally, upregulation of ganglioside-induced differentiation-associated protein 2 (GDAP2), a gene linked to brain function and metabolic regulation, suggests that modern breeding has not only optimized silk yield but may have also influenced behavioural traits related to resource utilization and energy efficiency (Zhu et al. 2022a).

Recent improvements in single-cell RNA sequencing (scRNA-seq) have further refined our understanding of silk gland development by providing a detailed transcriptional atlas of distinct silk gland cell types. Single-cell transcriptomics has revealed at least ten specialized cell types distributed across the anterior silk gland (ASG), middle silk gland (MSG), and PSG, offering insights into their unique roles in silk biosynthesis. By analysing gene expression at the single-cell level, researchers have been able to reconstruct the developmental trajectories of these cell types, uncovering key regulatory networks involved in fibroin protein synthesis. The integration of single-cell transcriptomic data with metabolic and proteomic analyses has paved the way for more targeted genetic interventions aimed at further optimizing silk production efficiency (Ma et al. 2022).

The use of RNA-seq has also contributed to understanding the genetic regulation of silk quality and fibroin biosynthesis. For instance, knockdown of the BmAbl1 gene in *B. mori* was found to suppress glutathione metabolism, leading to reduced free glycine and serine levels, which are critical for fibroin protein synthesis. Consequently, Fib-H, Fib-L, and P25 protein expression were all downregulated, resulting in a significant decrease in silk production (Qin et al. 2022). Another study focused on the impact of Fib-H gene knockout, revealing that mutations in this gene led to abnormal silk gland development, thin cocoons, and, in some cases, arrested metamorphosis. Transcriptomic comparisons between wild-type and Fib-H mutant *B. mori* demonstrated that genes associated with proteasome degradation, threonine-type endopeptidase activity, and autophagy were upregulated in the posterior silk glands of mutants, highlighting the essential role of Fib-H in normal silk gland function (Cui et al. 2018).

The continued application of RNA sequencing in *B. mori* research has significantly advanced our understanding of the molecular basis of silk production. Through transcriptomic analyses, researchers have uncovered key regulatory genes, metabolic pathways, and environmental influences that shape silk gland function. Comparative transcriptomics has highlighted the genetic adaptations associated with increased silk yield, while functional studies have elucidated how nutritional and environmental factors regulate fibroin biosynthesis.

### 3.6.2 *Role of Non-coding RNAs in* B. mori *Development and Silk Production*

Non-coding RNAs (ncRNAs) have emerged as key regulatory molecules in various biological processes, including development, metamorphosis, silk gland function, and silk protein synthesis in *B. mori*. These RNA molecules, which do not translate into proteins, were once considered "junk sequences". However, the functional role of ncRNAs is being explored in gene expression regulation, chromatin remodelling, transcription, and post-transcriptional modifications. These ncRNAs are categorized into small ncRNAs and long ncRNAs (lncRNAs), based on their length. Small ncRNAs include ribosomal RNA (rRNA), transfer RNA (tRNA), small nuclear RNA, small nucleolar RNA, and microRNA (miRNA). Among these, miRNAs are highly conserved molecules that modulate gene expression by inducing mRNA degradation or translational repression, playing an essential role in *B. mori* growth, silk gland development, and silk production efficiency (Gai et al. 2018; Chen et al. 2019; Chanda 2024). Additionally, lncRNAs are RNA molecules that exceed 200 nucleotides in length (Fu et al. 2022) and are transcribed by RNA polymerase II, similar to mRNAs. They typically feature a 5′ cap and a 3′ poly(A) tail, although they do not encode proteins (Ruan et al. 2021). Based on their genomic location relative to protein-coding genes, lncRNAs are categorized into sense, antisense, bidirectional, intronic, or intergenic types. These molecules play a crucial role in gene regulation by, modifying chromatin structure, influencing transcription, and interacting with DNA, RNA, and proteins at the post-transcriptional level (Zhang et al. 2021b). Some lncRNAs serve as precursors for miRNAs and small interfering RNAs (siRNAs), while others possess miRNA binding sites, acting as molecular decoys or competitors within miRNA-mediated regulatory pathways (Gai et al. 2018). Circular RNAs (circRNAs) represent a distinct class of ncRNAs, characterized by their covalently closed, continuous loop structures, which lack 5′–3′ polarity and a poly(A) tail. This structural conformation makes them highly stable and resistant to exonuclease degradation, allowing them to function as efficient miRNA sponges that influence gene expression. CircRNAs are involved in a wide range of cellular functions, particularly in response to environmental stimuli, where their ability to interact with DNA, RNA, and proteins enables them to regulate ribonucleoprotein complexes and cellular signalling pathways. As our understanding of lncRNAs and circRNAs continues to expand, their significance in gene regulation, stress responses, and cellular adaptation is becoming increasingly evident (Hu et al. 2018; Gai et al. 2018). An overview of the role of various non-coding RNAs in the development of *B. mori* and silk production is depicted in Fig. 3.4.

The importance of ncRNAs in *B. mori* development and silk production has been demonstrated through transcriptomic studies, particularly in silk gland regulation. During the larval-to-pupal transition, programmed cell death (apoptosis) leads to the degeneration of larval-specific tissues, including the silk gland. Transcriptomic analysis revealed that 40 apoptosis-related genes were differentially expressed, with 2.6% upregulated and 3.6% downregulated. Several lncRNAs were found to be involved in pathways regulating calcium signalling, metal ion binding, mitochondrial function, and protein phosphorylation, all of which play a role in silk gland degradation. Additionally, 285 miRNAs were identified as regulators of 124 apoptosis-related genes, demonstrating a complex interplay between ncRNAs and programmed cell death (Chen et al. 2019). Beyond developmental transitions, environmental stressors also impact ncRNA expression, as seen in *B. mori* exposed to dichlorvos, a widely used insecticide for mulberry crops. *B. mori* subjected to dichlorvos toxicity exhibited stunted growth, developmental deformities, erratic behaviour, and disrupted silk thread formation. RNA sequencing revealed significant alterations in miRNA expression within the PSG and MSG, with seventeen miRNAs upregulated and thirteen miRNAs downregulated. Notably, the upregulation of bmo-miR-190-5p resulted in the degradation of neuropeptide B2 (NGR B2), a neuroendocrine signalling regulator, causing abnormal spinning behaviour and irregular silk production patterns. (Chanda 2024).

Viral infections have also been shown to disrupt ncRNA regulatory networks in *B. mori*, affecting silk gland function and overall health. A study on *B. mori* cytoplasmic polyhedrosis virus (BmCPV) demonstrated that viral infection leads to significant differential expression of circRNAs, with 400 circRNAs being notably affected. Analysis of circRNA-miRNA interactions revealed that bmo-miR-3389-5p, bmo-miR-745-3p, and bmo-miR-3262 were central players in immune responses

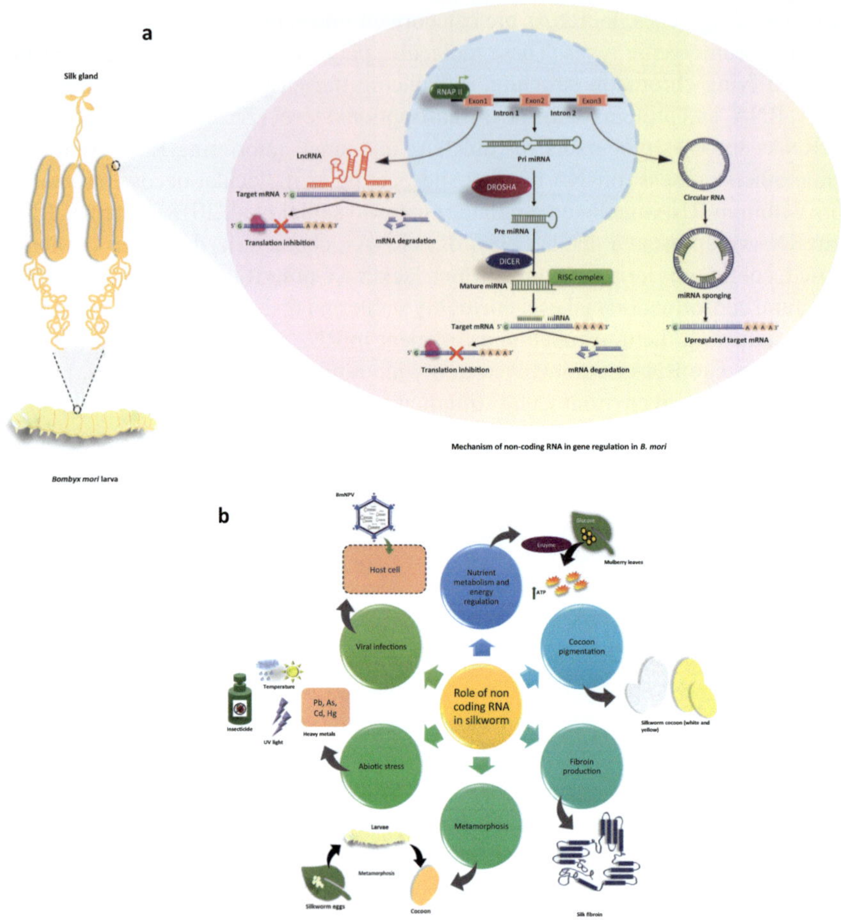


**Fig. 3.4** Schematic illustration of the functional roles of non-coding RNAs in *B.mori* growth and silk synthesis. **a** Gene regulatory mechanism mediated by various classes of non-coding RNAs including long non-coding RNAs (lncRNAs), microRNAs (miRNAs), and circular RNAs (circRNAs) within the silk gland. Adapted from Martinez-Useros et al. (2021) **b** Overview of non-coding RNAs influencing gene expression related to metamorphosis, fibroin synthesis, response to abiotic stress, and other physiological processes

and metabolic regulation, suggesting that viral infections interfere with silk gland homeostasis by altering ncRNA-mediated pathways (Hu et al. 2018). A similar study conducted by Zhu et al. showed that virus-mediated circDNA formation mediated by reverse transcriptase (RT) can potentially inhibit viral infection via the RNAi pathway by regulating the biogenesis of antiviral siRNAs (Zhu et al. 2022b). Similarly, lncRNAs have been found to participate in host-virus interactions, as seen in a transcriptomic study that identified 1,845 lncRNAs in BmCPV-infected *B. mori* larvae, with 41 showing significant differential expression. Further analysis of lncRNA-mRNA interaction networks revealed that lncRNAs modulated genes involved in autophagy, apoptosis, and immune response pathways, particularly autophagy-related gene (ATG3), programmed cell death 6 (PDCD6), and insulin-related peptide binding protein 2 (IBP2), which are essential for antiviral defence. These findings suggest that ncRNAs play a crucial role in maintaining silk gland function under biotic stress, making them potential targets for improving disease resistance and silk production efficiency in *B. mori* breeding programs. (Zhang et al. 2021b). Additionally, ncRNAs contribute to *B. mori* immune defence and silk gland protection, particularly against BmNPV, a major viral pathogen that affects silk yield. Research on Seroin2, a silk-associated antimicrobial protein, revealed that its expression was negatively regulated by two host miRNAs, bmo-miR-71 and bmo-miR-252, which suppressed Seroin2 mRNA levels. Overexpression of these miRNAs led to decreased Seroin2 expression, suggesting that ncRNA-mediated gene regulation plays a role in balancing silk production and immune response. This dual regulatory role underscores how ncRNAs influence both silk gland function and the *B. mori*'s ability to withstand environmental stressors, making them promising targets for genetic improvement in sericulture and disease-resistant *B. mori* breeding (Singh 2021).

Further research has highlighted the direct influence of ncRNAs on silk gland growth and fibroin synthesis, with several studies focusing on the lncRNA, lncR17454, which serves as a precursor for the let-7 miRNA cluster (let-7, miR-2795, and miR-100). Knockdown of lncR17454 led to moulting arrest and developmental failure, mirroring the effects of let-7 inhibition, indicating that this lncRNA plays a critical role in *B. mori* metamorphosis. Further investigation revealed that let-7 targets Abrupt (Ab), a transcription factor that negatively regulates the ecdysone signalling pathway, which is vital for silk gland differentiation and silk yield regulation. (Fu et al. 2022). Another study used CRISPR/Cas9-mediated knockdown of let-7 and observed that silk gland enlargement, enhanced endoreduplication, and increased silk yield occurred upon let-7 silencing, whereas overexpression resulted in silk gland atrophy and reduced silk fibres production. Mechanistic analysis showed that let-7 controls nutrient metabolism and energy regulation, with pyruvate carboxylase identified as a direct target, emphasizing its role in balancing silk gland growth and silk production efficiency (Wang et al. 2020b).

The domestication of *B. mori* has resulted in significant increases in silk yield compared to its wild ancestor, *B. mandarina*, yet the molecular mechanisms underlying these advancements remain largely unknown. A study investigating the role of lncRNAs in silk gland regulation identified 600 lncRNAs in wild and domesticated *B. mori*, with three differentially expressed lncRNAs (DELs) influencing silk protein synthesis. Specifically, lncRNA dw4sg_0483, which was highly expressed in domesticated *B. mori*, was linked to post-transcriptional regulation of fibroin production, potentially enhancing silk output. Conversely, lncRNAs dw4sg_0040 and dw4sg_0178, more abundant in wild *B. mori*, were associated with oxidative stress responses, indicating an adaptation to natural environments rather than silk production efficiency. These findings suggest that ncRNAs play a crucial role in optimizing silk gland metabolism, enabling domesticated *B. mori* to produce greater silk yields than their wild counterparts. (Zhou et al. 2018).

Further studies have explored the impact of ncRNAs on silk pigmentation and fibre quality, particularly in relation to cocoon colour determination in *B. mori*. Whole-transcriptome analysis of yellow and white cocoon strains revealed that the yellow blood inhibitor (I) gene is regulated by miRNAs and circRNAs, controlling carotenoid transport and pigment accumulation. miRNA-circRNA-mRNA interaction networks identified five key miRNAs, bmo-miR-2839-5P, bmo-miR-2808b, bmo-miR-2808a-5P, bmo-miR-745-5P, and bmo-miR-274-5P, that influence pigmentation genes, suggesting that ncRNAs modulate silk coloration through genetic and epigenetic mechanisms. These findings provide new insights for genetic engineering approaches in sericulture, offering possibilities for manipulating silk colour and improving fibre quality. (Ge et al. 2022).

Overall, ncRNAs play an indispensable role in *B. mori* growth, metamorphosis, silk gland differentiation, immune defence, and silk fibres production. As research continues to uncover their complex regulatory functions, targeting ncRNAs for genetic modification and selective breeding could revolutionize silk production, paving the way for more efficient, disease-resistant, and high-yielding *B. mori* strains.

## 3.7 *B. mori* Disease Resistance Through Genetic Improvements

### *3.7.1 Identification of Genes for Pathogen Resistance*

The domesticated *B. mori*, has undergone extensive breeding and domestication, leading to a weakened immune system that makes it highly vulnerable to various pathogens, including fungi, bacteria, viruses, and protozoa. The continuous rearing of *B. mori* under controlled environments has contributed to the reduction of innate immune traits and loss of natural resistance. However, despite this decline, *B. mori* have developed a complex genetic and immunological defence system that helps them combat infections. Multiple studies suggest that *B. mori* resistance is not confined to a single tissue, but is present across various tissues. Additionally, a range of genes, proteins, and metabolites play a significant role in this resistance. Previous studies have identified various genes, primarily in the *B. mori*'s midgut or midgut juice. The *B. mori* midgut is essential for both digestion and immune defence, serving as the first barrier against ingested pathogens. Its epithelial cells produce reactive oxygen species (ROS) and AMPs to neutralize harmful microorganisms, ensuring gut health and maintaining a balanced microbial community. This dual function supports both nutrient absorption and the *B. mori*'s overall resilience against infections (Deng et al. 2023).

Genetic resistance in *B. mori* is polygenic, meaning that multiple genes contribute to disease tolerance. These genes are primarily involved in midgut defence mechanisms, enzymatic activity, and immune responses, particularly against major pathogens such as BmNPV, BmCPV, and infectious flacherie virus (BmIFV). Resistance to these viruses is largely influenced by genes regulating midgut immunity, which includes antiviral activity in gut juices and the integrity of the peritrophic membrane. However, resistance to BmDNV is controlled by specific major genes, such as the recessive nuclear receptor binding SET domain protein 1(nsd-1) and nsd-2, and the dominant Nidogen-1 (Nid-1), which inhibit viral replication by interfering with essential enzymatic processes. Most *B. mori* strains exhibit resistance to BmDNV-1 due to the recessive nsd-1 gene, located on chromosome 21 at position 8.3. In contrast, resistance to BmDNV-2 is controlled by the recessive *nsd-2* gene, which is situated on a different chromosome. Additionally, the dominant *Nid-1* gene also confers resistance to *BmDNV-2*. Genetic studies suggest that resistance to *BmDNV-1* can be regulated by either a dominant or recessive major gene, facilitating the breeding of resistant *B. mori* strains (Buhroo 2018).

One of the most severe viral threats affecting *B. mori* is BmNPV, which causes substantial economic losses. Six genes, *dnapol* (which encodes the DNA polymerase), *p143* (which encodes a helicase), *ie-1, lef-1* (which encodes the primase), *lef-2* (which encodes a primase-associated factor), and *lef-3* (which encodes a single-stranded DNA binding protein), are essential for baculoviral DNA replication. Researchers have identified key genetic factors associated with viral resistance, particularly those involved in metabolic pathways and immune responses. A crucial discovery is the role of the Bmtret1 gene family, which regulates trehalose metabolism and transport, contributing to stress adaptation and antiviral defence. Trehalose transporters (TRET) help transfer trehalose from the fat body to the haemolymph, and while their function in viral resistance is not yet fully understood, studies suggest that Bmtret1-like and Bmtret1-X1 genes play significant roles in *B. mori* defence against BmNPV. Upregulation of Bmtret1 genes in resistant strains has been linked to inhibited viral gene expression, DNA replication, and virion assembly, whereas downregulation in susceptible strains increases vulnerability to infection (Kikawada et al. 2007; Kanamori et al. 2010; Lin et al. 2023). These findings position the Bmtret1 gene family as a potential target for breeding resistant *B. mori* strains to enhance sericulture productivity. Recent transcriptomic studies have provided additional insights into genes that are differentially expressed in resistant and susceptible *B. mori* strains upon BmNPV infection (Lin et al. 2023). Notably, the uncharacterized protein LOC101744317 isoform X2 (BGIBMGA004283) was found to be upregulated in the resistant BN variety, but downregulated in the susceptible B strain, suggesting its role in antiviral defence. Further characterization of this gene could provide deeper insights into *B. mori* immunity and resistance mechanisms.

Another major genetic factor associated with viral resistance is UDP glycosyltransferase (UGT41A3). Research has demonstrated that overexpression of UGT41A3 enhances *B. mori* resistance to BmNPV, emphasizing glycosylation as a crucial antiviral mechanism. These findings highlight the genetic complexity of *B.*

*mori* immunity and underscore potential targets for genetic modification or selective breeding to develop disease-resistant strains. Other key defence mechanisms include RNAi, AMPs, and epigenetic regulation. AMPs usually comprise 12–50 amino acids and exhibit antibacterial, antifungal, and antiviral activities (Yi et al. 2014). Research has shown that the *B. mori* midgut immune system can generate ROS and AMPs, which work together synergistically to defend against microbial infections. Analysis of endogenous AMPs in *B. mori* revealed a significant upregulation of Defensin A, Lebocin3, and Gloverin2. Notably, Lebocin3 was found to enhance the activity of Cecropin D, with their combined action significantly boosting antimicrobial effectiveness in *B. mori*.

AMPs, like gloverin, attacin, and moricin, help defend against bacterial and fungal infections. Gloverins are mainly active against *E. coli*, while Cecropins have a broad spectrum of activity against Gram-negative and Gram-positive bacteria (Kawaoka et al. 2008). Bacterial diseases, like flacherie, are typically induced by poor hygiene and environmental stress (Deng et al. 2023). Similarly, RNAi-related genes, such as *Dicer, Argonaute,* and *RdRP*, are upregulated in resistant strains, suggesting their role in silencing viral RNA. Recent studies also indicate that miRNAs play a role in post-transcriptional regulation of immune-related genes, contributing to *B. mori* disease resistance.

Heat Shock Proteins (HSPs), particularly HSP70 and HSP90, enhance cellular stress resistance during viral infections. *B. mori* is an economically significant insect that is highly susceptible to various environmental stresses and pathogens. Among the genes involved in pathogen resistance, the *GrpE* gene (*BmGrpE*) has been identified as a key player in protecting *B. mori* from high-temperature stress, which can weaken immunity and increase disease susceptibility. *BmGrpE* encodes a 24.1 kDa protein and contains a heat shock transcription factor (*Hsf1*) binding site, suggesting its role in cellular stress response. Given the strong link between environmental stress and disease susceptibility, understanding the function of *BmGrpE* provides valuable insights into the molecular mechanisms of pathogen resistance and stress adaptation in *B. mori*, contributing to developing more resilient strains through molecular breeding (Li et al. 2020).

In addition to genetic factors, cellular and humoral immune responses significantly influence *B. mori* resistance to viral infections. Viral infections lead to metabolic alterations in the host, with oxidative phosphorylation emerging as a key metabolic pathway associated with BmNPV resistance. Advances in biotechnology have enabled the identification of crucial genes and proteins involved in antiviral defence. Proteomic studies have linked metabolic pathways to immune responses, highlighting the importance of genetic markers in pathogen resistance. Additionally, transcriptomic analyses have identified CYP450s genes, contributing to immune priming and detoxification, thereby enhancing B. mori resilience to infections (Yi and Wu 2024).

Another essential defence mechanism is the JAK-STAT signalling pathway (Fig. 3.5), where Perilipin1 activates immune responses, particularly against the protozoan parasite *Nosema bombycis* (Nb), a microsporidian pathogen that infects the *B. mori*'s midgut cells and causes pebrine disease. Nb relies on lipids from *B.*

*mori* to facilitate its proliferation. Perilipin1, a lipid droplet-associated protein, has been identified as a crucial modulator of immune responses through the activation of the JAK-STAT signalling pathway. Perilipin1 influences the mobilization of lipids, which can be critical for the immune system's energy demands during an immune response. This pathway is activated when cytokines or immune-related signals bind to specific receptors, leading to the phosphorylation of Janus Kinase (JAK) proteins, which in turn activate STAT proteins. These activated STAT proteins translocate to the nucleus, where they regulate the expression of immune-related genes. Furthermore, Perilipin1 inhibits Nb proliferation by promoting the JAK-STAT pathway through increased expression of Domeless and Hop. The Domeless receptor is a key component of the JAK-STAT pathway in insects, functioning similarly to cytokine receptors in mammals, and the hopscotch gene codes for a JAK protein essential for transmitting signals from the Domeless receptor, facilitating the phosphorylation and activation of transcription factors in the JAK-STAT pathway (Su et al. 2024).

In summary, the immune defence mechanisms of *B. mori* are shaped by a combination of genetic, cellular, and biochemical factors, enabling resistance to various pathogens (Fig. 3.6). Despite the weakening of innate immunity due to domestication, *B. mori* have developed sophisticated defence strategies, including polygenic resistance, AMPs, RNAi, heat shock proteins, and metabolic adaptations. These insights into *B. mori* immunity not only contribute to the development of disease-resistant strains for sericulture, but also provide a broader understanding of insect immune systems, with potential applications in biotechnology and pest management.

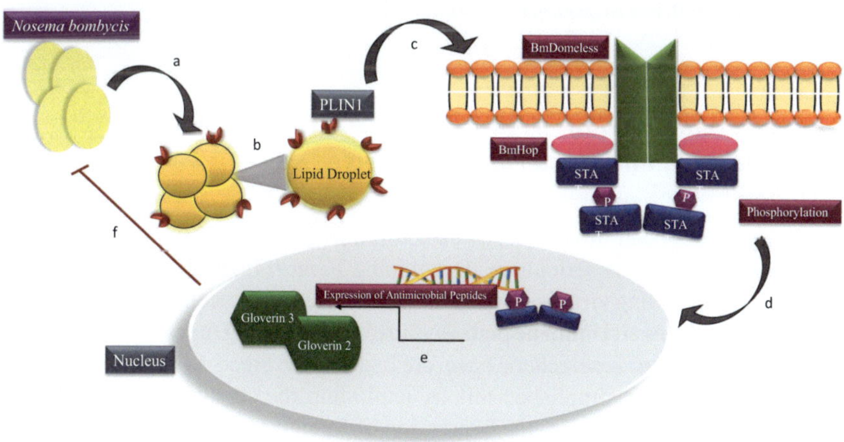


**Fig. 3.5** Illustration of Plin1-mediated suppression of *Nosema bombycis* (Nb) proliferation in *B. mori*. **a–b** Nb infection leads to lipid droplet accumulation and induces the expression of the *Plin1* gene. **c** *Plin1* activation promotes the expression of *Domeless* and *Hop*, thereby triggering the JAK-STAT signaling pathway. **d–e** *Plin1* further stimulates the production of antimicrobial peptides Gloverin 2 and Gloverin 3. **f** These peptides function to effectively limit Nb proliferation. Adapted from Su et al. (2024)

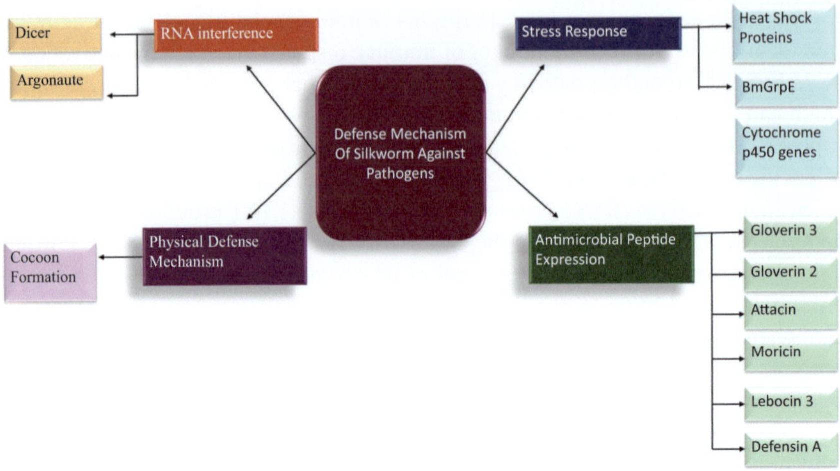


**Fig. 3.6** Immune defense mechanisms in *B. mori* against pathogenic infections

*B. mori* utilizes a range of protective responses to combat pathogens, including RNA interference pathways, activation of cellular stress responses, production of antimicrobial peptides, and structural defenses such as cocoon formation. These mechanisms function synergistically to strengthen the insect's immune system and prevent infection. The figure was designed using Microsoft® PowerPoint® Version 16.86 (Build 16.0.16924.20054) (Microsoft®, Redmond, WA, USA).

### 3.7.2 *Development of* B. mori *Lines Resistant to Viral, Fungal, and Bacterial Infections*

The *B. mori*, has been a model organism for studying insect immunity due to its economic importance in sericulture. Developing *B. mori* lines resistant to a variety of pathogens, including viruses, fungi, and bacteria, requires a comprehensive understanding of the genes and immune pathways involved in pathogen resistance. Through genetic modification, researchers have created a new breed of *B. mori* resistant to infections. They accomplished this by exploiting the so-called RNAi, a mechanism by which cells 'silence' the expression of foreign genes. Developing *B. mori* strains resistant to infections involves a combination of selective breeding, genetic engineering (CRISPR), and molecular biology techniques. The primary focus is on identifying resistance-related genes and utilizing them to develop robust *B. mori* strains.

Transgenic technology offers a promising approach to developing disease-resistant *B. mori*. Researchers identified *Bmlipase-1,* an enzyme found in the midgut juice of *B. mori,* which exhibits strong antiviral activity against BmNPV. To enhance resistance, they engineered a transgenic *B. mori* strain by introducing a vector that overexpresses the *Bmlipase-1* gene under the control of a viral promoter (IE1 promoter). This genetic modification was achieved through embryo microinjection, leading to a significant 27.3% increase in *Bmlipase-1* expression in the transgenic *B. mori* compared to the non-transgenic ones. Overexpression of endogenous antiviral genes, such as *Bmlispase-1* or *Bm-serine protease-2* (which encode proteins found in the digestive juice of *B. mori* larvae) or *BmAtlastin-n* (which encodes a member of the dynamin protein superfamily) or the exogenous antiviral gene *hycu-ep32,* enhance the resistance of *B. mori* to BmNPV. The knockdown of the BmNPV replication-related genes, such as "*ie-1, lef-1, lef-2, lef-3, lef-11,* and *gp64*", also significantly inhibit virus replication and are efficiently targeted for constructing anti-viral transgenic *B. mori* (Liu et al. 2021b).

Researchers have explored RNAi and CRISPR genome editing to enhance resistance against BmNPV in *B. mori.* While RNAi-based silencing of the *lef-1* gene only partially reduced viral replication, targeting the *ie-1* gene was more effective in inducing resistance. CRISPR/Cas9 genome editing has proven highly effective in inhibiting viral proliferation by targeting key genes like, *ie-1* and *me53.* Deleting large fragments of *ie-1* led to a 1,000-fold increase in antiviral resistance, and CRISPR/Cas9 has also been used for antiviral resistance breeding by modifying host and viral genes. These advancements highlight the potential of gene-editing technologies in developing virus-resistant *B. mori,* offering a sustainable approach to disease control in sericulture (Kanginakudru et al. 2007; Dong et al. 2018; Liu et al. 2021b).

CRISPR/Cas9-based methods have limitations in achieving strong, sustained resistance, including restricted antiviral resistance, limited target sites due to its G-rich PAM requirement, and lower editing efficiency. To overcome these issues, the researchers explored Cas12a (Cpf1), which offers higher editing efficiency, a T-rich PAM (expanding target sites), staggered DNA cuts (improving repair accuracy), and better multiplexing capabilities. The goal of this study was to develop virus-resistant *B. mori* using CRISPR/Cas12a (Cpf1) genome editing to protect them from BmNPV, a virus that causes major losses in the silk industry. The researchers compared three types of Cas12a enzymes (AsCas12a, FnCas12a, and LbCas12a) to see which worked best against BmNPV, and found that FnCas12a was the most effective. They then created genetically modified *B. mori* using FnCas12a and SpCas9, targeting the *ie-1* gene, which is essential for virus replication. Tests showed that FnCas12a provided stronger resistance to the virus than SpCas9, proving that Cas12a-based gene editing can be a powerful tool for breeding disease-resistant *B. mori* and improving silk production (Dong et al. 2020).

RNAi is a natural defence mechanism that helps cells silence harmful foreign genes, playing a crucial role in protecting against viruses. In *B. mori,* researchers aimed to harness this ability to create a stable antiviral defence against BmNPV. To achieve this, scientists selected four key viral genes, *ie1, lef1, lef3,* and *p74,* and introduced short sequences of these genes into the *B. mori* genome using piggyBac-derived vectors. These sequences were arranged in different orientations (sense, antisense, or inverted-repeat) to enhance gene-silencing efficiency. Among these, *B. mori* carrying the inverted-repeat sequences showed the strongest resistance to high doses of BmNPV, effectively stopping the virus from replicating and spreading. To ensure that this antiviral trait could be used in commercial silk production, the genetically modified *B. mori* were crossbred with CSR2, a high-yielding but virus-susceptible strain. The resulting hybrid *B. mori* showed a remarkable improvement in survival rates, over 75%, while maintaining their silk production quality. Notably, these *B. mori* produced dsRNA, which strongly suppressed viral replication and reduced the formation of infectious occlusion bodies (OBs), minimizing the risk of virus transmission (Subbaiah et al. 2013).

*B. mori* face significant threats from various pathogens, including viruses and microsporidia, which severely impact silk production. While RNAi-based strategies have shown promise in conferring resistance to viral infections, innovative approaches are also being explored to combat other pathogens, such as *Nb*, the causative agent of Pebrine disease. Pebrine disease, caused by the microsporidium *Nb*, poses a major threat to sericulture, necessitating innovative disease-resistant strategies. To combat this, a transgenic *B. mori* strain was developed expressing a single-chain fragment variable (scFv) antibody F12 (scFvF12) targeting a key pathogen protein. The engineered antibody, scFvF12, specifically binds to *N. bombycis* transmembrane protein NbTMP1 and is fused with an Fbox domain at the N-terminal of Slmb protein (NSlmb), to facilitate its degradation via the host's ubiquitin–proteasome system (UPS). Western blot analysis confirmed the effective recognition and degradation of NbTMP1, significantly suppressing *N. bombycis* proliferation in transgenic *B. mori* cells. The resulting transgenic *B. mori* exhibited enhanced resistance to *N. bombycis* while maintaining essential economic traits for silk production. This strategy introduces a novel method of utilizing the host's UPS for targeted pathogen protein degradation, offering potential applications in sericulture and broader host–pathogen interactions (Zhang et al. 2024).

The development of disease-resistant *B. mori* lines is crucial for ensuring the sustainability and productivity of the sericulture industry. Advances in selective breeding, molecular genetics, and biotechnology have enabled researchers to enhance resistance against major pathogens, including viruses, bacteria, and microsporidia. The identification of key resistance genes, immune-related genes involved in antiviral responses, has provided valuable genetic markers for breeding programs. RNAi and CRISPR-based genome editing have revolutionized disease resistance strategies, enabling precise genetic modifications to suppress pathogen replication. Transgenic approaches, such as the overexpression of antiviral proteins and engineered immune factors, have further strengthened *B. mori* immunity. By integrating these strategies, researchers have successfully developed robust *B. mori* strains capable of withstanding infections while maintaining economic traits essential for silk production. Continued advancements in genetic engineering, coupled with conventional breeding, will play a vital role in safeguarding *B. mori* populations against emerging diseases, ensuring the long-term viability of global sericulture.

## 3.8 Genetic Improvements in Mulberry Plants

Mulberry is a fast-growing, deep-rooted perennial plant that holds immense significance in sericulture, serving as the sole food source for the domesticated *B. mori*. Thought to have originated in the Himalayan foothills along the China–India border, mulberry has been cultivated for centuries, not just for silk production but also for its medicinal, nutritional, and environmental benefits. With around 68 recognized species, including 24 well-established ones, mulberry exhibits remarkable genetic diversity, which makes it an ideal candidate for scientific improvement. While species like *Morus alba, Morus indica, Morus bombycis, Morus latifolia,* and *Morus multicaulis* are primarily cultivated for *B. mori* rearing, *Morus nigra* is mainly grown for its fruit. One of the fascinating aspects of mulberry genetics is its variation in ploidy levels, ranging from diploids to high-polyploid forms, which adds complexity to breeding efforts but also provides opportunities for genetic advancements. (Dutta et al. 2023; Kannihalli et al. 2024). Mulberry leaf quality directly influences *B. mori* growth, metabolism, and silk yield, making genetic improvements in mulberry a crucial aspect of sericulture research. The biochemical composition of the leaves, rich in proteins, sugars, and essential micronutrients, determines the health and silk-producing capacity of *B. mori*. Beyond its role in sericulture, mulberry exhibits traits such as high biomass production, adaptability to diverse climatic conditions, and resilience to biotic and abiotic stresses, which make it a valuable crop in sustainable agriculture. Moreover, its ability to sequester carbon and contribute to phytoremediation highlights its ecological importance. However, despite its agronomic potential, traditional breeding methods have struggled to keep pace with the growing demands of sericulture, due to the plant's long juvenile phase, high heterozygosity, and complex inheritance of desirable traits (Sarkar et al. 2025). Recent advances in genetic improvement techniques have opened new possibilities for optimizing mulberry characteristics. Biotechnology-driven approaches, such as plant tissue culture, molecular marker-assisted breeding, recombinant DNA technology, and genetic transformation, have made it possible to accelerate breeding programs and introduce targeted traits with greater precision. Cryopreservation techniques have also facilitated the long-term conservation of valuable genetic resources, ensuring the preservation of elite germplasm for future breeding programs. Studies utilizing DNA markers, such as ISSRs have confirmed the genetic stability of in vitro-regenerated and cryopreserved mulberry plants, reinforcing the reliability of these approaches in preserving genetic integrity (Dutta et al. 2023).

### *3.8.1 Enhancing Mulberry Leaf Quality Through Genetic Modifications*

India, the second-largest producer of raw silk after China, faces a growing need to enhance mulberry leaf productivity to sustain its sericulture industry (Sajeevan et al. 2017). The quality and quantity of mulberry leaves directly influence the development, metabolism, and silk gland function of *B. mori,* ultimately shaping cocoon yield and silk quality. Several factors impact mulberry leaf quality, including genetic variability, agronomic practices, environmental conditions, and post-harvest management (Kannihalli et al. 2024). Among these, abiotic stressors, such as drought, salinity, and alkalinity, significantly reduce mulberry leaf yield by 50–60%, particularly in regions with dryland and irrigated sericulture. Additionally, moisture loss between leaf harvesting and *B. mori* feeding adversely affects leaf freshness and nutritional value, impairing *B. mori* growth and silk production efficiency (Sajeevan et al. 2017).

A critical determinant of mulberry leaf productivity is photosynthetic efficiency, which directly impacts chlorophyll content and $CO_2$ assimilation rate. A decline in relative chlorophyll content (SCMR) under drought stress is often indicative of oxidative damage, caused by pigment photo-oxidation and chlorophyll degradation. Engineering drought-tolerant mulberry genotypes with enhanced chlorophyll stability and oxidative stress resistance could help safeguard leaf productivity and maintain optimal leaf composition under environmental stressors (Sarkar et al. 2025).

Despite significant advancements in traditional breeding techniques, improving mulberry leaf productivity and quality remains challenging, due to the perennial nature of the plant, long juvenile phase, high heterozygosity, and limited genetic diversity within existing germplasm. The complex inheritance of key agronomic and yield traits further complicates selective breeding efforts. In contrast, genetic modification technologies, such as mutational breeding, plant tissue culture techniques and transgenic approaches, offer targeted, precise, and accelerated improvements in mulberry leaf traits. By modifying genes associated with drought tolerance, improved carbohydrate metabolism, enhanced protein synthesis, and delayed senescence, researchers can develop high-yielding, nutritionally superior mulberry varieties, tailored for optimal sericulture performance as depicted in Fig. 3.7 (Sarkar et al. 2025).

Genetic modifications have emerged as a powerful approach to improving mulberry leaf quality, directly impacting *B. mori* nutrition, cocoon formation, and overall silk production efficiency. Among various genetic improvement strategies, mutation breeding has proven to be an effective and relatively fast method for inducing desirable agronomic traits. Mutation, a heritable alteration in an organism's genetic structure, occurs naturally but can be accelerated using mutagens, which are physical or chemical agents that alter DNA or RNA sequences. These mutations create genetic variability, enabling the selection of superior plant traits that enhance growth patterns, nutrient composition, and stress resilience (Yali and Mitiku 2022). Mutation breeding has gained prominence in crop improvement, due to its ability to generate beneficial mutations in a shorter time frame compared to conventional breeding methods (Yali and Mitiku 2022; Türkoğlu et al. 2023). By employing chemical mutagens, such as ethyl methane sulfonate (EMS) or physical mutagens like ionizing radiation (X-rays and $\gamma$-rays), researchers can induce targeted genetic variations that lead to improved plant characteristics (Ichida et al. 2019).

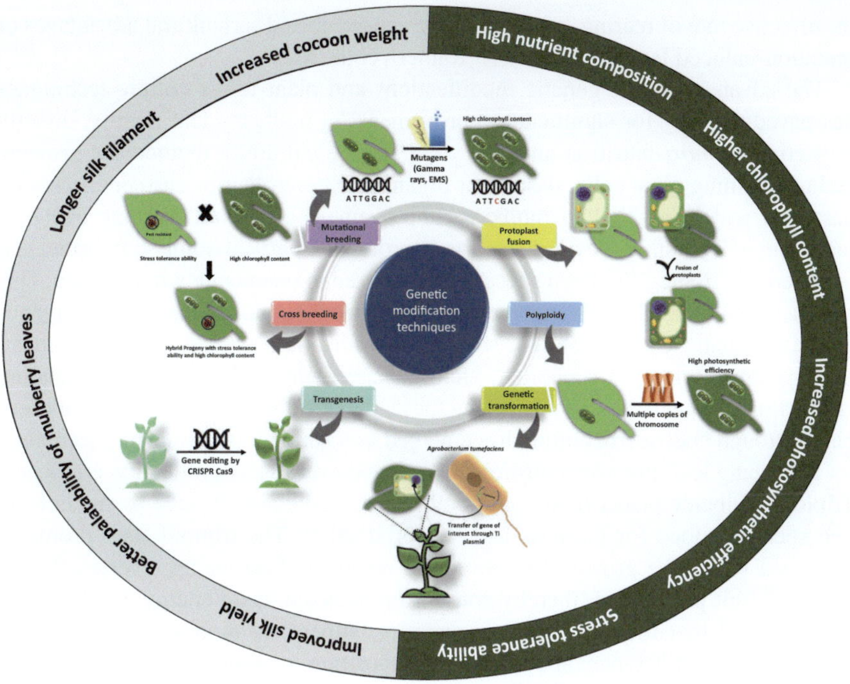


**Fig. 3.7 Overview of genetic modification strategies used to improve mulberry leaves for sericulture.** Various genetic enhancement approaches such as quality improvement, mutational breeding, genetic transformation, and transgenic techniques have been applied to optimize mulberry leaf traits. These modifications positively influence leaf quality, silkworm health, and overall silk yield. The figure was created using Microsoft® PowerPoint® Version 16.86 (Build 16.0.16924.20054) (Microsoft®, Redmond, WA, USA)

Among these, gamma rays, a highly penetrative form of ionizing radiation, possessing an energy level from 10 keV (kilo electron volts) to several hundred keV, have been extensively used in mulberry mutagenesis to enhance leaf quality and biochemical composition. Mutations caused by gamma irradiation can modify DNA at the molecular level, by introducing additions, deletions, or frame-shift mutations, which can positively influence plant physiology and metabolic activity (Ramesh et al. 2021). A study conducted by Ramesh H. L. et al. explored the impact of gamma irradiation on mulberry leaves and found that exposure to lower doses of gamma rays significantly improved total soluble sugar content, amino acid concentration, and chlorophyll levels in tender and medium-aged leaves. These biochemical enhancements increased the nutritional value of the leaves, making them more digestible and beneficial for *B. mori* growth. Additionally, the enhanced moisture retention capacity of irradiated leaves contributed to better palatability and prolonged post-harvest quality, ensuring efficient nutrient absorption and improved *B. mori* metabolism. When *B. mori* were fed with these genetically modified mulberry leaves, significant improvements were observed in cocoon weight, shell weight, filament length, and the effective rate of rearing, reinforcing the economic and sericultural advantages of mutation-induced leaf modifications (Ramesh et al. 2021).

The advancement of genetic modifications and plant tissue culture techniques has paved the way for significant improvements in mulberry leaf quality, directly impacting *B. mori* nutrition and silk yield. Tissue culture, a method for growing and maintaining plant cells, tissues, or organs in a controlled in vitro environment, has been widely applied in mulberry improvement programs. Various explants, including cells, anthers, ovules, embryos, protoplast-derived somatic hybrids, and meristematic tissues, has been successfully utilized to regenerate whole plants. This approach has facilitated advancements in micropropagation, callus culture, organogenesis, somatic embryogenesis, somatic hybridization, transgenic development, in vitro stress tolerance screening, yield enhancement, and secondary metabolite production (Sarkar et al. 2018). Among these, the development of triploid mulberry plantsthrough endosperm culture has emerged as a particularly promising technique for enhancing leaf biomass, nutritional value, and overall silk production efficiency. Triploid mulberry plants hold considerable agronomic significance particularly in tree species valued for biomass and leaf production. The triploid state promotes enhanced vegetative growth, by preventing resources from being diverted toward seed and fruit production, thereby conserving photosynthetic energy. Studies have demonstrated that triploid mulberry varieties outperform diploid counterparts in leaf yield, stress tolerance, and nutritive composition, making them highly suitable for sericulture applications. Researchers successfully regenerated triploid mulberry plants, which exhibited larger stomata, higher chloroplast density, and increased photosynthetic efficiency, contributing to better leaf biomass and nutrient composition. Triploid mulberry varieties have been shown to outperform diploids in leaf yield, nutritional quality, and stress resistance, highlighting how polyploidization, combined with genetic modifications, can optimize mulberry leaf traits, ensuring a more efficient and sustainable silk production system (Thomas et al. 2000).

Genetic transformation has become a fundamental technique in molecular breeding,, offering a means to introduce desirable traits into the mulberry genome while preserving its genetic identity. Among the various transformation methods, *Agrobacterium*-mediated transformation has proven to be one of the most efficient approaches for transferring target genes into plants, ensuring the expression of beneficial traits to enhance mulberry leaf quality (Ziemienowicz 2014). The bacterium *Agrobacterium tumefaciens* (*A. tumefaciens*), originally known for inducing crown gall disease, has been harnessed as a genetic tool due to its ability to transfer and integrate transfer DNA (T-DNA) into the plant genome via a type IV secretion system (T4SS) (Liu et al. 2021a). This process involves five key steps, including activation of bacterial virulence genes, formation of the T-DNA complex, transfer of the genetic material into the host nucleus, integration into the plant genome, and subsequent expression of the introduced genes. The tumour-inducing (Ti) plasmid carries the virulence (vir) region, which regulates transformation, along with chromosomal virulence loci such as chvA, chvB, and pscA, which play essential roles in DNA transfer (Ziemienowicz 2014). By modifying *A. tumefaciens*, researchers have developed a highly efficient system for plant genetic engineering, allowing for the stable introduction of genes that improve mulberry leaf attributes, including nutritional composition, moisture retention, and stress tolerance.

The advantages of *Agrobacterium*-mediated transformation over other genetic engineering techniques, such as particle bombardment, include higher efficiency, lower cost, and a lower copy number of T-DNA insertions, ensuring more stable and predictable genetic modifications (Hwang et al. 2019). This method has been successfully applied in mulberry to enhance post-harvest leaf quality and stress resilience, leading to improved *B. mori* nutrition and silk yield. A case study by Sajeevan et al. (2017) demonstrated the potential of this approach by introducing the *Arabidopsis* SHINE1 (AtSHN1) gene, a transcription factor known to regulate epicuticular wax biosynthesis, into Indian mulberry. The resulting transgenic plants exhibited increased surface wax deposition, reduced water loss, and improved chlorophyll retention, contributing to longer post-harvest leaf freshness and enhanced nutrient preservation. These modifications ensured that *B. mori* fed with the transgenic leaves experienced no adverse effects on larval development or cocoon formation, validating the feasibility of genetic modifications in improving mulberry leaf quality for sustainable sericulture (Sajeevan et al. 2017).

These studies exemplify how genetic engineering can enhance mulberry resilience to environmental stress, while maintaining the nutritional integrity of leaves, which is essential for *B. mori* growth and silk production. By integrating such genetic modification techniques with advanced molecular breeding approaches, such as CRISPR/Cas9 genome editing, future research can further refine targeted genetic modifications to optimize mulberry leaf traits, including increased drought tolerance, enhanced metabolic efficiency, and improved post-harvest longevity. As the demand for high-quality silk production continues to rise, leveraging genetic modifications to enhance mulberry leaf attributes presents a scientifically robust and economically viable strategy to support the sustainability and efficiency of sericulture.

### *3.8.2 Development of Drought- and Pest-Resistant Mulberry Varieties*

The increasing susceptibility of mulberry to pest infestations and drought stress necessitates the development of resilient varieties to sustain *B. mori* rearing and silk production. One of the most significant challenges in mulberry cultivation is pest-related leaf damage, which results in an estimated 34.24% loss in leaf yield, amounting to approximately 4500 kg per hectare annually. More than 300 insect and non-insect pests have been documented to infest mulberry at different growth stages and seasonal variations, significantly affecting leaf quality, biochemical composition, and overall plant vigor. *B. mori* fed with pest-infested leaves exhibit slower growth, reduced cocoon weight, and compromised silk quality, further reinforcing the need for genetic improvement in pest resistance. Among the major pests, Jassids (*Empoasca flavescens*), commonly known as leafhoppers, cause severe damage by sucking sap from the underside of leaves, leading to characteristic symptoms of "hopper burn." This manifests as triangular brown spots at leaf tips, which gradually spread along the veins, causing marginal curling, leaf crinkling, and eventual defoliation, significantly stunting plant growth. Similarly, mealybugs, another major pest in both irrigated and rain-fed mulberry gardens, cause tukra disease, which results in bunchy top formation, petiole thickening, and shortened internodes. The feeding mechanism of mealybugs, involving stylet penetration and sap extraction, deprives the plant of essential nutrients, leading to leaf deformation, yellowing, premature leaf drop, and weakened plant immunity. Additionally, mealybugs contribute to secondary infections by facilitating the spread of bacterial, fungal, and viral pathogens, as well as promoting black sooty mold growth through honeydew secretion, which further attracts ants. Studies indicate that feeding *B. mori* with mealybug-infested mulberry leaves results in reduced larval growth, decreased cocoon weight, and lower silk productivity, emphasizing the urgent need for resistant mulberry cultivars (Mahadeva 2017). To combat pest-induced losses, researchers have explored natural defence mechanisms in mulberry. A key discovery in *Morus multicaulis* identified MLX56, a chitin-binding protein, exhibits toxicity against lepidopteran pests while showing no harmful effects on *B. mori*, suggesting its potential for targeted pest resistance in mulberry cultivation (Gai et al. 2017).

The widespread use of imidacloprid, a common neonicotinoid insecticide, has been shown to reduce *B. mori* survival rates, lower cocoon quality, and cause metabolic stress due to its toxic effects on *B. mori*. When *B. mori* consume imidacloprid-contaminated mulberry leaves, their natural antioxidant defence system is disrupted, leading to increased oxidative stress, cell damage, and lipid peroxidation. A study conducted by Gao et al. (2018) explored a novel approach to mitigating imidacloprid-induced toxicity in *B. mori* through N-acetyl-L-cysteine (NAC) supplementation. NAC, a precursor to glutathione, is known for its antioxidant and detoxification properties. By pre-treating mulberry leaves with NAC, researchers observed improvements in *B. mori* survival rates, body weight, and feed efficiency, counteracting the negative effects of pesticide exposure. NAC also enhanced antioxidant enzyme activity, including superoxide dismutase (SOD), catalase (CAT), and glutathione peroxidase (GSH-Px), effectively reducing malondialdehyde (MDA) levels, a key marker of oxidative stress. The study demonstrated that *B. mori* reared on NAC-treated mulberry leaves exhibited superior cocoon weight, shell mass, and silk quality, reinforcing the potential of biochemical interventions to safeguard sericulture productivity (Gao et al. 2018).

Another major problem in mulberry cultivation is drought. Drought stress affects several important plant functions. It can reduce photosynthesis, lower chlorophyll levels, and reduce the efficiency of photosystem II (PSII). When plants don't get enough water, they also face oxidative stress, which leads to the production of harmful molecules called reactive oxygen species (ROS). These can damage enzymes, break down proteins, and harm cell membranes, eventually reducing crop yield. To cope with harsh conditions like drought and cold, plants make protective substances called osmoprotectants, such as proline, glycine betaine, glycerol, and mannitol. These are small, neutral molecules that help plants survive stress. Since water is essential for all living things, a lack of water (water scarcity) is a major environmental challenge that negatively impacts plant growth and health. Proline is a natural osmoprotectant, helps plants retain water, stabilize cell structures, and reduce damage from drought. A study on Thai mulberry varieties revealed how proline levels influence drought tolerance. The drought-resistant variety SRCM9809-34 maintained better water balance and strong antioxidant activity with lower proline levels, suggesting an efficient stress response. In contrast, Sakhonnakorn 85 accumulated excessive proline, signalling severe dehydration and leading to wilting and poor photosynthesis. This suggests that moderate proline regulation, not just high accumulation, is key to drought resilience in mulberry, offering valuable insights for developing hardier, drought-resistant varieties (Nutthapornnitchakul et al. 2024).

The role of the MmSK gene, a shaggy-like protein kinase involved in stress response has been widely explored. Transgenic mulberry plants with suppressed MmSK expression were developed using Virus-Induced Gene Silencing (VIGS), and their response to drought conditions was examined. The results showed that these plants had reduced drought tolerance, displaying lower levels of soluble proteins and proline, which are crucial for maintaining water balance. Additionally, the transgenic plants accumulated higher malondialdehyde (MDA), indicating increased oxidative stress, and exhibited weaker antioxidant enzyme activity (SOD, POD), leading to visible symptoms of drought damage, including wilting and leaf yellowing. In contrast, wild-type plants maintained better physiological stability under drought conditions. These findings highlight the potential of MmSK gene-based breeding strategies for developing drought-resistant mulberry varieties, ensuring more sustainable leaf production for sericulture in regions affected by water scarcity. (Li et al. 2018).

Similarly, to enhance mulberry's resilience, researchers developed transgenic mulberry lines co-expressing AtDREB2A and AtSHN1 genes through *Agrobacterium*-mediated genetic transformation. These genes, regulated by a constitutive promoter, significantly improved drought tolerance by enhancing cuticular resistance, reducing post-harvest water loss, and increasing relative water content (RWC) by 15–18%. The transgenic lines also exhibited higher proline accumulation (1.8–1.9-fold), which aids in osmotic adjustment under stress conditions. Additionally, the plants displayed reduced chlorophyll degradation and electrolyte leakage under drought and salinity stress, ensuring better photosynthetic efficiency (26–46%) and water use efficiency (9–15%). These findings suggest that genetic modifications targeting drought resistance in mulberry can significantly improve its adaptability, making it more sustainable for silk production in water-deficient regions (Sarkar et al. 2025).

### 3.8.3 *Impact of Mulberry Genetics on* B. mori *Nutrition and Silk Yield*

The genetic composition of mulberry plays a fundamental role in determining *B. mori* nutrition, metabolism, and silk production, influencing key physiological and biochemical processes such as nutrient assimilation, metabolic activity, and fibroin synthesis. The nutritional profile of mulberry leaves is governed by its water content, carbohydrates, proteins, amino acids, vitamins, and micronutrients, all of which contribute to *B. mori* growth, silk gland development, and cocoon quality. Among these, proteins, carbohydrates, and lipids are critical for larval development, while high soluble sugar content helps regulate osmotic balance and alleviate stress, ultimately enhancing *B. mori* health and rearing performance. Genetic interventions aimed at increasing sugar accumulation, optimizing protein synthesis, and improving nutrient bioavailability have led to improved mulberry varieties with superior nutritional properties, ensuring better digestion, enhanced silk filament length, and increased silk yield (Ramesh et al. 2021).

Further studies on the conversion efficiency of mulberry nutrients into silk production revealed that *B. mori* fed on Goshoerami leaves exhibited better nutrient assimilation, leading to higher cocoon yield and improved silk fibre quality. In contrast, those reared on SKM-33 showed reduced growth rates and lower filament production, emphasizing the need for genetic improvements in mulberry to enhance silk output. Moreover, mulberry leaf proteins such as urease have been shown to cross the *B. mori* midgut, hydrolysing urea into ammonia, which is then used for amino acid metabolism. Additionally, carotenoids, carotenes, and xanthophylls derived from mulberry leaves accumulate in the silk glands, contributing to cocoon pigmentation and strengthening silk fibres (E et al. 2022).

*B. mori* growth and silk productivity also depend on micronutrients such as calcium, potassium, magnesium, and phosphorus, which are essential for fibroin synthesis and silk filament development. Research has shown that *B. mori* reared on Jorhat and TR10 mulberry varieties, which have higher micronutrient content, produced longer filaments and superior cocoon traits. In contrast, while the Hmute variety contained higher levels of protein, carbohydrates, and sugars, it primarily supported larval growth without significantly improving silk output. Furthermore, *B. mori* hybrids such as CSR4 × CSR2 and FC1 × FC2 demonstrated greater cocoon weight and shell percentage when fed on nutritionally superior mulberry genotypes, reinforcing the importance of leaf quality in optimizing silk yield (Ruth et al. 2019).

Beyond macronutrients, secondary metabolites such as phenolic compounds and flavonoids significantly impact *B. mori* feeding behaviour, immunity, and silk fibre quality. Mulberry leaves contain chlorogenic acid, hydroxycinnamic acids, caffeoylquinic acid, and flavonols such as quercetin and kaempferol glycosides, which contribute to faster *B. mori* development, enhanced survival rates, and antimicrobial protection within the cocoon. Chlorogenic acid, in particular, plays a key role in stimulating feeding through chemosensory activation, ensuring higher ingestion rates and improved larval growth. These findings underscore the importance of genetic modifications aimed at increasing secondary metabolite content in mulberry leaves, ultimately enhancing the economic traits of silk production (Urbanek Krajnc et al. 2022).

Additionally, leaf moisture retention capacity (MRC) is a crucial factor in preserving post-harvest leaf quality, ensuring optimal feeding conditions for *B. mori*. A strong cuticular wax layer on the leaf surface has been shown to reduce water loss, prolong leaf freshness, and maintain nutrient density. Studies have demonstrated a positive correlation between wax load and moisture retention, highlighting the role of wax biosynthesis in improving leaf longevity. Since cuticular waxes are complex mixtures of long-chain fatty acids synthesized through intricate biochemical pathways, genetic modifications targeting wax biosynthetic genes can enhance mulberry leaf water retention, improve resistance to desiccation, and sustain nutrient composition even after harvest (Sajeevan et al. 2017).

Comparative studies assessing the impact of different mulberry varieties on *B. mori* nutrition and silk yield provide compelling evidence of genotypic influence on silk productivity. In an evaluation of S1635, V1, and Local varieties, S1635 emerged as the most nutritionally superior, with *B. mori* fed on its leaves exhibiting higher larval weight (4.88 g), increased cocoon and pupal weights, and an extended filament length of 1292 m. The higher protein content (147.03 μg/g) in S1635 enhanced silk gland function, leading to better fibroin synthesis and stronger silk filaments. Conversely, *B. mori* fed on the local variety exhibited lower protein intake, resulting in reduced cocoon weight and shorter filaments, further emphasizing the importance of selecting genetically superior mulberry varieties for sericulture (Lalfelpuii and Choudhury 2014).

Beyond conventional nutrients, mulberry leaves contribute to regulatory mechanisms in *B. mori* through miRNAs, which influence growth, metabolism, and reproductive capacity. A study by Chen et al. identified miR168a in mulberry leaves, which is absorbed by *B. mori* and transported via haemolymph to the fat body and ovary, where it binds to BmAGO1 and regulates gene expression. Functional analysis demonstrated that miR168a inhibits BmMthl1 mRNA expression, activating the JNK-FoxO signalling pathway, which in turn reduces oxidative stress, extends *B. mori* lifespan, and enhances reproductive efficiency. These findings highlight the critical regulatory role of mulberry-derived miRNAs in *B. mori* physiology, further underscoring the challenges of replacing mulberry leaves with alternative protein sources (Chen et al. 2024).

These findings strongly emphasize the need for targeted breeding programs to develop genetically superior mulberry varieties that maximize *B. mori* nutrition and silk productivity. By integrating molecular selection techniques, genetic engineering, and advanced breeding strategies, researchers can create mulberry strains with optimized protein content, improved moisture retention, and enhanced digestibility, ensuring consistent silk yield and superior filament properties. As demand for high-quality silk continues to grow, genetic advancements in mulberry cultivation will play a pivotal role in sustaining sericulture and enhancing its economic viability.

## 3.9 Challenges and Ethical Considerations

Genetic modification of *B. mori* and mulberry plants involves several potential risks, particularly concerning ecological stability, genetic integrity, and food safety. Modifications in *B. mori* have primarily focused on enhancing silk quality and producing therapeutic biomolecules. Advanced genetic engineering techniques, such as CRISPR/Cas9 and RNAi, have facilitated precise alterations, but concerns remain regarding genetic stability. Unintended heritable changes could introduce undesirable traits, potentially disrupting wild populations if genetically modified *B. mori* were to escape or cause unintended ecological consequences. Additionally, rigorous testing is required to ensure that engineered traits do not alter natural survival and reproductive dynamics, potentially leading to ecosystem imbalances (Dong et al. 2018; Aramwit et al. 2023).

The mulberry plant, as the exclusive food source of *B. mori*, plays a critical role in sericulture, making its genetic diversity essential for environmental adaptability. Traditional breeding and modern biotechnological techniques have been used to enhance stress resistance and improve productivity. (Thriveni et al. 2023). Mutation breeding using gamma irradiation has been employed to create genetic variability; however, this approach carries the risk of unintended phenotypic changes and the potential loss of beneficial alleles necessary for stress resilience. While genetic modifications can improve resistance to environmental stresses, they may also introduce unforeseen susceptibilities, emphasizing the need for a thorough assessment of their long-term impact. Overall, while genetic engineering holds promise for improving sericulture, careful evaluation and regulatory oversight are necessary to mitigate potential ecological and genetic risks.

One of the primary ethical considerations surrounding transgenic *B. mori* relates to the potential for unintended consequences stemming from their release into the environment. For instance, the introduction of genetically enhanced traits, such as increased silk yield or resistance to specific viruses, could lead to unforeseen ecological dynamics, potentially altering the food web and interactions among other plant and insect species (Thriveni et al. 2023).This concern is underscored by the role of *B. mori* as keystone species in mulberry cultivation, as changes in their population dynamics could disrupt existing agricultural systems and biodiversity. Furthermore, the creation of viral-resistant transgenic *B. mori* represents a significant breakthrough for the sericulture industry, as it can potentially mitigate economic losses due to diseases such as those caused by BmNPV. However, reliance on genetically modified organisms for pest management may reduce genetic diversity, making crops more susceptible to future pests or diseases that may not be addressed by existing genetic modifications. This highlights a broader concern associated with monoculture practices facilitated by genetic engineering: the long-term resilience of agricultural systems could become compromised as genetic homogeneity increases (Subbaiah et al. 2013).

Regulatory frameworks must therefore account for these multifaceted risks when evaluating genetically modified *B. mori* and mulberry. Genetic improvements should be rigorously assessed not just for their immediate benefits but also for their potential long-term impacts on ecosystems and agricultural practices. The regulation of GMOs in sericulture is a complex but essential process that ensures safety, sustainability, and ethical practices (Liu et al. 2022; Shen et al. 2023). On a global scale, frameworks like the Cartagena Protocol on Biosafety help regulate the movement of GMOs across borders, while the Codex Alimentarius (Food and Agricultural Organization/World Health Organization) sets food and agriculture safety standards. (Ma et al. 2012; Aramwit et al. 2023). Each country has its own approach to managing genetically modified *B. mori*. India's Genetic Engineering Appraisal Committee (GEAC) carefully evaluates approvals under its Environment Protection Act, while China's Biosafety Law ensures strict oversight through its Ministry of Agriculture and Rural Affairs (MARA). (Panthee et al. 2017). In the United States, organizations like the United States Department of Agriculture (USDA), Food and Drug Administration (FDA), and United States Environmental Protection Agency (EPA) work together to assess environmental and commercial risks. Meanwhile, the European Union follows some of the world's strictest GMO regulations, requiring thorough risk assessments by the European Food Safety Authority (EFSA). Japan, another key player in silk production, enforces careful monitoring through its Act on the Conservation and Sustainable Use of Biological Diversity. The key focus across all these regulations is to balance innovation with responsibility, ensuring that genetically modified *B. mori* contribute to progress in silk production while protecting the environment, farming communities, and consumer interests (Thriveni et al. 2023).As biotechnology continues to evolve, countries are working toward better harmonization of these rules, making sure that scientific advancements in sericulture are both safe and beneficial for everyone involved.

### 3.9.1 Future Directions in Genetic Research on Mulberry Silk

As the global silk industry advances, future genetic research on mulberry silk aims to enhance yield, quality, and sustainability by integrating advanced genomics, biotechnologies, and breeding practices. High-throughput sequencing and GWAS offer valuable insights into key traits of *B. mori*, such as silk yield, quality, and disease resistance (Dong et al. 2017; Ruth et al. 2019). Additionally, fundamental research on the genetic foundation of *Morus species*, particularly *Morus alba*, is essential for identifying resistance genes and characterizing metabolic pathways that influence silk production, which can be leveraged for selective breeding and genetic modification (Baranwal et al. 2016; Dutta et al. 2023). Epigenetics further contributes by uncovering non-genetic regulatory mechanisms affecting gene expression, particularly in response to environmental factors, thereby facilitating the development of resilient mulberry cultivars (Gao et al. 2020; Chen et al. 2022). The integration of genetic research with synthetic biology presents new opportunities for engineering *B. mori* to produce enhanced fibroin proteins, leading to specialized silk with improved tensile strength and elasticity for biomedical and bioengineering applications (Gu et al. 2020; Moraila-Martinez et al. 2024). Moreover, epigenomic studies can reveal additional regulatory layers influencing silk properties under environmental stress, allowing for precise manipulation of *B. mori* traits beyond direct genetic modifications (Rahman and Islam 2021). Increasing genetic diversity through biodiversity-rich breeding strategies could also strengthen resilience and productivity, contributing to the sustainability of silk production in response to climate change (Umarov et al. 2022; Umarov and Nasirillaev 2023). Advances in bioinformatics and data mining techniques will facilitate genomic data analysis, leading to the discovery of novel genes associated with silk fibres properties and optimizing breeding strategies through tools like CRISPR/Cas9 and transcription activator-like effector nuclease TALEN (Daniyarov et al. 2021; Dutta et al. 2023). Additionally, environmental genomics and population genomics will help assess the impact of cultivation practices on *B. mori* genetics and productivity, aligning research with sustainability goals in sericulture (Madhu et al. 2023). Collaborative efforts between research institutions, government bodies, and farmers will be crucial in driving innovation, ensuring economic growth, and enhancing silk production through cutting-edge genetic advancements (Raveesha et al. 2016).

### 3.9.2 Synthetic Biology and Epigenetics

The integration of synthetic biology and epigenetics in *B. mori* research presents transformative opportunities for enhancing silk production by optimizing genetic and environmental factors influencing silk yield and quality. Synthetic biology provides a platform for engineering biological systems with precision, enabling the modification of metabolic pathways to enhance silk protein synthesis. By redirecting cellular resources toward silk production rather than energy conservation, researchers can develop transgenic *B. mori* strains with tailored metabolic profiles that yield higher-quality silk (Shih et al. 2016, 2020; Turpal 2023). Additionally, the introduction of genes associated with high-strength fibres can result in silk with superior mechanical properties, expanding their applications in industries such as textiles, biotechnology, and medicine (Gu et al. 2020; Moraila-Martinez et al. 2024). A promising approach involves utilizing synthetic biology to engineer *B. mori* capable of biosynthesizing silk with unique functional attributes, including bioactive silk with medicinal properties for wound healing and tissue regeneration (Panja et al. 2017). The use of bioengineering techniques to create specialized silks, coupled with advances in synthetic gene circuits, could lead to the development of novel silk fibres with enhanced elasticity, durability, and resistance to environmental stressors. Moreover, bioreactor systems employing synthetic pathways to produce silk proteins in controlled environments may revolutionize sustainable silk production by reducing reliance on sericulture while preserving the biochemical integrity of natural silk fibres (Ghosh et al. 2017; Manjunath et al. 2023).

Epigenetics, which regulates gene expression without altering the underlying DNA sequence, plays a crucial role in modulating silk production traits. Understanding how environmental factors and nutritional inputs influence epigenetic modifications can provide new strategies for enhancing silk properties without permanent genetic changes (Leem et al. 2020). Epigenetic mechanisms such as DNA methylation, histone modification, and chromatin remodelling can be leveraged to fine-tune gene expression, thereby optimizing cocoon quality and resilience to environmental stress (Gao et al. 2020; Chen et al. 2022). Research on ncRNAs in lepidopterans suggests their involvement in growth and silk production signalling processes, which could be harnessed to improve *B. mori* efficiency in converting mulberry leaf nutrients into silk fibre (Li et al. 2023a). Additionally, epigenomic studies can uncover regulatory layers influencing silk yield under varying environmental conditions, enabling scientists to manipulate *B. mori* traits through controlled rearing practices (Rahman and Islam 2021). The ability to modulate silk properties using epigenetic control mechanisms presents a sustainable alternative to direct genetic modifications, offering practical applications for climate-resilient sericulture. Enhancing the genetic diversity of *B. mori* populations through biodiversity-rich breeding strategies may also improve their adaptability to environmental changes, supporting the long-term sustainability of silk production (Umarov et al. 2022; Umarov and Nasirillaev 2023).

Combining synthetic biology and epigenetics offers a novel strategy for developing disease-resistant and stress-tolerant *B. mori* strains. By manipulating epigenetic markers linked to stress responses, researchers can engineer *B. mori* with enhanced adaptability to diverse agricultural environments, ensuring stability in silk production despite climate fluctuations (Gao et al. 2020; Qayoom et al. 2023). The integration of bioinformatics and genetic engineering techniques, such as CRISPR/Cas9 and TALEN, will further refine breeding programs, allowing for precise genomic edits that enhance silk properties while preserving genetic diversity (Daniyarov et al., 2021). Additionally, developing genetically engineered mulberry plants enriched with specific phytochemicals can improve the nutritional value of *B. mori* diets, thereby boosting silk production efficiency (Ruth et al. 2019). The interconnectedness of plant and insect genetics underscores the potential of a holistic approach to sericulture, where genetic advancements in both mulberry and *B. mori* contribute to overall industry improvements. Furthermore, population genomics and environmental genomics will provide insights into how different cultivation practices impact *B. mori* genetics, supporting the development of innovative breeding strategies aligned with sustainability goals (Madhu et al. 2023).

Integrating synthetic biology and epigenetics in *B. mori* research presents groundbreaking possibilities for silk enhancement, improving productivity, resilience, and sustainability. The convergence of these fields enables precise genetic modifications, refined breeding strategies, and innovative environmental management practices that optimize silk yield and quality. As sericulture continues to evolve, fostering interdisciplinary collaborations and training researchers in advanced genetic techniques will be crucial for bridging long-established silk production methods with cutting-edge biotechnological advancements. Through these efforts, silk production can be revolutionized to meet future industrial and environmental challenges, ensuring its continued significance in the global market.

### *3.9.3 Conclusion*

The genetic developments in mulberry sericulture have significantly revolutionized sericulture, providing accurate and effective means to increase silk yield, fibre quality, and resistance to diseases. Sequencing of the genome of *B. mori* and identification of the major genes responsible for silk production have laid a solid groundwork for genetic enhancement. Traditional breeding practices, although still applicable, have been significantly accelerated by contemporary molecular tools, such as MAS, QTL mapping, and GWAS, to enable the selection of high-performance *B. mori* lines. The integration of gene-editing technologies, notably CRISPR-Cas9, has also transformed silk production by facilitating direct gene edits associated with silk production. This has led to the development of transgenic *B. mori* with enhanced fibre strength, elasticity, and environmental resilience, expanding their applications beyond textiles into biomedical and industrial fields. Moreover, genetic interventions aimed at improving B. mori immunity have played a crucial role in reducing losses due to pathogenic infections, ensuring stable and sustainable silk production. Beyond the B. mori, genetic improvements in mulberry plants have also contributed to the efficiency of sericulture. By developing mulberry varieties with superior nutritional profiles, drought resistance, and pest tolerance, researchers have ensured a reliable food source for B. mori, further optimizing cocoon production. These genetic modifications, combined with advancements in B. mori breeding and engineering, highlight the interconnected nature of genetic research in sericulture. Although these developments have greatly enhanced productivity, quality, and sustainability, they still face challenges. The ethical implications of genetic modifications, potential ecological effects, and regulatory issues associated with transgenic B. mori must be addressed to ensure the safe and proper application. Moreover, the incorporation of new technologies, such as synthetic biology, transcriptomics, and epigenetic mutations, will continue to improve the silk-making process, setting the stage for a new era of high-performance silk fibers. Ultimately, the convergence of conventional breeding methods with advanced genetic tools has ushered in a new era for silk production, aligning industrial needs with environmental and economic sustainability. As research continues to expand the horizons of genetic engineering and molecular breeding, mulberry silk is poised to become a significant and dynamic biomaterial, with the potential to impact numerous sectors in profound ways.

**Declaration** The authors acknowledge the use of AI tools during the preparation of this manuscript for the purposes of language editing and grammar checking. All content was subsequently reviewed, verified, and edited by the authors to ensure accuracy, originality, and compliance with ethical standards. The authors take full responsibility for the final content of the manuscript.

## References


Aramwit P, Jiang Q, Muppuri S, Reddy N (2023) Transgenic modifications of silkworms as a means to obtain therapeutic biomolecules and protein fibers with exceptional properties. Biotechnol Bioeng 120:2827–2839. https://doi.org/10.1002/Bit.28455

Attri K, Vakayil S, Shruthi GH et al (2024) Integrative approaches in sericulture from traditional practices to modern technologies sericulture and its contribution to biodiversity conservation. J Adv Biol Biotechnol 27:480–490. https://doi.org/10.9734/Jabb/2024/V27i81160

Baci G-M, Cucu A-A, Giurgiu A-I et al (2021) Advances in editing silkworms (Bombyx mori) genome by using the CRISPR-Cas system. Insects 13:28. https://doi.org/10.3390/Insects13010028

Baranwal VK, Negi N, Khurana P (2016) Genome-wide identification and structural, functional and evolutionary analysis of WRKY components of mulberry. Sci Rep 6:30794. https://doi.org/10.1038/Srep30794

Bhatti MF, Azizullah, Shahzadi N et al (2019) Effect of honey (*Apis dorsata* [hymenoptera: apidae]) on larval growth and silk cocoon yield of *Bombyx mori* (lepidoptera: bombycidae). J Insect Sci 19(6):11. https://doi.org/10.1093/Jisesa/Iez108

Bi H, Xu X, Li X et al (2019) CRISPR disruption of BmOvo resulted in the failure of emergence and affected the wing and gonad development in the silkworm *Bombyx mori*. Insects 10:254. https://doi.org/10.3390/Insects10080254

Brady D, Saviane A, Cappellozza S, Sandrelli F (2020) An efficient workflow for screening and stabilizing CRISPR/Cas9-mediated mutant lines in *Bombyx mori*. Methods Protoc 4(1):4. https://doi.org/10.3390/Mps4010004

Buhroo Z (2018) A review: disease resistance in mulberry silkworm *Bombyx Mori*.L. Asian J Sci Technol 4(11):157–166.

Bunnik EM, Le Roch KG (2013) An introduction to functional genomics and systems biology. Adv Wound Care (New Rochelle) 2:490–498. https://doi.org/10.1089/Wound.2012.0379

Cao J, Zheng H, Zhang R et al (2023) Dimmed gene knockout shortens larval growth and reduces silk yield in the silkworm, Bombyx mori. Insect Mol Biol 32:26–35. https://doi.org/10.1111/Imb.12810

Chanda S (2024) Effect of dichlorvos on mulberry silkworm and silk production in laboratory. Indian J Entomol 87(4):1–4. https://doi.org/10.55446/Ije.2024.2357

Chen S, Hou C, Bi H et al (2017) Transgenic clustered regularly interspaced short palindromic repeat/Cas9-mediated viral gene targeting for antiviral therapy of *Bombyx mori* nucleopolyhedrovirus. J Virol. https://doi.org/10.1128/Jvi.02465-16

Chen R, Xiao Y, Liu Z et al (2019) Three vital RNA functions and interactions in the process of silk gland apoptosis in silkworm *Bombyx mori*. Arch Insect Biochem Physiol 100(1):e21511. https://doi.org/10.1002/Arch.21511

Chen Y, Fan B, Yasen A et al (2022) YTHDF3 is involved in the diapause process of bivoltine *Bombyx mori* strains by regulating the expression of Cyp307a1 and Cyp18a1 genes in the ecdysone synthesis pathway. Biomolecules 12:1127. https://doi.org/10.3390/Biom12081127

Chen L, Liang J, Zhang Q et al (2024) Mulberry-derived miR168a downregulates BmMthl1 to promote physical development and fecundity in silkworms. Int J Biol Macromol 259:129077. https://doi.org/10.1016/J.Ijbiomac.2023.129077

Cui Y, Zhu Y, Lin Y et al (2018) New insight into the mechanism underlying the silk gland biological process by knocking out fibroin heavy chain in the silkworm. BMC Genomics 19:215. https://doi.org/10.1186/S12864-018-4602-4

Daniyarov U, Abdrimova G, Rajabov N et al (2021) Creation of fine silk fabric of mulberry silkworms through inbreed system mixing methods. E3S Web Conf 258:04046. https://doi.org/10.1051/E3sconf/202125804046

De La Peña PNO, Lao AGD, Bautista MaAM (2022) Global profiling of genes expressed in the silk glands of Philippine-reared mulberry silkworms (*Bombyx mori*). Insects 13:669. https://doi.org/10.3390/Insects13080669

Deepika I, Ramesh KV, Kumar I et al (2024) Molecular diagnostics in sericulture: a paradigm shift towards disease diagnosis in silkworms. Entomol Exp Appl 172:372–382. https://doi.org/10.1111/Eea.13419

Deng X, Liu L, Deng J, Zha X (2023) Specific expression of antimicrobial peptides from the black soldier fly in the midgut of silkworms (*Bombyx mori*) regulates silkworm immunity. Insects 14:443. https://doi.org/10.3390/Insects14050443

Dewi R, Minarningsih M, Andadari L (2021) Uji adaptasi hybrid ulat sutra asal tiongkok (adaptation test of hybrid silkworm from china). Jurnal Penelitian Hutan Tanaman 18:147–158. https://doi.org/10.20886/Jpht.2021.18.2.147-158

Dong H-L, Zhang S-X, Tao H et al (2017) Metabolomics differences between silkworms (*Bombyx mori*) reared on fresh mulberry (*Morus*) leaves or artificial diets. Sci Rep 7:10972. https://doi.org/10.1038/S41598-017-11592-4

Dong Z, Huang L, Dong F et al (2018) Establishment of a baculovirus-inducible CRISPR/Cas9 system for antiviral research in transgenic silkworms. Appl Microbiol Biotechnol 102:9255–9265. https://doi.org/10.1007/S00253-018-9295-8

Dong Z, Hu Z, Qin Q et al (2019) CRISPR/Cas9-mediated disruption of the immediate early-0 and 2 as a therapeutic approach to *Bombyx mori* nucleopolyhedrovirus in transgenic silkworm. Insect Mol Biol 28:112–122. https://doi.org/10.1111/Imb.12529

Dong Z, Qin Q, Hu Z et al (2019b) Construction of a one-vector multiplex CRISPR/Cas9 editing system to inhibit nucleopolyhedrovirus replication in silkworms. Virol Sin 34:444–453. https://doi.org/10.1007/S12250-019-00121-4

Dong Z, Qin Q, Hu Z et al (2020) CRISPR/Cas12a mediated genome editing enhances Bombyx mori resistance to BmNPV. Front Bioeng Biotechnol 8. https://doi.org/10.3389/Fbioe.2020.00841

Dutta H, Sawarkar A, Dutta S et al (2023) Genomic approaches to ensure a more sustainable and productive future of mulberry for sericulture industry. Pharma Innov 12:2001–2011. https://doi.org/10.22271/Tpi.2023.V12.I9v.22980

Fang S-M, Hu B-L, Zhou Q-Z et al (2015) Comparative analysis of the silk gland transcriptomes between the domestic and wild silkworms. BMC Genomics 16:60. https://doi.org/10.1186/S12864-015-1287-9

Fang S-M, Zhou Q-Z, Yu Q-Y, Zhang Z (2020) Genetic and genomic analysis for cocoon yield traits in silkworm. Sci Rep 10:5682. https://doi.org/10.1038/S41598-020-62507-9

Fu Y, Wang Y, Huang Q et al (2022) Long noncoding RNA lncR17454 regulates metamorphosis of silkworm through let-7 miRNA cluster. J Insect Sci 22. https://doi.org/10.1093/Jisesa/Ieac028

Fungmongkonsatean T, Jongjitwimol J, Paensuwan P et al (2022) Hemocompatibility evaluation of thai *Bombyx mori* silk fibroin and its improvement with low molecular weight heparin immobilization. Polymers (Basel) 14:2943. https://doi.org/10.3390/Polym14142943

Furdui EM, Mărghitaş LA, Dezmirean DS et al (2014) Genetic characterization of *Bombyx mori* (Lepidoptera: Bombycidae) breeding and hybrid lines with different geographic origins. J Insect Sci 14. https://doi.org/10.1093/Jisesa/Ieu073

Gai T, Tong X, Han M et al (2020) Cocoonase is indispensable for Lepidoptera insects breaking the sealed cocoon. Plos Genet 16:E1009004. https://doi.org/10.1371/Journal.Pgen.1009004

Gai Y-P, Zhao Y-N, Zhao H-N et al (2017) The latex protein MLX56 from mulberry (*Morus multicaulis*) protects plants against insect pests and pathogens. Front Plant Sci 8. https://doi.org/10.3389/Fpls.2017.01475

Gai Y-P, Yuan S-S, Zhao Y-N et al (2018) A novel lncRNA, MuLnc1, associated with environmental stress in mulberry (*Morus multicaulis*). Front Plant Sci 9. https://doi.org/10.3389/Fpls.2018.00669

Gao H, Sun Y, Song G et al (2018) N-acetyl-l-cysteine against imidacloprid intoxication on *Bombyx mori* larvae. Arch Insect Biochem Physiol 99. https://doi.org/10.1002/Arch.21497

Gao R, Li C-L, Tong X-L et al (2020) Identification, expression, and artificial selection of silkworm epigenetic modification enzymes. BMC Genomics 21:740. https://doi.org/10.1186/S12864-020-07155-Z

Ge Q, Fan Y, Xu J et al (2022) Whole transcriptome-based study to speculate upon the silkworm yellow blood inhibitor (i) gene and analyze the miRNA-mediated gene regulatory network. Processes 10:1556. https://doi.org/10.3390/Pr10081556

Ghosh A, Gangopadhyay D, Chowdhury T (2017) Economical and environmental importance of mulberry: a review. IJPE T 3:51–58. https://doi.org/10.18811/Ijpen.V3i02.10437

Giora D, Marchetti G, Cappellozza S et al (2022) Bibliometric analysis of trends in mulberry and silkworm research on the production of silk and its by-products. Insects 13:568. https://doi.org/10.3390/Insects13070568

Gowrisankar R, Dasari S, Anas M (2023) Silk reeling techniques: exploring traditional and advanced methods. J Exp Agric Int 45:91–99. https://doi.org/10.9734/Jeai/2023/V45i92179

Gu Y, Yu L, Mou J et al (2020) Mechanical properties and application analysis of spider silk bionic material. E-Polymers 20:443–457. https://doi.org/10.1515/Epoly-2020-0049

Gundarapu S, Gadde AK, Teja KSS et al (2024) A comprehensive review of the effect and mitigation of climate change on sericulture. Int J Environ Clim Change 14:776–788. https://doi.org/10.9734/Ijecc/2024/V14i74317

Gundi R, Vanitha C, Tulsi KSN et al (2023) Molecular marker assisted breeding and development of Bidensovirus resistant and thermo tolerant silkworm (*Bombyx mori*) hybrids suitable for tropical climatic conditions. Agric Res 12:428–438. https://doi.org/10.1007/S40003-023-00662-X

Guo Y, Shen Y-H, Sun W et al (2011) Nucleotide diversity and selection signature in the domesticated silkworm, *Bombyx mori*, and wild silkworm, *Bombyx mandarina*. J Insect Sci 11:1–16. https://doi.org/10.1673/031.011.15501

Gupta V, Rastogi S (2023) Engineering spider genes for high-tensile silk and patenting challenges. J Adv Zool 44:1004–1012. https://doi.org/10.17762/Jaz.V44is-3.404

Hanif U, Naveed S, Ul Huda N et al (2025) Silkworm interaction and its role in resistance to *Bombyx mori* nuclear polyhedrosis virus: a genetic and molecular review. Indus J Biosci Res 3:30–36. https://doi.org/10.70749/Ijbr.V3i2.603

Hong JW, Jeong CY, Yu JH et al (2020) *Bombyx mori* kynurenine 3-monooxygenase gene editing and insect molecular breeding using the clustered regularly interspaced short palindromic repeat/CRISPR associated protein 9 system. Biotechnol Prog 36. https://doi.org/10.1002/Btpr.3054

Hou CX, Sun PJ, Guo XJ et al (2013) Marker-assisted selection in breeding silkworm strains with high silk production and resistance to the densonucleosis virus. Genet Mol Res 12:4171–4178. https://doi.org/10.4238/2013.October.7.3

Hu Z, Zhu F, Chen K (2023) The mechanisms of silkworm resistance to the Baculovirus and antiviral breeding. Annu Rev Entomol 68:381–399. https://doi.org/10.1146/Annurev-Ento-120220-112317

Hu Y, Zhang Q, You R et al (2012) The relationship between secondary structure and biodegradation behavior of silk fibroin scaffolds. Adv Mater Sci Eng 2012:1–5. https://doi.org/10.1155/2012/185905

Hu X, Zhu M, Zhang X et al (2018) Identification and characterization of circular RNAs in the silkworm midgut following *Bombyx mori* cytoplasmic polyhedrosis virus infection. RNA Biol 15:292–301. https://doi.org/10.1080/15476286.2017.1411461

Hwang H-H, Wang C-H, Chen H-H et al (2019) Effective agrobacterium-mediated transformation protocols for callus and roots of halophyte ice plant (*Mesembryanthemum crystallinum*). Bot Stud 60:1. https://doi.org/10.1186/S40529-018-0249-3

Ichida H, Morita R, Shirakawa Y et al (2019) Targeted exome sequencing of unselected heavy-ion beam-irradiated populations reveals less-biased mutation characteristics in the rice genome. Plant J 98:301–314. https://doi.org/10.1111/Tpj.14213
Iizuka T, Sezutsu H, Tatematsu K et al (2013) Colored fluorescent silk made by transgenic silkworms. Adv Funct Mater 23:5232–5239. https://doi.org/10.1002/Adfm.201300365
Jiang L, Peng L-L, Cao Y-Y et al (2020) Transcriptome analysis reveals gene expression changes of the fat body of silkworm (*Bombyx mori* l.) in response to selenium treatment. Chemosphere 245:125660. https://doi.org/10.1016/J.Chemosphere.2019.125660
Kanamori Y, Saito A, Hagiwara-Komoda Y et al (2010) The trehalose transporter 1 gene sequence is conserved in insects and encodes proteins with different kinetic properties involved in trehalose import into peripheral tissues. Insect Biochem Mol Biol 40:30–37. https://doi.org/10.1016/J.Ibmb.2009.12.006
Kanginakudru S, Royer C, Edupalli SV et al (2007) Targeting ie-1 gene by RNAi induces baculoviral resistance in lepidopteran cell lines and in transgenic silkworms. Insect Mol Biol 16:635–644. https://doi.org/10.1111/J.1365-2583.2007.00753.X
Kannihalli S, Rayar SG, Mallapur CP et al (2024) Enhancing the mulberry leaf quality and yield through foliar application of nanofertilizers. Plant Arch 24. https://doi.org/10.51470/Plantarchives.2024.V24.No.1.138
Kant R, Islam T, Bali K et al (2024) Comparative analysis of rearing of silkworm, *Bombyx mori* L. under different conditions. Int J Vet Sci Anim Husbandry 9(5):520–528
Kanwal M, Younus W, Hussain M (2019) An insight into genomics of mulberry silkworm. Sci Inq Rev 3:08–16. https://doi.org/10.32350/Sir.32.02
Kawaoka S, Katsuma S, Daimon T et al (2008) Functional analysis of four Gloverin-like genes in the silkworm, *Bombyx mori*. Arch Insect Biochem Physiol 67:87–96. https://doi.org/10.1002/Arch.20223
Kayukawa T, Minakuchi C, Namiki T et al (2012) Transcriptional regulation of juvenile hormone-mediated induction of Krüppel homolog 1, a repressor of insect metamorphosis. Proc National Acad Sci 109:11729–11734. https://doi.org/10.1073/Pnas.1204951109
Kikawada T, Saito A, Kanamori Y et al (2007) Trehalose transporter 1, a facilitated and high-capacity trehalose transporter, allows exogenous trehalose uptake into cells. Proc National Acad Sci 104:11585–11590. https://doi.org/10.1073/Pnas.0702538104
Kim S-R, Choi K-H, Kim S-W, Park S-W (2016) Comparison of Gloverin gene expression patterns between domesticated and wild silkworms. Int J Indust Entomol 33:113–120. https://doi.org/10.7852/Ijie.2016.33.2.113
Kumar D (2016) The human genome. In: Medical and health genomics. Elsevier, pp 1–13
Kuwana Y, Sezutsu H, Nakajima K et al (2014) High-toughness silk produced by a transgenic silkworm expressing spider (*Araneus ventricosus*) dragline silk protein. Plos One 9:E105325. https://doi.org/10.1371/Journal.Pone.0105325
Lalfelpuii R, Choudhury BN (2014) Effect of different mulberry plant varieties on growth and economic parameters of the b. Mori *Bombyx Mori* in Mizoram. Sci Vis 14(1):34–38
Larina EA, Aqilov UH, Daniyarov UT (2021) Advantages of using parthenogenetic clones of the silkworm (*Bombyx mori* l.) in hybridization. J Sci Innov Dev 4:54–64. https://doi.org/10.36522/2181-9637-2021-3-5
Leem JW, Fraser MJ, Kim YL (2020) Transgenic and diet-enhanced silk production for reinforced biomaterials: a metamaterial perspective. Annu Rev Biomed Eng 22:79–102. https://doi.org/10.1146/Annurev-Bioeng-082719-032747
Levin M, Scheibe M, Butter F (2020) Proteotranscriptomics assisted gene annotation and spatial proteomics of *Bombyx mori* Bmn4 cell line. BMC Genomics 21:690
Li Y, Wang G, Tian J et al (2012) Transcriptome analysis of the silkworm (*Bombyx mori*) by high-throughput RNA sequencing. Plos One 7:E43713. https://doi.org/10.1371/Journal.Pone.0043713

Li Y, Chen X, Zhang C et al (2016) Effects of starvation and hormones on DNA synthesis in silk gland cells of the silkworm, Bombyx mori. Insect Sci 23:569–578. https://doi.org/10.1111/1744-7917.12199

Li R, Liu L, Dominic K et al (2018) Mulberry (*Morus alba*) MMSK gene enhances tolerance to drought stress in transgenic mulberry. Plant Physiol Biochem 132:603–611. https://doi.org/10.1016/J.Plaphy.2018.10.007

Li J, Lu Z, Mao T et al (2020) Identification of the nucleotide exchange factor Bmgrpe and its role under high-temperature stress in silkworm, *Bombyx mori*. Arch Insect Biochem Physiol 104(1):e21664. https://doi.org/10.1002/Arch.21664

Li D, Kan Y, Li Z (2023a) Non-coding DNAs in lepidoptera. In: Biodiversity and ecology of lepidoptera—insights and advances. Intechopen

Li J, Deng J, Deng X et al (2023b) Metabonomic analysis of silkworm midgut reveals differences between the physiological effects of an artificial and mulberry leaf diet. Insects 14:347. https://doi.org/10.3390/Insects14040347

Li S, Lao J, Sun Y et al (2024) CRISPR/Cas9-mediated editing of bmeckl1 gene sequence affected silk gland development of silkworms (*Bombyx mori*). Int J Mol Sci 25:1907. https://doi.org/10.3390/Ijms25031907

Lin M, Qian Y, Chen E Et Al (2023) The BMTRET1 gene family and its potential role in response to BmNPV stress in *Bombyx mori*. Int J Mol Sci 25:402. https://doi.org/10.3390/Ijms25010402

Liu T, Cao L, Cheng Y et al (2021a) MKK4/5-MPK3/6 cascade regulates agrobacterium-mediated transformation by modulating plant immunity in Arabidopsis. Front Plant Sci 12:731690. https://doi.org/10.3389/Fpls.2021.731690

Liu Y, Chen D, Zhang X et al (2021b) Construction of baculovirus-inducible CRISPR/Cas9 antiviral system targeting BmNPV in *Bombyx mori*. Viruses 14:59. https://doi.org/10.3390/V14010059

Liu Y, Zhang X, Chen D et al (2022) CRISPR/Cas9-mediated disruption of the lef8 and lef9 to inhibit nucleopolyhedrovirus replication in silkworms. Viruses 14:1119. https://doi.org/10.3390/V14061119

Long D, Liu R, Huang Y et al (2024) An efficient and safe strategy for germ cell-specific automatic excision of foreign DNA in F1 hybrid transgenic silkworms. Insect Sci 31:28–46. https://doi.org/10.1111/1744-7917.13219

Lu W, Zhang T, Zhang Q et al (2023) FIBH gene complete sequences (fibhome) revealed silkworm pedigree. Insects 14:244. https://doi.org/10.3390/Insects14030244

Lu H, Jian M, Liang X et al (2024) Strong silkworm silk fibers through CNT-feeding and forced reeling. Adv Mater 36. https://doi.org/10.1002/Adma.202408385

Luan Y, Zuo W, Li C et al (2018) Identification of genes that control silk yield by RNA sequencing analysis of silkworm (*Bombyx mori*) strains of variable silk yield. Int J Mol Sci 19:3718. https://doi.org/10.3390/Ijms19123718

Ma S, Zhang S, Wang F et al (2012) Highly efficient and specific genome editing in silkworm using custom talens. Plos One 7:E45035. https://doi.org/10.1371/Journal.Pone.0045035

Ma Y, Zeng W, Ba Y et al (2022) A single-cell transcriptomic atlas characterizes the silk-producing organ in the silkworm. Nat Commun 13:3316. https://doi.org/10.1038/S41467-022-31003-1

Madhu DM, Irfan SM, Prakash S et al (2023) Value chain analysis of cocoons, constraints faced by women in production and marketing of mulberry silk in Chikkaballapur district of Karnataka, India. Asian J Agric Extension Econ Sociol 41:36–43. https://doi.org/10.9734/Ajaees/2023/V41i51898

Mahadeva A (2017) Insect pest infestation, an obstacle in quality mulberry leaves production. Asian J Biol Sci 11:41–52. https://doi.org/10.3923/Ajbs.2018.41.52

Mandal BB, Grinberg A, Seok Gil E et al (2012) High-strength silk protein scaffolds for bone repair. Proc National Acad Sci 109:7699–7704. https://doi.org/10.1073/Pnas.1119474109

Manjunath KK, Harshitha BS, Anusha NM, Bhargavi HA (2023) Breeding for mulberry improvement: a review. Pharma Innov 12:477–481. https://doi.org/10.22271/Tpi.2023.V12.I2f.18461

Martinez-Useros J, Martin-Galan M, Florez-Cespedes M, Garcia-Foncillas J (2021) Epigenetics of most aggressive solid tumors: pathways, targets and treatments. Cancers (Basel) 13:3209. https://doi.org/10.3390/Cancers13133209

Meng G, Wang R, Chu Q (2020) Genome survey and mitochondrial genome analysis of wild silkworm, *Bombyx mandarina*. Mol Entomol. https://doi.org/10.5376/Me.2020.11.0002

Mieszawska AJ, Nadkarni LD, Perry CC, Kaplan DL (2010) Nanoscale control of silica particle formation via silk−silica fusion proteins for bone regeneration. Chem Mater 22:5780–5785. https://doi.org/10.1021/Cm101940u

Mirhoseini SZ, Rabiei B, Potki P, Dalirsefat SB (2010) Amplified fragment length polymorphism mapping of quantitative trait loci for economically important traits in the silkworm, *Bombyx mori*. J Insect Sci 10:1–21. https://doi.org/10.1673/031.010.14113

Mondal R, Shaw S, Mandal P et al (2024) Recent advances in the biosensors application for reviving infectious disease management in silkworm model: a new way to combat microbial pathogens. Arch Microbiol 206:206. https://doi.org/10.1007/S00203-024-03933-5

Moraila-Martinez CL, Rodríguez-Ortega A, Rodriguez N et al (2024) Biofabrication of silk fibers with enhanced conductivity through silkworm feeding with reduced graphene oxide: implications for smart textile innovations. ACS Appl Nano Mater 7:6229–6241. https://doi.org/10.1021/Acsanm.3c06082

Murali S, Singh S (2020) To workout nutrigenetic traits for silkworm, *Bombyx mori* l. by absorption/assimilation for determining growth and development for identifying parental breeds for future breeding during spring season under subtropical region of north west India. Curr J Appl Sci Technol 57–70. https://doi.org/10.9734/Cjast/2020/V39i2230843

Nutthapornnitchakul S, Sonjaroon W, Putthisawong N et al (2024) Effect of drought stress on proline gene expression, enzyme activity, and physiological responses in Thai mulberry (*Morus* spp.). Hayati 31:559–571. https://doi.org/10.4308/Hjb.31.3.559-571

Panja S, Behera S, Sc K, Halder M (2017) Optical spectroscopic and morphological characterizations of curcuminized silk biomaterials: a perspective from drug stabilization. ACS Omega 2:6755–6767. https://doi.org/10.1021/Acsomega.7b00809

Panthee S, Paudel A, Hamamoto H, Sekimizu K (2017) Advantages of the silkworm as an animal model for developing novel antimicrobial agents. Front Microbiol 8. https://doi.org/10.3389/Fmicb.2017.00373

Paudel A, Panthee S, Hamamoto H, Sekimizu K (2020) A simple artificial diet available for research of silkworm disease models. Drug Discov Ther 14:177–180. https://doi.org/10.5582/Ddt.2020.03061

Pavithra MR, Karur AS, Teja KSS et al (2024) Genetic improvements in silkworms: enhancing silk yield and quality. Uttar Pradesh J Zool 45:164–172. https://doi.org/10.56557/Upjoz/2024/V45i164297

Pm S, Liang Y, Loqué D (2016) Biotechnology and synthetic biology approaches for metabolic engineering of bioenergy crops. Plant J 87:103–117. https://doi.org/10.1111/Tpj.13176

Qayoom K, Manzoor S, Rafiqui AR, Bin AO (2023) Biotic and abiotic stress management under climate change in sericulture. Int J Agric Appl Sci 4:46–52. https://doi.org/10.52804/Ijaas2023.426

Qian H, Li G, Zhao G et al (2019) Fingerprint analysis of Bombyx mori local variety resources based on ssr markers. ScienceAsia 45:342. https://doi.org/10.2306/Scienceasia1513-1874.2019.45.342

Qian W, Yang Y, Li Z et al (2021) Enhanced MYC expression in silkworm silk gland promotes DNA replication and silk production. Insects 12:361. https://doi.org/10.3390/Insects12040361

Qin L, Xia H, Shi H et al (2012) Comparative proteomic analysis reveals that caspase-1 and serine protease may be involved in silkworm resistance to *Bombyx mori* nuclear polyhedrosis virus. J Proteomics 75:3630–3638. https://doi.org/10.1016/J.Jprot.2012.04.015

Qin S, Sun L, Zhang S et al (2022) Bmabl1 regulates silk protein synthesis via glutathione metabolism in *Bombyx mori*. Insects 13:967. https://doi.org/10.3390/Insects13110967

Qiu C-Z, Zhou Q-Z, Liu T-T et al (2018) Evidence of peripheral olfactory impairment in the domestic silkworms: insight from the comparative transcriptome and population genetics. BMC Genomics 19:788. https://doi.org/10.1186/S12864-018-5172-1

Rahman M, Islam S (2021) Studies on food, health and environmental perspectives in mulberry (*Morus* spp.)—a review. J Biosci (Rajshari) 163–179. https://doi.org/10.3329/Jbs.V29i0.54832

Rajawat D, Pandey JP, Saini T et al (2024) Overview of biotechnological interventions in tasar sericulture: prospects for bridging tradition to technology. Plant Arch 24:68–78. https://doi.org/10.51470/Plantarchives.2024.V24.Specialissue.013

Ramesh Hl, Shivashankarappa LH, Yoganandamurthy VN (2021) Gamma irradiation enhanced leaf bioactive components and bioassay parameters in m5 mulberry (*morus* sp.) mutant. Am J Plant Sci 12:1945–1962. https://doi.org/10.4236/Ajps.2021.1212134

Rao KJ, Korumilli T, Patni V (2019) Investigating silk yield and morphological changes in silk fibres obtained from silkworms fed with Ag and/or $TiO_2$ nanoparticles. Bull Mater Sci 42:222. https://doi.org/10.1007/S12034-019-1913-2

Rao Y, Yang X, Pan C et al (2022) Advance of clustered regularly interspaced short palindromic repeats-Cas9 system and its application in crop improvement. Front Plant Sci 13:839001. https://doi.org/10.3389/Fpls.2022.839001

Raveesha S, Kumar KA, Bai DS (2016) A socio-economic analysis of women's participation in sericulture. Adv Res J Soc Sci 7:55–61. https://doi.org/10.15740/Has/Arjss/7.1/55-61

Ruan J, Wu M, Ye X et al (2021) Comparative mRNA and lncRNA analysis of the molecular mechanisms associated with low silk production in *Bombyx mori*. Front Genet 11:592128. https://doi.org/10.3389/Fgene.2020.592128

Ruiz X, Almanza M (2018) Implications of genetic diversity in the improvement of silkworm *Bombyx mori* l. Chil J Agric Res 78:569–579. https://doi.org/10.4067/S0718-58392018000400569

Ruth L, Ghatak S, Subbarayan S et al (2019) Influence of micronutrients on the food consumption rate and silk production of *Bombyx mori* (lepidoptera: bombycidae) reared on mulberry plants grown in a mountainous agro-ecological condition. Front Physiol 10:878. https://doi.org/10.3389/Fphys.2019.00878

Saini P, Rohela GK, Kumar JS et al (2023) Cultivation, utilization, and economic benefits of mulberry. In: Gnanesh BN, Vijayan K (eds) The mulberry genome. Compendium of plant genomes. Springer, Cham, pp 13–56. https://doi.org/10.1007/978-3-031-28478-6_2

Sajeevan RS, Nataraja KN, Shivashankara KS et al (2017) Expression of Arabidopsis SHN1 in indian mulberry (*Morus indica* l.) increases leaf surface wax content and reduces post-harvest water loss. Front Plant Sci 8. https://doi.org/10.3389/Fpls.2017.00418

Sarkar T, Mogili T, Gandhi Doss S, Sivaprasad V (2018) Tissue culture in mulberry (*morus* spp.) intending genetic improvement, micropropagation and secondary metabolite production: a review on current status and future prospects. In: Biotechnological approaches for medicinal and aromatic plants. Springer Singapore, Singapore, pp 467–487

Sarkar T, Nataraja KN, Lalitha Kumari F et al (2025) Co-expression of AtDREB2A and AtSHN1 in transgenic mulberry (*morus* spp.) confers tolerance to drought and salinity and reduces post-harvest water loss. Plant Growth Regul. https://doi.org/10.1007/S10725-025-01306-6

Seong J, Kim M-J, Kim H-S et al (2011) Generation of transgenic silkworms for production of erythropoietin in *Bombyx mori*. Genes Genomics 33:237–243. https://doi.org/10.1007/S13258-011-0022-4

Shen G, Liu D, Xu H et al (2023) A study on the effect of energy on the development of silkworm embryos using an estrogen-related receptor. Int J Mol Sci 24:14485. https://doi.org/10.3390/Ijms241914485

Silva Pugina R, Almeida Caiut JM (2021) Silk: history, obtaining, structure and properties of an old material in the development of new technologies. Int J Adv Med Biotechnol 4:27–45. https://doi.org/10.52466/Ijamb.V4i1.88

Singh CP (2021) In vitro treatment of seroin proteins to bmnpv budded virions suppresses viral proliferation in *Bombyx mori* larvae and ectopic overexpression of host-mirnas downregulates the expression of Seroin2 mRNA in BMN cells. Int J Trop Insect Sci 41:1485–1491. https://doi.org/10.1007/S42690-020-00346-X

Srivastava PP, Vijayan K, Kar PK, Saratchandra B (2011) Diversity and marker association in tropical silkworm breeds of *Bombyx mori* (Lepidoptera: Bombycidae). Int J Trop Insect Sci 31:182–191. https://doi.org/10.1017/S1742758411000233

Stark R, Grzelak M, Hadfield J (2019) RNA sequencing: the teenage years. Nat Rev Genet 20:631–656. https://doi.org/10.1038/S41576-019-0150-2

Su Y, Qu Q, Li J et al (2024) Perilipin1 inhibits nosema bombycis proliferation by promoting domeless- and hop- mediated jak-stat pathway activation in *Bombyx mori*. Microbiol Spectr 12. https://doi.org/10.1128/Spectrum.03671-23

Subbaiah EV, Royer C, Kanginakudru S et al (2013) Engineering silkworms for resistance to baculovirus through multigene RNA interference. Genetics 193:63–75. https://doi.org/10.1534/Genetics.112.144402

Suetsugu Y, Futahashi R, Kanamori H et al (2013) Large scale full-length cDNA sequencing reveals a unique genomic landscape in a lepidopteran model insect, *Bombyx mori*. G3 Genes Genomes Genet 3:1481–1492. https://doi.org/10.1534/G3.113.006239

Tan A, Fu G, Jin L et al (2013) Transgene-based, female-specific lethality system for genetic sexing of the silkworm, *Bombyx mori*. Proc National Acad Sci 110:6766–6770. https://doi.org/10.1073/Pnas.1221700110

Tang X, Liu H, Shi Z et al (2020) Enhanced silk yield in transgenic silkworm (*Bombyx mori*) via ectopic expression of BmGT1-L in the posterior silk gland. Insect Mol Biol 29:452–465. https://doi.org/10.1111/Imb.12655

Tang X, Ye X, Wang X et al (2021) High mechanical property silk produced by transgenic silkworms expressing the spidroins PySp1 and ASG1. Sci Rep 11:20980. https://doi.org/10.1038/S41598-021-00029-8

Teramoto H, Kojima K (2015) Incorporation of methionine analogues into *Bombyx mori* silk fibroin for click modifications. Macromol Biosci 15:719–727. https://doi.org/10.1002/Mabi.201400482

Teulé F, Miao Y-G, Sohn B-H et al (2012) Silkworms transformed with chimeric silkworm/spider silk genes spin composite silk fibers with improved mechanical properties. Proc National Acad Sci 109:923–928. https://doi.org/10.1073/Pnas.1109420109

Thomas TD, Bhatnagar AK, Bhojwani SS (2000) Production of triploid plants of mulberry (*Morus alba* l) by endosperm culture. Plant Cell Rep 19:395–399. https://doi.org/10.1007/S002990050746

Thriveni MC, Deepa S, Thanavendan G et al (2023) Characterization and evaluation of mulberry genetic resources for the identification of promising accessions. Indian J Plant Genet Resour 36:85–95. https://doi.org/10.61949/0976-1926.2023.V36i01.12

Tomihara K, Andolfatto P, Kiuchi T (2022) Allele-specific knockouts reveal a role for apontic-like in the evolutionary loss of larval melanin pigmentation in the domesticated silkworm, *Bombyx mori*. Insect Mol Biol 31:701–710. https://doi.org/10.1111/Imb.12797

Trochez-Solarte JD, Ruiz-Erazo X, Almanza-Pinzon M, Zambrano-Gonzalez G (2019) Role of microsatellites in genetic analysis of *Bombyx mori* silkworm: a review. F1000res 8:1424. https://doi.org/10.12688/F1000research.20052.1

Tsubota T, Yoshioka T, Jouraku A et al (2021) Transcriptomic analysis of the bagworm moth silk gland reveals a number of silk genes conserved within lepidoptera. Insect Sci 28:885–900. https://doi.org/10.1111/1744-7917.12846s

Türkoğlu A, Haliloğlu K, Tosun M et al (2023) Ethyl methanesulfonate (ems) mutagen toxicity-induced DNA damage, cytosine methylation alteration, and ipbs-retrotransposon polymorphisms in wheat (*Triticum aestivum* l.). Agronomy 13:1767. https://doi.org/10.3390/Agronomy13071767

Turpal K (2023) Synthetic biology as a breakthrough direction in the creation of biological systems. Bio Web Conf 76:07002. https://doi.org/10.1051/Bioconf/20237607002

Umarov S, Nasirillaev B (2023) Creation of heat-resistant, high-product breeds and hybrids of mulberry silkworm (*Bombyx mori* l.). E3S Web Conf 371:01028. https://doi.org/10.1051/E3sconf/202337101028

Umarov S, Nasirillaev B, Batirova A (2022) Mulberry silkworm (*Bombyx mori* l.) hybrids in Uzbek selection and, viability, cocoon productivity, and reproductive indicators in summer and fall seasons. Iop Conf Ser Earth Environ Sci 1068:012021. https://doi.org/10.1088/1755-1315/1068/1/012021

Urbanek Krajnc A, Bakonyi T, Ando I et al (2022) The effect of feeding with central european local mulberry genotypes on the development and health status of silkworms and quality parameters of raw silk. Insects 13:836. https://doi.org/10.3390/Insects13090836

Veenstra JA (2019). The hormone neuroparsin seems essential in Lepidoptera but not in domesticated silkworms. bioRxiv 716746. https://doi.org/10.1101/716746

Vijayan K (2010) The emerging role of genomic tools in mulberry (*Morus*) genetic improvement. Tree Genet Genomes 6:613–625. https://doi.org/10.1007/S11295-010-0276-Z

Wang X, Zhao P, Li Y et al (2015) Modifying the mechanical properties of silk fiber by genetically disrupting the ionic environment for silk formation. Biomacromolecules 16:3119–3125. https://doi.org/10.1021/Acs.Biomac.5b00724

Wang M, Lin Y, Zhou S et al (2020a) Genetic mapping of climbing and mimicry: two behavioral traits degraded during silkworm domestication. Front Genet 11. https://doi.org/10.3389/Fgene.2020.566961

Wang W, Wang X, Luo C et al (2020b) Let-7 microrna is a critical regulator in controlling the growth and function of silk gland in the silkworm. RNA Biol 17:703–717. https://doi.org/10.1080/15476286.2020.1726128

Wei GQ, Yu L, Liu CL et al (2013) Linkage and mapping analyses of the normal marking gene +p in the silkworm (*Bombyx mori*) using SSR markers. Genet Mol Res 12:2351–2359. https://doi.org/10.4238/2013.January.22.1

Xia Q, Li S, Feng Q (2014) Advances in silkworm studies accelerated by the genome sequencing of *Bombyx mori*. Annu Rev Entomol 59:513–536. https://doi.org/10.1146/Annurev-Ento-011613-161940

Xiang H, Li X, Dai F et al (2013) Comparative methylomics between domesticated and wild silkworms implies possible epigenetic influences on silkworm domestication. BMC Genomics 14:646. https://doi.org/10.1186/1471-2164-14-646

Xiao R, Yuan Y, Xia H et al (2022) Comparative transcriptome and proteome reveal synergistic functions of differentially expressed genes and proteins implicated in an over-dominant silkworm heterosis of increased silk yield. Insect Mol Biol 31:551–567. https://doi.org/10.1111/Imb.12779

Xing W, Ma S, Liu Y, Xia Q (2020) CRISPR/dcas9-mediated imaging of endogenous genomic loci in living *Bombyx mori* cells. Insect Sci 27:1360–1364. https://doi.org/10.1111/1744-7917.12722

Xu H-M, Wei C-S, Tang Y-T et al (2011) A new mapping method for quantitative trait loci of silkworm. BMC Genet 12:19. https://doi.org/10.1186/1471-2156-12-19

Xu H-E, Zhang H-H, Xia T, Et Al (2013) BmTEdb: a collective database of transposable elements in the silkworm genome. Database 2013. https://doi.org/10.1093/Database/Bat055

Xu J, Dong Q, Yu Y et al (2018) Mass spider silk production through targeted gene replacement in *Bombyx mori*. Proc National Acad Sci 115:8757–8762. https://doi.org/10.1073/Pnas.1806805115

Xu X, Du X, Chen J et al (2024) Genetic diversity and differentiation of silkworm (*Bombyx mori*) local germplasm resources in China and Uzbekistan. Insects 15:1020. https://doi.org/10.3390/Insects15121020

Yali W, Mitiku T (2022) Mutation breeding and its importance in modern plant breeding. J Plant Sci 10:64. https://doi.org/10.11648/J.Jps.20221002.13

Yamamoto K, Nohata J, Kadono-Okuda K et al (2008) A bac-based integrated linkage map of the silkworm *Bombyx mori*. Genome Biol 9:R21. https://doi.org/10.1186/Gb-2008-9-1-R21

Yanagisawa S, Zhu Z, Kobayashi I et al (2007) Improving cell-adhesive properties of recombinant *Bombyx mori* silk by incorporation of collagen or fibronectin derived peptides produced by transgenic silkworms. Biomacromolecules 8:3487–3492. https://doi.org/10.1021/Bm700646f
Yang C, Yokoi K, Yamamoto K, Jouraku A (2021) An update of kaikobase, the silkworm genome database. Database 2021. https://doi.org/10.1093/Database/Baaa099
Yi Y, Wu G (2024) RNA-seq analysis of LPS-induced immune priming in silkworms (*Bombyx mori*) and the role of cytochrome p450 detoxification system in the process. Int J Biol Macromol 283:136551. https://doi.org/10.1016/J.Ijbiomac.2024.136551
Yi H-Y, Chowdhury M, Huang Y-D, Yu X-Q (2014) Insect antimicrobial peptides and their applications. Appl Microbiol Biotechnol 98:5807–5822. https://doi.org/10.1007/S00253-014-5792-6
Zhan S, Huang J, Guo Q et al (2009) An integrated genetic linkage map for silkworms with three parental combinations and its application to the mapping of single genes and qtl. BMC Genomics 10:389. https://doi.org/10.1186/1471-2164-10-389
Zhang J-Y, Pan M-H, Sun Z-Y et al (2010) The genomic underpinnings of apoptosis in the silkworm, *Bombyx mori*. BMC Genomics 11:611. https://doi.org/10.1186/1471-2164-11-611
Zhang GX, Ju HM, Zhang FX, Jing LX (2012) Study on increasing strength of silkworm silk. Adv Mat Res 472–475:614–617. https://doi.org/10.4028/Www.Scientific.Net/Amr.472-475.614
Zhang R, Cao Y-Y, Du J et al (2021a) Transcriptome analysis reveals the gene expression changes in the silkworm (*Bombyx mori*) in response to hydrogen sulfide exposure. Insects 12:1110. https://doi.org/10.3390/Insects12121110
Zhang Z, Zhao Z, Lin S et al (2021b) Identification of long noncoding rnas in silkworm larvae infected with *Bombyx mori* cypovirus. Arch Insect Biochem Physiol 106:1–12. https://doi.org/10.1002/Arch.21777
Zhang J, Zhang X, Zhang H et al (2023) Overexpression of BmJHBPD2 repressed silk synthesis by inhibiting the JH/KR-H1 signaling pathway in *Bombyx mori*. Int J Mol Sci 24:12650. https://doi.org/10.3390/Ijms241612650
Zhang R, Zheng S, Yu B et al (2026) Developing pébrine-resistant silkworms through targeting the transmembrane protein NbTMP1 in *Nosema Bombycis*. J Invertebr Pathol 214:108438. https://doi.org/10.1016/j.jip.2025.108438
Zhao Q, Han M-J, Sun W, Zhang Z (2014) Copy number variations among silkworms. BMC Genomics 15:251. https://doi.org/10.1186/1471-2164-15-251
Zhao Q, Zhu Z, Kasahara M et al (2013) Segmental duplications in the silkworm genome. BMC Genomics 14:521. https://doi.org/10.1186/1471-2164-14-521
Zhao L, Sun X, Wang X et al (2022) *Bombyx mori* VPS13D is a key gene affecting silk yield. Plos One 17:E0270840. https://doi.org/10.1371/Journal.Pone.0270840
Zhou Q, Fang S, Zhang Q et al (2018) Identification and comparison of long non-coding rnas in the silk gland between domestic and wild silkworms. Insect Sci 25:604–616. https://doi.org/10.1111/1744-7917.12443
Zhou Q-Z, Fu P, Li S-S et al (2020) A comparison of co-expression networks in silk gland reveals the causes of silk yield increase during silkworm domestication. Front Genet 11: 225. https://doi.org/10.3389/Fgene.2020.00225
Zhu K, Chen Y, Chen L, Xiang H (2022a) Comparative silk transcriptomics illuminates distinctive impact of artificial selection in silkworm modern breeding. Insects 13:1163. https://doi.org/10.3390/Insects13121163
Zhu Z, Hu Q, Tong C et al (2021a) Transcriptomic analysis reveals the regulation network of bmkrüppel homolog 1 in the oocyte development of *Bombyx mori*. Insect Sci 28:47–62. https://doi.org/10.1111/1744-7917.12747
Zhu Z, Tong C, Qiu B et al (2021b) 20e-mediated regulation of BmKR-H1 by BmKRP promotes oocyte maturation. BMC Biol 19:39. https://doi.org/10.1186/S12915-021-00952-2
Zhu M, Pan J, Tong X et al (2022b) BmCPV-derived circular DNA vcdna-s7 mediated by *Bombyx mori* reverse transcriptase (rt) regulates bmcpv infection. Front Immunol 13:861007. https://doi.org/10.3389/Fimmu.2022.861007

Ziemienowicz A (2014) Agrobacterium-mediated plant transformation: factors, applications and recent advances. Biocatal Agric Biotechnol 3:95–102. https://doi.org/10.1016/J.Bcab.2013.10.004

Zou Y, Ding X, Zhang L et al (2021) BmMP influences wing shape by regulating anterior-posterior and proximal-distal axis development. Insect Sci 29(6):1569–1582. https://doi.org/10.1111/1744-7917.12998

# Chapter 4
# Mulberry Silk Sericulture: Historical Perspectives and Modern Practices

**O. S. K. Paavana, S. P. Sruthi, and M. S. A. Muthukumar Nadar**

## 4.1 Introduction

Sericulture, the ancient art of raising silkworms for silk production, has been practiced for thousands of years (Thrilekha et al. 2024). The word "silk" comes from the Chinese word "Su (Si)" (Taufique and Hoque 2021). Silk fibers have been utilized in the textile industry for clothing purposes since ancient times, extracted from silkworms (Girijappa et al. 2019). Silk is a textile that is yarn made out of silk filaments with fine, soft, and silky properties. Silk was a fine-class product used exclusively by royals in the past, but it later became accessible to various social classes. Silk is considered an upper-class product due to its high price range (Le et al. 2025).

### *4.1.1 Historical Origins of Mulberry Silk Sericulture*

#### 4.1.1.1 The Legend of Lady Hsi-Ling-Shih

The history of sericulture (silk production) is associated with legend and mythology (Pugina and Caiut 2021). According to Chinese customs, including Confucian scriptures, Empress Leizu, wife of Huang Di (or Huang-Ti), the "Yellow Emperor," who is believed to have ruled China from 2697 to 2597 BC, before the Xia dynasty, is said to have had this story (Fig. 4.1). Legend says that around 3000 B.C., the cocoon of a silkworm fell into her teacup. The 14-year-old Leizu began unraveling the thread as she was tearing off the cocoon (Butler 1888). Noticing the long fibers, she gathered more cocoons and wove them into cloth. Her husband, the Yellow Emperor, guided her in studying the life cycle of silkworms and taught her attendants how to spin

O. S. K. Paavana · S. P. Sruthi · M. S. A. M. Nadar (✉)
Division of Biotechnology, Karunya Institute of Technology and Sciences, Coimbatore, India
e-mail: research.muthukumar74@gmail.com



D. Kumar and M. Shahid (eds.), *Science and Technology of Silk*, Textile Science and Clothing Technology, https://doi.org/10.1007/978-3-032-16230-4_4

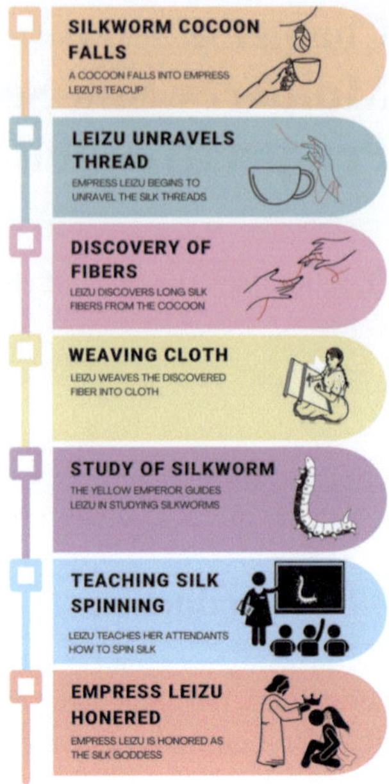


**Fig. 4.1** Historical origin of sericulture (Butler 1888)

silk. Hence, Empress Leizu was revered as the goddess of silk throughout Chinese history (Butler 1888). According to legend, silk production traveled from China to Khotan with a Chinese princess, the heir to the throne, who was probably the wife of the prince of Khotan, possibly in the early first century AD. The princess, who would not be deprived of her beloved silk fabric, defied the imperial order against the exportation of silkworms (Pugina and Caiut 2021; Butler 1888; Kirby 1888). Despite the Chinese exporting silk in copious quantities, they were very secretive about sericulture. Consequently, there were variations of myths invented by other cultures to explain the origins of silk. The ancients, particularly the Romans, who greatly admired silk and were aware that it came from trees, believed it grew out of tree leaves (Seneca the Elder and Virgil both believed this) (Robert 2007). Pliny the Elder was right about the source, though, writing in his "Natural History" that the *Bombyx* or silk moth "weaves webs, like spiders, that become a luxurious clothing material for women, called silk" (Pliny et al. 1855).

### *4.1.2 Archaeological Evidence of Early Sericulture*

Although mythological stories color the history of silk production, archaeological evidence of its history is much stronger. Recent findings have extended the timeline for silk.

A study conducted by the Department of History of Science and Scientific Archaeology at the University of Science and Technology of China, Hefei, China, utilized soil proteomics methods to identify the invisible products of buried silk degradation in tomb soils at the Jiahu archaeological site in Henan Province. Two samples were detected with silk protein special peptides (GAGAGAGY, which is part of the crystalline region of the fibroin), which can be regarded as new and reliable evidence of the earliest silk fiber in human history. Prehistoric biomolecular evidence of silk in the early Neolithic Age, discovered in the tombs of Jiahu, indicates that silk was used more than 8500 years ago, although such burial objects were not ubiquitous (Gong et al. 2016). Another research study utilized woven silk fabric, dating back to 3630 BC, to wrap the body of a child. The textile originates from a Yangshao site at Qingtaicun in Rongyang, Henan (Vainker 2004). Another exploration at the Shuanghuaishu site revealed a bone-carved silkworm figurine measuring 6.4 cm in length and <1 cm in width, indicating it is China's oldest known silkworm carving. Its shape and posture suggest that it depicts the silk-spinning stage, indicating that ancient people were not only familiar with but also capable of large-scale domestication of silkworms. In the core area of the site, a barbican-type structural enclosure was constructed, and simultaneously, the earliest bone silkworm carving artwork in China was discovered on the northern wall of the shelter (Chen et al. 2022).

Textiles can be effectively fossilized through mineralization, when the fibers of a textile are partially or completely replaced by minerals, and are commonly found alongside metal artifacts during excavation (Jia et al. 2024). In another study, textile fragments on a ring-shaped iron sword were discovered from the Maoling Mausoleum in Shanxi province, China, which have preserved the fiber shapes intact, but the textile surface is mineralized. For the textile, the morphology of the fiber and the mineral element information were observed and analyzed with a three-dimensional microscope, scanning electron microscope–energy disperse spectroscopy (SEM–EDS), and Raman spectroscopy, respectively (Zheng et al. 2020). Two immunology testing techniques have been used to study whether the textile fragments contained the silk protein. The liquid chromatography–tandem mass spectrometry analysis showed that the hilt string imprint (sample A) was a twill braiding with cinnabar remaining on the surface, while the blade textile imprint (sample B) was a tabby fabric, which was mineralized by iron ions to form a cavity structure. Sample B exhibited a silk-like morphology, characterized by a surface similar to that of mulberry silk and a triangular cross-section (Zheng et al. 2020). Meanwhile, detection of silk protein in several milligrams of sample B was achieved by enzyme-linked immunosorbent assay (ELISA) and immunofluorescence method (IFM) using fibroin polyclonal antibody. Therefore, it can be understood that the iron sword was previously covered or wrapped in silk. This study provides technical support for identifying silk residues in mineralized textiles, which is of great significance for the study of early organic fabrics (Zheng et al. 2020).

### 4.1.3 Development of Sericulture Techniques

Since the fresh cocoons have a limited life, the reeling process must be performed within seven or eight days of coming down from the cluster. If the pupa in the cocoon has developed into a moth, reeling cannot be performed. Therefore, the reeling production process should be conducted directly in the silkworm breeding area (Ju 2022). Due to the specialized nature of cocoon and silk production, cocoons are not suitable for transportation to independent manual systems for centralized processing (Ju 2022). Silk reeling and sericulture were typically inseparable in medieval China. Consequently, farmers who cultivate mulberries and rear silkworms, as well as artisans who extract silk from the cocoons, are collectively referred to as sericulturists, or practitioners of sericulture (Taufique and Hoque 2021; Ju 2022).

### 4.1.4 Spread of Sericulture

It's impossible to pinpoint the exact date when sericulture started (Takeda 2009). According to most historical sources, silkworms were bred from the earliest times by people in East Asia, primarily in China. The secret of sericulture was hidden by the Chinese emperors for several centuries (Butler 1888). Those who violated this secret or took silkworm eggs out of China were punishable by death (Czaplicki et al. 2021).

Sericulture originated in China and spread along the Silk Road (Takeda 2009). The Silk Road was a vast trade network that spanned over 12,000 km (nearly 7500 miles) (Mishra 2020). For centuries, the roads that linked China to the Middle East and Europe, now known as the Silk Road, were a long-standing trade route of immense economic significance. Merchants took gold, precious stones, perfumes, and arable crops to China using this route (Blaydes and Paik 2021). Silk held significant importance as a trade commodity along the Silk Road, where it was bartered for its equivalent weight in gold (Takeda 2009). Silk fabrics made their way to the Mediterranean for the first time in the second century B.C. (Czaplicki et al. 2021).

In the East, sericulture was introduced into Korea and subsequently to Japan around the third century B.C. (Takeda 2009). By the first half of the nineteenth century, Japan's production was far greater than China's (Ma 2005). For many years, Japan had a tradition of labor-focused technology. Japan surged ahead because it invested in nationwide public infrastructure and institutions, whereas China made it more difficult for businesses to grow (Ma 2004).

The Chinese silk monopoly lasted for several centuries until the Byzantine Emperor Justinian (482–565) was informed that silk thread came from caterpillars raised in China. He then sent a team of spies, disguised as monks, to learn about silk production (Yilmaz et al. 2015). Upon learning of the hatching, the monks carried an egg from the mulberry tree to the sticks among the pilgrims from China. This process of production took the silk industry to the Byzantine Empire (Yilmaz et al. 2015;

Weber 2008). Sericulture reached Europe in 550 A.D. (Takeda 2009). Silkworm cultivation commenced in Turkey in 552 A.D. on the shores of Anatolia, the Asian region of the country. Bursa, Turkey, became renowned for its silk trade (Yilmaz et al. 2015, 2016).

In the Western region, the spread occurred to Central Asia and India during the fourth century and subsequently to Persia and then to Rome (Takeda 2009; Islam 2024). India is the second-largest producer of silk in the world, next to China, where sericulture has become a tradition and a thriving culture (Bharathi 2016). In the past, there have been unsuccessful attempts to establish India as the origin of sericulture. They have endeavored to refute these assertions in light of the empirical evidence available in the literature. Hutton made a claim as early as 1864 that all the domesticated forms of the mulberry-feeding insects originated in China (Alam 1999). In Roman society, some of the most valuable silk products were exclusively owned by the Persians until the sixth century A.D. (Islam 2024).

Sericulture gradually spread to Italy by the end of the ninth century. Later, sericulture reached France in 1340 (Islam 2024). Silkworm farming started in America in the year 1522 when Hernán Cortés (the person credited with introducing silkworms to Mexico and thus, indirectly, to America) brought mulberry trees and an egg of a silkworm to Mexico. In North America, silkworm breeding, in 1609 (Czaplicki et al. 2021). King James I of England dispatched mulberry seedlings and a grain of silkworm to Virginia to get breeding going there. The reason it never took off was that silkworm farming was a labor-intensive process that offered less profit than farming cotton or tobacco. At the start of the nineteenth century, the northeastern parts of the USA were developing the production of silk fabric (Czaplicki et al. 2021). By 1870, Louis Pasteur had successfully prevented the disease known as Pebrine, thereby preventing the extinction of sericulture (Islam 2024). An indicative historical timeline of the spread of sericulture is shown in Fig. 4.2.

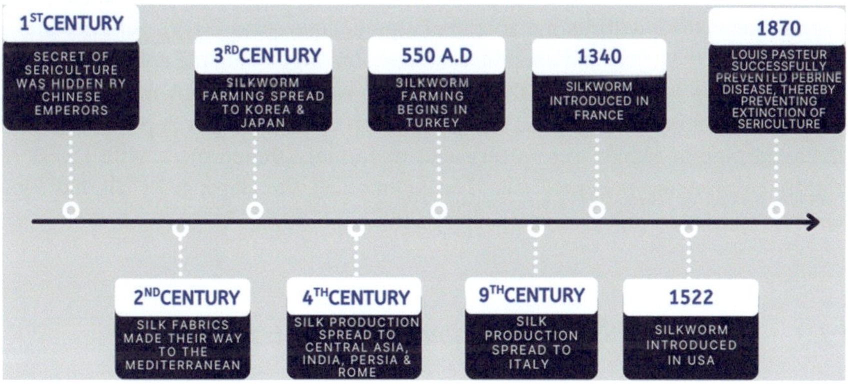


**Fig. 4.2** Spread of sericulture

## 4.2 Cultural Significance of Silk in Ancient Civilizations

### 4.2.1 The Role of Silk in Traditions and Festivals

Silk has been an essential component of cultural traditions and festivals throughout civilizations for millennia. Silk, a carrier of cultural influences, was further integrated into transactions between diverse cultures over a span of three millennia due to its role as a currency (Schoeser 2007). Silk holds a sacred place in India's cultural heritage, deeply intertwined with its lives and culture, a distinction that no other country can claim (Savithri et al. 2013). Silk bridal garments hold significant importance in Hindu weddings, and their presence is also documented in historical records dating back to medieval times (Schoeser 2007). In China, silk clothing, especially in the form of elaborate qipaos and brocaded robes, has long been worn on other major celebrations, including the Lunar New Year and weddings, as an emblem of abundance and good fortune (Garrett 2020). More decorative covers made of red silk, padded, and embroidered with auspicious symbols, some with fur lining and hoods, were given as gifts at the Lunar New Year (Garrett 2020). For 2000 years, in Japan, people have worn kimonos. The kimono is a floor-length silk robe embroidered with intricate designs (Kalman 2000).

### 4.2.2 Symbolism of Silk in Art and Literature

It shouldn't be surprising, then, that silk is a common theme in art and literature that explores beauty, fragility, and power. Silk is a frequent metaphor in Chinese poetry and painting, suggesting elegance and scholarly refinement. Classical poets such as Li Bai mentioned silk and the Silk Road to evoke imagery of delicacy and longing, resonating with its metaphorical significance (Han 2019). Silk became a vital fabric for theatrical and operatic costumes, distinguishing the nobility from the commoners and lending it visual and symbolic weight in the performance arts. A geisha is a professional Japanese artist and entertainer. Young or apprentice geisha are called maiko. Maiko were a furisode, or formal kimono made with layers of brightly colored and patterned silk. This symbolizes the elegance of silk has long been represented in Japanese art (Kalman 2000).

### 4.2.3 The Use of Silk in Religious Ceremonies

One of the reasons silk is considered a sacred fabric in various religious traditions is its perceived symbolism of purity, devotion, and divine connection. Silk is associated with the deities, as silk garments are offered to them (Young 2013). The ruling class and folk silkworm farmers had become many vivid silkworm deities in medieval

China for praying for a good harvest and the success of sericulture activities. The silkworm deity was believed to be both the inventor of sericulture and the protector of silkworm affairs. Between different classes and regions, the deity worship of silkworm can achieve integration and maintenance of sericulture organizations through the belief in the ancestor God and the offering ceremony (Ju 2022). The Empress performed state rituals before feeding imperial silkworms (Young 2013). The silkworm deity is worshipped by sericulture practitioners, while the Dyeing-vat-God is worshipped by people in the printing and dyeing trade, among others. Silk weavers worship the Huang-Dao-Po and so on (Ju 2022). Even though the evolution of the silkworm deity in medieval China reflected a religious consciousness with a polytheistic tendency, it was promoted by Taoism, Buddhism, and Confucianism, which had been continuously integrated into medieval Chinese society and folk religion (Ju 2022). Silk fabrics, crafted into liturgical vestments and altar textiles, were used in Christian religious ceremonies in Norman Sicily and its Ifrīqiyan-connected churches. Their use signified the sacredness of Christian worship and the wealth/power of the church community (Dolezalek 2013).

## 4.3 The Silk Road: Trade Routes and Economic Impact

### *4.3.1 Key Trade Cities Along the Silk Road*

The Silk Road was remarkable not just for trade, but also for the other cultural goods that spread along it, leading to the emergence of several significant cities:

**Samarkand**: Now, modern Uzbekistanis known today for its strategic location, and it remains a vibrant center of trade, culture, and education (Vichare 2024).

**Kashgar**: Located at the junction of the northern and southern Silk Roads, and Kashgar (in present-day China) served as a crucial trading hub, bringing together merchants from all corners of the world (Vichare 2024).

**Baghdad**: As the heart of the Abbasid Caliphate, Baghdad (now in Iraq) grew into a significant hub of international commerce and knowledge, linking the East to the West (Vichare 2024).

**Constantinople**: Istanbul, the ancient capital of the Eastern Roman Empire, remains a modern hub to this day, connecting Europe and Asia through sea routes and land traffic, with the two continents spied on from across the Bosphorus Strait (Vichare 2024).

### 4.3.2 The Exchange of Goods Beyond Silk

Silk was a main commodity; however, the Silk Road enabled the trading of a diverse range of goods:

**Spices and Herbs**: Spices, valued for their culinary and medicinal properties, such as pepper, cinnamon, and curative herbs, were traded extensively (Haksöz and Uşar 2011).

**Precious Metals and Stones**: Gold, silver, and precious stones, such as rubies and sapphires, were transported between continents and enriched many civilizations (Blaydes and Paik 2021).

**Other Textiles**: Along with silk, other fabrics, such as wool and cotton, were traded, showcasing the diverse weaving traditions of various cultures (Haksöz and Uşar 2011).

**Cultural Exchange**: Paper-making techniques that crossed the Silk Road assisted in the development of written letters and double-stick tapes. The compass replaced the sun, allowing for navigation by both the sky and the land. While the ground is making gears, the sky is making stars (Haksöz and Uşar 2011).

### 4.3.3 Cultural Exchanges Facilitated by the Silk Road

The Silk Road also played a critical role in enabling cultural exchange:

**Spread of Religions**: This movement of people facilitated the spread of religions, as Buddhism traveled from India into Central Asia, China, and Japan. Both Islam and Christianity also utilized these paths to expand their reach, creating diverse religious landscapes (Khezri and Zandieh 2024).

**Cultural Exchange**: Different cultures were also heavily influenced by each other through trade and conquest (Khezri and Zandieh 2024).

**Astrology**: Ideas of the heavens and later medicine were learned from each other (Khezri and Zandieh 2024).

### 4.3.4 Decline of the Silk Road and Its Causes

Several factors contributed to the fall of the Silk Road:

**Trade Route Fragmentation**: Political fragmentation along the trade route significantly decreased overnight trade for merchants (Blaydes and Paik 2021).

**Sea Routes**: As more efficient routes developed across the ocean, merchants increasingly turned to water transport as a safer and quicker means to move their goods, thereby decreasing their reliance on overland paths (Blaydes and Paik 2021).

**Fragmentation of Security**: Powerful empires that had once threatened or supported trade along the Silk Road became weaker, resulting in less secure and less protected trade routes (Blaydes and Paik 2021).

To summarize, the Silk Road was much more than a network of trade routes; it laid the framework for economic vibrancy, cultural exchange, and technological progress. Its legacy still shapes the interconnectedness that is felt in our modern world.

## 4.4 Techniques of Traditional Sericulture

### *4.4.1 Methods of Silkworm Rearing and Care*

The biological cycle of the silkworm essentially includes the metamorphosis of egg to larva, which is commonly referred to as "Silkworm," a stage where the larva is reared and is actually enclosed in a cocoon. Silkworms, or *Bombyx mori*, are used almost exclusively for the production of silk, so they undergo four distinct growth stages: egg, larva, pupa, and adult moth (Buhroo & Zafar 2019).

**Environmental Conditions**: Additionally, silkworm rearing is significantly influenced by temperature and humidity. In order for the growth of silkworms to be perfect, an ideal temperature of 24–27 °C and a relative humidity of 75–80% are needed. Deviations from these conditions can result in reduced growth or higher rates of larval mortality (Rahmathulla 2012).

**Feeding and Nutrition**: The primary food for silkworms is mulberry leaves (*Morus alba*). Leaf nutritional composition affects both silk yield and quality. To feed the larvae, mulberry leaves harvested at the right maturity ensure maximum silk gland development (Alipanah et al. 2020).

**Disease Prevention and Hygiene**: The rightful principle to be adopted in disease control is "Prevention is better than cure." Silkworms can also suffer from viral, bacterial, and fungal diseases, including grasserie, flacherie, and muscardine. Hukuhara (1987) mentioned that the different stages of silkworm diseases can be characterized by enzootic and epizootic states. There have been ingenious methods for silkworm crop protection, but prevention technology is the primary one. Consequently, considerable focus has been placed on improving disinfectants and their formulations (Gupta et al. 2016). To that end, a room disinfectant, a 5% bleaching powder solution, and a dust formulation of silkworms whose bed is disinfected (Labex—a commercial name) have been developed to inactivate all kinds of pathogens (Gupta et al. 2016; Frobisher and Crabtree 1974). In three epidemiological studies from India, this infectant has shown to be unique, effective, and economic aspects that reduced the disease mortality from 35 to 14% on average in six years during 1986–1991 in West Bengal, but little costlier than bleaching powder solution is the room disinfectant, Chlorine dioxide (Gupta et al. 2016). These practices would reduce the prevalence of diseases, such as regular sanitation of rearing trays, disinfection of the rearing houses with lime or formalin, and proper aeration (Trivedy et al. 2011).

### *4.4.2 Harvesting and Processing Raw Silk*

**Cocoon Collection and Sorting**: Silkworms spin cocoons made of fibroin protein secreted from their silk glands after they complete their larval stage. To prevent moths from causing silk fibers to break, cocoons are harvested within 7–10 days of spinning. Silk cocoons are sorted by size, shape, and color to ensure uniform processing (Babu 2012; Murugesh-Babu 2019).

**Stifling and Reeling**: Cocoons are treated with steam or a hot bath in a process called stifling to kill the pupae, so the silk fiber won't break. Once stifled, the filaments get unwound in a process called reeling, in which raw silk threads are removed and twisted into yarn (Babu 2012; Murugesh-Babu 2019).

**Degumming and Weaving**: Degumming—this process removes sericin, a protective protein that coats fibers of fibroin. This is used to increase silk's shine and smoothness. After degumming, silk yarns are woven into fabrics on traditional or mechanized looms (Babu 2012; Murugesh-Babu 2019).

### *4.4.3 Regional Variations in Sericulture Techniques*

Different techniques of sericulture, which have been influenced by climate, resources, and culture, are adopted in various regions.

**Chinese Sericulture**: As the top silk-producing country, China has incorporated effective technologies into its sericulture. The artificial diets not only enabled the year-round breeding and rearing of silkworms but also played a crucial role in promoting uniform quality and increased productivity (Pan et al. 2024).

**Uzbekistan Sericulture**: In Uzbekistan, the primary challenge in sericulture is maintaining a stable hygrothermal environment (i.e., a balance of heat and moisture). To address this, farmers employ innovative technologies, such as special devices equipped with fabric and polyethylene film, to retain humidity. This approach has proven to be an effective practice (Ubaydullaevich and Sodikugli 2021).

**Indian Sericulture**: Indian Sericulture is practiced in various places where silk is collected in different forms, such as Muga, Eri, and Tasar. These are widespread traditions in these territories, and they have adapted to local climate conditions (Sharma et al. 2020a).

**Japanese Sericulture**: The practice of sericulture in China originated in 2600 B.C., and the first introduction of the silkworm eggs to Japan occurred in 195 A.D. (Singh and Bhargava 2023). Japanese sericulture has seen numerous advancements, such as the "*Seiryo-iku*" rearing method, which utilizes heating and ventilation to enhance silkworm rearing (Cheng and Ono 2020). Silkworm contributed to Japan's modernization. Japan has been the global leader in silkworm bioresource storage. Japan, known for its labor-focused technology, surged ahead by investing in public infrastructure and institutions (Ma 2004).

### *4.4.4 Role of Family Units in Traditional Sericulture*

Sericulture is a rural-based industry that is labor-intensive and provides employment to both skilled and unskilled laborers (Bukhari et al. 2019). Sericulture is an ancillary occupation; yet with the help of technology, its cultivation can be turned into a high-intensity, even high-yielding activity. Sericulture may be called the cottage industry, as family members of all ages and sexes can participate in it. Sericulture has great potential to increase family income. Sericulture has long been a family-based industry, particularly in the rural areas of Asia. Silk-rearing touched families in every aspect, from mulberry cultivation to the collection of cocoons. We know this type of men-and-women-labor arrangement from the past, where men ran the farm and its marketing and the women fed silkworms, spun, and weaved. In numerous places, silk-making methods have been passed down through the generations to preserve the ancient art of sericulture (Kasi 2013).

### *4.4.5 Training Programs for Traditional Sericulturists*

Extension teaching is primarily a form of informal education (Alabi 2024). The labor-intensive, non-skilled cocoon production and reeling operations of sericulture tend to keep sericulturists subordinated, so that they are alienated from knowledge and skill dissemination processes and left with only underpaid traditional work. There is a requirement to enhance the skills and knowledge of sericulturists through training on various activities to sustain the annual cocoon crop income (Attri et al. 2024).

A study Capacity Building and Training (CBT) studies on sericulture farmers in the Rajouri district of Jammu and Kashmir. The data on registered farmers was collected from the Department of Sericulture, Government of Jammu and Kashmir. The contraceptive effect of CBT training on the silkworm was very obvious. They also emphasize the importance of training in new technology, integrated pest management, and disease management for sericulture production (Bukhari et al. 2024).

A study was conducted in the Kolar district of Karnataka to determine the knowledge and adoption of mulberry sericulture technologies among farmers. Primary data were collected from a selected sample of 80 sericulturists trained using a structured schedule through personal interviews before and after training. Motivation and carrying out extension programs like group discussion, result demonstration could enhance the adoption of technology. Study tours, refresher training programs, and confidence-building steps would result in greater technology adoption, which would have a positive ripple effect in the study area (Srinivasa et al. 2013).

To increase productivity and preserve traditional sericulture, various training programs taught by government and non-government organizations introduce modern techniques, disease management, and market integration. A few programs include the training modules offered by the Central Silk Board of India and the Sericulture Research Institutes in China, which provide hands-on training in hybrid silkworm breeding and silk reeling technology. At the international level, organizations such as the International Sericulture Commission and close-knit local cooperative networks conduct workshops to impart knowledge of sustainable sericulture practices and organic silk production, and provide access to fair-trade markets. Sericulture education has now become accessible to a broader audience through online training modules. Online courses on disease management, sustainable cultivation, and silk entrepreneurship are available on various digital platforms.

## 4.5 Geographic Distribution of Sericulture Practices

### *4.5.1 Major Global Producers of Mulberry Silk*

Mulberry silk is mainly produced in Asia, with China producing the most (India is the next largest producer) (Bharathi 2016). India is not only the second-largest producer of silk in the world but also the largest consumer of the product, with production originating from states such as Karnataka, Andhra Pradesh, Tamil Nadu, Jammu and Kashmir, and West Bengal (Kumar et al. 2019a). Other major producers include Thailand, Japan, Vietnam, and North Korea, but their output is significantly less than that of China and India (Bharathi 2016). The top mulberry silk-producing nations are listed in Table 4.1.

Global raw silk production has fluctuated year on year. It declined by 30% over that span compared to 2011, falling to 91,221 metric tons in 2022. Production then peaked in 2015 at 202,073 metric tons, followed by a decrease. China, India, Uzbekistan, Vietnam, Thailand, and Brazil are the top silk-producing countries. China and India collectively account for 94% of the global silk production (Popescu et al. 2024).

**Table 4.1** Top-producing mulberry silk nations by country (Bharathi 2016; Popescu et al. 2024)

| Country | Production status |
|---|---|
| China | Largest producer |
| India | Second-largest producer and largest consumer |
| Thailand | Significant producer in Southeast Asia |
| Japan | Known for high-quality silk production |
| Vietnam | Growing producer in Southeast Asia |
| North Korea | Minor but notable producer |

### *4.5.2 Climate Requirements for Successful Sericulture*

Sericulture, the rearing of silk, requires specific climate conditions to facilitate the growth of the worm and maximize yield (Table 4.2).

**Optimal Temperature Range**: The ideal temperature for silkworm larvae is 24–28 °C (Rahmathulla 2012). The host plant, mulberry, on which the silkworm feeds, has an ideal temperature range of 24–28 °C, although a range of 18–24 °C can be tolerated.

**Temperature Sensitivity**: High temperatures increase the development of the silkworm, promote premature larval emergence, and affect cocoon quality. Development can be slowed by low temperatures (Sarkar 2018; Hariharan et al. 2024).

**Relative Humidity**: Relative humidity of 75–90% is the best condition for the rearing of silkworm, especially in the initial stages (Rahmathulla 2012).

**Mulberry Plant Watering**: Watering is an essential aspect of growing a mulberry plant. Climate change can have direct implications on mulberry supply, making it harder for farmers to produce their crop, meaning less food for silkworms to feed on (Hariharan et al. 2024).

**Table 4.2** Environmental factors for various stages of silkworm (Rahmathulla 2012)

| Environmental factors | Temperature (°C) | Humidity (%) |
|---|---|---|
| Incubation | 25 | 75–80 |
| I instar | 28 | 85–90 |
| II instar | 27 | 85 |
| III instar | 26 | 80 |
| IV instar | 25 | 70–75 |
| V instar | 24 | 65–70 |
| Spinning | 25 | 70 |
| Cocoon preservation | 25 | 80 |

**Extreme Weather Events**: Silkworms and mulberry plants are affected by extreme weather conditions, including storms and heatwaves. As these events occur, protective measures to save against them are required (Hariharan et al. 2024).

**Adaptation Strategies**: Several adaptation strategies can be adopted to sustain sericulture under altered climate conditions (Sarkar 2018; Hariharan et al. 2024).

- **Climate-Tolerant Varieties**: Creating silkworm and mulberry varieties to be tolerant to heat, pests, and diseases.
- **Temperature Control**: Ensure temperature control measures are in place in rearing facilities.
- **Pest and Disease Management**: Biological, cultural, and chemical control.
- **Ease of Mulberry Cultivation**: Intensifying the mulberry-growing method to control water and food resources efficiently.

### *4.5.3 Impact of Geography on Mulberry Cultivation*

The cultivation of mulberry is largely dependent on geography, which not only influences the conditions under which mulberry grows but also its bioactive properties.

**Climate and Temperature**: Mulberry flourishes across tropical to temperate zones (ideal, 24–28 °C) and tolerates a wide range of temperatures, growing up to altitudes of 1000 m (Mulberry cultivation and utilization in India. https://www.fao.org/4/x9895e/x9895e04.htm). In regions with less temperature disparity, high levels of bioactive compounds (rutin and morusin) are increased, improving antiapoptotic potential. High-temperature changes decrease valuable components, such as chlorogenic acid and resveratrol (Bai et al. 2023).

**Rainfall and Irrigation**: The annual rainwater requirement varies between 600 and 2500 mm, particularly in low-rainfall areas, where irrigation is important (Mulberry cultivation and utilization in India. https://www.fao.org/4/x9895e/x9895e04.htm). Bioactive components are affected by precipitation patterns, such as more total annual precipitation raises quercitrin and scopoletin, enhancing the antioxidative capacity (Bai et al. 2023). Crops in areas like Karnataka and Tamil Nadu (India) depend on irrigation in dry weather to sustain yields (Mulberry cultivation and utilization in India. https://www.fao.org/4/x9895e/x9895e04.htm).

**Altitude**: Elevation has a direct impact on phytochemical profiles. Higher altitudes are correlated with higher levels of chlorogenic acid, cryptochlorogenic acid, and resveratrol. Higher-altitude areas where S9 (Zhejiang) occurred reflected a comparatively richer bioactive content due to the larger temperature swings observed in low-altitude areas (Bai et al. 2023).

**Soil and Sunlight**: Loamy or clayey soils with good drainage are preferred by mulberry, which requires 340 $m^3$/ha of water every 10–15 days, depending on the type of soil. Sunlight (9–13 h/day) contributes to the quality and growth of leaves, especially in tropical regions (Mulberry cultivation and utilization in India. https://www.fao.org/4/x9895e/x9895e04.htm).

**Regional Adaptability**: In India, most cultivation occurs in tropical states like Karnataka, which has a favorable climate and adequate irrigation infrastructure. Nonetheless, geographical dispersion (e.g., temperature gradients, rainfall distribution) adds variability to the medicinal and economic functions of mulberry, defining region-specific crops, such as in high-precipitation zones, where mulberry serves as an antioxidant-rich crop (Bai et al. 2023).

Such factors underscore the importance of site-specific cultivation strategies to enhance yield and bioactive potential, tailored to the local geography.

## *4.5.4 Comparative Analysis: Sericulture*

A country-wise comparative analysis of sericulture aspects is given in Table 4.3.

### 4.5.4.1 Sericultural Practices

**Silkworm Races**: Different climatic regions grow different silkworm races. The Chinese silkworm breed from Kashmir is famous for its high yield and adaptability to Pakistan's climate. Japan and Korea, by contrast, have a wide variety of silkworm strains collected for research and production (Bhatti et al. 2023).

**Mulberry Cultivation**: Breeding silkworms by rearing mulberry trees is essential. The productivity of sericulture is greatly influenced by the quality and quantity of

**Table 4.3** Comparative analysis of sericulture country-wise (Ma 2004; Bharathi 2016; Rahmathulla 2012, Popescu et al. 2024)

| Aspect | China | India | Japan/Korea | Pakistan |
|---|---|---|---|---|
| Silkworm races | Diverse strains maintained for research and production | Focus on bivoltine and polyvoltine strains | Extensive germplasm resources | Chinese race preferred for local conditions |
| Economic impact | Largest producer with significant exports | Major producer with substantial rural employment | Smaller scale but high- quality silk production | Declining industry due to challenges |
| Environmental considerations | Advanced practices but high resource use | Mulberry cultivation impacts biodiversity | Emphasis on sustainable practices | Environmental concerns due to intensive farming |

these leaves. Seasonal variations and the replacement of mulberry trees with other crops hinder regions such as India and Pakistan (Bai et al. 2023; Bhatti et al. 2023).

**Raising Methods**: More advanced raising methods, such as controlled environments and specialized feeding practices, are more commonly used in countries like China and Japan. These techniques improve the quality and yield of silk (Bhatti et al. 2023).

#### 4.5.4.2 Economic Impacts

**Employment and Income**: Sericulture offers substantial employment opportunities in rural areas, thereby contributing to local economies. It is an important source of income for the farming community in India (Kasi 2013).

**Production and Export**: India and China are the world leaders in silk production and export revenue. Japan and Korea also produce silk, but to a lesser degree, and do contribute to the global silk market (Bharathi 2016; Popescu et al. 2024).

**Challenges**: Despite its economic advantages, the benefits of sericulture face several challenges, including high investment costs, a lack of knowledge and exposure among farmers about sericulture farming, and stiff competition from synthetic fibers. For instance, due to these, the industry in Pakistan has declined (Bhatti et al. 2023).

#### 4.5.4.3 Environmental Considerations

**Mulberry Plantations**: Sericulture encourages the planting of mulberry trees, a beneficial approach to combating deforestation and promoting biodiversity. However, the intensive farming practices required for mulberry cultivation can have adverse consequences for soil health and water resources (Rohela et al. 2020).

**Resource Efficiency**: Growing silk can also be an efficient use of resources, as the process utilizes land that may be underutilized. However, the industry's dependence on water and pesticides has significant environmental implications (Attri et al. 2024).

**Sustainability**: A growing number of sustainable practices are emerging in sericulture, including organic farming and the development of disease-resistant silkworm varieties. Such initiatives aim to reduce the environmental footprint of the sector while enhancing productivity (Altman and Farrell 2022).

## 4.6 Disease Management in Silkworm Cultivation

The oldest accounts of silk production originate from China. While no scientific documentation exists about the disease during that era, conventional practice indicated awareness of the disease, as management was undertaken to improve the cocoon yield.

The great explorer Alexander von Humboldt (1769–1859) once said: "*The frequent sequence of reactions to an important discovery is first a denial of its veracity, then a denigration of its importance, and finally usurpation of credit for it.*" This statement aligns entirely with Pasteur's contributions to the fight against silkworm disease (Cavaillon and Legout 2022; Raichvarg 2022). Louis Pasteur's role in combating silkworm disease exemplifies an observation from Alexander von Humboldt about the typical responses to significant discoveries: denial, denigration, and appropriation of credit. In 1865, Pasteur was asked by J.B. Dumas to help the silk industry due to an epidemic affecting silkworms. Pasteur, who had not previously turned a microscope toward entomology, took years to come to the conclusion that the disease, pebrine, was infectious: despite Emmanuel Bénèdict-à-l'Asso's evidence about the disease's infectious nature, and work by contemporaries such as A. Béchamp, who had argued for a parasitic origin. Pasteur only then conceded a parasitic cause, rejecting Béchamp's earlier work with contempt. Although Béchamp's treatment methods failed to achieve any notable success, Pasteur's segregation technique (which he built upon Emilio Cornalia's ideas) helped contain the epidemic and earned him widespread acclaim. However, the silkworm industry never really recovered (Cavaillon and Legout 2022).

### 4.6.1 *Common Diseases Affecting Silkworms*

Silkworms (*Bombyx mori*) are affected by multiple diseases that significantly threaten silk production (Tables 4.4 and 4.5). The principal diseases are Grasserie, Flacherie, Muscardine, and Pebrine (Tayal and Chauhan 2017). These diseases are distributed at different prevalences depending on the region, season, and environmental conditions.

India suffers an average silkworm crop loss of 15–47% due to various diseases, compared to 10–15% in countries such as Japan, China, and Italy. In India, Flacherie (33.88%), one of four diseases, is the leading cause of death. In North West India, the incidence of silkworm diseases is higher in the autumn season (35–47%) compared to the spring season (15–20%) (Tayal and Chauhan 2017).

#### 4.6.1.1 Grasserie

Grasserie disease has a strong prevalence, representing around 33% of the total diseases in the worms. This is a disease that is seen widely in all the silk rearing

**Table 4.4** Common diseases affecting silkworms

| Disease | Causative agent (few examples) | Type | Symptoms | Mode of transmission | Control measures | References |
|---|---|---|---|---|---|---|
| Pebrine | *Nosema bombycis* | Protozoan | Black spots on body, irregular molting, deformed pupae | Transovarial (from mother to egg), contaminated food | Mother moth examination, surface sterilization of eggs | Gu et al. (2024) |
| Grasserie | *Bombyx mori nucleopolyhe drovirus* (BmNPV) | Viral | Swollen, shiny body, skin rupture, white body fluid oozes out | Ingestion of infected leaves, contact with infected | Disinfection, early detection, feeding antiviral plant extracts | Reddy et al. (2024), Chandrasekharan et al. (2006) |
| Flacherie | *Streptococcus*, *Staphylococcus*, *Serratia marcescens* | Bacterial | Vomiting, foul-smelling diarrhea, soft body, quick death | Ingestion of contaminated food or water | Hygiene, feeding fresh mulberry leaves, lime treatment | Gadwala et al. (2021) |
| Muscardine | *Beauveria bassiana* | Fungal | Hard, mummified body covered in white fungal growth | Contact with fungal spores in air or rearing bed | Formalin spray, maintaining low humidity | Kodli and Anita (2025) |
| Aspergillosis | *Aspergillus flavus*, *A. niger* | Fungal | Greenish/blackish mold growth, lethargy, mortality | Spore contamination from environment | Dry and clean rearing rooms, antifungal treatment | Kumar et al. (2019b) |

**Table 4.5** Viral diseases of *B. mori* (Chopade et al. 2021; Sharma et al. 2020b)

| Disease | Causative virus | Virus type | Genome type | Replication site | Main symptoms | Transmission |
|---|---|---|---|---|---|---|
| Grasserie (nuclear polyhedrosis) | *B. mori Nucleopolyhedrovirus* (BmNPV) | Baculovirus | Double- stranded DNA | Nucleus | Swelling, oily shine, sluggishness, rupture with milkyooze, hanging posture | Oral (contaminated leaves), contact |
| Cytoplasmic polyhedrosis | *B. mori cytoplasmic polyhedrosis virus* (BmCPV) | Cypovirus (reoviridae) | Double- stranded RNA | Cytoplasm (midgut) | Pale, thin larvae, chalky white midgut, vomiting, sluggishness | Oral, fecal contaminati on |
| Infectious flacherie | *B. mori infectious flacherie virus* (BmIFV) | Picornavirus-like | Single-stranded RNA | Cytoplasm (goblet cells) | Flaccid body, vomiting, diarrhea, translucent thorax, foul odor after death | Oral, feces, environmental stress |
| Densonucleosis | *B. mori densonucleosis virus* (BmDNV-1 and BmDNV-2) | Parvovirus (bidnaviridae) | Single- stranded DNA (bipartite for BmDNV-2) | Nucleus | Shrinking, dull/ red skin, gut swelling, high mortality in early instars | Oral, feces, persistent in environment |

farms. Slugging movement has been seen in the larval stage. The skin of the larva becomes shiny prior to molting, and it is unable to molt, thus dying. The intersegments of the larva swell, the skin color of the larva becomes yellowish, the integument of the larva gets fragile, and it ruptures easily, exuding turbid milky white retrograde hemolymph. When it passes the anal region, the grasping power of the larva is reduced or completely lost. On occasion, the worm becomes restless and a rectal protrusion occurs (Chopade et al. 2021).

***Nuclear Polyhedrosis Virus (BmNPV)***: is a destructive disease caused by Nuclear Polyhedrosis Virus (BmNPV). It forms polyhedra within the nuclei of the host cells to produce symptoms such as larval swelling and a shiny, yellowish skin, which eventually leads to death. Transmission is primarily achieved through ingestion of infected mulberry leaves. BmNPV can be contagious and persistent in the environment (Chopade et al. 2021; Sharma et al. 2020b).

***Cytoplasmic Polyhedrosis Virus*** **(BmCPV)**: This virus first affects the midgut and brings about polyhedral in the cytoplasm. Lethargy, vomiting, chalky white midgut, and retarded growth are the common symptoms of this disease. It is transmitted orally through contaminated food; therefore, it affects digestion and metabolism (Chopade et al. 2021; Sharma et al. 2020b).

***Infectious Flacherie Virus*** **(BmIFV)**: This infection can also develop into flacherie disease. The virus is characterized by atony of the body, vomiting, diarrhea, and translucent skin. Frequently, coinfections also occur, exacerbating the outcome. This is an RNA virus, infecting the midgut (goblet cells), impairing digestive efficiency and enabling microbe overgrowth (Chopade et al. 2021; Sharma et al. 2020b).

***Densonucleosis Virus*** **(BmDNV-1 and BmDNV-2)**: The infection affects early instar, producing chronic symptoms. Swelling is observed in the infected midgut cells, followed by rupture, and finally, the virus is released in the feces (Chopade et al. 2021; Sharma et al. 2020b).

#### 4.6.1.2 Flacharie

The disease is another type of disease that occurs due to the presence of both virus and bacteria. It accounts for ~57.22% of all other silkworm diseases (Chopade et al. 2021). From laboratory notebook entries kept by Pasteur in 1866, the rare presence of sick noncorpuscular larvae is feasible to be associated with a disease other than pebrine. After the failure of his pebrine-free eggs, however, Pasteur definitively separated from both pebrine and from muscardine and grasserie the non-corpuscular disease that he called flacherie. Whenever Pasteur turned his attention to flacherie with a view to allaying disbelief in his method of prophylaxis against pebrine, he found that in larvae suffering from flacherie, these fermentative microbes, which had been taken in with the leaves, multiplied far beyond the normal in the midgut. He made a credible case in connection therewith: flacherie developed from fermentation of the intestinal contents, which had been caused by several disparate factors, including constitutional debility, adverse rearing conditions, the two fermentative microbes Vibrio (*vibrion*) and chaplet-arranged ferment (*ferment en chaplet*) (Hukuhara 2014).

Perhaps unsurprisingly, other studies have uncovered a range of bacteria associated with silkworm bacteriosis. Flacherie was thought by Cuboni and Garibini (1890) to be caused by *Bacillus cubonianus*. The pathogen causing Sotto disease was isolated in Japan as *Bacillus sotto*, presently reclassified as *Bacillus thuringiensis* var. Sotto (Nazar et al. 2020). As early as 1960, Sawamura isolated *Bacillus megaterium* and *Bacillus ellenbachi* from infected silkworms. Some other pathogens like *Serratia marcescens, Bacillus bombysepticus, Bacillus bombycoides*, and *Streptococcus bombycis, Bacillus mucoides, Bacillus latrosporus,* and *Bacillus cereus* var. *Aeromonas alesti* (and *Aerobacter cloaceae* and *Achromobacter superficialis*) have also been identified as pathogenic to silkworms with flacherie. Other pathogens were mentioned by Kodama and Nakasuji, including *Streptococcus faecalis, Serratia piscatorium*, species of *Proteus beta,* and *Micrococcus flavus*. Flacherie is a syndrome characterized by the enlargement of the host's nerve system and nearly flaccid body of the host, harboring the larvae. It is referred to as bacterial flacherie when caused by bacterial infection. Although the symptoms of different strains of bacteria are generally atypical, there are commonalities—larvae affected are lethargic, lose their appetite, and are able to vomit the contents of their gut. Slow development, with a limp appearance, is being observed. The pigmentation of the larvae, which itself varies with the species causing the disease, gives the dead worms a black appearance, inducing a putrid smell. Bacteria are classified into three types according to the site of infection and their pathogenicity to the silkworm. These are as follows: Bacteremia, Septicemia, and Toxicosis (Manimegalai 2009).

**Bacteremia**

The disease of silkworms, which is characterized by bacteremia or digestion-dwelling bacteria, is the result of bacterial multiplication in the gut. Those bacteria primarily infect the silkworms through the mouth and gut, causing a more severe, chronic form of flacherie. Initially, the larvae have opaque cephalothoracic regions, but as the disease (presumably parasitism) progresses, even the posterior part can become translucent. Other symptoms include anorexia, lethargy, stunted growth, and diarrhea of infected larvae (Selvi et al. 2023). The gram-positive bacteria known to be most associated with these diseases of the digestive organs in *B. mori* are *S. faecalis* and *S. faecium*, both of which are 0.7–0.9 μm in diameter. *Streptococcus faecalis* can reproduce, forming large colonies that attach to the larval peritrophic membrane, and they can remain in this state for 2–3 days. This membrane dissolves after ~6 days, causing the goblet and cylindrical gut cells to vacuolate. The larva, therefore, dies as digestive tract fluid and bacteria flow from the vacuolated midgut epithelium into the hemolymph. These bacteria also secrete different enzymes, including chitinase, lecithinase, and proteinase that act on the midgut epithelium within the digestive system and also promote entry of the bacteria into the hemocoel. The damage occurs through the actions of bacterial exotoxins and endotoxins, which also primarily aid different bacteria in invading the digestive tract by harming the gut wall to gain access to the hemocoel. Significantly, fifth instar larvae infected with *S. faecium* experience a gut pH shift from the alkaline range of 8.0–9.5 to a more permissive 7.5–7.6 that helps support many septicemic microbes (Selvakumar 2013; Schleifer and Kilpper-Balz 1984).

**Septicemia**

Bacteria entering the silkworm's blood system cause septicemia. The silkworm is attacked by bacteria, and then there is a multiplication of bacteria throughout the silkworm's body, leading to septicemia. Silkworms frequently throw up fluid from their guts before they die. Dead and diseased larvae are soft, shrunken, and swollen in the thoracic region. In this state, most bacteria are recoverable from hemolymph. The body wall of the infected larva ruptures readily, and there is an escape of foul-smelling fluid filled with bacterial cells, and so death follows rapidly upon infection. Several bacteria have been isolated from silkworms suffering from septicemia, viz., *Serratia marcescens, Pseudomonas species, Bacillus proteus, Bacillus allergens, Bacillus prodigiousus, Bacillus pyocyanes,* and many *Streptococcus* species.

But *Serratia marcescens* and *Bacillus* species are two of the most frequent offenders. If the cause of the septicemia is *Bacillus* species, a dark greenish color appears on the dorso-thoracic region of dead silkworms within a short period, which gradually extends to their whole body. When *Serratia marcescens* leads to septicemia, larvae appear with a reddish coloration that also englobes their entire body, but also dark brown spots. Infection with around $6–8 \times 10^2$ bacterial cells may result in death in 18 h. *Serratia* given orally is more pathogenic under high humidity conditions, but mortality rates differ according to the number of bacteria added to their feed (Park et al. 2024; Enomoto et al. 1987).

Septicemia caused by bacteria during the pre-pupal or pupal stages of silkworms reduces the quality of cocoons and grainage productivity. Rod-shaped ($1–1.5 \times 3$ μm), with sub-terminal spores and peritrichate flagella, *Bacillus* species cause septicemia. On the other hand, *Serratia marcescens* are $0.6–1.0 \times 0.5$ μm small rods, non-sporulating, and with peritrichate flagella. *Serratia marcescens* is generally non-pathogenic when localized in the intestine of silkworms, but is pathogenic and proliferative when allowed to translocate into the hemocoel, and can induce acute disease and death in 1–3 days. Septic bacteria invade silkworms through their wounds or injuries, in addition to damaging the gut wall, and proliferate rapidly in the hemolymph. *Serratia piscatorium* has an increased invasiveness in the hemocoel in the presence of *S. faecalis* or *S. faecium* in the larval midgut. The decrease of pH by such bacteria allows *S. piscatorium* to grow and invade, causing deadly septicemia (Park et al. 2024; Hanh 2025).

**Toxicosis**

The main organism to cause bacterial toxicosis is *B. thuringiensis*, especially the strain var. sotto. Symptoms of the bacterial toxicosis include loss of appetite and lethargy in larvae. Their dorsal vessel beats rapidly, and they may thrash as if in great pain. From *Bacillus thuringiensis*, four toxins (α-exotoxin, β-exotoxin, γ-exotoxin, and δ-endotoxin) were identified. When the larvae ingest these toxins, the digestive fluids first dissolve the toxins, followed by partial absorption into their bodies, and lead to toxicity affecting the nervous system, causing convulsions or paralysis. There are two forms of bacterial toxicosis (Sivaprasad et al. 2021; Watanabe et al. 1966).

i. **Acute Bacterial Toxicosis**

In acute bacterial toxicosis, the δ-endotoxins, represented by parasporal crystal bodies, are released from *B. thuringiensis* and enter the bodies of the organisms. The main symptoms of this toxicosis include an immediate cessation of feeding, head escalation, spasm and tremor, dominion, agony, and final collapse with death. Death can occur from 10 min to several hours after the onset of the symptoms. The body becomes stiff, and the head curls back into a hook shape after death. The dead body then blackens and begins to decompose, putting forth a nasty-smelling dark brown liquid (Watanabe et al. 1966; Takizawa and Iizuka 1963).

ii. **Chronic Bacterial Toxicosis**

Small amounts of *B. thuringiensis* crystal toxin give rise to chronic bacterial toxicosis. In this case, silkworms have reduced mulberry leaf intake, feces are irregularly shaped, and they vomit occasionally, which results in muscle paralysis. The thorax and the abdominal end become translucent, and the larvae become motionless. Roughly 10–12 h post-infection, there are signs of lethargy, as well as loss of quadrupedal grip strength of the legs, flaccidity, and then death (Watanabe et al. 1966; Tojo 1986).

#### 4.6.1.3 Muscardine

"White Muscardine" is a fungal disease commonly observed in India. It is called an occasional disease, but if it occurs, it can even cause total crop failure, and it is particularly prevalent in the eri silkworm. These disease symptoms can be observed even in the eggs and continue to the pupae stage. A brief overview of symptoms seen in different life cycles is as follows (Chopade et al. 2021).

**Egg Stage**: Conidia will grow on the surface of eggs which get infected by Muscardine. Fungal mycelium can also be seen inside and outside the infected egg. Those eggs with these symptoms of disease will not hatch.

**Larval Stage**: A less severe lethargy and anorexia, vomiting, and diarrhea are the symptoms. Black marks like dull border and oily stains are visible on the whole body. The dead larva hardens, dries out, and mummifies itself over time, and whitens due to the formation of conidia and mycelial fungus.

**Pupa Stage**: The pupa may have black spots, mummified inside the cocoon, hard, light, and white fungus oozes out from the intersegmental region.

**Moth Stage**: In the moth stage, we can see the improper growing of the wings, and the wings are slugged. This disease-killed moth dies at an early stage and becomes hardened, light, and whitish, eventually mummifying (Chandrasekharan and Nataraju 2008; Vanitha and Seethapathy 2017).

#### 4.6.1.4 Pebrine

This is a protozoan disease of silkworm, usually seen in the muga silkworm, but the muga silkworm is one of the most harmful diseases of the silk moth. This disease can only be controlled through the examination of the mother moth. The symptoms of this disease can also be observed in the different phases of the silkworm's lifecycle. The infected moth will lay a very limited number of eggs, and these eggs will not adhere well to the egg sheets. An infected moth will lay eggs that are mainly unfertile and/or dead. Irregular hatching, and a few of the eggs have hatched. The larva becomes apathetic, and larval development may be delayed, or molting may be impaired. Infected worms develop black dark spots on their body, and growth is retarded. When the larva is observed, it appears pale and translucent, and the spores are shiny and oval in shape when viewed under a microscope. It is rubbery for a period of time after death, but later becomes black. The pupa develops black spots on its surface. The moth's stunted development and crippled wings is observed (Chopade et al. 2021).

### *4.6.2 Integrated Pest Management in Silkworm Cultivation*

Pest and disease incidence during the silkworm rearing period is very frequent and occasionally causes complete crop loss. Various insect pests infest the mulberry silkworm, including the uzifly (*Exorista bombycis*), earwig, dermestid beetle, and ants. Uzifly is the most serious pest among them in Karnataka, Andhra Pradesh, Tamil Nadu, and West Bengal. Vijayan et al. (2014) reported 10–40% silkworm crop loss due to uzi infestation. However, this method of controlling pests has some limitations, including pollution due to toxic residues, resistance in the pest, destruction of the natural enemy complex, and hazardous effects on the silkworm and humans. Thus, Integrated Pest Management (IPM) measures involving physical, chemical, and biological means have evolved and been publicized for the management of mulberry and silkworm diseases. The aim of IPM is to maximize pest control in a manner that balances overall economic, social, and environmental values (Sakthivel 2012).

However, the adoption level was low despite the successful results. The highest technology gap (87.60%) was in the adoption of biological control measures against the mulberry pest, whereas other measures, cultural/mechanical, were 33.30%, and the least, 9.80%, in the adoption of chemical measures. No technological gap was observed for the mechanical control methods of Uzi menace on silkworm, whereas 77.00 and 71.50% technological gaps were observed for the biological and chemical control methods, respectively. Hence, the application of IPM practices, with a special focus on biocontrol methods, also needs to be popularized among farmers through vigorous extension activities for wider adoption at the farmer's level (Bhat et al. 2024).

## 4.7 Role of Mulberry Trees in Silk Production

Table 4.6 lists key aspects of mulberry cultivation, which play a crucial role in determining the productivity and quality of silk in *B. mori* sericulture.

### *4.7.1 Different Species of Mulberry Trees Used*

Mulberry (*Morus spp*), essentially the only feed of the silkworm *Bombyx mori*, is an economically significant genus in sericulture. The genus was initially proposed and described by Linnaeus with seven species; however, a number of other species have subsequently been added to the genus. Species of the genus have a broad geographic range. Due to its wide adaptability, the species grows from sea level to an elevation of 3000 m and does well under all types of climates (Dandin and Kumari 2021).

Over 68 species of *Morus* are widely recognized, of which *M. alba, M. latifolia, and M. mulicauilis* are cultivated extensively for their leaves, *M. nigra* for its fruit, and *M. serrata* for its timber. In mulberry, various cytomorphs are present, including diploids (*Morus alba*), triploids (*M. alba, M. indica*), tetraploids (*M. laevigata, M. cathayana, and M. boninensis*), hexaploids (*M. serrata and M. tiliaefolia*), octaploids (*M. cathayana*), and docosahexaploids (*M. nigra*). However, diploids and triploids are widely predominant. Mulberry is thought to have originated in the northern

**Table 4.6** Role of mulberry trees in silk production (Sangha et al. 2024; Peng et al. 2023; Hăbeanu et al. 2023; Giacomin et al. 2017)

| Aspect | Key points |
|---|---|
| Exclusive feed for *Bombyx mori* | Silkworms are obligate mulberry feeders, approximately. 1 ton of fresh leaves is required to raise the larvae that spin cocoons |
| Determinant of cocoon and filament quality | Leaf-nutrient profile (protein, aminoacid balance, minerals) directly shapes larval growth rate, shell ratio, filament length, and tensile strength of silk. Fortifying leaves with alanine or selecting high protein cultivars significantly boosts commercial traits |
| Cultivar selection and breeding | Numerous of mulberry varieties exist; breeders target high leaf yield per hectare, seasonal sprouting, drought tolerance, and balanced amino acid spectra to match specific silkworm races |
| Postharvest handling | Leaf quality deteriorates quickly. Cold storing leaves beyond 24 h depresses antioxidant content and reduces silk protein synthesis, even if larval weight looks normal |
| Agroecological services | Mulberry is a fast-growing perennial that sequesters carbon, stabilizes soil on slopes, and fits well in agroforestry belts- lowering the overall footprint of sericulture compared with annual fodder crops |
| Economic ratios | Approx. 25 mature trees sustain 5000 larvae, yielding approx. 1.1 kg raw reeled silk. Efficient plantations can reach 20–40 tons leaves each year |

hemisphere, especially in the Himalayan foothills, and it eventually spread into the tropics of the southern hemisphere (Vijayan et al. 2014).

## 4.8 Conclusion

Mulberry silk sericulture is an area where tradition, science, and sustainable development truly strengthen one another. From its storied beginnings, when an empress observed the cocoon, to the vibrant exchanges of the Silk Roads between India, Japan, and Europe, each generation has added layers of agronomic insight, genetic enhancement, and cultural significance to *B. mori* husbandry. The accumulated practices of today intersect with twenty-first century technologies, molecular breeding, precision IoT-enabled rearing, climate-smart agroforestry, circular bioeconomy models that valorize every byproduct, and immersive training that upskills a new generation of farmers and researchers. What doesn't change is the enterprise's core proposition: a labor-intensive, high-value rural industry that transforms modest landholdings and leafy mulberry canopies into livelihoods, luxury fiber, and downstream innovations in biomedicine, cosmetics, and sustainable textiles. The historical trajectory of mulberry sericulture thus provides more than the history of silk. Cultivating that continuum that pays tribute to extremis wisdom while upholding contemporary practices will facilitate an efficient thread of mulberry silk to uphold economic opportunity, environmental stewardship, and cultural vibrancy in the fabric of our collective future.

**Declaration for AI Usage** AI tools were utilized solely for language improvement and formatting purposes. The authors take full responsibility for the accuracy, originality, and final content of the chapter.

## References

Alabi M (2024) Experiential learning: fostering deep learning through active engagement. Ladoke Akintola University of Technology

Alam I (1999) History of sericulture in India to the 17th century. Proceed Indian Hist Cong 60:339–352

Alipanah M, Abedian Z, Nasiri A, Sarjamei F (2020) Nutritional effects of three mulberry varieties on silkworms in Torbat Heydarieh. Psyche 2020:1–4. https://doi.org/10.1155/2020/6483427

Altman GH, Farrell BD (2022) Sericulture as a sustainable agroindustry. Clean Circ Bioecon 2:100011. https://doi.org/10.1016/j.clcb.2022.100011

Attri K, Sujatha S, Jekinakatti B et al (2024) Advancements in sericulture: innovations and sustainability in silk. Uttar Pradehs J Zool 45:305–317

Babu KM (2012) Silk production and the future of natural silk manufacture. Handbook of natural fibres. Elsevier, Amsterdam, pp 3–29

Bai H, Jiang S, Liu J et al (2023) Planting conditions can enhance the bioactivity of mulberry by affecting its composition. Front Plant Sci 14:1133062. https://doi.org/10.3389/fpls.2023.1133062

Bharathi PD (2016) Sericulture industry in India: a source of employment generation. Int J Adv Eng Res Sci 3:144–147

Bhat MR, Brahma D et al (2024) Role of integrated pest management in agriculture. Int J Zool Appl Biosci 9:40–46

Bhatti MF, Tahir HM, Ali A et al (2023) Adaptability and comparative analysis of economic and biological traits of various silkworm races (*B. mori* L.). Punjab Univ J Zool 38(2):191–200

Blaydes L, Paik C (2021) Trade and political fragmentation on the silk roads: the economic effects of historical exchange between China and the Muslim east. Am J Pol Sci 65:115–132. https://doi.org/10.1111/ajps.12541

Buhroo & Zafar (2019) Latest trends in entomology and zoology studies. Akinik Publications

Bukhari R, Kour H, Aziz A (2019) Women and the Indian sericulture industry. Int J Curr Microbiol Appl Sci 8:857–871

Bukhari R, Hussan F, Kousar F et al (2024) Evaluating the effects of capacity building and training (CBT) on sericulture practices of farmers in Rajouri district of Jammu and Kashmir, India. J Sci Res Rep 30:78–87. https://doi.org/10.9734/jsrr/2024/v30i102433

Butler EA (1888) Silkworms. Swan, Sonnenschein, Lowry & Co, London

Cavaillon J-M, Legout S (2022) Louis Pasteur: between myth and reality. Biomolecules 12:596. https://doi.org/10.3390/biom12040596

Chandrasekharan K, Nataraju B (2008) Studies on white muscardine disease of mulberry silkworm *B. moril* in India—a review. Indian J Sericult 47:136–154

Chandrasekharan K, Nataraju B, Balavenkatasubbaiah M et al (2006) Effect of BmNPV infection during the later instars on the larval and cocoon characters of silkworm, *B. mori* L. Indian J Sericult 45:104–109

Chen G, Yang R, Lu P et al (2022) How can we understand the past from now on? Three-dimensional modelling and landscape reconstruction of the Shuanghuaishu site in the central plains of China. Remote Sens 14:1233. https://doi.org/10.3390/rs14051233

Cheng SY, Ono S (2020) Preservation and management of a World Heritage site Tajima Yahei sericulture farm and its buffer zone. Built Herit 4:13. https://doi.org/10.1186/s43238-020-00013-7

Chopade P, Raghavendra CG, Kumar M, Bhaskar S (2021) Assessment of diseases in *B. mori* silkworm—a survey. Glob Transit Proceed 2:133–136. https://doi.org/10.1016/j.gltp.2021.01.019

Cuboni G, Garbini A (1890) Sopra una malattia del gel—so in rapporto colla flaccidezza. R Acc Lincei 2(4):26–27

Czaplicki Z, Gliścińska E, Machnowski W (2021) Natural silk—an unusual fibre: origin, processing and world production. Fibres Text East Eur 29:22–28. https://doi.org/10.5604/01.3001.0014.9291

Dandin SB, Kumari V (2021) Mulberry (*Morus* sps.) cultivation for sustainable sericulture. In: Mulberry: genetic improvement in context of climate change. CRC Press, Boca Raton, pp 188–207

Dolezalek I (2013) Textile connections? Two ifrīqiyan church treasuries in Norman Sicily and the problem of continuity across political change. Al-Masāq 25:92–112. https://doi.org/10.1080/09503110.2013.767009

Enomoto S, Moriyama H, Iwanami S (1987) Septicemia occurrence in cocoons as related to silkworm rearing conditions. Jpn Agri Res Quart 21:117–121

Frobisher H, Crabtree G (1974) Fundamentals of microbiology, 9th edn. Thomson Learning, London

Gadwala M, Vootla S, Tewary P, Babulal S (2021) Immune mechanism in silkworm *B. mori* L. In: Methods in microbiology. Elsevier, Amsterdam, pp 115–141

Garrett V (2020) Chinese dress: from the Qing dynasty to the present day. Tuttle Publishing, Clarendon, VT

Giacomin AM, Baruque-Ramos J, Zonatti WF et al (2017) Silk industry and carbon footprint mitigation. IOP Confer Ser Mater Sci Eng 254:192008
Girijappa YGT, Rangappa SM, Parameswaranpillai J, Siengchin S (2019) Natural fibers as sustainable and renewable resource for development of eco-friendly composites: a comprehensive review. Front Mater 6(6):226. https://doi.org/10.3389/fmats.2019.00226
Gong Y, Li L, Gong D et al (2016) Biomolecular evidence of silk from 8500 years ago. PLoS ONE 11:e0168042. https://doi.org/10.1371/journal.pone.0168042
Gu H, Cao Z, Chen K, Lü P (2024) Detection and control of pébrine disease in the silkworm (*Bombyx mori*). ISJ Invert Surv J 21:82–98
Gupta SK, Mukhopadhay SK, Bhattacharyya H, Modak BK (2016) Integrated management of diseases and pests of silkworm. J Env Sociobiol 13:147–155
Hăbeanu M, Gheorghe A, Mihalcea T (2023) Nutritional value of silkworm pupae (*B. mori*) with emphases on fatty acids profile and their potential applications for humans and animals. Insects 14(3):254
Haksöz Ç, Uşar D (2011) Silk road supply chains: a historical perspective. Managing supply chains on the silk road. CRC Press, New York, pp 3–26
Han Y (2019) Li Bai and silk road culture. In: Proceedings of the 2019 international conference on art design, music and culture. Xi'an
Hanh NT (2025) Research on the causative agents of septicemia in silkworms in northern Vietnam. In: Contemporary research and perspectives in biological science, vol 11. BP International, pp 71–86
Hariharan S, Monisha U, Logeshwari SM et al (2024) Impact of climate change on sericulture: adaptation strategies and future directions. Uttar Pradesh J Zool 45:570–580
Hukuhara T (1987) Epizootiology of insect diseases. Wiley Interscience, New York
Hukuhara T (2014) The etiology of flacherie, one of the great scourges of sericulture. J Insect Biotechnol Sericol 83:25–31
Islam TU (2024) The historical journey of sericulture: insights into sustainability from past to present. Int J Emerg Knowl Stud 03:719–726
Jia R, Zheng H, Chen H et al (2024) Archaeological textiles preserved by copper mineralization. Herit Sci 12:312. https://doi.org/10.1186/s40494-024-01418-8
Ju F (2022) Polytheism tendency in the trend of integration of the three major religions: worship of silkworm deity art of medieval China. Religions 13:1047. https://doi.org/10.3390/rel13111047
Kalman B (2000) Japan, the culture, 2nd edn. Crabtree Publishing, New York
Kasi E (2013) Role of women in sericulture and community development: a study from a south Indian Village. SAGE Open 3:215824401350298. https://doi.org/10.1177/2158244013502984
Khezri S, Zandieh M (2024) Threads of culture (role of silk road in intercultural exchanges). Tour Cult 5(18):26–35
Kirby WF (1888) Silkworms. Nature 38:386–387. https://doi.org/10.1038/038386a0
Kodli SB, Anita S (2025) A novel approach for silkworm disease detection and classification by using CNN and image processing techniques. J Sci Res Technol 3(3):21–29
Kumar GNA, Reddy BS, Goudappa SB et al (2019a) Growth performance of silkworm cocoon production in Karnataka, India. Int J Curr Microbiol Appl Sci 8:674–682
Kumar S, Ravindra MA, Ramya MN (2019b) Aflotoxin: a higher threat of health hazards for the bivoltine silkworms in temperate zones—a review. Int J Curr Microbiol Appl Sci 8:3144–3153
Le THT, Singyabuth S, Nguyen TTT, Tran TTT (2025) Quang Nam Silk, Vietnam: material culture and storytelling in the process of cultural revitalization. ArtsEduca 42:1–15
Ma D (2004) Why Japan, not China, was the first to develop in east Asia: lessons from sericulture, 1850–1937. Econ Dev Cult Change 52:369–394. https://doi.org/10.1086/380947
Ma D (2005) Between cottage and factory: the evolution of Chinese and Japanese silk-reeling industries in the latter half of the nineteenth century. J Asia Pac Econ 10:195–213. https://doi.org/10.1080/13547860500071451
Manimegalai L (2009) Bacterial pathogens of mulberry silkworm, *B. mori* L. and their management strategies. Sericologia 49:401–420

Mishra MK (2020) The silk road growing role of India. EconStor Preprints

Mulberry cultivation and utilization in India. https://www.fao.org/4/x9895e/x9895e04.htm

Murugesh-Babu K (2019) Silk: processing, properties and applications. Woodhead Publishing, Cambridge

Nazar MZ, Umer A, Sarwar ZM et al (2020) Haemolymph assay and morphological changes of silkworm, *B. mori* L. (Bombycidae: Lepidoptera) Infected with *B. thuringiensis* var. sotto. Pak J Agric Res 33(2):212–216

Pan M, Jiang K, Jin Y et al (2024) Study on the structure and properties of silk fibers obtained from factory all-age artificial diets. Int J Mol Sci 25:6129. https://doi.org/10.3390/ijms25116129

Park JW, Park SK, Jeong CY et al (2024) Microbial community changes in silkworms suspected of septicemia and identification of *Serratia* sp. Int J Mol Sci 25(7):39–57. https://doi.org/10.3390/ijms25073957

Peng M, Wang G, Zhu S (2023) Cold-stored mulberry leaves affect antioxidant system and silk proteins of silkworm (*B. mori*) larva. J Sci Food Agric 103:7673–7682. https://doi.org/10.1002/jsfa.12849

Pliny S, Bostock J, Riley HT (1855) The natural history of Pliny. In: Translated, with copious notes and illustrations by John Bostock and H.T. Riley. H.G. Bohn, London

Popescu A, Şerban V, Ciocan HN (2024) New trends in the global silk production in the period 2011–2022. In: Scientific papers series management, economic engineering in agriculture and rural development, vol 24

Pugina RS, Caiut JMA (2021) Silk: history, obtaining, structure and properties of an old material in the development of new technologies. Int Adv Med Biotechnol 4:27–45

Rahmathulla VK (2012) Management of climatic factors for successful silkworm (*B. mori* L.) crop and higher silk production: a review. Psyche 2012:1–12. https://doi.org/10.1155/2012/121234

Raichvarg D (2022) Silkworm, science worm. C R Biol 345:35–50. https://doi.org/10.5802/crbiol.94

Reddy YP, Kumar UA, Parasuramudu M, Nandhini K (2024) Evaluation of influenced antiviral medicinal plant extracts for the control of grasserie disease and estimation of biochemical and hematological changes in silkworm, *B. mori* L. BioResToday 5:844–849

Robert J-N (2007) Les relations entre le monde Romain et la Chine: La tentation du Far East. https://www.clio.fr/BIBLIOTHEQUE/pdf/pdf_les_relations_entre_le_monde_romain_et_la_chine_la_tentation_du_far_east.pdf

Rohela GK, Shukla P et al (2020) Mulberry (*Morus* spp.): an ideal plant for sustainable development. Trees For People 2:100011

Sakthivel N (2012) Adoption of integrated pest management practices in sericulture—a case study in Tamil Nadu. J Biopesticides 5:212–215

Sangha MN, Barwani DK, Xavier C et al (2024) We are what we eat: implications of host plant suitability on sustainable production of silkworm pupae as novel ingredient with dietary and health benefits. PLoS ONE 19:e0316290. https://doi.org/10.1371/journal.pone.0316290

Sarkar K (2018) Management of nutritional and climatic factors for silkworm rearing in west Bengal: a review. Int J Agric Environ Biotechnol 11(5):769–780

Savithri G, Sujathamma P, Ramanamma CH (2013) Glory of Indian traditional silk sarees. Int J Text Fashion Technol 3:61–68

Schleifer KH, Kilpper-Balz R (1984) Transfer of *S. faecalis* and *S. faecium* to the Genus *Enterococcus* nom. rev. as *Enterococcus faecalis* comb. nov. and *Enterococcus faecium* comb. nov. Int J Syst Bacteriol 34:31–34. https://doi.org/10.1099/00207713-34-1-31

Schoeser M (2007) Silk. Yale University Press, New Haven, CT

Selvakumar T (2013) Prevalence of flacherie disease and pathogenicity of isolated pathogensin silkworm, *B. mori* (L.) under different environmental conditions. Agric Sci Dig Res J 33:253

Selvi GSA, Chelladurai D, Kumar VP, Chandran MN (2023) Bacterial flacherie in *B. mori* (L.): isolation and identification of pathogenic organism. Uttar Pradesh J Zool 44:52–61

Sharma A, Chanotra S, Gupta R, Kumar R (2020a) Influence of climate change on cocoon crop loss under subtropical conditions. Int J Curr Microbiol Appl Sci 9:167–171

Sharma A, Sharma P, Thakur J et al (2020b) Viral diseases of mulberry silkworm *B. mori* L.—a review. J Pharmacognosy Phytochem 9:415–423

Singh K, Bhargava J (2023) Tasar sericulture: a sustainable economical booster. Ann Sci Allied Res 1:97–107. https://doi.org/10.5281/ZENODO.10479569

Sivaprasad V, Rahul K, Makwana P (2021) Immunodiagnosis of silkworm diseases. Methods in microbiology. Elsevier, Amsterdam, pp 27–46

Srinivasa G, Gope M, Manjula A, Somireddy J (2013) Impact of training on knowledge and adoption of sericulturists in Kolar district of India. Agric Sci Dig Res J 33:294. https://doi.org/10.5958/j.0976-0547.33.4.028

Takeda S (2009) Sericulture. Encyclopedia of insects. Elsevier, Amsterdam, pp 912–914

Takizawa Y, Iizuka T (1963) Studies on the toxicity of some *Bacillus* upon the silkworm, *B. mori* L. J Sericult Sci Jpn 32:343–347

Taufique M, Hoque M (2021) Current scenario of sericulture production in India: a spatio-temporal analysis. Int Res J Ed Technol 2(4):12–23

Tayal MK, Chauhan TPS (2017) Silkworm diseases and pests. Industrial entomology. Springer, Singapore, pp 265–289

Thrilekha D, Mala H, Reddy NC et al (2024) Sericulture in the 21st century: global trends and future prospects. Int J Adv Biochem Res 8:1057–1061

Tojo A (1986) Mode of action of Bipyramidal delta-endotoxin of *B. thuringiensis* subsp. kurstaki HD-1. Appl Environ Microbiol 51:630–633. https://doi.org/10.1128/aem.51.3.630-633.1986

Trivedy K, Kumar SN, Vinutha N, Qadri SMH (2011) In vitro testing of common disinfectants used in sericulture to control the growth of fungi in rearing houses. Res J Microbiol 6:439–465. https://doi.org/10.3923/jm.2011.439.465

Ubaydullaevich NB, Sodikugli SD (2021) Leaf consumption in the rearing of silkworm with high humidity method. EPRA Int J Res Dev 21:135–139

Vainker SJ (2004) Chinese silk: a cultural history. British Museum Press, London

Vanitha S, Seethapathy P (2017) Isolation and characterization of white muscardine fungi *B. bassiana* (Bals) Vulii: a causative of mulberry silkworm. J Entomol Zool Stud 14:512–515

Vichare J (2024) The impact of the silk road on global trade: a historical perspective. Int J Novel Res Dev 9:92–96

Vijayan K, Saratchnadra B, Raju PJ, Tikader A (2014) Biotechnology of mulberry (*Morus* L.)—a review. Emirates J Food Agricult 26(6):472–496

Watanabe T, Tsutsui R, Iwahana H (1966) Toxic crystals produced by *B. thuringiensis* var. sotto. J Sericult Sci Jpn 35:418–422

Weber RW (2008) On the cover: silkworm. Ann Allergy Asthma Immunol 101(4):A4. https://doi.org/10.1016/s1081-1206(10)60306-7

Yilmaz O et al (2015) History of sericulture in Turkey. Asian J Agri Food 3(2):237–242

Yilmaz O, Erturk YE, Coskun F, Ertugrul M (2016) Sericulture in provinces of bursa, Amasya and mugla (Turkey). J Agric For 62(1):241–247

Young SH (2013) For a compassionate killing: Chinese Buddhism, sericulture, and the silkworm god aśvaghoṣa. J Chin Relig 41:25–58. https://doi.org/10.1179/0737769x13z.0000000002

Zheng H, Yang H, Zhang W et al (2020) Insight of silk relics of mineralized preservation in Maoling Mausoleum using two enzyme-linked immunological methods. J Archaeol Sci 115:105089. https://doi.org/10.1016/j.jas.2020.105089

# Chapter 5
# Non-mulberry Silks: Diversity, Origins, and Production Techniques

**Farhan Khan, Mst. Romana Akter, Nadim Ibn Sayed, Saima Mahjabin, A. Z. M. Mofasser, and Rony Mia**

## 5.1 Introduction

Silk is one of the most essential raw materials for the textile manufacturing industry because of its distinctive attributes, including endurance, shimmer, silky texture, and breathable properties, which make it ideal for delicate textiles and high-end attire. According to the philosopher Confucius, over three thousand years ago in China, the Chinese princess Xi Lin Shi serendipitously invented silk reeling, paving the way for the identification of the fiber (Üzümcü et al. 2019). The wild silk moth or worm species *Bombyx mandarina* (Moore), which thrived on white mulberry trees, is speculated to be the primary source for silk in the past (Üzümcü et al. 2019; Üzümcü and Kadoğlu 2016; Hyde 1984; Kundu et al. 2012; Peigler 1993; Colomban et al. 2018). Many insect species use a protein having a fibrous structure to create external structures known as cocoons, and that fibrous structure is called silk. Mulberry silk, derived from the *Bombyx* species, and non-mulberry silk, obtained from the *Antheraea* species, are its two principal varieties of natural silk. Non-mulberry silks are frequently referred to as untamed or vanya silks (Das 2020).

F. Khan
Department of Textile Engineering, National Institute of Textile Engineering and Research, Nayarhat, Savar, Dhaka, Bangladesh

Mst. R. Akter
College of Textile and Clothing, Qingdao University, Qingdao, Shandong, China

N. Ibn Sayed · A. Z. M. Mofasser
Department of Textile Engineering, European University of Bangladesh, Mirpur, Dhaka, Bangladesh

S. Mahjabin
Department of Textile Engineering, Fareast International University, Banani, Dhaka, Bangladesh

R. Mia (✉)
Wilson College of Textiles, North Carolina State University, Raleigh, NC, USA
e-mail: mroni_mia@yahoo.com



D. Kumar and M. Shahid (eds.), *Science and Technology of Silk*, Textile Science and Clothing Technology, https://doi.org/10.1007/978-3-032-16230-4_5

Non-mulberry silk moths, frequently referred to as wild silk moths, are usually individuals not raised in houses. Alternatively, indigenous tribes collect the cocoons from naturally occurring moth populations. There are occasions when rearing takes place outside, where the larvae have little to no protection. In the United States of America, the Saturniidae family, which comprises the great majority of these lepidopterans, is commonly referred to as "enormous silk caterpillars". Nearly half of the more than 1500 species in this complex group pupate underground without creating cocoons. The widely recognized tent caterpillars (*Malacosoma* hübner) are a kind of non-mulberry silk moth that is a member of the family Lasiocampidae (Peigler 1993). Apart from that, the Notodontidae family encompasses two distinct groups of non-mulberry silk moths that can be found all along Africa, whereas the Pieridae family, well known for the widely recognized whites and sulfurs among butterflies, has numerous subspecies in Mexico (Peigler 2020). Non-mulberry silks, produced by eri, tasar, and muga silkworms, are gaining appeal in industries like textiles and medicine because of their ecologically friendly manufacturing procedures (Das et al. 2020). The increased interest in non-mulberry silkworms has encouraged further research on the sustainable use and preservation of wild silks, ensuring their survival in both commercial and non-commercial circumstances (Dandin and Prabhu 2022).

Industries are reconsidering the materials they employ as well as how they handle production to process and produce in response to growing environmental concerns. In order to lessen their effects on the environment, this change has led producers, researchers, and textile fiber manufacturers to concentrate on developing biodegradable and sustainable fibers. The negative consequences of material and energy exchanges between the biological sphere, or the natural environment, and the technosphere, or the artificial technological environment, are known as environmental repercussions. The textile sector in particular has drawn criticism for its role in environmental deterioration, which has increased demand for environmentally friendly substitutes (Babu and Sinclair 2015). Non-mulberry silk is one such substitute that has drawn interest and has been investigated as a novel biomaterial. It is a viable choice for many applications due to its remarkable mechanical properties, compatibility, and biodegradability. These include technological applications in biosensors, membranes, filters (of various types), and the medical and biomedical domains (Guo et al. 2017; Chand et al. 2023). Silk's attraction is further enhanced by its unique ability to interact with cells, especially in medical applications where biocompatibility is essential. Broader themes such as ecological responsibility, the ecology of industry, environmental effectiveness, and ethical chemistry are also driving the development of sustainable materials. The development of next-generation materials, goods, and procedures aimed at reducing environmental harm is guided by these frameworks. New laws and increased social consciousness are accelerating the pursuit of environmentally friendly materials and techniques as awareness of the textile industry's negative environmental impact grows. The fiber has spurred innovation, encouraging scientists and producers to prioritize biodegradable and sustainable fibers as a more sensible and efficient way to mitigate the ecological impact of the sector (Babu and Sinclair 2015).

The non-mulberry silk possesses a tougher quality and improved thermal and mechanical properties compared to mulberry silk, due to its growth in an unfavorable environment (Guo et al. 2017; Fang et al. 2016). The silk cocoon membrane (SCM) obtained from the non-mulberry silk works as an insulator in the dry state and is able to induce electricity by absorbing moisture. This discovery enhances its application in various technological devices, including batteries, biosensors, humidity sensors, waste heat management, and steam engines (Tulachan et al. 2014). Hence, the use of semi-domesticated or wild-type silk, known as non-mulberry silk in the modern era, is expanding alongside domesticated silk.

Figure 5.1 depicts the progressive growth in India's silk production over a 17-year period. Starting at 18.32 thousand metric tons in fiscal year (FY) 2008, the production steadily increased, reaching an estimated 38.91 thousand metric tons by FY 2024.

This constant growth demonstrates India's increasing interest in both domesticated (mulberry) and wild (non-mulberry) silk farming. As one of the world's leading silk producers, India benefits from different agricultural settings that sustain a wide range of silk varieties, most notably mulberry silk in the south and non-mulberry varieties such as eri, tasar, and muga in the eastern and northeastern region. The findings show that rather than just agricultural and technical breakthroughs, there is also an acceleration of local and worldwide demand for silk. This emerging trend highlights

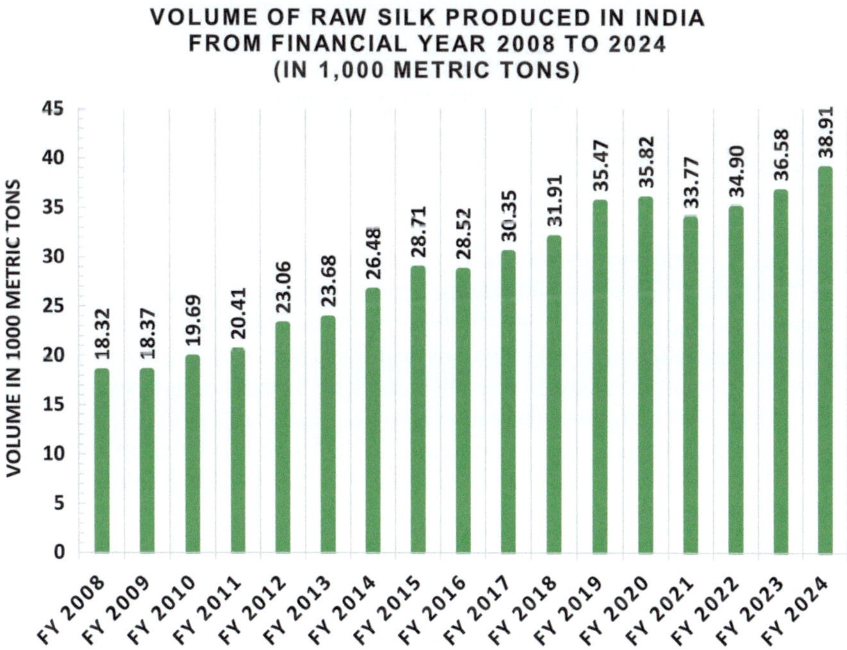


**Fig. 5.1** Volume of raw silk production in India from financial year 2008–2024 (in 1000 metric tons). *Source* central silk board, adapted from Statista (https://www.statista.com/statistics/622914/silk-production-india/)

India's prominent position in the global silk industry, where tradition and modern innovation are closely intertwined.

Therefore, in this chapter, an elaborate discussion about non-mulberry silkworms, the species type found worldwide, is presented, along with the distinction between wild and domesticated mulberry silkworms. Furthermore, the diversity of these wild silkworms, together with their evolutionary background, is also described. Additionally, the comparative characteristics of wild silks, including their physical, mechanical, and thermal properties, are presented alongside their cultural significance and early uses in human society. The habitat and ecology of non-mulberry silkworms, as well as their production methods and related difficulties, are also covered in this chapter, along with potential applications of non-mulberry silk-based biomaterials.

## 5.2 Diversity of Non-mulberry Silks

The allocation of some of the world's wild silk fauna has been described by Peigler (1993) and Jolly et al. (1979). He has documented about 25 species of wild silk moths that have been utilized to make non-mulberry silk. Some of the silks' names reflect the fact that tribal people have been the primary exploiters of these resources (Kioko 1998). Silkworms belong to the Lepidoptera order and are divided into two types: mulberry silkworms and non-mulberry silkworms. Silkworms from the Bombycidae family create mulberry silk, whereas silkworms from the Saturniidae and Lasiocampidae families generate non-mulberry silk. The predominant producer of mulberry silk is *Bombyx mori*, a domesticated species found worldwide, forming the backbone of the sericulture industry (Jasmine et al. 2014). Except for mulberry silk, there are five major forms of economically relevant silk obtained from distinct species of silkworms, each of which feeds on a different host plant. All other silk variants are commonly known as non-mulberry silk. Non-mulberry silk has recently been referred to as "Vanya Silk" due to its wild origins. South Asia is the only region in the world that produces economically viable silk (Elumalai et al. 2021). The classification of silkworms has traditionally relied on physical traits, such as follicular patterns on eggshells, chromosome analysis, and tubercular setae located arrangement on larvae (Ahmad et al. 2004). However, these morphological characteristics can be influenced by environmental factors, making classification challenging. As a result, the development of DNA markers has become essential for accurate classification (Jasmine et al. 2014; Mahendran et al. 2006).

In sharp contrast to their cultivated mulberry counterparts, non-mulberry silkworms, sometimes referred to as Emperor moths, are wild members of the Saturniidae family. This large family comprises ~1100 species worldwide. Voltinism, or the number of generations a silkworm produces annually, divides them into three categories: univoltine (producing just one brood), bivoltine (producing two broods), and multivoltine (making more than two broods) (Dey et al. 2023). These silkworms have been the subject of rigorous scientific scrutiny for decades, with documented studies dating back to 1907 and continuing through 1975, as recorded in the Memoirs of

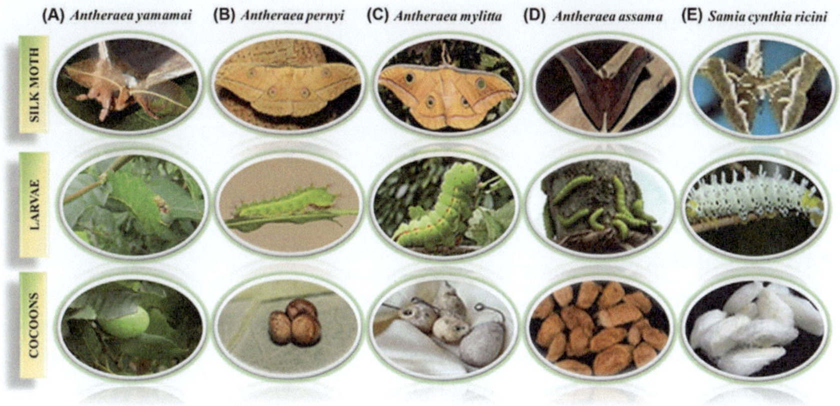


**Fig. 5.2** Diagrammatic representation of the variety of non-mulberry silk types found worldwide **a** *Antheraea yamamai* [Japanese Tasar silkworm], **b** *Antheraea pernyi* [temperate tussar or oak tasar], **c** *Antheraea mylitta* [tropical tasar silkworm], **d** *Antheraea assama* [muga silkworm], and **e** *Samia cynthia ricini* [Eri silkworm]. Reproduced with permission from (Jaiswal et al. 2024), Copyright 2024 Elsevier

the Zoological Survey of India. The non-mulberry silkworm population is primarily dominated by the *Antheraea* and *Philosamia* species. A visual representation showcasing the diverse range of non-mulberry silkworms is illustrated in Fig. 5.2 (Fazal and Latief 2018; Kim et al. 2018; Jaiswal et al. 2024). Besides, the history of Indian silk production traces its origins as far back as 2450–2000 BCE, with evidence emerging predominantly from the Indus Valley civilization.

### *5.2.1 Key Species and Types of Non-mulberry Silks*

Wild silks include Eri, Muga, Tasar, Fragaria, and cricula, which are produced by silkworms from the *Saturniidae* family, while shashe silk is derived from those belonging to the *Lasiocampidae* family (Peigler 2020). These silkworm species are distributed across various Asian nations, including India, China, Japan, Indonesia, and Thailand, as well as in Namibia (Jasmine et al. 2014).

Muga, eri, and tasar silkworm from the kingdom up to the super family and family level is given below (Elumalai et al. 2021).

Kingdom: Animalia.

Phylum: Arthropoda.

Sub phylum: Mandibulata.

Class: Insecta.

Division: Endo Pterygota.

Order: Lepidoptera.

Sub order: Ditrysia.

Super family: Bombycoidea.

Family: Saturniidae.

#### 5.2.1.1 Tasar Silkworm

Multivoltine non-mulberry silkworms are cultivated in both temperate and tropical regions. The commercial production of tasar silk primarily involves four species from the genus *Antheraea* Hubner listed in Table 5.1.

*Antheraea* has more species than any other category of silk-producing insects. A total of 35 species have been identified, comprising 31 pertaining to the Indo-Australian biogeographic area, three found in the Palearctic realm, and one discovered in the USA. Furthermore, more than fifty variations, aberrant forms, or different races have been discovered (Elumalai et al. 2021).

Tasar silk can be divided into two types: tropical tasar and temperate tasar. Tropical tasar silkworms are widely distributed across the Indian subcontinent. The tropical tasar silkworm *A. mylitta* cocoons produce the greatest amount of silk. These silkworms can be bivoltine or trivoltine, depending on the variety and habitat (Jasmine et al. 2014). The worm, moth, and cocoons of tasar silk are depicted in Fig. 5.3 (Babu and Bhat 2017).

*Antheraea mylitta*, often known as the tasar non-mulberry silkworm, is typically reared on host plants such as Sal, Saaj, and Arjuna. Spanning the Indian subcontinent, from Bihar to West Bengal and all the way to the southern state of Hyderabad, Maharashtra, several regional varieties of these silkworms are raised. Numerous secondary metabolites, some of which serve as organic insect repellents, have been identified as coming from tasar silkworms (Ahamad et al. 2021). The proteins sericin and silk fibroin, which are obtained from these silkworms, have also been investigated for possible uses in tissue remodeling. The tasar silkworm is one of them that is important for the production of silk (Jaiswal et al. 2024).

*Antheraea pernyi*, an individual of the *Saturniidae* family, is also known as temperate tussar moths or oak tasar silkworms and is native to China. The oak tasar

**Table 5.1** Species of the genus, *Antheraea* (hubner), their commercial names and countries of origin (Elumalai et al. 2021)

| Scientific name | Commercial name | Country of origin |
|---|---|---|
| *Antheraea proylei* J. | Temperate tasar silkworm | India |
| *A. mylitta* D. | Tropical tasar silkworm | India |
| *A. yamamai* G.M. | Japanese tasar silkworm | Japan |
| *A. pernyi* G.M. | Chinese tasar silkworm | China and Union of Soviet Socialist Republics (USSR) |

**Fig. 5.3** Worm, moth, and cocoons of tasar silk. Adapted from WIKIMEDIA COMMONS (https://commons.wikimedia.org/)

**Fig. 5.4** Oak Tasar- **a** worm, **b** moth and **c** cocoons. Reproduced with permission from (Babu and Bhat 2017), Copyright 2017 Elsevier

silkworm, moth, and cocoons are shown in Fig. 5.4 (Babu and Bhat 2017). They are thought to have developed from their original species, *Antheraea roylei*, based on their significant similarities in nucleosome structure. Wild silkworms have been widely studied for their potential in bacterial resistance (Qu et al. 1982). Additionally, silk films developed from their silk sericin have demonstrated promising applications in mineralization and tissue engineering (Jaiswal et al. 2024; Yang et al. 2014).

*Antheraea yamamai* is often known as the Japanese oak silk caterpillar. These kinds of insects are native to subtropical Eastern Asia and parts of Europe. They were initially reported around 1861 as a counterfeit hybrid of *Antheraea polyphemus*, a species endemic to North America. The proteins produced by these silkworms were tested for their ability to synthesize a compound known as biliverdin, a powerful antioxidant and therapeutic agent (Leem et al. 2018).

#### 5.2.1.2 Muga Silkworm

Muga is a golden-yellow silk derived from *A. assama*, primarily found in the north-eastern region of India (Jasmine et al. 2014). More than 95% of the world's muga silk is produced in Assam. Other regions that border Assam also engage in this silk growing in places like Meghalaya, West Bengal, Manipur, Nagaland, and Mizoram.

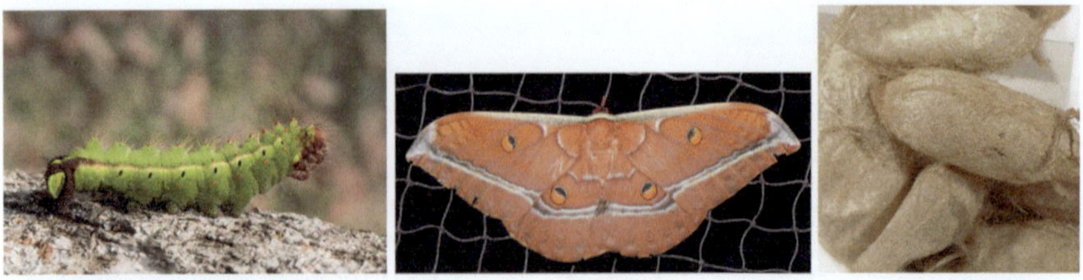


**Fig. 5.5** Worm, moth and cocoons of Muga. Adapted from WIKIMEDIA COMMONS (https://commons.wikimedia.org/)

The golden-yellow silk is produced by a semi-domesticated, multivoltine species known as the muga silkworm (Elumalai et al. 2021). It is partially domesticated and follows a multivoltine life cycle (Jasmine et al. 2014). This breed is extensively cultivated and found in Eastern India Assam civilization; moreover, it is widely grown in the Brahmaputra canyon (Elumalai et al. 2021). The picture of muga silkworm, moth, and cocoons is given in Fig. 5.5 (Babu and Bhat 2017).

#### 5.2.1.3 Eri Silkworm

Eri silk is obtained from *Philosamia ricini* (also known as *Samia ricini* or *Cynthia ricini*), a species belonging to the family *Saturniidae*, (Babu and Bhat 2017). It is also referred to as endi. Eri is a multivoltine wild silk that is spun from cocoons or sheaths from an open end which distinguishes it from other silk varieties. Eri silk primarily feeds on castor leaves. Being polyphagous, the eri silkworm consumes a diverse range of host plants, including kessaru, papaya, barkessuru, tapioca (cassava), payam, and among others. Eri culture is a traditional domestic practice that is mostly carried out in the Northeastern area, where the native tribal groups view the protein-rich pupae as a delicacy. Consequently, eri cocoons spin into silk while maintaining their open mouths (Elumalai et al. 2021). The Brahmaputra canyon of Assam, especially the tribally dominated areas, was historically the center of the Eri culture. Mizoram, Manipur, Meghalaya, Nagaland and Arunachal Pradesh were next in line. States like Tamil Nadu, Chhattisgarh, Andhra Pradesh, Madhya Pradesh, Odisha, and West Bengal have recently implemented ericulture on an experimental basis (Elumalai et al. 2021). The picture of eri silkworm, moth, and cocoons is given in Fig. 5.6 (Babu and Bhat 2017).

**Fig. 5.6** Worm, moth, and cocoons of Eri. Adapted from WIKIMEDIA COMMONS (https://commons.wikimedia.org/)

#### 5.2.1.4 Other Lesser-Known Varieties

##### 5.2.1.4.1 Fagara Silkworm

Along with a few other similar species that are present in China, Sudan, and the Indo-Australian biogeographic area, Fagara silk is obtained from the gigantic silk moth *Attacus atlas L.* With a wingspan of up to eleven inches, this moth is the biggest insect currently in existence. With peduncles that range in length from two to ten cm, it weaves light brown cocoons that are about 6 cm long (Elumalai et al. 2021). These silkworms' yarn has a higher tensile strength than that of muga and eri (Jaiswal et al. 2024).

##### 5.2.1.4.2 Anaphe Silkworm

The *Anaphe* silkworm is found in southern and central Africa. *Anaphe* silk is produced by silkworms belonging to the *Anaphe* genus, which includes *Anaphe venata* (Butler), *Anaphe reticulata* (Walker), *Anaphe ambrizia* (Butler), *Anaphe moloneyi* (Druce), *Anaphe panda* (Boisduval), *Anaphe infracta*, and *Anaphe carteri* (Walsingham) (In brackets, the names of the scientists who first described these moth species are mentioned) (Elumalai et al. 2021). Among these, the *A. panda* has the most potential for non-mulberry silk cultivation due to its large cocoon size and the abundance of its primary host plant, *Bridelia micrantha* (Walsingham). The cocoons weigh between 0.03 kg and 0.04 kg. Encased in a thin coating of silk, these silkworms spin their cocoons in groups of cells. There are seven larval instars for the species. The entire developing period lasts approximately 4 months during the rainy season and 3 months during the dry season. During the wet season, the phase known as pupal lasts approximately 5–6 months, whereas during the summertime, it lasts 3.5–4.5 months. The weight of the silk nest varies based on the number of silkworms involved in spinning it. These nests are brown in color and can be classified into two types: a medium-sized nest, which contains around 5105 worms, and a smaller nest, which consists of approximately twenty worms. At the pre-pupal phase, the larvae weigh an average of 1.09 g. The moths' cream-colored wings with black transverse streaks make them very appealing. These cocoons are harvested from the forest by tribal people, who then twirl the silk fibers to create raw silk. The resultant silk has a modest sheen and is supple. While the silk from *A. moloneyi* is called koko and tissnian-tsamia besides, the silk from *A. infracta* is called boko (Elumalai et al. 2021).

#### 5.2.1.4.3 *Gonometa postica* (Shashe) Silkworm

African non-mulberry silk moths, or *G. postica*, are found throughout eastern and southern Africa and are renowned to produce silk of superior quality (Teshome et al. 2014). Females of this species are twice as big as males, and they exhibit dimorphism throughout their cocoons and moths. The spines on the Shashe silkworm protective shell (cocoons) immobilize the pupae and shield the adult larvae (Kundu et al. 2012; Jaiswal et al. 2024).

#### 5.2.1.4.4 *Cricula trifenestrata* Silkworm

*Cricula* is the smallest non-mulberry silkworm, producing perforated cocoons with a golden tint (Kundu et al. 2012; Hridya et al. 2021). These moths are extensively spread in South Asia, with an enormous presence in India, the Vietnam, Andaman Islands, Bangladesh, Myanmar, Singapore, Malaysia, Cambodia, the Philippines, Thailand, and Java (Jaiswal et al. 2024).

#### 5.2.1.4.5 Coan Silk

Coan silk is produced by the Mediterranean biogeographic area's *Pachypasa otus D.*, a polyphagous silkworm species that feeds on a range of trees, including oak, juniper, ash, pine, cypress, and louisiana. The larvae of this species spin white cocoons that measure about 0.089 by 0.076 m. Historically, this silk was used to create crimson gowns for Roman officials, but production is limited (Elumalai et al. 2021).

### *5.2.2 Comparison of Characteristics of Wild Silkworms*

Silk is a naturally occurring structural protein that falls under a specific class of environmentally beneficial biopolymers. A filamentous integral protein termed silk fibroin (SF) and a sticky outer layer made up of water-soluble glycoproteins called sericin (SS) make up the majority of the original form of silkworm silk (Bandyopadhyay et al. 2019). The following parameters are associated with variations among different types of silk: (i) variations in the presence of repetitive polypeptide sequences; (ii) differences in the presence and configuration of the glycoprotein, light chain (L-chain), and heavy chain (H-chain); and (iii) variations in the sericin content among various types of silk (Zou et al. 2022; Konwarh et al. 2017). About 20–30% of the cocoon's bulk is made up of the amorphous and easily soluble in water glycoprotein designated as silk sericin (SS) (Jasmine et al. 2014). Serine (32–34%) makes up the majority of its composition, followed by other amino acids including glutamic acid, tyrosine, histidine, and threonine, as well as aspartic acid (14–16%) (Mandal and Kundu 2010; Kumar and Mandal 2017). Sericin mostly appears in an arbitrary coil mold and has a molecular weight between 10,000 and 350,000 Da. The isolation algorithm implemented dictates the proportion of sericin that is removed (Kumar and Mandal 2017; Kumar et al. 2018). The biochemical and immunological properties of non-mulberry sericin differ from those of mulberry silk variants (Ahmad et al. 2004). In comparison to *B. mori*, *A. mylitta* has less glycine (Bandyopadhyay et al. 2019). *A. mylitta* sericin's secondary structure contains random coils and β-sheets, which are similar to *B. mori* sericin (Dash et al. 2006). Sericin from both species lacks α-helical structures. Refer to mulberry silk variants (*B. mori* and *B. mandarina*) and non-mulberry silk kinds (*A. yamamai*, *P. ricini*, *A. pernyi*, *A. assama* and *A. mylitta*). Table 5.2 points out the protein molecular weights of silk sericin (SS) along with silk fibroin (SF) and recurring sequences as well as tensile strength data.

It is evident that the characteristics of silk fibers derived from various non-mulberry species varies significantly. Nonetheless, insects of the same species can also generate fibers with quite different and special qualities. According to studies by Reddy and Yang (2010b), the qualities of wild silk fibers obtained from a range of uncommon saturniidae non-mulberry silk moths varies substantially across insects in the same family. Wild moths of the same species produced fibers varied in elongation from 4 to 20% and tenacity from 0.9 to 5 g/den as indicated in Table 5.3 (Reddy et al. 2015). Variations in the wild silk moths physical characteristics, environmental environments, and the kind and quantity of amino acids were cited as the causes of differences in the fiber qualities (Reddy and Yang 2010b). Fibers from various layers of a single cocoon may have unique qualities, much as fiber properties can change across species and even within the same species. The fibers from the outside, middle, and inner layers, however, have comparable compositions and tensile characteristics, according to studies on *H. cecropia* cocoons (Reddy and Yang 2010a). These fibers have comparable tensile qualities to *B. mori* silk and have better tensile qualities than wild silk fibers, as given in Table 5.3. Because *H. cecropia* silkworms had lower food requirements than *B. mori*, it was proposed that they were simpler to raise. Their bigger cocoons further suggested that they might constitute a better option than traditional silk.

When silk threads are heated, they rapidly lose weight at 175 °C and become black at 250 °C. Silk is highly resistant to fire, emitting an odor akin to burned horn and leaving a black carbon deposit. Silk outperforms wool in terms of heat resistance and thermal insulating properties. According to a thermal gravimetric analysis, wild silk (*P. ricini*) decomposes in at least two phases, but cultured silk (*B. mori*) typically

**Table 5.2** Molecular weight of silk sericin, silk fibroin, tensile strength, and the repetitive sequences of various mulberry and non-mulberry silk varieties (Jaiswal et al. 2024)

| Family | Silkworm species | Silk fibroin | Silk sericin | Tensile strength of cocoon fiber (GPa) | Repetitive sequences in silk fibroin | References |
|---|---|---|---|---|---|---|
| Saturniidae | *A. mylitta* | H-chain: 197 kDa | High-molecular-weight fractions of. 200, 200, and 70 kDa | 0.250 ± 0.5 | AAAAAAAAAAAASS | Gupta et al. (2015), Chattopadhyay et al. (2005), Reddy and Yang (2010a) |
| | *A. assama* | H-chain: 230 kDa | High-molecular-weight fractions of 100 and 66 kDa<br>Low-molecular-weight fractions of 3650 kDa | 0.36 ± 0.10 | AAA(A)515 | Gupta et al. (2015), Kundu et al. (2008), Malay et al. (2016) |
| | *A. pernyi* | H-chain: 220 kDa | High-molecular-weight fraction between 200 and 70 kDa | 0.43 ± 0.08 | AAAAAAAAAAAAAGS | Ho et al. (2012), Reddy and Yang (2012), Nayak and Kundu (2016), Tao et al. (2007) |
| | *A. yamamai* | H-chain: 218 kDa | High-molecular-weight fractions of 240, 200, and 180 kDa | 0.39 ± 0.07 | AAAAAAAAAAAASS | Gupta et al. (2015), Malay et al. (2016), Kobayashi et al. (2009) |

(continued)

**Table 5.2** (continued)

| Family | Silkworm species | Silk fibroin | Silk sericin | Tensile strength of cocoon fiber (GPa) | Repetitive sequences in silk fibroin | References |
|---|---|---|---|---|---|---|
| | *P. ricini* | H-chain: 230 kDa | A single fraction of 66 kDa | $0.47 \pm 0.11$ | YGGDGG(A)12GGAG | Ahmad et al. (2004), Mai-Ngam et al. (2011), Pal et al. (2013), Suzuki et al. (2015) |
| Bombycidae | *B. mori*<br>*B. mandarina* | H-chain: 350 kDa<br>L-chain: 26 kDa<br>P25 (glycoprotein): 30 kDa | Four polypeptide fractions of 400, 250, 150, and 24 kDa | $0.57 \pm 0.12$ | GAGAGSGAGAGY | Kundu et al. (2012), Malay et al. (2016), Takasu et al. (2002), Yang et al. (2016), Tanaka and Mizuno (2001) |

**Table 5.3** Comparative analysis of the tensile characteristics of silk strands made by several non-mulberry silk moths in the Saturniidae family (Reddy et al. 2015)

| Wild silk moths | Fiber fineness (denier) | Breaking elongation (%) | Tenacity (g/den) | Young's modulus (g/den) | References |
|---|---|---|---|---|---|
| *B. mori* | 0.4–1.1 | 10.0–23.4 | 4.3–5.2 | 84–121 | Reddy and Yang (2010c) |
| *A. mylitta* | 4.7–10.7 | 26–39 | 2.5–4.5 | 66–70 | |
| *P. ricini* | 1.3–2.7 | 24–28 | 1.9–3.5 | 29–31 | |
| *H. cecropia* | 2.0 | 14.8 ± 6.8 | 3.8 ± 0.6 | 68 ± 9.9 | Reddy and Yang (2010a) |
| *A. atlas* | 2.0 | 18.7 ± 9.3 | 4.3 ± 0.8 | 48 ± 18 | Reddy et al. (2013a) |
| *A. mittrei* | 12.1 | 10.7 ± 0.9 | 1.6 ± 0.4 | 46 ± 7.1 | Reddy et al. (2012) |
| *E. calleta* | 2.0 | 11.8 ± 5.5 | 2.8 ± 0.7 | 56 ± 18 | Reddy et al. (2013b) |
| *R. lebeau* | 2.7 | 12.3 ± 4.4 | 3.3 ± 1.2 | 64 ± 17 | Reddy et al. (2013c) |

decomposes in one. The divergence between *B. mori* and *P. ricini* is most likely caused by structural and amino acid changes. Moreover, Eri silk's breakdown temperature is quite similar to that of other wild silks, especially *A. pernyi.* Some characteristics of Eri silk are different from those of *A. pernyi*, such as the endothermic maxima at around 310 and 380 °C (Hutmacher et al. 2023). Furthermore, it has been shown that Eri silk has a greater maximal decomposition temperature than *B. mori* silk. The findings suggest that the constituents, composition, and silk strains of various silks influence their distinct characteristics (Horan et al. 2006). Nonetheless, the differential scanning calorimetry (DSC) the curves for wild silkworm silks show discrete exothermic peaks, showing a phase change from α-helix to β-form. The fluid silk comes from multiple silkworm species undergoes a unique, irreversible phase transition (Van Eijk et al. 2004).

### 5.2.3 Geographic Distribution of Species

Wild silk moths, also known as non-mulberry silk moths, are typically not bred in captivity. Instead, indigenous communities gather their cocoons from natural populations. In some instances, partial rearing is practiced, traditionally in the natural environment, with little or no shelter for the larvae. The majority of these lepidopterans pertain to the Saturniidae family, which includes over 1500 species commonly referred to as giant silk moths in the United States (Peigler 1993). Yet, over half of the Saturniidae subspecies pupate underground, creating cocoons. The Lasiocampidae family includes other wild silk moths, which encompass the well-known tent caterpillars (*Malacosoma* Hubner). Furthermore, the Notodontidae family includes two genera of wild silk moths, which are used in Africa. In contrast, the Pieridae family, best known for its white and sulfur butterflies, has a species in Mexico whose silk has been used to produce textiles and paper (Peigler 1993). Species-wise family distribution along with their region are given in Table 5.4.

However, it is important to first clarify the distinction between Tussah and Tasar silk. Tasar refers to the silk fiber obtained from the cocoons of *A. mylitta*, a species predominantly found in India (Fig. 5.7), where the fiber is also known as tropical Tasar (Üzümcü et al. 2019). Tussah, alternatively spelled as "Tusser" according to Peigler (Peigler 1993), is derived from *A. pernyi*, commonly known as the Chinese oak silk moth. In the 1970s, a hybrid species was developed by crossing *A. roylei* and *A. pernyi*, initially named *A. proylei*. The silk from this hybrid is referred to as oak Tasar or temperate Tasar. In various studies, the terms "Tasar" and "Tussah" have been used interchangeably or separately, with some works specifying their distinctions (Üzümcü et al. 2019).

Other amazing types of untamed silk fibers are created worldwide, each displaying the distinctive characteristics of its natural environment. Many of these unusual silk-producing insects have histories that are completely distinct from the well-known wild silks from the Saturniidae family. Members of the Notodontidae family contribute to the lesser-known silk history of Western Africa's thick woods. In Madagascar and other parts of Africa, the Lasiocampidae family is also important because these lepidopterans produce tough, textured fibers. The great ecological and biological diversity that exists across continents is reflected in these many wild silk sources

**Table 5.4** Different species of silkworms, their families, and the regions they are found (Peigler 1993)

| Species | Family | Region |
|---|---|---|
| *Antheraea assamensis* (Muga) | Saturniidae | Assam |
| *Anaphe* | Notodontidae | Western Africa |
| *A. pernyi* (Temperate Tussaer) | Saturniidae | China |
| *A. yamamai* (Japanese Oak Silk) | Saturniidae | Japan |
| *S. cynthia* (Eri) | Saturniidae | China |
| *S. ricini* (Eri) | Saturniidae | India, Japan, China |
| *G. postica* (Sashe Silkworm) | Lasiocampidae | Eastern and Southern Africa |
| *P. otus* | Lasiocampidae | Southeastern Europe |
| *Gonometa rufobrunnea* | Lasiocampidae | Botswana |
| *Borocera cajani* | Lasiocampidae | Madagascar |
| *A. polyphemus* | Saturniidae | Massachusetts |
| *Saturnia pyretorum* | Saturniidae | Southern China |
| *Eucheira socialis* | Pieridae | Mexico |
| *Antheraea paphia* | Saturniidae | India |

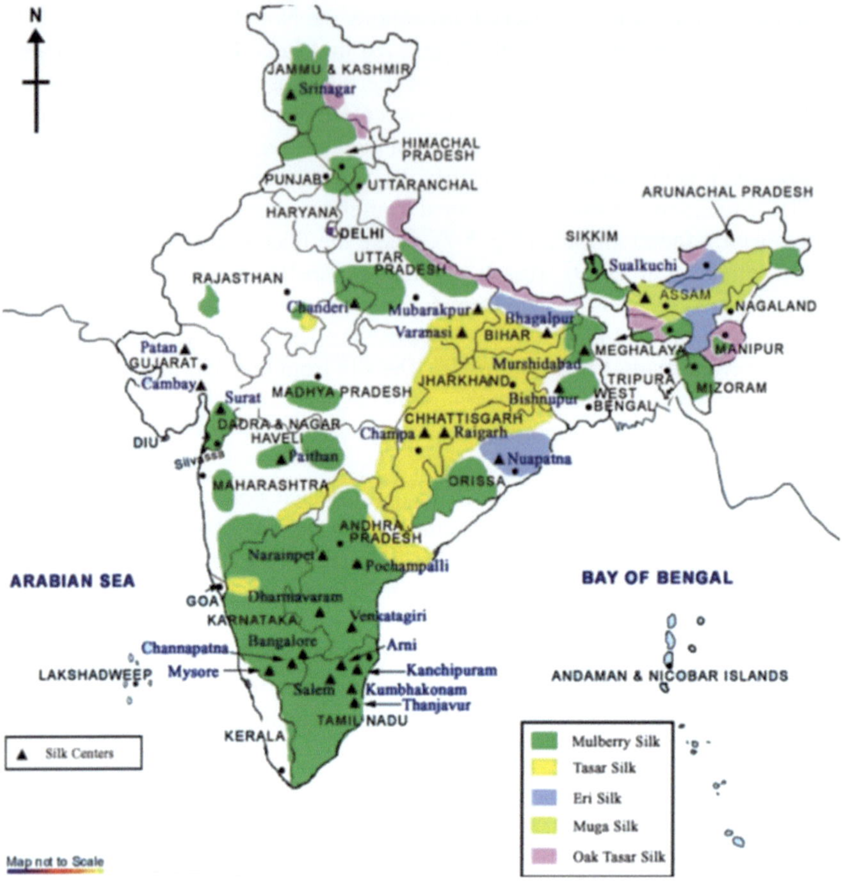


**Fig. 5.7** Geographic distribution of non-mulberry silkworms along with mulberry silkworms in India. Reproduced from (Üzümcü et al. 2019), Copyright CC-BY Terms, 2019

(Peigler 1993, 2020). In addition, the production volume of Tasar, Muga, and Eri silk in India during a 16-year period (FY 2008–FY 2023) is depicted in Fig. 5.8.

From that Pi Chart, it is being seen that silk output increased steadily from FY 2008 (2.075 thousand metric tons) to FY 2020 (10.577 thousand metric tons), according to the chart. Following this high, there is a slow decrease that ends in FY 2023 with 5.929 thousand metric tons.

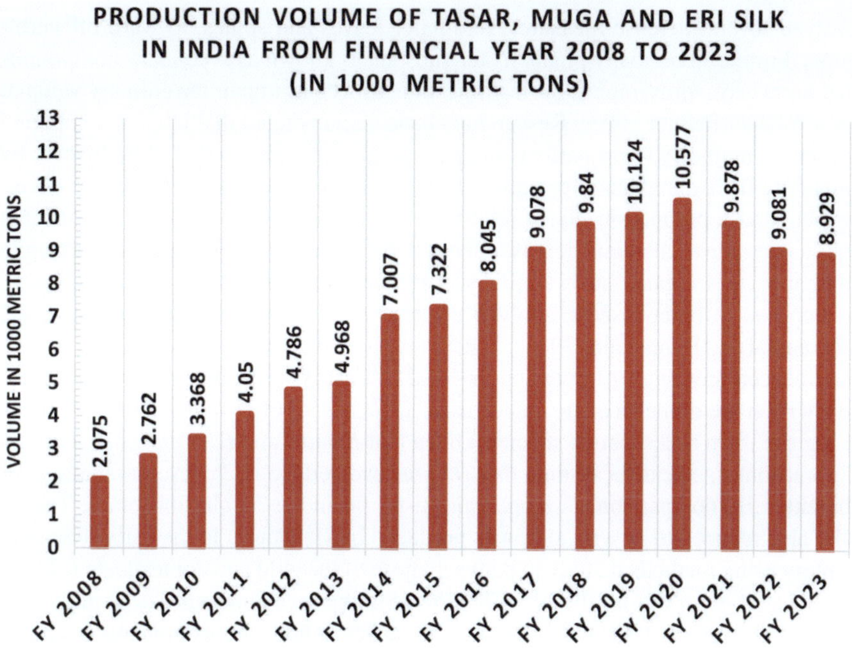


**Fig. 5.8** Production volume of Tasar, Muga, and Eri silk in India from financial year 2008–2023 (in 1000 metric tons). *Source* Central silk board, adapted from Statista (https://www.statista.com/)

## 5.3 Origins and Historical Context of Non-mulberry Silk

### 5.3.1 *Evolutionary Background of Non-mulberry Silkworm Species*

Evolution is defined as a shift in the frequency of hereditary features within a population. Whenever the incidence of a characteristic changes, whether or not it leads to the formation of a new species, the population evolves. These characteristics might be physical or behavioral. Multiple systems govern how organisms inherit characteristics. Genes encode features, but culture (ideas about behavior, sometimes known as memes) also plays a role. In addition, epigenetic factors can be inherited. Silk, according to biologists, is a kind of water-insoluble filament produced by animals' specialized glands. Animals store silk precursors inside their bodies in the form of a gel. When these gels leave the body, a chemical process converts them into fibers. Silk is produced by invertebrates using several glands, demonstrating that silk manufacture developed independently many times (Russell 2017). The *B. mandarina* is one species of wild silk moth that may be found in China and Japan. Although it may appear easy, this species has evolved to consume the leaves of several mulberry trees (genus *Morus*). Because they are unable to flee from their predators, plants have evolved several defense measures, like hairy leaves and spines, to ward off herbivores. Furthermore, many plants foster substances known as secondary compounds that harm herbivores impede during digestion, as well as impair the culinary value of their foliage (Peigler 1993). Researchers have demonstrated that latex, a compound present in mulberry leaves, affects the emergence and reproduction of herbivores by impeding their metabolism of carbohydrates which in turn slows herbivore development and reproduction. Since relatively few bug species are known to consume mulberries, these chemical defenses seem to be successful. This suggests that while mulberry trees created such auxiliary substances to deter insects, *B. mandarina*, as opposed to other insects, evolved a tolerance to these safeguarding chemicals (Rosenthal and Berenbaum 2012). The captive breeding of *B. mandarina* by humans commenced thousands of years beforehand in East Asia. It might be more precise to envision the domesticated silkworms as a subspecies of *B. mori*, but taxonomists designate them as a separate species. Given that *B. mori* and *B. mandarina* share 56 chromosomes, scientists assume that *B. mori* evolved from Chinese populations of the latter. By comparison, *B. mandarina* from Japan has 54 chromosomes. Tussah silk, also referred to as wild silk, was produced in Manchuria in the early twentieth century using methods that lay in between those of the wild and domesticated. There could have been a similar shift with *Bombyx*. Similar to members of the Bombycidae family (which includes *Bombyx*), moths in the Saturniidae family spin silk cocoons. The oak silk moth (*A. pernyi*), a member of the Saturniidae family, is capable of producing silk. It eats oak leaves, as the name suggests. These caterpillars received nourishment on a variety of untamed oak trees by Manchurian farmers circa 1920 (Russell 2017). However, in order to make up for the lack of native oak trees, farmers started growing oaks. In order to separate the silk in their cocoons, moths frequently discharge a liquid at one end of their shelter. This process breaks the continual silk fiber into several shorter strands, diminishing its market value. In addition, the tussah silk-producing terrain made it challenging to differentiate what was cultivated and what was wild. Although some oaks were intentionally evolved for fodder, native trees stipulated nearly every bit of the nutrition for the oak silk caterpillars. In order to facilitate the diversity of cocoons, they also pruned off primeval oaks, which were neither quite tamed nor extremely wild. It is unclear exactly when *B. mandarina* was tamed into *B. Mori*, although it could have transpired thousands of years ago. A Prehistoric archeological site at Hemudu lies in China's Yellow River Canyon. The results of the genetic study are consistent with archeological findings and the similar situation of tussah silk. Geneticists estimate that silkworm breeding originated around 7500 years ago, thus being essentially the same age as the 7000-year-old hemudu bowl with the silkworm-like etchings. According to their viewpoint, domestication ceased around 4000 years ago, which means that for over 3500 years, both domestic and wild silkworms interbred before parting to form discrete reproductive species. Because domestic populations can often be sprung from a small number of wild individuals, domesticated plant and animal groupings usually only maintain a small portion of the genetic diversity found in their wild counterparts. Scientists call this a genetic bottleneck or sampling effect since these ancient organisms included less genetic variety than the overall wild population. Like other domesticated animals, *B.*

*mori* shows evidence of a somewhat large genetic bottleneck. *B. mori* exhibits 83% of the genetic variability observed in *B. mandarina* due to substantial domestication and persistent hybridization between wild and semi-domestic silkworms (Russell 2017). In essence, natural plants and animals developed characteristics that humans illicitly deemed useful. Insects developed silk to aid in their survival and reproduction, and humans eventually found efficient use for it. Like other animals, humans ate and disseminated the fruit that mulberry trees had adapted to help propagate seeds. Humans have probably been eating mulberries since the hunter-gatherer age, and it is possible that silk was first used at the same time.

### 5.3.2 Regional Traditions and Cultural Significance

Native communities have been retrieving cocoons from non-mulberry silk moth populations since prehistoric times. In other circumstances, rearing happens, mostly outside, with little to no covering for the larvae. While most of these silk-producing species are quite ancient and have limited historical value, several were in use well into the early twentieth century and are now rarely used. Despite this, a number of cultivars remain useful in tiny sectors of the Chinese and Japanese industries, as well as a significant portion of India. As fewer components of humanity's progress become outmoded in growing economies, it is possible that old ways for rearing and producing silk are no longer sustainable in today's quickly changing technology environment. As a result of global commerce reaching practically all human groups, there is now a decline in desire for attractive or unique handcrafted things. In the Indian subcontinent, Africa, and Japan, untamed silks continue to be produced and marketed mostly because of traditional practices and ceremonial value (Peigler 1993). For instance, ancient methods of silk caring for and processing have been forgotten in the middle of today's fast technological advancement, when tiny aspects of human civilization are rapidly disappearing in emerging nations. Furthermore, there is less demand for exquisite or unique handmade goods since commerce worldwide has expanded to the majority of human nations. In Africa, India, and Japan, certain wild silks remain produced and pitched with success, mostly because of cultural customs and ceremonial significance. In Africa, India, and Japan, some wild silks are still made and sold with success, mostly because of their socioeconomic and ritualistic relevance. In contrast to the mulberry silkworm, which is an industry worth billion that creates silk throughout Asia with an assortment of digital transformation, contrived diets, and poor wage levels for workers, wilder silks continue to exist as a mainly independent sector. In certain areas, individuals cultivate or collect a restricted amount of their cocoons or larvae and afterward market the raw cocoons to boost their income. The cultural and artistic value of raw silk merchandise becomes insignificant when their production is merely assessed through a materialistic and profit-driven perspective. According to Indian history, eri, muga, alongside tasar are all four fundamental techniques for farming that are exclusive to India. Furthermore, India has produced more wild silk than almost any other country in history, both monetarily and artistically. The Indian government vigorously encourages small enterprises by developing and sponsoring research institutes, colleges, and universities around the country. In similar conditions, the Korean government stresses the monetary importance of the old tussah silk business and promotes it (Peigler 1993).

### 5.3.3 *Early Uses of Non-mulberry Silks in Human Societies*

The usage of silk by people in the past and now is confirmed by historical documentation. Assamese rulers and Ahom nobility once wore the white chapa, also known as champa, of the several varieties of muga silk. Silkworms were raised on *Michelia champaca* L. plants to create the aforementioned variation. Nevertheless, almost a century ago, it ceased producing them. The resulting garments were renowned for their sturdiness and dependability (Peigler 1993). In the latter part of the decade and before, Burma's middle and upper classes wore silk muga. However, in the recent years, costs have risen to the point that many individuals cannot afford it. Throughout the 1960s, India earned significant foreign exchange earnings from the global trade in tasar silk, despite the fact that muga and eri silk are harvested and used only in the Indian subcontinent, with annual profits of hundreds of millions of dollars. The most common export markets are Germany, Japan, and the United States. The bulk of the shipments in question are finished items, such as scarves, neckties, saris, and other garments (Peigler 1993). Eri silk is highly prized and widely used in the weaving of traditional clothing, such as chaddars, mekhalas, and sarees. The cultivation of Muga silk is distinctive to the region of Assamese and has a deep cultural and traditional heritage. Eri silk is currently gaining international recognition as "Ahimsa Silk". Fabric has become progressively more common due to its adiabatic features, which render it ideal for manufacturing jackets, shawls, and blankets (Elumalai et al. 2021).

## 5.4 Habitat and Ecology of Non-mulberry Silks

### 5.4.1 *Preferred Habitats of Non-mulberry Silkworms*

Silkworm categorization is based on the eating habits of silk-producing insects in the Lasiocampidae and Saturniidae families (Mahendran et al. 2006). Most of these species are wild, except for *P. ricini*, which is the only non-mulberry species fully domesticated (Nagaraju 2008). These silkworms exist in polymorphic forms and can be found on various host plants in different geographic areas (Naskar et al. 2014).

### 5.4.2 Ecological Roles in Natural Ecosystems

In their native habitats, non-mulberry silkworms are vital players who serve to maintain the equilibrium of the surroundings. By eating different plants, they assist in regulating plant development and stop any one species from dominating. Maintaining healthy ecosystems depends on a more varied spectrum of plants growing, hence, this makes it possible (Das et al. 2020). Furthermore, important for many predators, including birds and other insects, non-mulberry silkworms sustain the food chain and enable the survival of other species by means of their consumption (Ahmed et al. 2015). Beyond their function as consumers, non-mulberry silkworms greatly help to cycle nutrients in their environments. Their waste products break down organic materials as they eat plant material, therefore, returning vital nutrients to the soil (Konwarh et al. 2017). Another vital ecological function is that of the silk they create during pupation. Once the larvae spin their cocoons, these structures serve as homes for various creatures like bacteria and tiny insects colonizing the silk fabric (Ramappa et al. 2017). The silk's biodegradability guarantees that it breaks down gradually, therefore adding organic material to the ground. By means of these linked ecological roles, non-mulberry silkworms support sustainability and resilience in the natural environment by helping to preserve the fragile equilibrium of their habitats (Das et al. 2020).

### 5.4.3 Effect of Environmental and Climate Factors on Production

The production and expansion of non-mulberry silkworms are much influenced by environmental and climate elements. These are very sensitive to variations in temperature, humidity, and other environmental factors that will directly affect their growth and silk output. For example, the duration of the larval stage and the total number of life-cycles a silkworm may finish in a year depend much on temperature. Extreme temperatures—too hot or too cold—can disrupt their molting, eating, and spinning mechanisms, therefore lowering the silk output (Gahukar 2014). Likewise, humidity conditions impact the health of the silkworms, as low humidity may lead to dehydration and stress in the larvae (Gautam et al. 2022), while excessive humidity can let dangerous infections proliferate. Furthermore, the survival and development of non-mulberry silkworms depend on the availability of appropriate host plants, which are shaped by temperature. A change in climatic pattern, like extended droughts or heavy rain, might influence the number and quality of these plants, therefore influencing silkworm productivity (Konwarh et al. 2017). Therefore, the stability of their habitat and any significant changes in climatic circumstances directly affect the health of non-mulberry silkworm populations; therefore, any such changes may have long-lasting consequences on their growth and silk production (Sujatha et al. 2024).

## 5.5 Production Techniques of Non-mulberry Silkworms

### *5.5.1 Silkworm Rearing Practices*

#### 5.5.1.1 Life Cycle and Biology of Key Non-mulberry Silkworm Species

Non-mulberry silkworms show a wide range of variations in their life cycle patterns. The number of life cycles per year, or voltinism, is influenced by both the species and environmental factors, particularly climatic conditions. These silkworms can be grouped into different categories based on their breeding cycles, such as univoltine (one cycle per year), bivoltine (like *A. mylitta* and *A. atlas*), trivoltine (e.g., *A. mylitta*), or multivoltine (like *A. assama*), depending on the conditions they live in (Jaiswal et al. 2024; Janani et al. 2019). A non-mulberry silkworm's life cycle consists of four basic stages: egg, larva, pupa, and adult, as depicted in n Fig. 5.9 (Jasmine et al. 2014; Jaiswal et al. 2024; Silva et al. 2016).

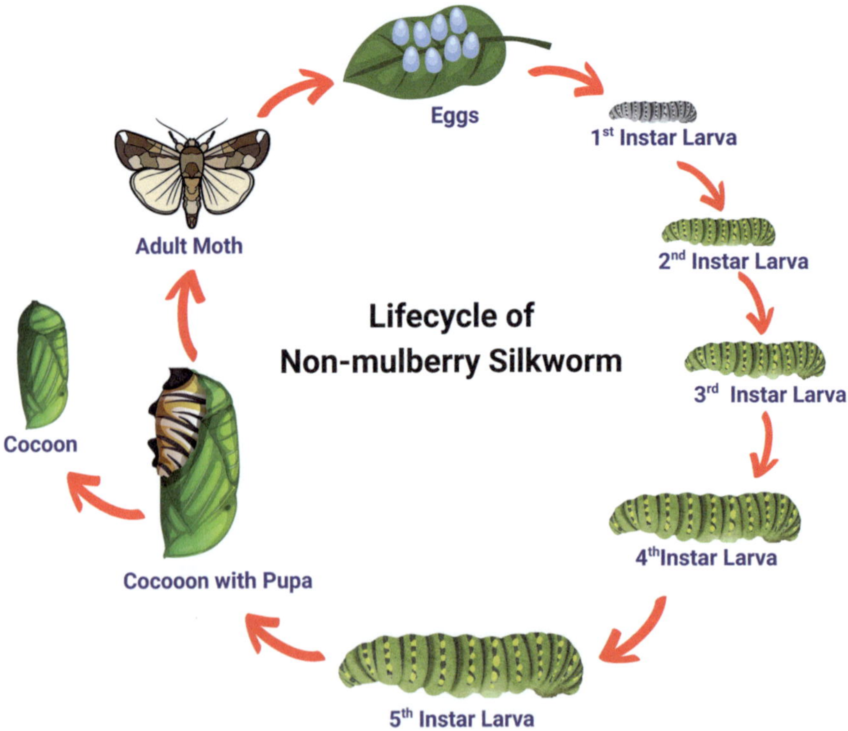


**Fig. 5.9** Various life cycle phases of the non-mulberry silkworm

I. **Eggs**: A female moth typically lays between 200 and 800 eggs during her lifetime, though the number can differ between species. The freshly laid eggs are yellow, with a sticky outer layer that helps them cling to surfaces. When conditions are right, the female moth pledges its fertilized offspring on the leafy parts of the host plant (Naskar et al. 2014).

II. **Larva**: After the hatching process from the eggs, the pupae begin eating on the host plant's leaves, which marks the beginning of their growth. Throughout their larval development, the larvae go through several molting processes, and the interval between these molts is known as "instar". Non-mulberry silkworms generally go through four molts and five instars before they start spinning silk. Their feeding behavior is limited to the larval stage. The length of each instar phase can vary from 4 to 15 days, with the molting period lasting between 24 and 36 h. Once the larvae reach the fifth instar, they cease eating and started to spun silk, which hardens into a cocoon when in contact with air. This cocoon serves as a protective barrier for the larvae. *Antheraea mylitta* is particularly known for its large cocoon, which includes a distinctive stalk (Jaiswal et al. 2024; Mandal and Kundu 2010).

III. **Pupa**: The transformation into a pupa happens inside the cocoon. Under favorable conditions, the larva undergoes transformation into a pupa and stays inside the cocoon until the environment is suitable for further development (Jaiswal et al. 2024).

IV. **Adult Moth**: When conditions are ideal, the pupa secretes an enzyme called "cocoonase", which helps break down the cocoon and allows the moth to emerge. Once the moths reach adulthood, they mate and lay eggs, starting the cycle for the next generation (Jasmine et al. 2014; Jolly et al. 1974). Adult male and female moths are depicted in Fig. 5.10 (Dingle et al. 2005).

**Fig. 5.10** Male and female adult moths of silkworm. Reproduced from (Dingle et al. 2005), Copyright CC-BY terms, 2005

#### 5.5.1.2 Host Plants and Their Management

Wild silkworms, like the Tasar, Eri, and Muga silkworms, depend on different types of plants for food, and these plants play a key role in their growth and the quality of the silk they produce. For example, Eri silkworms mainly eat castor leaves, while Tasar silkworms prefer oak, sal, and arjun trees. Muga silkworms, found in Assam, India, mostly feed on the leaves of the Som tree. The type of plant they feed on is incredibly important, as the nutritional quality of the leaves affects both the health of the silkworms and the quality of the silk they create (Elumalai et al. 2021).

Managing host plants is essential to have a good silkworm farming system. Good plant management guarantees a consistent supply of nutritious leaves for the silkworms, helps to avoid overgrazing and increases farm sustainability (Beirne 1985).

**Site Selection and Planting**: The first stage in effective non-mulberry silkworm farming is choosing suitable surroundings for the plants. Eri silkworms perform well in dry, arid environments; tasar silkworms flourish in temperate and tropical places with plenty of sunshine and rain. Proper spacing of the plants will provide excellent airflow and sunshine, thereby promoting healthy growth free from diseases (Tikader and Kamble 2010).

**Soil and Fertilization**: Healthy soil is thus vital; it is rich in organic matter, well-drained, and somewhat acidic to neutral. Although regular fertilizing helps encourage development, it is crucial not to overdo it as too much fertilizer might produce leaves not optimal for the silkworms. Preference is for organic fertilizers, such as animal dung or compost, which liberate the leaves from dangerous residues (Nath et al. 2023).

**Pruning and Leaf Management**: Ensuring the silkworms have soft, nutritious leaves depends on pruning and leaf management. Tough or bitter leaves generated by overgrown trees might be difficult for silkworms to break down. Frequent cutting promotes new leaf development; however, harvesting should be done in moderation to avoid overgrazing and provide a continuous supply of food (Ahmed et al. 2015).

**Pest and Disease Management**: Host plants, much like other crops, may attract pests and illnesses. Farmers have to be alert for issues such as aphids or fungal development and respond fast. Commonly utilized organic pest management techniques to keep the plants and silkworms safe from damage. Include neem oil or introducing natural predators (Singh and Saratchandra 2002).

**Watering and Irrigation**: Consistent irrigation is necessary, especially in dry conditions, to keep the plants healthy and productive. However, overwatering should be avoided to prevent root rot. Drip irrigation is an efficient method to provide just the right amount of moisture while conserving water (Chauhan and Tayal 2017).

#### 5.5.1.3 Rearing Methods

The rearing of non-mulberry silkworms, such as Eri, Tasar, and Muga, has evolved over time. While traditional methods are still common, modern techniques are being used to improve productivity, silk quality, and sustainability. Both approaches are influenced by the biology of silkworms and their host plants.

**Traditional Rearing Methods**: Traditional silkworm farming is hands-on and based on generations of local knowledge. It relies heavily on the natural environment and local resources. Traditional rearing uses local knowledge of silkworm behavior and environmental conditions to maximize silk production (Sivaprasad et al. 2024).

- Natural Habitat: Tasar and Muga silkworms are often reared in forested areas or open spaces, where host plants, like oak or Som trees, grow naturally. For direct feeding from these trees, the larvae need little infrastructure but careful observation of both plants and silkworms (Chauhan and Tayal 2017).
- Manual Labor: Daily silkworm monitoring and fresh leaf harvesting are among the tasks mostly done by hand. To avoid crowding, farmers also hand separate the silkworms (Narzary et al. 2022).
- Local Resources: To save expenses and encourage sustainability, traditional raising makes use of locally accessible resources such as bamboo, thatch, or wood for shelters (Sujatha et al. 2024).

**Modern Rearing Methods**: Modern methods of breeding silkworms use sustainable practices and technology to increase output and silk quality.

- Controlled Environment: Usually housed in silkworm homes, silkworms are grown under regulated temperature (25–28 °C) and humidity to guarantee good development and avoid illnesses like flacherie (Hăbeanu et al. 1209).
- Artificial Diets: Eri silkworms may be fed synthetic diets created from soybeans or other plant components in places lacking natural host plants. This guarantees food, but it might affect the quality of silk relative to natural leaves (Singh and Saratchandra 2002).
- Automation: Technological developments provide for automated feeding, temperature and humidity control, thereby lowering manual work and increasing efficiency. Additionally helping with leaf collecting and silkworm washing are machines (Singh et al. 2023).
- Management of Pests and illnesses: Eco-friendly pest management strategies like neem extracts and biological agents help to control pests and illnesses thus less dependence on dangerous chemicals (Buhroo et al. 2021).
- Sustainability: Modern farming stresses sustainability by means of organic fertilizers, water-saving irrigation techniques and energy-efficient ways. Priority also is conservation of native silkworm species and habitats (Dandin and Prabhu 2022).

## 5.5.2 *Extraction Processes of Non-mulberry Silks*

### 5.5.2.1 Cocoon Harvesting and Preparation

Harvesting and processing the silk from non-mulberry silkworms is an artistic as well as a scientific endeavor that calls for great attention to detail at every level. Non-mulberry silkworms are known for producing silk with a unique texture and strength, and the quality of the final product depends heavily on how the cocoons are handled.

**Harvesting the Cocoons**: Usually, within three to four days, harvesting starts after the silkworms have finished spinning and the cocoons are completely developed. Harvest time is very important:

- Early Harvesting: Should the silk be taken too soon, it could not have developed completely, resulting in shorter, weaker threads (Elumalai et al. 2021).
- Late Harvesting: Should the larvae start to break out from staying too long, the silk will suffer (Konwarh et al. 2017).

The cocoons are gathered delicately to prevent fraying the threads. For species that naturally spin on trees, care must be taken to harvest without disturbing the surrounding environment (Chauhan and Tayal 2017).

**Sorting the Cocoons**: Once collected, the cocoons are sorted by quality. Non-mulberry silkworms tend to produce cocoons in a variety of shapes, sizes, and colors:

- High-Quality Cocoons: These are smooth, uniform, and free from defects such as holes or blemishes. They yield the best silk (Ramappa et al. 2017).
- Lower-Quality Cocoons: While still usable, these may have imperfections that result in silk of lesser quality, often used for more industrial or rustic applications (Jayaram et al. 2024).

Sorting ensures that only the best cocoons are processed for premium silk products (Tuigong et al. 2015).

**Boiling or Steaming the Cocoons**: The next crucial step is softening the silk by boiling or steaming the cocoons to dissolve the sericin, the protein that binds the silk threads together. Non-mulberry silk tends to be tougher and coarser than mulberry silk, requiring extra care during this stage:

- Boiling: Cocoons are boiled in hot water for 30–45 min. The heat loosens the sericin and allows the silk to be unwound more easily (Hd and Reddy 2024). Given their greater sericin, non-mulberry silkworms may need a longer boiling time or higher temperature (Ahmed et al. 2015).
- Steaming: Alternatively, steaming is a gentler method that preserves the natural texture of the silk, which is important for higher-end products (Ahmed et al. 2015).

The cocoons are washed in cold water to eliminate any residual sericin after boiling or steaming, therefore preventing the silk from feeling sticky (Sahu et al. 2016).

**Unwinding the Silk**: Once the sericin is removed, the silk threads are ready to be unwound. Often shorter and thicker than mulberry silk, non-mulberry silk makes the unwinding process somewhat more labor-intensive:

- Reeling: Under conventional techniques, the cocoon is submerged in warm water, and the silk thread is manually unwound using a tiny hook or reel. The shorter, coarser threads in non-mulberry silk might cause this stage to take more time (Sahu et al. 2016).
- Spinning: Mechanical spinning devices may also be used to yarn the silk. When working with shorter threads, this especially helps as the fibers are twisted together to create longer, continuous strands (Sujatha et al. 2024).

**Dyeing and Finishing**: Non-mulberry silk's natural color spectrum runs from earthy tones to rich, deep colors, which makes it ideal for dyeing. The silk is colored once it has been spun:

- Dyeing: Non-mulberry silk absorbs dyes well, allowing for vibrant, unique colors (Babu 2018). Natural to synthetic dyes may be used in the dyeing process, depending on the intended result (Chauhan and Tayal 2017).
- Finishing: The silk can be treated in various ways to enhance its durability and texture after dying. Calendering—pressing the fabric—helps to enhance the finish, thus smoothing the silk and increasing its shininess. Another rich feel motif might be added using silk softeners (Datta and Nanavaty 2005).

**Storage and Packaging**: At last, after the silk is treated, it is ready for sale or storage. Maintaining the quality of the silk depends on appropriate packing. To reduce damage from insects like moths, it should be kept cold and dry. Correctly wrapped silk guarantees it arrives at makers or consumers in the best shape (Manjunath et al. 2020).

#### 5.5.2.2 Degumming and Spinning Methods Related to Non-mulberry Silks

Processing non-mulberry silk depends especially on the degumming and spinning phases because they help release the full potential of this special fiber. Although these techniques resemble those used for mulberry silks, the rougher, harder character of non-mulberry fibers sets important distinctions.

**Degumming of Non-mulberry Silks**: Degumming eliminates sericin, the protein that binds the silk strands, therefore softening them for spinning. But since non-mulberry silk's stronger structure calls for more thorough degumming:

- Boiling: Non-mulberry silks often need longer or hotter boiling (90–100 °C) for 45 min to an hour to break down the sericin. This method lets the silk strands be untangled more quickly (Ghodke and Chavan 2023).
- Steaming: A milder substitute, steaming at temperatures between 70 and 85 °C, gradually softens the sericin, thereby maintaining the natural texture of the silk (Babu 2018).
- Chemical Degumming: Though this might compromise the inherent qualities of the silk, in certain situations, sodium carbonate or soap solutions are used to hasten the process. The silk is washed in cold water after degumming to eliminate any remaining sericin (Reddy 2019).

**Spinning Methods for Non-mulberry Silks**: Once degummed, non-mulberry silk is spun into yarn. The fibers are shorter and coarser, so the spinning methods differ slightly from those for mulberry silk:

- Hand Reeling: For high-quality silk, traditional hand reeling is used. The cocoon is soaked in warm water, and silk workers reel the threads by hand, combining them into longer strands. This method preserves the silk's organic texture but is labor-intensive (Sonwalkar 1993).
- Mechanical Spinning: For mass production, mechanical spinning machines are employed. These machines twist the short fibers together to form continuous threads, offering consistency but often losing some of the natural texture (Konwarh et al. 2017).
- Spinning on Spinning Frames: Using spinning frames or ring spindles, non-mulberry silk is twisted into finer, more uniform yarns while retaining strength and luster. This is common for high-quality fabric production (Biswal et al. 2022).
- Blending with Other Fibers: Occasionally, non-mulberry silk is blended with other fibers, such as cotton or wool, to enhance texture or improve strength. This can help reduce the natural coarseness, making it easier to spin and weave (Chartvivatpornchai and Okahisa 2025).

#### 5.5.2.3 Challenges in Production of Non-mulberry Silks

Collecting silkworm samples presents significant challenges for the silk industry, particularly since many varieties are restricted to specific geographic regions. Most silkworms are raised outdoors, and the production of seeds is often a bottleneck in the process of producing wild silk. Additionally, many of the worms are prone to various diseases, which negatively impact the yield. It is also challenging to protect silkworms from predators, as they are raised in their natural environments. Furthermore, large-scale deforestation poses a threat to the cultivation of these non-mulberry silk species (Takasu et al. 2002; Naskar et al. 2014).

## 5.6 Perspectives on Future Trends Regarding Silk Production

Some research projects were conducted, which led to the creation of innovative technologies and the subsequent distribution of modern technologies. The wireless personal area network (WPAN) systems, Internet of Things (IoT), image processing techniques, smart sensors, biosensor technology, public–private partnerships (PPP), the establishment of cocoon banks, and startups in the sericulture industry are some of the technologies that fall under this category. In addition, the Cluster Promotion Program (CPP) has resulted in a change in both the techniques used to raise silkworms and the production of gradable raw silk, both of which have seen considerable advances. The upshot of these technologies is that they have contributed to a rise in the production of silk as well as the beginning and implementation of contemporary raw silk production. This has been accomplished through the enhancement of silkworm or sericulture activities on farms as well as in an industrial setting (Chand et al. 2023; Geetha et al. 2020).

## 5.7 Applications of Non-mulberry Silk

Research is being conducted to investigate the potential of protein fibers for use in various technological applications, including filters, membranes, paper, textiles, and leather. The recent developments include the application of these materials for nanoscale patterning. In the fields of medicine and biomedicine, protein fibers are being researched for their potential applications as surgical threads and sutures, as well as for the production of biological membranes and scaffolds that would facilitate the formation of cells and the function of tissues. In addition, they are utilized in the treatment of wounds, the replacement of ligaments, the installation of implants, cosmetic surgery, and the distribution of pharmaceuticals. Additionally, protein fibers have the potential to act as carriers in medicinal applications and can be utilized as stabilizers in the food industry. Contact lens materials and cosmetics are two further applications that could be considered. Collagen, fibrin, and elastin are examples of natural protein fibers that are utilized in hydrogels for the purposes of tissue engineering and medication delivery. Due to the fact that it is both biodegradable and compatible with cells, silk, on the other hand, is currently being examined as a potential biomaterial that could be useful in the future. This is because these fibers have limited mechanical properties, which is the reason behind this (Babu and Sinclair 2015).

### 5.7.1 Biomedical Applications of Non-mulberry Silk

For more than five thousand years, silk has been utilized in textiles and for several medical purposes. Silk's biocompatibility, biodegradability, and self-assembly, in addition to its one-of-a-kind physical, chemical, and mechanical qualities, have recently sparked a growing interest in the use of silk for biomedical applications. The fundamental unit of silk, which is referred to as fibroin fibers, is composed of repeated sequences of amino acids that have an effect on the structure of the protein chain as well as the qualities of the substance. According to Jastrzebska et al. (2015), these amazing features of silk have notable implementations in the field of biomedicine. In addition, the hydrophobic amino acids and antioxidant characteristics of sericin make it beneficial not only to cosmetics but also to the food industry. As a result of its high moisture content, silk is also used as a medicinal material for tissue regeneration, encouraging cell growth, shielding against ultraviolet radiation, and being incorporated into the creation of lotions and shampoos. The antioxidant activity of sericin, in conjunction with its low digestibility, contributes to the enhancement of its medical applications. These applications include the potential for sericin to operate as an anticancer, antibacterial, anti-inflammatory, and anticoagulant agent, as well as to foster colon health (Kunz et al. 2016). The inherent tripeptide regions of non-mulberry silks enable their utilization in various biomedical applications, including drug delivery systems (Baruah et al. 2020), bone regeneration (Bhattacharjee et al. 2015), skin repair (Chouhan et al. 2017), vascular grafts (Gupta et al. 2020), 3D disease models (Singh et al. 2022), and cancer models (Talukdar and Kundu 2012). Table 5.5 presents various formats of the materials and other applications of non-mulberry silks as biomaterials.

**Table 5.5** Species of non-mulberry silks and their biomedical applications (Jasmine et al. 2014)

| Species of non-mulberry silk | Format of material | Applications | References |
|---|---|---|---|
| *A. mylitta* | Gland fibroin 3D scaffold, sericin film, gland fibroin 3D scaffold, gland fibroin 3D scaffold, gland fibroin and sericin nanoparticles, gland fibroin and sericin nanoparticles, gland fibroin 3D scaffold, films, micropatterned surfaces | Bone tissue engineering, skin tissue engineering, adipose tissue engineering, cartilage tissue engineering, delivery vehicles, model matrices | Mandal and Kundu (2009a, b), Dash et al. (2008), Talukdar et al. (2011a, b) |
| *A. pernyi* | Braided fibroin | Tendon engineering | Fang et al. (2009) |

### *5.7.2 Therapeutic Applications of Silk*

There have been a great number of studies carried out on different kinds of silk for the purpose of determining its therapeutic applications in a variety of disciplines. Research has demonstrated that silk sericin, because of its protein makeup, is vulnerable to the action of digestible proteolytic enzymes in the body. As a result, silk sericin is a material that is both biocompatible and biodegradable. The gelling qualities, moisture-retention capacity, and skin adhesion that it possesses allow it to be utilized in a wide variety of applications within the medical, pharmaceutical, and cosmetic industries. There have been additional studies that have brought attention to the anti-proliferative effects of sericin. These studies have also emphasized the impact of $\gamma$-irradiated silk fibroin on the cell cycle and its anti-tumor activity in mouse peritoneal macrophages. These findings suggest that the proliferative effects of sericin are dependent on the concentration of the substance (Byun et al. 2010; Kaewkorn et al. 2012). Furthermore, researchers have investigated the potential in vivo and in vitro medicinal applications of non-mulberry silk sericin, including its antibacterial qualities, wound-healing capacities, and minimal inflammation effects with no infections (Sapru et al. 2018). Additionally, silk sericin mixed with silk fibroin is utilized in the cosmetics industry for the purpose of nail care, hair care, and skin care. It has been put into lotions and ointments, and it has been shown to increase the suppleness of the skin, in addition to providing anti-aging and anti-wrinkle advantages (Joseph and Raj 2012; Padamwar et al. 2005). Specifically, for the antioxidant properties of non-mulberry silks (*A. mylitta*, *A. assama*, and *Aernthaea ricini*), their usage in the potential cosmetic sector is also remarkable, along with the mulberry silk (Kumar and Mandal 2017).

### *5.7.3 Silk Fiber as a Possible Industrial and Engineering Material*

Fibrin and sericin are the two principal protein components that are found in the cocoons surrounding silkworms. Using its role as a glue protein, sericin is responsible for bringing together two fibroin filaments. During the process of removing fibroin from the cocoon in the textile industry, sericin is often discarded as waste. This is despite the fact that fibroin is widely utilized in a variety of applications, including those in the automotive, industrial, and medical sectors. It is possible for sericin to be digested due to its protein structure, which renders it vulnerable to proteolytic enzymes within the organism's body. Due to this feature, it is recognized as a material that is both biocompatible and biodegradable (Chand et al. 2023). However, these attributes are influenced by the variability in the reproducibility of the silk. Scanning electron microscopy has been utilized in research projects that have focused on the diversity of tensile properties (Pérez-Rigueiro et al. 1998). In addition, silkworm sericins isolated from mulberry (*B. mori*) and non-mulberry silks (tropical tasar,

*A. mylitta*, muga, *A. assama*, and eri, *S. ricini*) have been shown to be effective in a variety of applications, including cryopreservation, tissue engineering, culture media, hydrogels, and films (Sahu et al. 2016; Kunz et al. 2016).

### 5.7.4 *Electricity and Optical Devices from Silkworm Silk*

The increasing application of silk in biomedicine has led to the enhancement of silk fibroin-based materials for targeted functions, such as sensing, UV light resistance, cell visualization, and antibacterial properties (Chand et al. 2023; Saleh 2021; Seifipour et al. 2020). Silk cocoon membranes, engineered by insects, have been investigated for their electrical properties. Investigations have been conducted on the electrical properties of mulberry (*B. mori*) and non-mulberry silk cocoon membranes (SCM), including those derived from tussar and *A. mylitta*. Research indicates that dry SCMs function as insulators, absorb moisture, and produce electrical currents that are affected by temperature. The results indicate that the electrical properties of SCMs, which are influenced by temperature and humidity, may be applicable in various fields such as battery technology, biosensors, humidity sensors, steam engines, and waste heat management (Tulachan et al. 2014). The previous studies indicate that functionalized silk is extensively utilized as an electrode material in supercapacitors and resistive switching memory devices due to its cost-effectiveness, natural availability, and high carbon content. Metal salts such as $FeCl_2$ and $ZnCl_2$, obtained from natural silk, function as efficient activation-graphitization agents during the dissolution process, leading to the formation of hierarchical porous nitrogen-doped carbon nanosheets (Zhu et al. 2016). Fluorescent silk has garnered considerable interest due to its potential applications in cell visualization and the monitoring of scaffold performance, representing a novel modification of natural biomaterials. Advancements in genetic engineering have enabled the production of fluorescent silks in multiple colors, such as red, green, and orange (Fan et al. 2019; Li et al. 2017; Zhou et al. 2017). Fibroin and sericin possess diverse applications, and their recovery and reuse in the textile industry contribute to addressing environmental challenges while also offering significant scientific and commercial benefits.

## 5.8 Conclusion

Non-mulberry silkworms, comprising varieties such as tasar, muga, and eri, represent a valuable and biologically diverse group of silk-producing worms that extend beyond the commonly known mulberry silk. These are not only integral to regional economies and cultural heritage, particularly in parts of India and Southeast Asia, but also possess unique physical and chemical properties that set them apart from mulberry silk. The structural proteins in these silks, primarily fibroin and sericin, exhibit varied amino acid compositions that influence mechanical strength, biodegradability, and compatibility in biomedical applications. The recent research into the molecular structure, mechanical behavior, and processing techniques of non-mulberry silks has opened up innovative applications, especially in tissue engineering, drug delivery, and biodegradable composites. Despite their potential, non-mulberry silks are underutilized at an industrial scale due to limited availability, rearing challenges, and processing complexities. With continued research and investment, non-mulberry silks can contribute significantly to sustainable materials science, rural development, and high-performance biomaterials.

**Declaration** The authors declare that they have no known competing financial interests or personal relationships that could have appeared to influence the work reported in this paper. Generative AI and AI-assisted technologies have been used in the writing process to improve the readability and language of the work. After using this tool, the author(s) reviewed and edited the content as needed and take full responsibility for the content of the publication.

## References

Ahamad SI, Neetha K, Vootla SK (2021) Bioactive secondary metabolites of wild *A. mylitta* silkworm cocoons. In: Moths and caterpillars. IntechOpen

Ahmad R, Kamra A, Hasnain SE (2004) Fibroin silk proteins from the nonmulberry silkworm *P. ricini* are biochemically and immunochemically distinct from those of the mulberry silkworm *B. mori*. DNA Cell Biol 23(3):149–154

Ahmed S, Singh N, Sarkar C (2015) Role of forest biodiversity in conservation of non-mulberry (vanya) silk in India. Munis Entomol Zool 10(1):342–357

Babu K (2015) Natural textile fibres: animal and silk fibres. In: Sinclair R (ed) Textiles and fashion. Elsevier, Amsterdam, pp 57–78

Babu KM (2017) Silk from silkworms and spiders as high-performance fibers. In: Bhat G (ed) Structure and properties of high-performance fibers. Woodhead Publishing series in textiles, Elsevier, pp 327–366

Babu KM (2018) Silk: processing, properties and applications. Woodhead Publishing

Bandyopadhyay A, Chowdhury SK, Dey S, Moses JC, Mandal BB (2019) Silk: a promising biomaterial opening new vistas towards affordable healthcare solutions. J Indian Instit Sci 99(3):445–487. https://doi.org/10.1007/s41745-019-00114-y

Baruah RR, Kalita MC, Devi D (2020) Novel non-mulberry silk fibroin nanoparticles with enhanced activity as potential candidate in nanocarrier mediated delivery system. RSC Adv 10(15):9070–9078. https://doi.org/10.1039/C9RA08901B

Beirne BP (1985) Avoidable obstacles to colonization in classical biological control of insects. Can J Zool 63(4):743–747

Bhattacharjee P, Naskar D, Kim H-W, Maiti TK, Bhattacharya D, Kundu SC (2015) Non-mulberry silk fibroin grafted PCL nanofibrous scaffold: promising ECM for bone tissue engineering. Eur Polym J 71:490–509. https://doi.org/10.1016/j.eurpolymj.2015.08.025

Biswal B et al (2022) Extraction of silk fibroin with several sericin removal processes and its importance in tissue engineering: a review. J Polym Environ 30(6):2222–2253. https://doi.org/10.1007/s10924-022-02381-w

Buhroo ZI, Bhat MA, Aziz A (2021) Chapter-5 sericulture. In: Latest trends in zoology and entomology sciences, p 69

Byun E-B et al (2010) Enhancement of anti-tumor activity of gamma-irradiated silk fibroin via immunomodulatory effects. Chem Biol Interact 186(1):90–95. https://doi.org/10.1016/j.cbi.2010.03.032

Chand S, Chand S, Raula B (2023) Usage of silkworm materials in various ground of science and research. J Nat Fibers 20(1):2139328. https://doi.org/10.1080/15440478.2022.2139328

Chartvivatpornchai N, Okahisa Y (2025) Structural and mechanical comparison of Eri and Mulberry silk fibroin nanofibers films through advanced mechanical treatments for sustainable applications. Int J Biol Macromol 295:139554. https://doi.org/10.1016/j.ijbiomac.2025.139554

Chattopadhyay SDR, Gulrajani M, Sen K (2005) Study of property and structural variants of mulberry and tasar silk filaments. AUTEX Res J 5(2):81–86

Chauhan T, Tayal MK (2017) Mulberry sericulture. Ind Entomol 21:197–263

Chouhan D, Chakraborty B, Nandi SK, Mandal BB (2017) Role of non-mulberry silk fibroin in deposition and regulation of extracellular matrix towards accelerated wound healing. Acta Biomater 48:157–174. https://doi.org/10.1016/j.actbio.2016.10.019

Colomban P, Jauzein V (2018) 5—silk: fibers, films, and composites—types, processing, structure, and mechanics. In: Bunsell AR (ed) Handbook of properties of textile and technical fibres, 2nd edn. Woodhead Publishing, New York, pp 137–183

Dandin S, Prabhu N (2022) Silkworm biodiversity, conservation, threats, and sustainable utilization in Indian subcontinent. Biodiversity. CRC Press, New York, pp 157–192

Das S (2020) Effect of drying on tensile properties and structure of tasar and muga silk. Indian J Fibre Text Res 45(2):211–214

Das SK, Sahu BK, Singh D (2020) Host plant diversity of non-mulberry silkworms: a review. J Pharmacogn Phytochem 9(3):109–113

Dash R, Mukherjee S, Kundu S (2006) Isolation, purification and characterization of silk protein sericin from cocoon peduncles of tropical tasar silkworm, *A. mylitta*. Int J Biol Macromol 38(3–5):255–258. https://doi.org/10.1016/j.ijbiomac.2006.03.001

Dash R, Mandal M, Ghosh SK, Kundu S (2008) Silk sericin protein of tropical tasar silkworm inhibits UVB-induced apoptosis in human skin keratinocytes. Mol Cell Biochem 311:111–119. https://doi.org/10.1007/s11010-008-9702-z

Datta RK, Nanavaty M (2005) Global silk industry: a complete source book. Universal-Publishers

Dey S et al (2023) Photocrosslinkable silk-based biomaterials for regenerative medicine and healthcare applications. Regener Eng Transl Med 9(2):181–201. https://doi.org/10.1007/s40883-022-00277-8

Dingle JG, Hassan E, Gupta M, George D, Anota L, Begum H (2005) Silk production in Australia: a report for the rural industries research and development corporation

Elumalai D, Raj PM, Ramamoorthy R, Mohan C, Poovizhiraja B (2021) Introduction to non-mulberry silkworms. CRC Press, New York

Fan S et al (2019) Silk materials for medical, electronic and optical applications. Sci China Technol Sci 62:903–918. https://doi.org/10.1007/s11431-018-9403-8

Fang Q, Chen D, Yang Z, Li M (2009) In vitro and in vivo research on using *A. pernyi* silk fibroin as tissue engineering tendon scaffolds. Mater Sci Eng C 29(5):1527–1534. https://doi.org/10.1016/j.msec.2008.12.007

Fang G et al (2016) Exploration of the tight structural–mechanical relationship in mulberry and non-mulberry silkworm silks. J Mater Chem B 4(24):4337–4347. https://doi.org/10.1039/C6TB01049K

Fazal N, Latief N (2018) *Bombyx mori* derived scaffolds and their use in cartilage regeneration: a systematic review. Osteoarth Cartil 26(12):1583–1594. https://doi.org/10.1016/j.joca.2018.07.009

Gahukar R (2014) Impact of major biotic factors on tropical silkworm rearing in India and monitoring of unfavourable elements: a review. Sericologia 54:150–170

Gautam M, Singh D, Singh S, Singh S, Kumar M, Singh S (2022) A review on silkworm (*B. mori* Linn.) an economic important insect. Biol Forum Int J 14(4):482–491

Geetha G, Dasari JR, Saha A (2020) Impact of cluster promotion programme on socio-economic aspects on women sericulture farmers in Karnataka, India. Curr Agricult Res J 8(1):58–68

Ghodke PB, Chavan RJ (2023) Research trends in silk spinning process: a review. J Entomol Res 1:200–204

Guo C, Zhang J, Wang X, Nguyen AT, Liu XY, Kaplan DL (2017) Comparative study of strain-dependent structural changes of silkworm silks: insight into the structural origin of strain-stiffening. Small 13(47):1702266. https://doi.org/10.1002/smll.201702266

Gupta A, Mita K, Arunkumar KP, Nagaraju J (2015) Molecular architecture of silk fibroin of Indian golden silkmoth, *A. assama*. Sci Rep 5(1):12706

Gupta P et al (2020) Bioresorbable silk grafts for small diameter vascular tissue engineering applications: in vitro and in vivo functional analysis. Acta Biomater 105:146–158. https://doi.org/10.1016/j.actbio.2020.01.020

Hăbeanu M, Gheorghe A, Mihalcea T (2023) Silkworm B mori—sustainability and economic opportunity, particularly for Romania. Agriculture 13(6):1209

Hd YK, Reddy NN (2024) Sericulture—soil to silk. Advances 69:85

Ho M-P, Wang H, Lau K-T (2012) Effect of degumming time on silkworm silk fibre for biodegradable polymer composites. Appl Surf Sci 258(8):3948–3955. https://doi.org/10.1016/j.apsusc.2011.12.068

Horan RL, Collette AL, Lee C, Antle K, Chen J, Altman GH (2006) Yarn design for functional tissue engineering. J Biomech 39(12):2232–2240. https://doi.org/10.1016/j.jbiomech.2005.07.007

Hridya H, Guha L, Mazumdar M, Sarkar B, Vijayakumar S, Borpuzari P (2021) Probing the potentiality of the defoliator *C. trifenestrata* Helfer silk: a revisit. Bull Natl Res Centre 45:1–8. https://doi.org/10.1186/s42269-021-00669-w

Hutmacher DW, Tandon B, Dalton PD (2023) Scaffold design and fabrication. In: Jan De Boer CAVB, Uquillas JA, Malik N (eds) Tissue engineering (3rd edn). Elsevier, Amsterdam, pp 355–385

Hyde N (1984) The queen of textiles. Natl Geogr 165(1):3–48

Jaiswal C et al (2024) Nonmulberry silk-based biomaterials: biomedical applications, current status, and future perspective. In: Subhas RLR, Kundu C (ed) Silk-based biomaterials for tissue engineering, regenerative and precision medicine. Woodhead Publishing series in biomaterials, pp 55–87

Janani G et al (2019) Insight into silk-based biomaterials: from physicochemical attributes to recent biomedical applications. ACS Appl Bio-Mater 2(12):5460–5491. https://doi.org/10.1021/acsabm.9b00576

Jasmine S, Mandal B (2014) Types and properties of non-mulberry silk biomaterials for tissue engineering applications. In: Kundu SC (ed) Silk Biomaterials for tissue engineering and regenerative medicine. Elsevier, Amsterdam, pp 275–298

Jastrzebska K, Kucharczyk K, Florczak A, Dondajewska E, Mackiewicz A, Dams-Kozlowska H (2015) Silk as an innovative biomaterial for cancer therapy. Rep Pract Oncol Radiother 20(2):87–98. https://doi.org/10.1016/j.rpor.2014.11.010

Jayaram H, Mahadevegowda LG, Boregowda MH (2024) Seri-entrepreneurship: current status and potential opportunities. Curr Agricult Res J 12(1):31

Jolly M, Sen S, Sonawalkar T, Prasad G (1979) Non mulberry silks. Food and agriculture organization. United Nat Serv Bull 29:178

Jolly MS, Sen S, Ahsan MM (1974) Tasar culture. Ambika publishers

Joseph B, Raj SJ (2012) Therapeutic applications and properties of silk proteins from *B. mori*. Front Life Sci 6(3–4):55–60. https://doi.org/10.1080/21553769.2012.760491

Kaewkorn W, Limpeanchob N, Tiyaboonchai W, Pongcharoen S, Sutheerawattananonda M (2012) Effects of silk sericin on the proliferation and apoptosis of colon cancer cells. Biol Res 45(1):45–50. https://doi.org/10.4067/S0716-97602012000100006

Kim S-R et al (2018) Genome sequence of the Japanese oak silk moth, *A. yamamai*: the first draft genome in the family Saturniidae. Gigascience 7(1):gix113. https://doi.org/10.1093/gigascience/gix113

Kioko EN (1998) Biodiversity of wild silkmoths (Lepidoptera) and their potential for silk production in East Africa. Kenyatta University, Nairobi

Kobayashi I, Kojima K, Sezutsu H, Uchino K, Tamura T (2009) Expression of the Japanese oak silkworm *A. yamamai* fibroin gene in the domesticated silkworm *B. mori*. Insect Sci 16(6):465–473. https://doi.org/10.1111/j.1744-7917.2009.01293.x

Konwarh R, Bhunia BK, Mandal BB (2017) Opportunities and challenges in exploring Indian nonmulberry silk for biomedical applications. Proc Indian Natn Sci Acad 83(1):85–101

Kumar JP, Mandal BB (2017) Antioxidant potential of mulberry and non-mulberry silk sericin and its implications in biomedicine. Free Rad Biol Med 108:803–818. https://doi.org/10.1016/j.freeradbiomed.2017.05.002

Kumar JP, Alam S, Jain AK, Ansari KM, Mandal BB (2018) Protective activity of silk sericin against UV radiation-induced skin damage by downregulating oxidative stress. ACS Appl Bio-Mater 1(6):2120–2132. https://doi.org/10.1021/acsabm.8b00558

Kundu SC, Dash BC, Dash R, Kaplan DL (2008) Natural protective glue protein, sericin bioengineered by silkworms: potential for biomedical and biotechnological applications. Prog Polym Sci 33(10):998–1012. https://doi.org/10.1016/j.progpolymsci.2008.08.002

Kundu S et al (2012) Nonmulberry silk biopolymers. Biopolymers 97(6):455–467. https://doi.org/10.1002/bip.22024

Kunz RI, Brancalhão RMC, Ribeiro LFC, Natali MRM (2016) Silkworm sericin: properties and biomedical applications. BioMed Res Int 2016(1):8175701

Leem JW et al (2018) Visible light biophotosensors using biliverdin from *A. yamamai*. Opt Exp 26(24):31817–31828. https://doi.org/10.1364/OE.26.031817

Li Q et al (2017) Sub-micron silk fibroin film with high humidity sensibility through color changing. RSC Adv 7(29):17889–17897. https://doi.org/10.1039/C6RA28460D

Mahendran B, Ghosh S, Kundu S (2006) Molecular phylogeny of silk-producing insects based on 16S ribosomal RNA and cytochrome oxidase subunit I genes. J Genet 85:31–38. https://doi.org/10.1007/BF02728967

Mai-Ngam K, Boonkitpattarakul K, Jaipaew J, Mai-Ngam B (2011) Evaluation of the properties of silk fibroin films from the non-mulberry silkworm *S. cynthia ricini* for biomaterial design. J Biomater Sci Polym Edn 22(15):2001–2022. https://doi.org/10.1163/092050610X530964

Malay AD et al (2016) Relationships between physical properties and sequence in silkworm silks. Scie Rep 6(1):27573

Mandal BB, Kundu SC (2009a) Osteogenic and adipogenic differentiation of rat bone marrow cells on non-mulberry and mulberry silk gland fibroin 3D scaffolds. Biomaterials 30(28):5019–5030. https://doi.org/10.1016/j.biomaterials.2009.05.064

Mandal BB, Kundu S (2009b) Self-assembled silk sericin/poloxamer nanoparticles as nanocarriers of hydrophobic andhydrophilic drugs for targeted delivery. Nanotechnology 20(35):355101. https://doi.org/10.1088/0957-4484/20/35/355101

Mandal BB, Kundu SC (2010) Biospinning by silkworms: silk fiber matrices for tissue engineering applications. Acta Biomater 6(2):360–371. https://doi.org/10.1016/j.actbio.2009.08.035

Manjunath R, Kumar A, Kumar KA (2020) Utilisation of sericulture waste by employing possible approaches. In: Naeem M, Ansari A, Gill S (eds) Contaminants in agriculture: sources, impacts and management. Springer, Cham, pp 385–398

Nagaraju J (2008) Silk of India, grace and luster. Biotechnol News 3(5):4–7

Narzary PR et al (2022) Recent trends in seri-bioscience: its prospects in modern sericulture. Pharma Innov 11(1):604–611

Naskar D, Barua R, Ghosh A, Kundu S (2014) Introduction to silk biomaterials. In: Kundu SC (ed) Silk biomaterials for tissue engineering and regenerative medicine. Elsevier, Amsterdam, pp 3–40

Nath I, Dutta PL, Gadge AS, Narzary PR (2023) Role of plant extracts on growth and development of mulberry and non-mulberry silkworms. Sci World Monthly E Mag 3(4):494–501. https://doi.org/10.5281/zenodo.7827408

Nayak S, Kundu SC (2016) Silk protein sericin: promising biopolymer for biological and biomedical applications. In: Nuno RLR, Neves M (eds) Biomaterials from nature for advanced devices and therapies, 1st edn. Wiley, Amsterdam, pp 142–158

Padamwar MN, Pawar AP, Daithankar AV, Mahadik K (2005) Silk sericin as a moisturizer: an in vivo study. J Cosm Dermatol 4(4):250–257. https://doi.org/10.1111/j.1473-2165.2005.00200.x

Pal S, Kundu J, Talukdar S, Thomas T, Kundu SC (2013) An emerging functional natural silk biomaterial from the only domesticated non-mulberry silkworm *S. ricini*. Macromol Biosci 13(8):1020–1035. https://doi.org/10.1002/mabi.201300013

Peigler RS (1993) Wild silks of the world. Am Entomol 39(3):151–162. https://doi.org/10.1093/ae/39.3.151

Peigler RS (2020) Wild silks: their entomological aspects and their textile applications. Handbook of natural fibres. Elsevier, Amsterdam, pp 715–745

Pérez-Rigueiro J, Viney C, Llorca J, Elices M (1998) Silkworm silk as an engineering material. J Appl Polym Sci 70(12):2439–2447

Qu XM, Steiner H, Engström Å, Bennich H, Boman HG (1982) Insect immunity: isolation and structure of cecropins B and D from pupae of the Chinese oak silk moth, *A. pernyi*. Eur J Biochem 127(1):219–224. https://doi.org/10.1111/j.1432-1033.1982.tb06858.x

Ramappa V, Dev P, Ranjan B, Kumar Y, Sangeeta S, Gopal R (2017) Comparative study of antioxidant potential of mulberry and non-mulberry silkworm pupal oil

Reddy N (2019) Silk: materials, processes, and applications. Woodhead Publishing

Reddy N, Yang Y (2010a) Structure and properties of cocoons and silk fibers produced by *Hyalophora cecropia*. J Mater Sci 45:4414–4421. https://doi.org/10.1007/s10853-010-4523-3

Reddy N, Yang Y (2010b) Morphology and tensile properties of silk fibers produced by uncommon Saturniidae. Int J Biol Macromol 46(4):419–424. https://doi.org/10.1016/j.ijbiomac.2010.02.013

Reddy N, Yang Y (2010c) Structure and properties of ultrafine silk fibers produced by *Theriodopteryx ephemeraeformis*. J Mater Sci 45:6617–6622. https://doi.org/10.1007/s10853-010-4752-5

Reddy N, Yang Y (2012) Investigation of the structure and properties of silk fibers produced by *Actias lunas*. J Polym Environ 20:659–664. https://doi.org/10.1007/s10924-012-0482-x

Reddy N, Jiang Q, Yang Y (2012) Biocompatible natural silk fibers from *Argema mittrei*. J Biobased Mater Bioenergy 6(5):558–563. https://doi.org/10.1166/jbmb.2012.1255

Reddy N, Zhao Y, Yang Y (2013a) Structure and properties of cocoons and silk fibers produced by *A. atlas*. J Polym Environ 21:16–23. https://doi.org/10.1007/s10924-012-0549-8

Reddy N, Jiang Q, Yang Y (2013b) Investigation of the properties and potential medical applications of natural silk fibers produced by *Eupackardia calleta*. J Biomater Sci Polym Edn 24(4):460–469. https://doi.org/10.1080/09205063.2012.690712

Reddy N, Jiang Q, Yang Y (2013c) Properties and potential medical applications of silk fibers produced by *Rothischildia lebeau*. J Biomater Sci Polym Edn 24(7):820–830. https://doi.org/10.1080/09205063.2012.719756

Reddy N, Yang Y, Reddy N, Yang Y (2015) Non-mulberry silk fibers: natural protein fibers. Innovative biofibers from renewable resources. Springer, Berlin, pp 165–174

Rosenthal GA, Berenbaum MR (2012) Herbivores: their interactions with secondary plant metabolites: ecological and evolutionary processes. Academic press, New York

Russell E (2017) Spinning their way into history: silkworms, mulberries and manufacturing landscapes in China. Global Environ 10(1):21–53. https://doi.org/10.3197/ge.2017.100102

Sahu N et al (2016) Non-mulberry and mulberry silk protein sericins as potential media supplement for animal cell culture. BioMed Res Int 2016(1):7461041. https://doi.org/10.1155/2016/7461041

Saleh TA (2021) Protocols for synthesis of nanomaterials, polymers, and green materials as adsorbents for water treatment technologies. Environ Technol Innov 24:101821. https://doi.org/10.1016/j.eti.2021.101821

Sapru S, Das S, Mandal M, Ghosh AK, Kundu SC (2018) Prospects of nonmulberry silk protein sericin-based nanofibrous matrices for wound healing—in vitro and in vivo investigations. Acta Biomater 78:137–150. https://doi.org/10.1016/j.actbio.2018.07.047

Seifipour R, Nozari M, Pishkar L (2020) Green synthesis of silver nanoparticles using *Tragopogon collinus* leaf extract and study of their antibacterial effects. J Inorg Organometal Polym Mater 30(8):2926–2936. https://doi.org/10.1007/s10904-020-01441-9

Silva SS et al (2016) Fabrication and characterization of Eri silk fibers-based sponges for biomedical application. Acta Biomater 32:178–189. https://doi.org/10.1016/j.actbio.2016.01.003

Singh R, Saratchandra B (2002) An integrated approach in the pest management in sericulture. Int J Ind Entomol Biomater 5(2):141–151

Singh YP, Moses JC, Bandyopadhyay A, Mandal BB (2022) 3D bioprinted silk-based in vitro osteochondral model for osteoarthritis therapeutics. Adv Healthcare Mater 11(24):2200209. https://doi.org/10.1002/adhm.202200209

Singh S, Jigyasu DK, Patidar OP, Kumar A, Shabnam AA (2023) Eri silk insect (*S. ricini*, Donovan). In: Commercial insects. CRC Press, New York, pp 103–125

Sivaprasad V, Sangannavar PA, Lingaiah K (2024) An introduction to biotechnology driven advances for silkworm improvement and sustainable perspectives. In: Suresh RV, Saha S, Alam K (eds) Biotechnology for silkworm crop enhancement: tools and applications. Springer, Singapore, pp 1–20

Sonwalkar TN (1993) Hand book of silk technology. Taylor & Francis

Sujatha G, Kumar GA, Teja K, Devi DL, Rupali J, Gautam SK (2024) A comprehensive review of the effect and mitigation of climate change on sericulture. Int J Environ Clim Change 14(7):776–788. https://doi.org/10.9734/ijecc/2024/v14i74317

Suzuki Y, Kawanishi S, Yamazaki T, Aoki A, Saito H, Asakura T (2015) Structural determination of the tandem repeat motif in *S. cynthia ricini* liquid silk by solution NMR. Macromolecules 48(18):6574–6579. https://doi.org/10.1021/acs.macromol.5b01717

Takasu Y, Yamada H, Tsubouchi K (2002) Isolation of three main sericin components from the cocoon of the silkworm, *B. mori*. Biosci Biotechnol Biochem 66(12):2715–2718. https://doi.org/10.1271/bbb.66.2715

Talukdar S, Kundu SC (2012) A non-mulberry silk fibroin protein based 3D in vitro tumor model for evaluation of anticancer drug activity. Adv Funct Mater 22(22):4778–4788. https://doi.org/10.1002/adfm.201200375

Talukdar S, Nguyen QT, Chen AC, Sah RL, Kundu SC (2011a) Effect of initial cell seeding density on 3D-engineered silk fibroin scaffolds for articular cartilage tissue engineering. Biomaterials 32(34):8927–8937. https://doi.org/10.1016/j.biomaterials.2011.08.027

Talukdar S, Mandal M, Hutmacher DW, Russell PJ, Soekmadji C, Kundu SC (2011b) Engineered silk fibroin protein 3D matrices for in vitro tumor model. Biomaterials 32(8):2149–2159. https://doi.org/10.1016/j.biomaterials.2010.11.052

Tanaka K, Mizuno S (2001) Homologues of fibroin L-chain and P25 of *B. mori* are present in *Dendrolimus spectabilis* and *Papilio xuthus* but not detectable in *A. yamamai*. Insect Biochem Mol Biol 31(6–7):665–677. https://doi.org/10.1016/S0965-1748(00)00173-9

Tao W, Li M, Zhao C (2007) Structure and properties of regenerated *A. pernyi* silk fibroin in aqueous solution. Int J Biol Macromol 40(5):472–478. https://doi.org/10.1016/j.ijbiomac.2006.11.006

Teshome A, Raina S, Vollrath F (2014) Structure and properties of silk from the African wild silkmoth *G. postica* reared indoors. J Insect Sci 14(1):36. https://doi.org/10.1093/jis/14.1.36

Tikader A, Kamble C (2010) Seri-biodiversity with reference to host plants in India. Asian Austr J Plant Sci Biotechnol 4(1):1–11

Tuigong DR, Kipkurgat TK, Madara DS (2015) Mulberry and silk production in Kenya. J Text Sci Eng 5(6):1

Tulachan B et al (2014) Electricity from the silk cocoon membrane. Sci Rep 4(1):5434. https://doi.org/10.1038/srep05434

Üzümcü MB, Borazan I, Kaplan M (2019) Wild silk fibers: types, properties and utilization areas. In: International congress on wool and luxury fibres Iconwoolf, Tekirdağ

Üzümcü MB, Kadoğlu H (2016) The effects of spinning system and blending ratio on quality of silk/cotton blended yarns. Text Apparel 26(3):244–248

Van Eijk F et al (2004) Tissue engineering of ligaments: a comparison of bone marrow stromal cells, anterior cruciate ligament, and skin fibroblasts as cell source. Tissue Eng 10(5–6):893–903. https://doi.org/10.1089/1076327041348428

Yang M, Mandal N, Shuai Y, Zhou G, Min S, Zhu L (2014) Mineralization and biocompatibility of *Antheraea pernyi* (*A. pernyi*) silk sericin film for potential bone tissue engineering. Bio-Med Mater Eng 24(1):815–824. https://doi.org/10.3233/BME-130873

Yang G, Wu M, Yi H, Wang J (2016) Biosynthesis and characterization of a non-repetitive polypeptide derived from silk fibroin heavy chain. Mater Sci Eng C 59:278–285. https://doi.org/10.1016/j.msec.2015.10.023

Zhou Z et al (2017) The use of functionalized silk fibroin films as a platform for optical diffraction-based sensing applications. Adv Mater 29(15):5471

Zhu B et al (2016) Silk fibroin for flexible electronic devices. Adv Mater 28(22):4250–4265. https://doi.org/10.1002/adma.201504276

Zou S, Yao X, Shao H, Reis RL, Kundu SC, Zhang Y (2022) Nonmulberry silk fibroin-based biomaterials: impact on cell behavior regulation and tissue regeneration. Acta Biomater 153:68–84. https://doi.org/10.1016/j.actbio.2022.09.021

# Chapter 6
# Silk Yarn and Its Quality Evaluation

**Tithi Mitra and Sankar Roy Maulik**

## 6.1 Introduction

Very few natural fibres possess the balance of strength and elegance, and silk is among the finest. Throughout history, silk has been one of the most renowned natural fibres, admired for its smooth texture, exceptional sheen, elasticity, strength, natural thermo-regulating properties, and biocompatibility. Different silkworm species produce silk through unstructured fluid secretion and solidify it into a continuous filament. Thus, it is regarded as the longest natural fibre. The silkworm's feeding culture primarily influences the quality of silk (e.g. texture, colour, fineness). Two major varieties emerge when categorising silk: cultivated silk and wild silk. The former is primarily sourced from the *Bombyx mori* species of silkworm, which generally feeds on the leaves of mulberry trees. This variety of silk production relies on various factors, including controlled weather (i.e. controlled humidity, optimal temperature, etc.) for growth and protection from other environmental influences. Producing good-quality silk yarn involves multiple stages, including silkworm rearing, cocoon formation, harvesting, reeling, spinning, and many others, each playing a crucial role. An in-depth analysis of the quality evaluation process of silk fibre and yarn is mentioned in this chapter.

T. Mitra · S. Roy Maulik (✉)
Department of Silpa-Sadana, Visva-Bharati (A Central University), Sriniketan, India
e-mail: sankar.raymaulik@visva-bharati.ac.in; s_r_moulik@yahoo.co.in

T. Mitra
e-mail: tithimitra999@gmail.com

 
D. Kumar and M. Shahid (eds.), *Science and Technology of Silk*, Textile Science and Clothing Technology, https://doi.org/10.1007/978-3-032-16230-4_6

## 6.2 Properties of Silk

Silk fibre possesses a distinct morphological structure; from its longitudinal view, it reveals an irregular surface, while its cross-sectional view is elliptical. Silk is produced through the extrusion process by the silkworm in the form of a double filament (bave) that sticks together with the help of sericin or silk gum. The core of the silk fibre is known as fibroin, and the outer layer is the silk gum, known as sericin, which protects the fibre. Around 70–80% of silk is composed of fibroin, whereas the remaining 20–30% consists of silk gum or sericin. The silk gum gets dissolved entirely through a process commonly known as degumming. The polar and non-polar ratios of silk, as well as the ratio of crystalline to amorphous regions, notably affect the degumming process. Silk fibroin comprises around 76% of amino acids with non-polar side chains (the main ones are glycine and alanine) and only about 21% of polar groups. Whereas sericin consists of 25% non-polar groups and about 75% of polar side chains (the main among these are serine, aspartic acid, glycine, etc.), this difference in composition helps remove sericin more effectively.

## 6.3 Raw Silk Preparation

The raw silk preparation follows a few crucial stages, including sorting, stifling, cooking and reeling. The primary goal of the stifling stage is to destroy the pupa inside the cocoon to prevent it from coming out as an adult moth, ensuring the silk filament remains continuous. This process also helps dry the cocoons and store them for a long period. Some standard methods of stifling are—sun drying, steam stifling and hot air drying (Babu 2019).

The reeling of silk is an essential step. The process generally involves unwinding filaments from the cocoons to produce raw silk. It can be performed in three different ways, i.e. (i) the charka reeling system, a traditional and manually operated process of reeling, (ii) the cottage basin reeling system, which involves a semi-mechanised process, and (iii) the filature reeling system, which is a fully power-driven reeling process (Mishra 2000). The charka reeling system is a manual process that requires skilled labour and is performed in a single cooking basin, whereas the cottage basin and filature reeling systems rely on mechanised processes and are carried out in two separate basins, one with boiling water and another with hot water (Kiyokawa 1993).

China is the leading raw silk suppliers of the world and is the global hub for silk production. Some major companies are Zhejiang, Jiangsu, Sichuan, Tongxiang Heshan Weiye Textile Co. Ltd., Zhejiang Golden Eagle Spun Silk Co. Ltd., Guangdong Wintex Supply Chain Co. Ltd., etc. These companies produce the best quality mulberry filament yarn and supply raw silk yarn to the global market. Raw silk testing and classification are done to assess the quality and assign a grade based on test results. The grading of mulberry silk is based on various factors such as fibre length, uniformity, colour, strength, etc. 6A to F is the most common grading range, which was first adopted by China and then, gradually became used worldwide. Indian silk industries commonly source grades 5A and 4A from China.

Before further processing, the raw material undergoes an initial quality inspection, focusing on parameters such as yarn appearance, tensile strength, elongation, size deviation, moisture regain and the occurrence of winding breaks.

## 6.4 Automated Testing Methods

### *6.4.1 Test Volume*

The unit of testing and grading is the lot, consisting of about 300 or 600 kg of homogeneous raw silk in the form of skeins or cones, which can be termed ‘package’. The sample packages for automated testing consist of 40 original packages representatively drawn from one lot. All tests are carried out in standard atmospheric conditions.

### *6.4.2 Size Test*

The primary goal of this test is to evaluate average size, size variation and the largest deviation in size.

*Sizing reel*—Applicable to both methods, i.e. traditional and semi-automated size testing. The measuring machine for making the skeins has a reel of 1.25 m in circumference, a uniform velocity and an automatic stop motion, which actuates when the skein is completed. For automated size testing, a sample length of 100 m is often used, and this can be checked simultaneously along with the evenness testing operation.

*Balance*—The sensitivity and capacity of the balance are 5 mg or better, and 50 g or more, respectively.

### *6.4.3 Evenness, Defects and Impurities Tests*

Traditional testing using *Seriplane* boards involves visual comparison of various aspects, e.g. evenness, cleanness and neatness with standard photographs (recognised and adopted by the International Silk Association). Standardised numerical values are assigned to evaluate these three parameters. As this method is subjective and the result depends on the operator’s expertise, it may introduce differences in the result (Chattopadhyay et al. 1997).

- Evenness variation refers to those defective portions of raw silk threads on an inspection board which show stripes caused by denier variation of raw silk noticed during visual inspection.
- The cleanness defects are waste, large and small slugs, bad casts, long knots, heavy corkscrews and long loops.
- Neatness defects refer to minor imperfections in the raw silk threads, smaller than those classified as minor cleanness defects. Other commonly observed defects in raw silk are filament stains and contaminations.
- Filament stains are commonly observed in raw silk when the stained and melted cocoons are not appropriately sorted before the boiling and reeling operation. These stains may not be removed entirely during degumming and bleaching processes and become visible in white, ivory and light colour taffeta or satin quality fabrics. Such fabric defects are noticeable and unacceptable according to international product quality standards. Similarly, contamination from dirty filament fragments can leave unwanted marks on the final fabric.

Since no international norms or standards are available for the *Seriplane* test, it has been largely replaced by automated ones for assessing evenness variation, cleanness and neatness. For this test, electronic equipment is used, which allows a neutral inspection of a representative length of raw silk and gives reproducible results. Due to the differences between the electronic mass monitoring and the visual *Seriplane* inspection, the test results obtained by these two test methods may vary.

*Uster Tester-3* (UT3) can be used for automatic testing. The tests of evenness, defects and impurities consist of 40 tests from 40 packages. Every test has to be made on 1000 m of yarn with a testing speed of 400 m/min. The yarn pre-tension during testing is important to prevent the yarn from vibrating because it could lead to faulty results. The UT3 test generates the following parameters for the inspection certificate, all obtained through a single operation.

- Evenness (CV%)
- Defects (Thin/1000 m)
- Defects (Thick/1000 m)
- Impurities/1000 m
- CV size 50 m%
- CV size 450 m%.

### 6.4.4 Tenacity and Elongation Tests

These tests aim to determine the single-end tenacity and elongation of raw silk along with their respective variations. These parameters are essential for the efficient performance of high-speed weaving machines to improve productivity and fabric quality. The test is performed using a 'Constant Rate of Extension' (CRE) type instrument, and 'USTER TENSORAPID' is one such instrument. The test requires a speed of 5 m/min and a gauge length of 50 cm, and 10 tests from each of 40 packages should be performed, i.e. the total number of tests to be done is 400. The following parameters are required in the inspection certificate.

Tenacity in (cN/dtex), CV $_{tenacity}$ %, elongation (%) and CV $_{elongation}$ %.

### 6.4.5 Cohesion Test

This is a crucial parameter mainly for weavers; the main objective of this test is to determine the degree of adhesion of cocoon filaments. It is assessed using the Duplan Cohesion Tester, which measures the number of strokes.

### 6.4.6 Winding Test

This test aims to determine the number of breaks which occur in the silk threads during a specific period of winding. The sample size for this test is 40 skeins. The winding breaks are calculated for a total yarn length of 300 km. The winding breaks are expressed as a total number of breaks/300 km. The winding speed is at least 165 m/min or more.

### 6.4.7 Visual Inspection

The primary purpose of this inspection is to evaluate the uniformity visually, overall finish and various characteristics of raw silk. The evaluation process is performed in a standard visual inspection room under controlled settings. Key parameters such as colour, lustre and hand feel are evaluated and indicated as follows:

*Colour*: Light, Medium and Dark
*Lustre*: Bright, Medium and Dark
*Hand*: Smooth, Medium and Rough.

### 6.4.8 Conditioned Weight Test

Conditioned weight of raw silk is the moisture-free weight plus 11% of the oven-dry weight.

### *6.4.9 Exfoliation Test*

This additional test is primarily conducted on raw silk to assess a specific undesirable property. Exfoliation refers to the splitting of individual silk filaments into very fine fibrils. These fibrils initially resist dye absorption, creating a white fuzz on the surface of silk yarn or fabric, affecting its overall quality.

## 6.5 Silk Yarn Preparatory Process

Figure 6.1 presents a generalised flowchart for the silk yarn preparatory process.

**Fig. 6.1** Silk yarn preparatory process flowchart

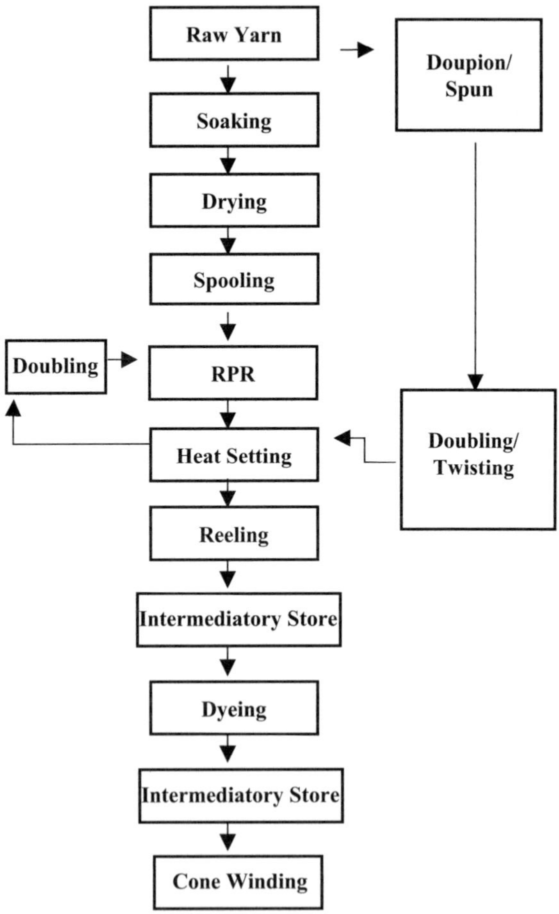

### 6.5.1 Soaking

This is a preparatory process in which filature raw silk yarn is immersed in a bath containing a soaking or softening agent consisting of neutral and vegetable oils, fatty acids and anionic surfactants. Doupion, spun and noil varieties of silk yarn do not require soaking, as their inherent properties differ from filature yarns. The soaking process helps reduce the risk of yarn breakage in further processing stages, ensuring smoother handling. The process involves two tubs where water is heated to 45 °C. Once heated, the required soaking solution is added to ensure optimal treatment of the yarns. The amounts of the solution vary for warp and weft yarns and are optimised to meet the differing mechanical stresses they undergo. To distinguish different lots, a colour tint is added to the solution and is thoroughly stirred with a stirrer to ensure uniformity. The tints are generally colours with no affinity for silk fibre; thus, they can be easily washed away during degumming. The process consists of multiple cycles, with the initial ones involving a bath. The soaking process is automatic once the prepared solution and yarns are transferred to the machine tub. After soaking, the yarns need to be appropriately dressed by hand, and any gum spots present on the yarn hanks must be gently removed manually to prevent any damage to the fibre. 'Gum spot' is a defect in the raw yarn at different places where yarns are glued together, causing problems during winding. This defect arises due to improper cooking of bivoltine cocoons, imperfect adjustment of process parameters in reeling, etc. This defect can be confirmed by comparing the staining behaviour of the defective and the normal portion of hank with a staining reagent.

After the soaking process, the yarns are properly dried and assessed with a moisture metre to confirm the required moisture content before moving on to the following processing stage, i.e. spooling. The duration of drying varies, typically ranging between 24 and 48 h.

*Key Considerations for the Soaking Process*

- Water should not contain acid, alkali or salts, since soap will not dissolve in hard water.
- The tint colour should not be fast as it may dye the yarn.
- The silk hanks should not be exposed directly to sunlight because it affects the strength of the silk and turns it yellow.
- It is advisable not to increase the temperature beyond 45 °C.

### 6.5.2 Spooling (Winding)

This process converts hanks into double-flanged bobbin form. For filature, doupion and spun silk, the standard winding speed ranges from 220 to 300 m/min, depending on factors such as yarn quality, type of yarn and specific production requirements.

*Key Considerations for This Process*

- A weaver's knot is advised for fixing broken yarn, ensuring that the tail end does not exceed 2–3 mm in length. This is a common practice to avoid slippage of knots during warping and weaving.
- Gauge setting is another important parameter during the spooling operation. The gauge should not be too tight or too loose. Tight gauge results in rupture in the single yarn, whereas loose gauge allows slugs, knots and other foreign matters to pass along with the yarn, which is also not advisable.
- It is also necessary to avoid certain steps during this operation, such as moist or semi-dried skeins should not be taken for winding. It may result in the stretching of silk yarn. The skeins should not be mounted haphazardly, since it becomes difficult to get the two ends, resulting in more breakages.
- Varying yarn contents on the bobbins is not recommended because it leads to variation in unwinding tension during doubling, resulting in more wastage.
- Traverse should not be less or more than the bobbin length, and there should not be any jerks in the traverse, since it leads to improper bobbin build-up.
- There should not be any scratches on the thread and guide plates because the yarn gets abraded, leading to rupture and end breakages.

### *6.5.3 Doubling*

This is a crucial stage in the yarn preparatory process, primarily aimed at producing plied yarn and minimising irregularities in yarn. To achieve high-quality yarn, several considerations need to be followed during the doubling process, such as:

- It is recommended to maintain low tension in individual yarns.
- The operating speed should be around 400 m/min.
- It is essential to maintain a relative humidity of 65–70% for optimal yarn performance.
- A tensioner should be used to achieve uniform tension on the yarn.
- The spindle must stop automatically and instantly when a single yarn breaks to avoid production and quality loss.
- It is advisable to maintain uniform yarn tension during the doubling process, as any variation can form loose and tight ends, adversely leading to the corkscrew effect, a significant fabric defect.
- For 2 and 3-ply warp yarns, a 'group knot' is recommended and a 'component knot' for thicker yarns to reduce knot visibility. Also, avoiding the long tail end of the knot is recommended, which adversely leads to defects during twisting and weaving.
- To avoid the corkscrew effect, it is essential that the yarn packages stay in place and have the same amount of yarn during the doubling process, as any difference can create tension variation.
- Crossing in spooling packages intended for the doubling process must be avoided, as it can also result in tension variation while unwinding, affecting the final quality of yarn and fabric.

### 6.5.4 Twisting

Yarn twisting improves its suitability to meet consumer requirements. A Two-For-One twister (T.F.O.) is commonly used for twisting filature yarns. The key concept behind T.F.O. is that every single rotation of the spindle results in two twists, increasing productivity over traditional twisting methods. Figure 6.2 presents a process for producing warp yarn with specifications of 2 and 3 ply, 20/22 denier.

Figure 6.3 presents a sequence of steps in a recommended process for producing a weft yarn with specifications of 8 and 12 ply.

The speed of the spindle should be adjusted based on the yarn type and desired twist level. It is generally advisable to maintain the speed of the spindle within 8000–8500 rpm; this range ensures better yarn quality. To achieve high-quality yarns, key considerations include:

- Any abrasions on the pots or yarn path must be avoided, or else they can cause yarn breakage or they may form weak spots in woven fabrics, which is a defect. Inspecting the pots and yarn paths is necessary to avoid further defects.
- To prevent uneven twisting, spindle speed must be monitored regularly to maintain efficient rotation and minimise twist variation.
- In this process, knotting during yarn breakage is not advisable, as it may cause twist irregularities due to speed differences. It may result in over-twisted yarn or low-twisted yarn.
- Weekly machine cleaning and maintenance are essential to avoid excess yarn accumulation and keep the spindles rust-free.
- Package hardness should be checked and optimised as loose packages can result in yarn slippage, causing unnecessary wastage.

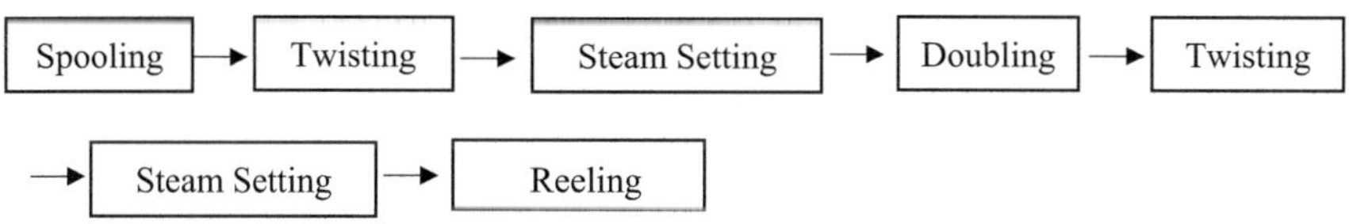


**Fig. 6.2** Process to produce warp yarn with a specification of 2 and 3 ply, 20/22 denier

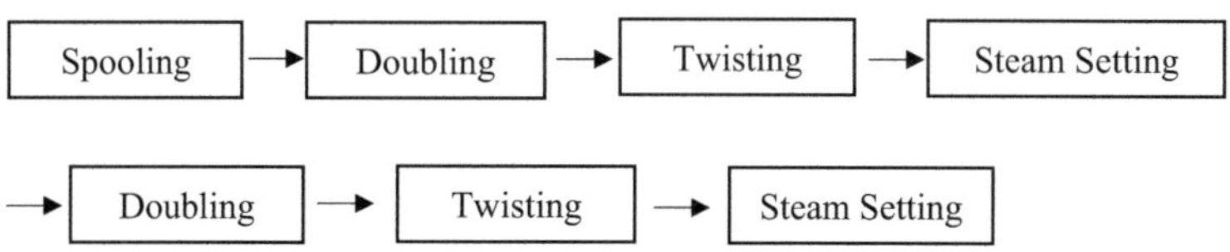


**Fig. 6.3** Process to produce weft yarn having specifications of 8 and 12 ply

### 6.5.5 Steam-Setting

Steaming is a crucial stage that removes the stress from twisted yarns and evens the twist, making it more suitable for further processing. Based on yarn type, the recommended steaming temperature is generally 65 °C, with a duration of 70–100 min. Key considerations during this process include:

- It is essential to maintain the right level of steam. Over-steaming reduces the yarn's strength, and less steaming leads to a defect known as yarn snarling.
- In the steam chamber, the package must be set in a proper sequence to allow uniform steam distribution.
- It is advisable to use soft water for steaming, and the chamber must be rust-free.
- Autoclaves are better for steam-setting than open-bath steamers as the former offer more uniform and controlled steaming.

### 6.5.6 Reeling

Efficient reeling is crucial for producing high-quality yarn, which is important for the degumming and dyeing. The primary objective of this stage is to convert yarns into hank form, which is suitable for degumming and dyeing. During this stage, only draw knots are used in case of yarn breakage for both filature warp and weft yarns. Key considerations for better quality reeled yarns include:

- It is essential to maintain equal length of hanks, as variation in length can lead to problems during degumming, dyeing, cone winding and warping.
- It is always advisable for operators to use bags to collect yarn waste, as loose yarn can get entangled with the hanks and may cause processing disruptions in later stages.

## 6.6 Technological Development in Silk Yarn Quality Evaluation

Recent technological advancements have incorporated automated systems to overcome the limitations of traditional systems, which mainly rely on subjective visual inspection while evaluating silk yarn quality. These automated inspection systems are capable of providing more consistent and reliable results. A notable contribution in this field is the *Silk Check—an* optical sensing system which offers instant analysis of hand-reeled silk yarns. Traditional grading of silk yarns to assess colour uniformity is subjective and labour-intensive, requiring experienced inspectors who may not be readily available in remote production locations. This optical sensing system provides high accuracy in measuring linear mass density and colour shade, which are difficult to assess manually. This advancement effectively enhances grading quality and transparency (Chaitavon et al. 2021). Researchers have introduced a digital image processing-based method for yarn colour analysis. This process involves automated image capture and colour analysis, followed by hierarchical grouping using Principal Component Analysis (PCA) (Pal et al. 2012). These approaches align with the perspectives of Industry 4.0, which integrates analytics and improved systems into the textile industry. In recent years, the incorporation of artificial intelligence into textiles has introduced new dimensions in the silk yarn quality prediction field. Researchers have explored new insights into how machine learning can characterise and forecast silk fibroin's mechanical and biochemical behaviour, which is the main building block responsible for the strength, lustre, elasticity, biocompatibility and durability of silk yarn. By simulating fibroin behaviour under different conditions, the AI system can forecast yarn characteristics such as tenacity, elongation and uniformity. These predictions are effective in silk yarn production, where quality control is essential (Shanmughan and Kandasubramanian 2025).

Researchers have compared the traditional *Seriplane* visual grading system and the Uster evenness tester to analyse key quality parameters such as evenness, cleanness and neatness. The result emphasised that although *Seriplane* visual assessment is widely practised, it is a subjective and operator-dependent test. On the contrary, the Uster system provides reproducible and quantifiable data on mass variation, which includes defects such as loops, knots and stripes, thus offering a more reliable standard for evaluating yarn quality (Chattopadhyay et al. 1997). Latest advancements in artificial intelligence have transformed fibre and yarn quality assessment processes. A review discusses how deep learning techniques such as artificial neural networks, conventional neural networks, principal component analysis and support vector machines are gaining popularity for predicting yarn properties (Ingle and Jasper 2025). While the use of these frameworks is noticeable mainly in cotton, jute and alpaca fibres, they also offer promising potential for evaluating silk yarn quality.

## 6.7 Conclusion

The yarn preparation process, from the initial raw material selection to the reeling stages, is crucial for producing high-quality and value-added silk fabrics that meet international market standards. Therefore, maintaining a consistent high-grade raw silk and optimal process parameters is paramount. Relying on inspection certificates provided by the suppliers may not be adequate; to maintain transparency and standards, it is advisable that all parameters must be evaluated through standardised testing. Every stage in the yarn preparation process has its considerations; it is important to adhere to these for obtaining good-quality end products. Raw silk users mainly rely on identifying the raw silk quality for suitability with optimum efficiency on high-speed machinery involved in twisting, weaving and knitting.

It is always recommended that during the soaking process, the bath temperature must be below 45 °C to prevent damage. Moreover, it is advisable to avoid direct sunlight exposure during the drying stage; instead, controlled drying is recommended to preserve the strength of the material. Compared to open-bath steaming, autoclave steaming is more preferable. Reeling is the final stage of the yarn preparation, and it is crucial to maintain equal hank length. By maintaining proper raw material selection, standardised processing parameters and reliable testing methods, producing good quality silk yarn can improve global market performance. The traditional system of evaluating quality parameters mainly relies on subjective visual inspection. The technological advancements have incorporated automated systems to overcome these limitations in evaluating silk yarn quality.

**Declaration** The significant portions of this article are based on the practical experience of the author (Dr. Sankar Roy Maulik) while working as Manager Quality Assurance in an export-oriented silk industry.

## References


Babu KM (2019) Silk: processing, properties and applications. Elsevier—Woodhead Publishing, Duxford

Chaitavon K, Sumriddetchkajorn S, Kamtongdee C, Chanhorm S (2021) Optical sensing system for real-time physical quality evaluation of hand reeled silk yarn. IEEE J Sel Top Quantum Electron 27:1–8. https://doi.org/10.1109/JSTQE.2021.3096418

Chattopadhyay R, Das S, Gulrajani ML, Sen K (1997) A comparison between the quality attributes of silk yarn using the Uster and Seriplane test systems. J Text Inst 88:92–98. https://doi.org/10.1080/00405009708658533

Ingle N, Jasper WJ (2025) A review of deep learning within the framework of artificial intelligence for enhanced fiber and yarn quality. Text Res J 95:904–933. https://doi.org/10.1177/00405175241265510

Kiyokawa Y (1993) Possibilities for producing high-quality raw silk and fostering a skilled workforce in the Indian silk-reeling industry. Hitotsubashi J Econ 34:87–109

Mishra SP (2000) A textbook of fibre science and technology. New Age International, New Delhi

Pal A, Dey T, Chopra P, Akuli A, Ray M, Bhattacharvva N (2012) A new method for grading of silk yarn using electronic vision. In: Proceedings of the 2012 sixth international conference on sensing technology (ICST). IEEE, Kolkata, pp 387–392

Shanmughan B, Kandasubramanian B (2025) Artificial intelligence techniques to predict the behavior of silk fibroin. In: Kandasubramanian B, Jaya-Prakash N (eds) Engineering natural silk. Springer, Singapore, pp 193–206

# Chapter 7
# Advances in Extraction and Characterization of Silk Proteins for Future Uses

**Shubhajit Shaw, Ritwik Acharya, Pankaj Mandal, Amit Kumar Mandal, and Debnirmalya Gangopadhyay**

## 7.1 Introduction

Silk is a natural protein fiber largely produced by the domesticated silkworm *B. mori* and has long been valued for its aesthetic appeal, mechanical strength, and thermal stability. Although traditionally used in the textile industry, its unique molecular structure and biodegradability have recently drawn increasing attention in diverse fields, such as biomedicine, environmental engineering, and materials science (Seo et al. 2023; Nguyen et al. 2019). Silk is composed of two primary proteins such as silk fibroin (SF), which forms the structural core, and silk sericin (SS), a hydrophilic protein that envelopes fibroin filaments and functions as a biological adhesive. SS is often discarded during the degumming process, which serves as a preparatory stage for the extraction of pure fibroin, primarily due to its perceived low value and the challenges associated with its efficient extraction (Hassan et al. 2024; Beniwal et al. 2024). However, recent studies revealed that SS possesses notable biofunctional properties, including antioxidant, antimicrobial, anti-inflammatory, and moisturizing effects, positioning it as a promising candidate for diverse biomedical, pharmaceutical, and industrial applications (Silva et al. 2022).

The growing scientific interest in silk is driven not only by its complementary structural and functional characteristics but also by the increasing need to develop sustainable and eco-friendly biomaterials. As environmental concerns surrounding synthetic polymers continue to rise, naturally derived polymers from silk offer a sustainable and attractive alternative through their extraction and functional utilization (Seo et al. 2023). However, challenges persist in realizing this potential. Conventional extraction techniques for silk proteins often employ harsh chemical reagents

S. Shaw · R. Acharya · P. Mandal · A. K. Mandal · D. Gangopadhyay (✉)
Department of Sericulture, Raiganj University, Raiganj, Uttar Dinajpur, West Bengal, India
e-mail: debganguly.rgu@gmail.com



D. Kumar and M. Shahid (eds.), *Science and Technology of Silk*, Textile Science and Clothing Technology, https://doi.org/10.1007/978-3-032-16230-4_7

like sodium carbonate, urea, or lithium bromide, which can degrade protein structure, reduce biocompatibility, and pose ecological hazards due to effluent toxicity (Huang et al. 2023). In response, contemporary research has increasingly focused on greener alternatives, such as enzymatic, ultrasound-assisted, microwave-assisted, and ionic liquid-based methods, which primarily aim to enhance extraction efficiency while preserving protein integrity and minimizing environmental impact (Ward 2019; Biswal et al. 2022; Vyas and Shukla 2016; Pereira et al. 2024).

In addition to extraction, the structural and functional optimization of SF and SS requires comprehensive characterization. Analytical techniques such as Fourier transform infrared spectroscopy (FTIR), X-ray diffraction (XRD), nuclear magnetic resonance (NMR), scanning electron microscopy (SEM), and transmission electron microscopy (TEM) play a crucial role in revealing the secondary structures, crystallinity, thermal stability, and surface morphology of these silk proteins (Koczoń et al. 2023; Kanoujia et al. 2023; Chandra and Chandra 2022; Reizabal et al. 2021). These insights are crucial for tailoring silk-based biomaterials for targeted applications, ranging from tissue scaffolding and wound healing to cosmetics, drug delivery systems, and biosensors (Nguyen et al. 2019; Silva et al. 2022). For instance, SF's β-sheet-rich domains lend it remarkable mechanical strength and slow biodegradability, ideal for structural biomedical implants. On the contrary, the amorphous nature of SS, combined with its high water solubility and bioactivity, renders it suitable for use in hydrogels, wound dressings, and skincare formulations (Luo et al. 2015; Li and Sun 2022; Silva et al. 2022).

This chapter presents a focused scientific examination of mulberry silk, the most commercially important and extensively researched form of silk. It offers a comprehensive evaluation of its two principal protein components, such as SF and SS, with particular focus on their sustainable extraction, advanced characterization, and broad spectrum of applications. The review critically assesses both conventional and emerging extraction techniques, considering their efficiency, environmental sustainability, and effects on the structural and functional integrity of silk proteins. By systematically comparing traditional and innovative methods and analyzing their physicochemical outcomes, the chapter addresses key challenges related to scalability, reproducibility, and ecological impact. It further explores the multifunctionality of SF and SS within contemporary material science, underscoring their cross-sector relevance and potential for sustainable use in next-generation applications.

## 7.2 Silk: Composition and Structure

Silk production in the mulberry silkworm is a complex and well-regulated process involving a series of coordinated physiological and biochemical steps. This process results in the formation of silk fibers known for their strength, flexibility, and ability to break down naturally in the environment. These features are essential for the protection of the silkworm during its cocoon stage and are now being explored for use in fields like biotechnology, medicine, and material science (Kunz et al. 2016). The silk fiber is made of two main proteins: SF and SS. SF forms the strong, inner part of the silk thread, while SS covers SF and acts as a protective and sticky outer layer. Their production is carefully controlled by hormones and genes, and it occurs during the transformation of the larva into a pupa (Mondal et al. 2007; Wang et al. 2014). The silk gland in *B. mori* is divided into three parts: the posterior, middle, and anterior regions. Each part is responsible for producing different parts of the silk proteins. The posterior gland produces the heavy chains of SF, while the middle part makes the light chains and P25 protein, which help stabilize the SF structure. SS is mainly secreted by the middle gland and coats the SF. The structure of the silk gland also plays a role in adjusting the final qualities of the fiber, which makes it a useful model for designing new materials (Liu and Zhang 2014).

At the molecular level, SF and SS have different structures and roles, but both are necessary to create a high-quality silk fiber. SF contains amino acids like glycine, alanine, and serine. These help form special folded structures called β-sheets that give silk its strength and crystal-like organization (Biswal et al. 2022). SS, on the other hand, contains more polar amino acids such as serine, threonine, and asparagine. These make it water-soluble and suitable for coating and protecting the SF during fiber spinning (Rising and Harrington 2023).

SF is highly organized and forms a stable structure held together by hydrogen bonds. This gives it strong and durable qualities (Qiu et al. 2019). It also has a balanced structure that allows it to be both stiff and flexible, making it ideal for use in biomedical and textile products (Liu et al. 2022). SS is less organized, with random coil and α-helix structures. This flexibility makes it useful as a natural adhesive and moisture-retaining layer (Aad et al. 2024).

The interaction between SF and SS is very important for forming stable silk threads. SS sticks to SF through hydrogen bonds and other weak forces, helping keep the thread together as it solidifies in air (Reizabal et al. 2023; Seo et al. 2023). The special combination of SF and SS gives mulberry silk its unique properties. These include high strength, flexibility, resistance to heat, and compatibility with living tissues. SS makes the fiber hydrophilic, meaning it can absorb and hold water, which improves moisture retention and breathability (Zhao and Zhang 2020; Hu et al. 2023). SF is more hydrophobic, helping the fiber resist water. The balance between these two qualities makes silk useful in a wide range of environments and applications. SF is also known for its thermal stability, as it can resist breaking down at temperatures above 300 °C due to its β-sheet structures (Yuan et al. 2024). While SS is not as heat-resistant, it still works well at moderate temperatures and adds to the overall strength of the silk fiber (Kunz et al. 2016). Together, SF and SS make mulberry silk a highly valuable natural material with many potential uses in science and industry.

## 7.3 Extraction Techniques for SF and SS

Extraction of SF and SS begins with degumming, a crucial step in silk processing. It involves the removal of SS, a hydrophilic glue-like protein that binds SF filaments, through thermochemical treatments that disrupt peptide bonds (Rastogi and Kandasubramanian 2020; Biswal et al. 2022). Although SS offers antimicrobial activity and moisture retention (Lin et al. 2024), its removal is essential for applications requiring pure SF (Fig. 7.1, Table 7.1).

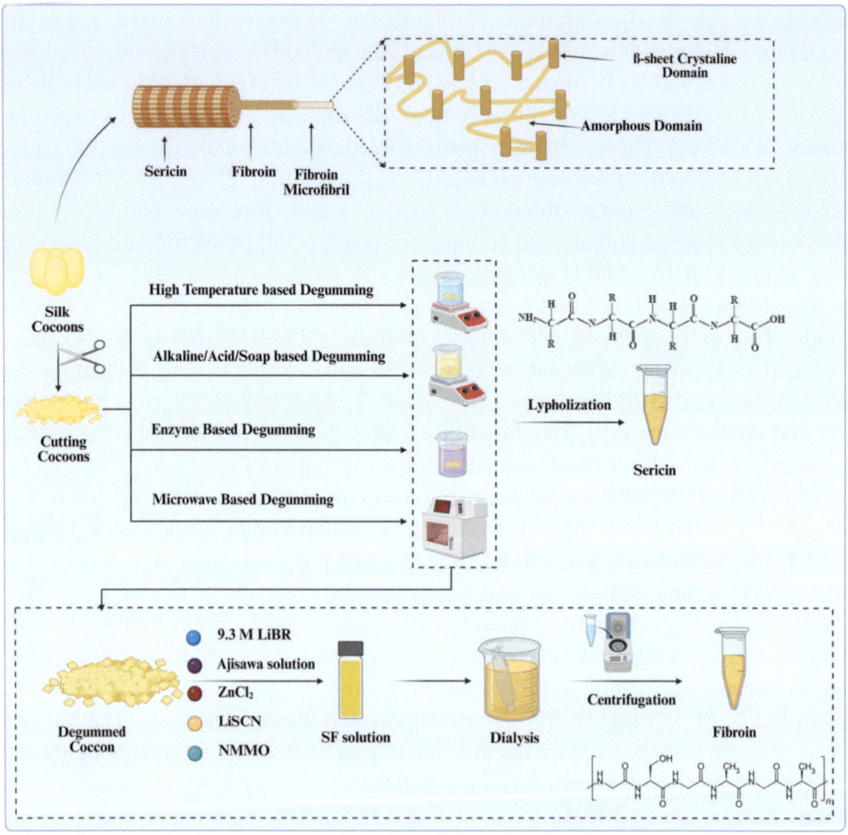


**Fig. 7.1** Overview of silk degumming processes for SS and SF, highlighting structural domains and common extraction methods

**Table 7.1** Different types of degumming process, advantages and disadvantages of SS

| Extraction method | Description | Advantages | Disadvantages | References |
|---|---|---|---|---|
| Water degumming | Uses hot water (50–60 °C) to soften and partially dissolve SS | Environmentally friendly, cost-effective | Requires high temperature and long processing times, may damage *SF*, incomplete SS removal | (Sharma et al. 2019; Rastogi and Kandasubramanian 2020) |
| Soap degumming | Uses sodium-based soaps (e.g., sodium stearate, sodium oleate) in mildly alkaline conditions | Mild processing, improved fiber softness, whiteness, and luster | Requires careful control of variables like temperature, pH, and soap concentration | (Gulrajani 1992; Gore et al. 2019; Biswal et al. 2022; Wang et al. 2023) |
| Alkali degumming | Uses alkalis (e.g., $Na_2CO_3$, NaOH) to hydrolyze SS | Cost-effective, high efficiency | Can cause yellowing and rough texture; often combined with soap for better results | (Franck 2001; Cui et al. 2021; Lin et al. 2024) |
| Acid degumming | Uses dilute acids (e.g., citric acid, sulfuric acid) to hydrolyze SS | Rapid degumming, no need for high temperatures | Less commonly used due to potential damage to *SF* | (Gulrajani 1992; Wang et al. 2023) |
| Enzymatic degumming | Uses enzymes (e.g., trypsin, papain) to break down SS into amino acids | Gentle process, preserves fiber quality, results in softer texture and whiteness | Inefficient for industrial applications, does not fully remove hydrophobic impurities | (Dijkstra 2010; Ward 2019) |
| Microwave-assisted degumming | Uses microwave energy to enhance SS removal at lower temperatures | Faster process, reduced water and energy consumption | Requires specialized equipment | (Gupta et al. 2013; Rastogi and Kandasubramanian 2020; Biswal et al. 2022) |
| Amine-assisted degumming | Uses organic amines (e.g., ethylamine, diethylamine) to remove SS | Works well even in hard water conditions | Slow process, requires precise control | (Gulrajani and Malik 1993; Wang et al. 2023) |

(continued)

**Table 7.1** (continued)

| Extraction method | Description | Advantages | Disadvantages | References |
|---|---|---|---|---|
| Ultrasound-based degumming | Uses ultrasonication with degumming agents for rapid and uniform SS removal | Shortens processing time, produces high-quality silk fibers | May require additional chemicals for complete degumming | (Vyas and Shukla 2016; Lo 2020) |
| Ionic liquid-based degumming | Uses ionic liquids (e.g., 1-butyl-3-methylimidazolium chloride) to dissolve SS selectively | Environmentally friendly, preserves silk integrity, reusable solvents | High cost of ionic liquids, process optimization needed | (Wang et al. 2020; Pereira et al. 2024) |

## *7.3.1 Extraction of SS*

### 7.3.1.1 Conventional Degumming Methods

#### 7.3.1.1.1 Water Degumming

Water degumming is a simple, eco-friendly technique that excludes chemical additives (Sharma et al. 2019). Typically performed by immersing silk cocoons in water at 50–60 °C, this method softens SS to aid in filament unwinding during reeling. However, it suffers from incomplete degumming, requires extended exposure to high temperatures and pressures, and may damage SF, limiting its use in industrial applications (Rastogi and Kandasubramanian 2020).

#### 7.3.1.1.2 Soap Degumming

Soap degumming is widely adopted due to its effectiveness and relatively mild processing conditions. It uses alkaline soaps derived from oils (e.g., olive, coconut, tallow) such as sodium stearate, sodium oleate, and others (Wang et al. 2023). These soaps hydrolyze to release alkali, which reacts with SS to form soluble soda salts. Efficiency depends on factors like soap concentration (20–30% w/w), temperature (90–95 °C), time (1.5–2 h), liquor ratio (30–40:1), and pH (9.7–10.5) (Gulrajani 1992; Gore et al. 2019). The method enhances the softness, whiteness, absorbency, and luster of SF.

#### 7.3.1.1.3 Alkali Degumming

Alkaline agents such as sodium carbonate, sodium hydroxide, and potassium carbonate are commonly used for their low cost and high effectiveness (Lin et al. 2024). These substances efficiently hydrolyze SS, but may alter SF's appearance and texture if used alone. Thus, alkalis are often combined with soaps or detergents to mitigate harsh effects (Franck 2001). Key parameters like concentration and pH must be carefully controlled to preserve SF quality (Cui et al. 2021).

#### 7.3.1.1.4 Acid Degumming

Acidic solutions including tartaric, citric, and sulfuric acids are traditional degumming agents. Operating within a pH of 1.5–2, they hydrolyze peptide bonds in SS, particularly between aspartic and glutamic acids (Wang et al. 2023). Although effective and rapid, acid treatments are less commonly used due to their potential to degrade SF and the greater safety profile of alkaline methods (Gulrajani 1992).

### 7.3.1.2 Advanced Green Extraction Methods

#### 7.3.1.2.1 Enzymatic Degumming

Enzymatic degumming utilizes proteolytic enzymes (e.g., trypsin, papain) to cleave specific peptide bonds in SS under mild temperature (37–60 °C) and pH conditions (Ward 2019; Dijkstra 2010). This specificity minimizes damage to SF, yielding fibers with superior texture and whiteness. However, it is less efficient in removing hydrophobic impurities like waxes and oils, and is not yet ideal for large-scale processing due to limited yield and high cost.

#### 7.3.1.2.2 Microwave-Assisted Degumming

This technique employs microwave energy to uniformly and rapidly heat water molecules within the degumming solution, accelerating SS removal and reducing processing time and temperature requirements (Biswal et al. 2022). Studies have shown microwave-assisted degumming to be more efficient and resource-conserving compared to high-temperature high-pressure (HTHP) methods (Rastogi and Kandasubramanian 2020; Gupta et al. 2013).

#### 7.3.1.2.3 Amine-Assisted Degumming

Organic amines such as methylamine and triethylamine act similarly to alkalis by generating hydroxyl ions in aqueous solution, facilitating SS removal (Gulrajani and Malik 1993). This method is less affected by water hardness and can yield results comparable to soap degumming. However, the degumming rate is relatively slow, which limits its practicality in commercial settings (Wang et al. 2023).

#### 7.3.1.2.4 Ultrasound-Assisted Degumming

Ultrasonication enhances the penetration of degumming agents into silk fibers through high-frequency sound waves, resulting in uniform SS removal and reduced processing times (Vyas and Shukla 2016). An alternative involving supercritical $CO_2$ under acidic conditions demonstrated superior SF quality and recyclability of $CO_2$, offering an energy- and water-efficient process (Lo 2020).

#### 7.3.1.2.5 Ionic Liquid-Based Degumming

Ionic liquids (ILs) such as 1-butyl-3-methylimidazolium chloride offer a sustainable approach to SS extraction by dissolving SS under mild conditions without compromising SF integrity (Pereira et al. 2024). ILs are reusable, reducing environmental impact. However, their high cost and need for process optimization currently limit broader adoption (Wang et al. 2020).

### *7.3.2 Extraction of SF*

Once dried and degummed, SF fibers can be dissolved using various solvents to form aqueous solutions. Due to their crystalline structure and high β-sheet content, SF requires strong solvents such as concentrated acids or high-molarity chaotropic salts for effective dissolution. Commonly used solvents include LiBr, Ajisawa's reagent ($CaCl_2/H_2O$/ethanol), $ZnCl_2$, NMMO, LiSCN, HFIP, and $Ca(NO_3)_2$ (AJISAWA 1998; Yamada et al. 2001; Sashina et al. 2006; Rockwood et al. 2011). These solvents differ in solubility efficiency, requiring specific conditions for dissolution. The resulting SF solutions contain high salt concentrations, necessitating desalting, typically by dialysis against ultrapure water. While effective, dialysis is time-consuming and water-intensive. Faster alternatives, such as HPLC and desalting columns, have been explored; however, they often reduce SF yield and increase processing costs (Yeo et al. 2006).

### *7.3.3 Comparative Analysis of Extraction Methods*

#### 7.3.3.1 Efficiency

Conventional alkaline and acid degumming methods are well-established and effective, but often require high temperatures, extended durations, and large volumes of water (Wang et al. 2018; Feng et al. 2020). In contrast, microwave and ultrasound-assisted methods offer faster, more uniform SS removal, with reduced energy and water usage (Vyas and Shukla 2016). Supercritical $CO_2$ and IL-based techniques provide high efficiency under mild conditions, further preserving SF quality (Biswal et al. 2022).

#### 7.3.3.2 Purity of Extracted Components

Alkaline and acid methods can degrade SF, leading to reduced fiber strength and luster. They also yield SS with lower purity due to residual SF contamination (Seo et al. 2023). In contrast, modern methods like microwave and ultrasonic-assisted degumming yield higher purity SF and SS due to their controlled action (Carissimi et al. 2019).

#### 7.3.3.3 Environmental and Economic Considerations

Traditional techniques have significant environmental impacts, including high chemical and water use and toxic effluents. In contrast, advanced methods like microwave, ultrasound, supercritical $CO_2$, and IL-based degumming reduce water and chemical use, minimize environmental footprint, and improve process sustainability (Rastogi and Kandasubramanian 2020). However, higher costs associated with equipment and materials (e.g., ILs, $CO_2$ systems) can be a barrier to large-scale industrial implementation.

## 7.4 Characterization of SF and SS

Comprehensive characterization of SF and SS is essential to understand their structural, chemical, thermal, and morphological properties. A variety of analytical techniques are employed to reveal their conformational states, crystallinity, thermal stability, and micro- to nanoscale structures (Fig. 7.2). Below is an overview of the major methods used:

### *7.4.1 Fourier Transform Infrared Spectroscopy (FTIR)*

FTIR is a powerful technique for analyzing silk proteins, offering insights into their functional groups and secondary structures. The method detects characteristic amide bands, amide I (~1600–1700 $cm^{-1}$) associated with C=O stretching, amide II (~1500–1600 $cm^{-1}$) corresponding to N–H bending and C–N stretching, and amide III (~1200–1300 $cm^{-1}$), which further elucidates protein conformations (Chatterley et al. 2022; Koczoń et al. 2023). β-sheet structures, critical for mechanical integrity, are typically identified by peaks at 1620–1630 $cm^{-1}$, while random coil formations appear near 1645–1650 $cm^{-1}$, indicating amorphous characteristics. The combination of FTIR with ATR-FTIR allows for non-destructive surface characterization of silk films, fibers, and hydrogels, making it especially suitable for biomedical and textile applications (Kazarian and Chan 2013). Moreover, deconvolution of the amide I band enables quantification of secondary structural elements, facilitating the optimization of silk protein processing and functionality (Lee et al. 2018; Fellows et al. 2020; Seo et al. 2023).

### *7.4.2 X-ray Diffraction (XRD)*

XRD is used to analyze crystallinity in silk proteins by detecting diffraction patterns from crystalline domains. SF displays sharp peaks at ~20° and 24° (2θ), reflecting

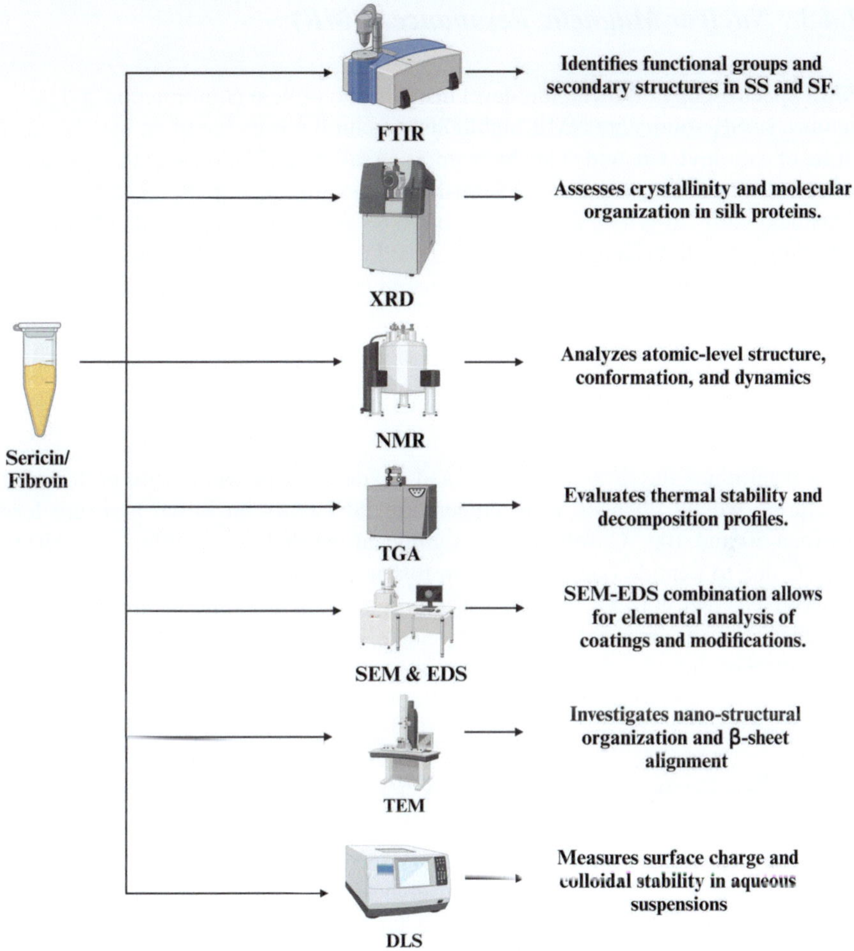


**Fig. 7.2** Overview of analytical techniques employed for the comprehensive characterization of silk proteins

ordered β-sheet structures (Narita et al. 2020). In contrast, SS shows a broad peak at ~20.5° (2θ), indicating amorphous or semi-ordered structure (Kanoujia et al. 2023; Gupta et al. 2013). Processing parameters like degumming, solvent exposure, and post-treatments influence the crystalline–amorphous transition (Kanoujia et al. 2023; Johnston et al. 2018). XRD also reveals nanoscale features through peak broadening (Holder and Schaak 2019).

### 7.4.3 Nuclear Magnetic Resonance (NMR)

NMR spectroscopy offers atomic-level insights into protein conformation and interactions. In SF, solid-state NMR highlights structural motifs based on the chemical shifts of carbonyl, Cα, and side-chain atoms (Asakura 2024). It helps track alanine, glycine, and serine residues, which form the backbone of SF. For SS, 1H NMR detects downfield shifts in tyrosine signals—from 6.68 and 6.96 to 7.09 and 7.26 ppm—reflecting structural changes (Cho et al. 2003; Diehl 2008). NMR is thus crucial for mapping conformational stability in silk-based biomaterials (Hu et al. 2021).

### 7.4.4 Thermogravimetric Analysis (TGA)

TGA determines the thermal stability and degradation profile of silk proteins by measuring weight changes with temperature. SF shows an initial moisture loss between 50 and 100 °C, followed by significant thermal degradation from 250 to 350 °C due to peptide bond cleavage in β-sheets (Zhao et al. 2021). SS degrades between 210 and 400 °C, reflecting its amorphous, hydrophilic nature (Castrillón Martínez et al. 2017). Analysis of TGA and derivative curves provides information on purity, molecular breakdown, and thermal resilience (Chandra and Chandra 2022).

### 7.4.5 Scanning Electron Microscopy (SEM)

SEM provides high-resolution imaging to assess surface morphology. SF fibers appear smooth and continuous, with differences arising from solvent or regeneration conditions (Reizabal et al. 2021). SS typically exhibits irregular, aggregated globules when dried from solution (Gupta et al. 2014). SEM analysis is valuable for structural evaluation in silk scaffolds, hydrogels, and nanofibers (Gupta and Ghosh 2016; Goldstein et al. 2017).

### 7.4.6 Transmission Electron Microscopy (TEM)

TEM provides nanoscale imaging of internal structure and crystallinity. In SF, TEM identifies β-sheet nanocrystals that enhance strength and rigidity (Drummy et al. 2007; Liang et al. 2024). SS often forms pine-like or aggregated morphologies influenced by pH and ionic conditions (WU et al. 2010). TEM is instrumental in analyzing silk-based nanocomposites for biomedical and industrial applications (Akhtar et al. 2018; Nilogal et al. 2021).

## 7.5 Properties of SF and SS

Silk proteins, SF and SS, exhibit distinct physicochemical and biological characteristics that underpin their diverse applications in biomedical, cosmetic, and material science fields. Their key properties, including amino acid composition, antioxidant activity, biodegradability, and mechanical strength, are discussed below and summarized in Table 7.2.

### 7.5.1 *Amino Acid Composition*

The amino acid profile critically defines silk protein behavior. Fibroin (SF), the core structural protein of silk, is rich in glycine (~45%), alanine (~30%), and serine (~12%), supporting the formation of repetitive GAGAGS motifs that drive β-sheet crystallinity and high tensile strength (Zhao et al. 2015; De Giorgio et al. 2024). Sericin (SS), by contrast, is hydrophilic and rich in polar amino acids such as serine (~30–40%), aspartic acid, and glycine, contributing to its water solubility and emulsifying properties (Seo et al. 2023). Tyrosine present in both proteins allows for chemical crosslinking and functional modification. SF's structure makes it ideal for

**Table 7.2** Properties of SS and SF

| Properties | SS | SF | References |
|---|---|---|---|
| Amino acid profile | Rich in serine (30–40%), aspartic acid, and glycine. Hydrophilic and bioactive | Predominantly glycine (~45%), alanine (~30%), serine (~12%). Forms β-sheet structures, providing mechanical strength | (Zhao et al. 2015; Seo et al. 2023; De Giorgio et al. 2024) |
| Antioxidant activity | Strong antioxidant due to tyrosine, tryptophan, histidine, and cysteine. Reduces oxidative stress and enhances stability | Moderate antioxidant activity. Can be enhanced with polyphenols or nanoparticles. Used in therapeutic applications | (Kunz et al. 2016; Lozano-Pérez et al. 2017; Seo et al. 2023; Qian et al. 2024) |
| Biodegradability | Rapid degradation due to amorphous structure. Releases bioactive peptides that aid in wound healing and tissue repair | Enzymatically degraded by protease XIV and collagenase. Higher crystallinity leads to slower degradation, making it ideal for long-term biomedical use | (Zhu et al. 2011; Luo et al. 2015; Lee et al. 2023) |
| Mechanical strength | Weak mechanical properties due to amorphous nature and hydrophilicity. Used in hydrogels and coatings | High tensile strength (610–690 MPa) due to β-sheet structures. Used in sutures, bone scaffolds, and artificial ligaments | (Partlow et al. 2016; Zhang et al. 2018; Li and Sun 2022) |

load-bearing roles, while SS's composition supports its use in moisture-retentive and bioactive systems like drug delivery and skincare.

### 7.5.2 *Antioxidant Activity*

SS exhibits strong intrinsic antioxidant activity due to its content of tyrosine, tryptophan, histidine, and cysteine, which neutralize reactive oxygen species (ROS) (Seo et al. 2023). This makes SS suitable for biomedical and cosmetic uses targeting oxidative stress (Kunz et al. 2016). SF has lower native antioxidant capacity but shows enhanced activity when hydrolyzed or combined with polyphenols or nanoparticles (Lozano-Pérez et al. 2017; Qian et al. 2024). Modified SF scaffolds have shown promise in regenerative medicine. Thus, while SS offers stronger inherent antioxidant properties, SF can be engineered for similar outcomes.

### 7.5.3 *Biodegradability*

Both proteins are biodegradable but differ in degradation rates and pathways. SF degrades via proteases such as collagenase and protease XIV, with β-sheet content governing its degradation rate, more crystalline SF degrades slowly, ideal for long-term implants (Zhu et al. 2011; Luo et al. 2015). SS degrades faster due to its amorphous and hydrophilic structure, releasing bioactive peptides beneficial for wound healing and tissue regeneration (Lee et al. 2023). The differing degradation profiles of SF and SS allow for complementary or staged biomaterial designs.

### 7.5.4 *Mechanical Properties*

SF is known for its remarkable tensile strength (610–690 MPa), attributed to its β-sheet-rich crystalline structure and strong hydrogen bonding. Its mechanical properties can be tailored through processing methods for applications such as bone scaffolds and artificial ligaments (Partlow et al. 2016; Li and Sun 2022). SS, due to its amorphous nature and low β-sheet content, has lower mechanical strength. However, advances in polymer blending and crosslinking have improved its robustness, expanding its application range beyond soft biomaterials to more durable composites (Zhang et al. 2018).

## 7.6 Insights into Applications of SF and SS

SF and SS, the two principal proteins extracted from silkworm cocoons, have emerged as multifunctional biomaterials with applications across diverse sectors including biomedicine, textiles, nanotechnology, cosmetics, and environmental sustainability. Their favorable biological properties, such as biocompatibility, biodegradability, antioxidant activity, mechanical strength, and moisture retention, contribute to their increasing demand in both academic research and industrial development (Table 7.3; Fig. 7.3).

### *7.6.1 Biomedical Applications*

In regenerative medicine, SF and SS play complementary roles due to their distinct structural and functional properties. SF possesses high tensile strength, flexibility, and low immunogenicity, which makes it suitable for developing scaffolds, films, and hydrogels that promote cell adhesion and proliferation. These structures support collagen synthesis, fibroblast migration, and formation of a moist environment that accelerates tissue regeneration (Liu et al. 2022; De Giorgio et al. 2024). SS, a water-soluble glycoprotein, contributes significantly by providing antioxidant, anti-inflammatory, and cell-adhesive properties. It stimulates keratinocyte and fibroblast activity and supports re-epithelialization and tissue remodeling (Vidya and Rajagopal 2021; Li et al. 2023). Together, SF and SS provide an effective platform for healing wounds and regenerating skin, bone, cartilage, and neural tissues.

Both proteins are also extensively studied for their drug delivery potential. SF's ability to form stable β-sheet structures allows it to be processed into films, nanoparticles, and hydrogels capable of controlled and sustained drug release, particularly for anticancer agents and bioactive peptides (Crivelli et al. 2018; Yu et al. 2023). SS enhances the solubility and bioavailability of hydrophobic drugs and facilitates their cellular uptake. Moreover, its antioxidant nature can enhance therapeutic outcomes (Yu et al. 2023). The integration of SF and SS in delivery systems improves biocompatibility, drug stability, and release kinetics, contributing to more efficient and patient-friendly therapies (Crivelli et al. 2018).

Although SF lacks inherent antibacterial activity, it can be functionalized with antimicrobial agents such as silver nanoparticles, peptides, or antibiotics to prevent bacterial colonization. Its slow degradation supports long-term antimicrobial activity (Schäfer et al. 2023; Hassan et al. 2024). SS naturally exhibits antibacterial effects by disrupting microbial membranes and chelating essential ions like iron. This activity is effective against both Gram-positive and Gram-negative bacteria and is further enhanced by its antioxidant properties, which help reduce inflammation and promote wound healing (Silva et al. 2022; Li et al. 2023). SF-SS composites are thus valuable in producing antibacterial wound dressings and medical textiles.

**Table 7.3** Different silk-based biomaterials and their key properties, production methods, and applications

| Silk-based nano material | Key properties | Production method | Applications | References |
|---|---|---|---|---|
| Sponges | 3D porous, tunable pore size, macroporous | Salt leaching, freeze casting, HFIP solvent | Bone tissue regeneration, vascularized bone formation | (Karageorgiou and Kaplan 2005; Rockwood et al. 2011; Bhattacharjee et al. 2017) |
| Electrospun mats | Nanofibrous, high surface area, tunable fiber diameter | Electrospinning with voltage-induced fiber formation | Wound dressings, textiles, wearable electrodes, nerve guides | (Rockwood et al. 2011; Yukseloglu et al. 2015; Belanger et al. 2018) |
| Microspheres | Encapsulation, controlled degradation, drug stabilization | Emulsion, phase separation, potassium phosphate precipitation | Drug delivery, enzyme stabilization, controlled release | (Lammel et al. 2010; Rockwood et al. 2011; Li et al. 2015, 2017) |
| Hydrogels | Soft tissue-like, high water retention, tunable crosslinking | Physical, photochemical, chemical gelation | Drug delivery, tissue engineering, regenerative medicine | (Fernández-García et al. 2016; Frauchiger et al. 2017; Vidal et al. 2019; Li et al. 2020) |
| HRP-crosslinked hydrogels | Enzyme-mediated crosslinking, biocompatible, in situ gelation | HRP catalysis with phenol/aniline derivatives | Biomedical applications, bioactive factor encapsulation | (Partlow et al. 2014; Bang et al. 2017; Bi et al. 2019; Hoang Thi et al. 2019) |
| Aerogels | Ultra-lightweight, highly porous, thermal insulating | Freeze drying, ice templating | Fire retardants, pollution control, biomaterials | (Mandal and Kundu 2009; Xiong et al. 2018; Maleki and Huesing 2021) |
| 3D printed structures | Complex, customizable, patient-specific designs | 3D bioprinting with silk-based bioinks | Tissue engineering, controlled macroporosity scaffolds | (Zheng et al. 2018; Rodriguez et al. 2018; Mu et al. 2020) |
| Hard silk materials | High mechanical strength, machinable, resorbable | Biomimetic silk processing, thermal molding | Orthopedic devices, resorbable bone screws and plates | (Li et al. 2016; James et al. 2019; Guo et al. 2020) |

(continued)

**Table 7.3** (continued)

| Silk-based nano material | Key properties | Production method | Applications | References |
|---|---|---|---|---|
| Films | Transparent, flexible, customizable surface properties | Air drying, annealing, electrogelation | Drug delivery, corneal tissue models, vaccine stabilization | (Lawrence et al. 2009; Zhou et al. 2014; Stinson et al. 2020) |
| Silk foams | Injectable, compressible, expandable | Pre-loaded enzyme degradation tuning | Soft tissue regeneration, drug delivery | (Bellas et al. 2015; Brown et al. 2017; Ornell et al. 2020) |
| Microneedles | Biodegradable, mechanical strength, controlled drug release | Molded *SF* with post-processing adjustments | Drug, vaccine, and insulin delivery | (Tsioris et al. 2012; DeMuth et al. 2014; Wang et al. 2019) |

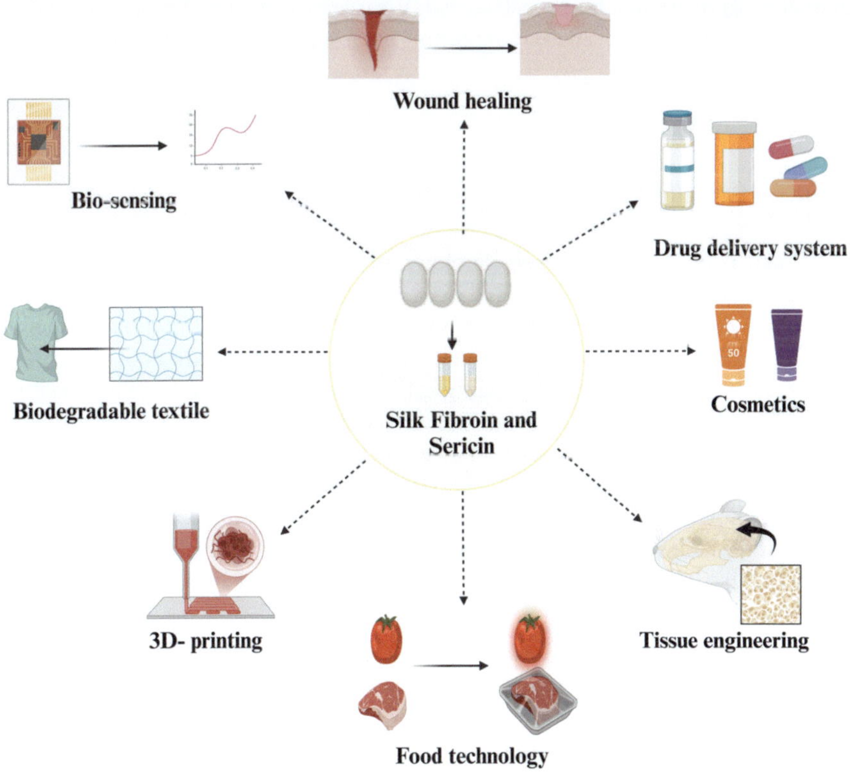


**Fig. 7.3** Diverse applications of silk fibroin and sericin in healthcare, textiles, food, and advanced technologies

### 7.6.2 Environmental and Textile Industry Applications

In the textile industry, SF is recognized for its biodegradability, strength, and comfort-enhancing properties such as breathability and moisture regulation. It decomposes into non-toxic amino acids through microbial or enzymatic action, reducing environmental impact (Zhang 2002; Vidya and Rajagopal 2021). SS, previously discarded as waste, is now appreciated for improving textile functionality. It enhances moisture retention, provides UV protection, and increases thermal stability. SS coatings also promote fiber biodegradability and impart antioxidant and antibacterial properties to fabrics (Silva et al. 2022; Yang et al. 2024). The combined use of SF and SS in textiles contributes to circular economy initiatives by creating sustainable, multifunctional fabrics.

In eco-friendly dyeing, SF and SS offer sustainable alternatives to chemical-intensive methods. SF fabrics can be dyed using natural colorants or biosynthetic techniques, reducing synthetic dye use (Wray et al. 2011; Lee et al. 2020). SS improves dye absorption and fixation due to its hydrophilicity and serves as a natural alternative to mordants. It also enhances UV resistance and dye fastness in textiles (Silva et al. 2022; Zare 2023), supporting environmentally responsible finishing processes.

### 7.6.3 Nanotechnology and Cosmetics

SF is increasingly used in nanotechnology as a matrix for producing nanofibers, microspheres, and nanoparticles for drug delivery, biosensors, and tissue scaffolding. Its composites with inorganic materials offer additional properties such as antimicrobial activity, fluorescence, and magnetism (Belda Marín et al. 2020). SS further enhances these systems by adding antioxidant and antimicrobial functionality, making it suitable for biomedical nanomaterials (Silva et al. 2022). In cosmetics, SF acts as a carrier for active ingredients and protects skin structure in formulations like anti-aging creams and skin-repair gels (Hassan et al. 2024). SS provides moisturizing, UV-protective, and antioxidant effects due to its rich amino acid composition, making it valuable in sunscreens and personal care products (Tengattini et al. 2020). Their combined use promotes the development of skin-compatible and eco-friendly cosmetic formulations.

## 7.7 Challenges and Future Perspectives

### 7.7.1 *Challenges*

#### 7.7.1.1 Technical Challenges in Extraction

SF and SS, derived from silkworms, are highly valued for their biocompatibility, biodegradability, and multifunctionality, enabling their use across biomedical, textile, and material science sectors (Hassan et al. 2024). However, their extraction poses key challenges related to environmental sustainability, efficiency, and scalability (Beniwal et al. 2024). SF extraction typically involves dissolving SS using harsh solvents like calcium chloride ($CaCl_2$), lithium bromide (LiBr), or urea, followed by purification. While effective, these chemicals can damage protein structure, reduce biofunctionality, and raise environmental concerns due to toxicity and waste disposal (Huang et al. 2023). SS extraction is more complex due to its strong affinity for SF. It is commonly removed with hot water or alkaline treatments, but these approaches often degrade SS's molecular integrity, limiting its utility (Seo et al. 2023). Moreover, SS composition varies among silkworm species and developmental stages, complicating standardization for large-scale processing.

#### 7.7.1.2 Optimization of Green Methods

Eco-friendly approaches such as ionic liquids (ILs) and enzymatic extraction are being explored to reduce environmental impact. ILs like BMIMCl effectively dissolve silk proteins but require careful control of concentration and temperature (Heng et al. 2022). Enzymatic methods offer a gentler, more selective extraction, though they demand precise optimization of enzyme type, concentration, and reaction time (Feng et al. 2020). Supercritical $CO_2$ also shows promise as a green solvent, but its high operational cost limits large-scale use (Kopilovic et al. 2023).

#### 7.7.1.3 Scalability for Industrial Use

Scaling up silk protein extraction remains a major bottleneck (Bitar et al. 2024). Parameters like solvent concentration, temperature, and time must be tightly controlled to ensure consistency and yield, easily managed in lab conditions, but challenging at an industrial scale (Sachin et al. 2024). Green methods, although sustainable, often involve high operational costs. ILs and supercritical $CO_2$ systems require costly reagents and equipment (Ying et al. 2022). Moreover, these eco-friendly approaches may yield lower purity or quantity compared to traditional chemical methods (Kabir et al. 2024).

#### 7.7.1.4 Environmental Impact

Traditional methods using $CaCl_2$, LiBr, and urea generate hazardous waste that can harm ecosystems (Bitar et al. 2024). This has led to increased interest in green technologies. SS-rich silk waste, usually discarded, holds potential for repurposing in cosmetics, pharmaceuticals, and biodegradable materials (Yuan et al. 2024). Enzymatic extraction reduces chemical use, supporting eco-sustainability (Bitar et al. 2024). Solvent recycling, particularly using ionic liquids and supercritical carbon dioxide, can significantly reduce waste. Ionic liquids are both efficient and recyclable, while supercritical carbon dioxide leaves minimal residues and produces no toxic byproducts. (Almohasin et al. 2023).

### 7.7.2 *Future Perspectives*

#### 7.7.2.1 Development of Hybrid Biopolymer Materials

Combining SF and SS with other biopolymers like collagen, chitosan, and alginate results in hybrid materials with enhanced biological and mechanical properties (Hassan et al. 2024). These composites, often embedded with nanoparticles or graphene, are being developed for applications like wound healing, drug delivery, and bone regeneration due to their bioactivity, cell adhesion, and biodegradability (Wu et al. 2021).

#### 7.7.2.2 Applications in Regenerative Medicine

SF has shown excellent promise in tissue engineering, supporting various cell types including stem cells, and offering tunable degradation rates (Pritchard et al. 2014). SF-based scaffolds facilitate the regeneration of bone, cartilage, and nerve tissues through enhanced cell migration and differentiation (Ma et al. 2024). Meanwhile, SS is emerging as a functional biomaterial with bioactive properties like promoting angiogenesis and cell adhesion. It is under exploration for wound healing, skin grafts, and drug delivery systems, especially in neuroregeneration and stem cell therapies (Liu et al. 2022; Yang et al. 2023). In orthopedics, SF scaffolds combined with growth factors have demonstrated success in directing stem cell differentiation toward bone and cartilage formation, offering efficient support for tissue repair.

## 7.8 Conclusion

In conclusion, the extraction and characterization of silk proteins sericin and fibroin represent an active area of research with wide-ranging implications in biomedical, environmental, textile, and nanotechnological sectors. This review has critically analyzed traditional and advanced green extraction techniques, highlighting their efficiency, environmental impact, and suitability for large-scale applications. The integration of advanced analytical tools such as FTIR, SEM, XRD, and NMR has enabled a deeper understanding of the molecular structure, thermal behavior, and functional properties of SF and SS, revealing their complementary strengths; SF for its mechanical robustness and biocompatibility, and SS for its hydrophilicity, antioxidant, and antimicrobial properties. Despite current challenges, such as the variability in extraction outcomes, the environmental burden of conventional solvents, and scalability issues, innovative approaches involving ionic liquids, enzymatic methods, and hybrid biopolymer systems offer promising alternatives. As research continues to bridge knowledge gaps and refine sustainable methodologies, silk proteins are poised to play a pivotal role in next-generation biomaterials, offering eco-friendly, high-performance solutions that align with circular economy principles and advanced material science.

**Acknowledgements** Shubhajit Shaw, would like to acknowledge the University Grant Commission, Government of India, for the UGC-JRF fellowship (UGC-JRF; NTA Ref. No. 201610120606).

**Declaration of AI Usage** The AI tools were used for purposes such as language refinement, clarity, and coherence. The authors has thoroughly reviewed all material, and take full responsibility for the accuracy, integrity, and originality of the final content.

## References

Aad R, Dragojlov I, Vesentini S (2024) Sericin protein: structure, properties, and applications. J Funct Biomater 15:322. https://doi.org/10.3390/jfb15110322

Ajisawa A (1998) Dissolution of silk fibroin with calciumchloride/ethanol aqueous solution. J Sericultural Sci 67(2):91–94. https://doi.org/10.11416/kontyushigen1930.67.91

Akhtar K, Khan SA, Khan SB, Asiri AM (2018) Scanning electron microscopy: principle and applications in nanomaterials characterization. In: Sharma SK (ed) Handbook of materials characterization. Springer, Cham, pp 113–145

Almohasin JA, Balag J, Miral VG, et al (2023) Green solvents for liquid–liquid extraction: recent advances and future trends. In: The 4th international electronic conference on applied sciences. MDPI, p 174

Asakura T (2024) Structural analysis of silk using solid-state NMR. Magn Reson Lett 4:200111. https://doi.org/10.1016/j.mrl.2024.200111

Bang S, Ko Y-G, Kim WI et al (2017) Preventing postoperative tissue adhesion using injectable carboxymethyl cellulose-pullulan hydrogels. Int J Biol Macromol 105:886–893. https://doi.org/10.1016/j.ijbiomac.2017.07.103

Belanger K, Schlatter G, Hébraud A et al (2018) A multi-layered nerve guidance conduit design adapted to facilitate surgical implantation. Health Sci Rep 1:e86. https://doi.org/10.1002/hsr2.86

Belda Marín C, Fitzpatrick V, Kaplan DL et al (2020) Silk polymers and nanoparticles: a powerful combination for the design of versatile biomaterials. Front Chem 8:604398. https://doi.org/10.3389/fchem.2020.604398

Bellas E, Lo TJ, Fournier EP et al (2015) Injectable silk foams for soft tissue regeneration. Adv Healthcare Mater 4:452–459. https://doi.org/10.1002/adhm.201400506

Beniwal A, Mahapatara D, Das M et al (2024) A comprehensive review on the multifaceted applications of mulberry silkworm pupae: a sustainable resource. J Asia-Pac Entomol 27:102312. https://doi.org/10.1016/j.aspen.2024.102312

Bhattacharjee P, Kundu B, Naskar D et al (2017) Silk scaffolds in bone tissue engineering: an overview. Acta Biomater 63:1–17. https://doi.org/10.1016/j.actbio.2017.09.027

Bi B, Liu H, Kang W et al (2019) An injectable enzymatically crosslinked tyramine-modified carboxymethyl chitin hydrogel for biomedical applications. Colloids Surf B Biointerf 175:614–624. https://doi.org/10.1016/j.colsurfb.2018.12.029

Biswal B, Dan AK, Sengupta A et al (2022) Extraction of silk fibroin with several sericin removal processes and its importance in tissue engineering: a review. J Polym Environ 30:2222–2253. https://doi.org/10.1007/s10924-022-02381-w

Bitar L, Isella B, Bertella F et al (2024) Sustainable *B. mori's* silk fibroin for biomedical applications as a molecular biotechnology challenge: a review. Int J Biol Macromol 264:130374. https://doi.org/10.1016/j.ijbiomac.2024.130374

Brown JE, Moreau JE, Berman AM et al (2017) Shape memory silk protein sponges for minimally invasive tissue regeneration. Adv Healthcare Mater 6:1600762. https://doi.org/10.1002/adhm.201600762

Carissimi G, Lozano-Pérez AA, Montalbán MG et al (2019) Revealing the influence of the degumming process in the properties of silk fibroin nanoparticles. Polymers 11:2045. https://doi.org/10.3390/polym11122045

Castrillón Martínez DC, Zuluaga CL, Restrepo-Osorio A, Álvarez-López C (2017) Characterization of sericin obtained from cocoons and silk yarns. Proced Eng 200:377–383. https://doi.org/10.1016/j.proeng.2017.07.053

Chandra A, Chandra A (2022) Ion conduction mechanism of nanoparticle-based polymer composites. Nanoparticle-based polymer composites. Elsevier, Amsterdam, pp 277–306

Chatterley AS, Laity P, Holland C et al (2022) Broadband multidimensional spectroscopy identifies the amide II vibrations in silkworm films. Molecules 27:6275. https://doi.org/10.3390/molecules27196275

Cho KY, Moon JY, Lee YW et al (2003) Preparation of self-assembled silk sericin nanoparticles. Int J Biol Macromol 32:36–42. https://doi.org/10.1016/S0141-8130(03)00023-0

Crivelli B, Perteghella S, Bari E et al (2018) Silk nanoparticles: from inert supports to bioactive natural carriers for drug delivery. Soft Matter 14:546–557. https://doi.org/10.1039/C7SM01631J

Cui H, Shi C, Duan J et al (2021) Study on biochemical degumming process of mulberry fiber, part 1:single factor analysis. J Nat Fibers 18:1533–1544. https://doi.org/10.1080/15440478.2019.1691699

De Giorgio G, Matera B, Vurro D et al (2024) Silk fibroin materials: biomedical applications and perspectives. Bioengineering 11:167. https://doi.org/10.3390/bioengineering11020167

DeMuth PC, Min Y, Irvine DJ, Hammond PT (2014) Implantable silk composite microneedles for programmable vaccine release kinetics and enhanced immunogenicity in transcutaneous immunization. Adv Healthcare Mater 3:47–58. https://doi.org/10.1002/adhm.201300139

Diehl B (2008) Principles in NMR spectroscopy. NMR spectroscopy in pharmaceutical analysis. Elsevier, Amsterdam, pp 1–41

Dijkstra AJ (2010) Enzymatic degumming. Euro J Lipid Sci Tech 112:1178–1189. https://doi.org/10.1002/ejlt.201000320

Drummy LF, Farmer BL, Naik RR (2007) Correlation of the β-sheet crystal size in silk fibers with the protein amino acid sequence. Soft Matter 3:877–882. https://doi.org/10.1039/B701220A

Fellows AP, Casford MTL, Davies PB (2020) Spectral analysis and deconvolution of the amide I band of proteins presenting with high-frequency noise and baseline shifts. Appl Spectrosc 74:597–615. https://doi.org/10.1177/0003702819898536

Feng Y, Lin J, Niu L et al (2020) High molecular weight silk fibroin prepared by papain degumming. Polymers 12:2105. https://doi.org/10.3390/polym12092105

Fernández-García L, Marí-Buyé N, Barios JA et al (2016) Safety and tolerability of silk fibroin hydrogels implanted into the mouse brain. Acta Biomater 45:262–275. https://doi.org/10.1016/j.actbio.2016.09.003

Franck RR (2001) Silk, mohair, cashmere and other luxury fibres. Woodhead Publishing

Frauchiger D, Tekari A, Wöltje M et al (2017) A review of the application of reinforced hydrogels and silk as biomaterials for intervertebral disc repair. Eur Cells Mater 34:271–290

Goldstein JI, Newbury DE, Michael JR et al (2017) Scanning electron microscopy and X-ray microanalysis. Springer, Berlin

Gore PM, Naebe M, Wang X, Kandasubramanian B (2019) Progress in silk materials for integrated water treatments: fabrication, modification and applications. Chem Eng J 374:437–470. https://doi.org/10.1016/j.cej.2019.05.163

Gulrajani ML (1992) Degumming of silk. Rev Prog Color Relat Top 22:79–89. https://doi.org/10.1111/j.1478-4408.1992.tb00091.x

Gulrajani ML, Malik R (1993) Degumming of silk with methylamine. Ind J Fiber Text Res 18(2):72–78

Guo C, Li C, Vu HV et al (2020) Thermoplastic moulding of regenerated silk. Nat Mater 19:102–108. https://doi.org/10.1038/s41563-019-0560-8

Gupta SS, Ghosh M (2016) Advanced nanocarriers for nutraceuticals based on structured lipid and nonlipid components. Nutraceuticals. Academic Press, pp 271–304

Gupta D, Agrawal A, Chaudhary H et al (2013) Cleaner process for extraction of sericin using infrared. J Clean Prod 52:488–494. https://doi.org/10.1016/j.jclepro.2013.03.016

Gupta D, Agrawal A, Rangi A (2014) Extraction and characterization of silk sericin. Ind J Fibre Text Res 39(4):364–372

Hassan MA, Basha AA, Eraky M et al (2024) Advancements in silk fibroin and silk sericin-based biomaterial applications for cancer therapy and wound dressing formulation: a comprehensive review. Int J Pharm 662:124494. https://doi.org/10.1016/j.ijpharm.2024.124494

Heng H, Deng Q, Yang Y, Wang F (2022) Recent research progress of ionic liquid dissolving silks for biomedicine and tissue engineering applications. Int J Mol Sci 23:8706. https://doi.org/10.3390/ijms23158706

Hoang Thi TT, Lee Y, Le Thi P, Park KD (2019) Engineered horseradish peroxidase-catalyzed hydrogels with high tissue adhesiveness for biomedical applications. J Ind Eng Chem 78:34–52. https://doi.org/10.1016/j.jiec.2019.05.026

Holder CF, Schaak RE (2019) Tutorial on powder X-ray diffraction for characterizing nanoscale materials. ACS Nano 13:7359–7365. https://doi.org/10.1021/acsnano.9b05157

Hu Y, Cheng K, He L et al (2021) NMR-based methods for protein analysis. Anal Chem 93:1866–1879. https://doi.org/10.1021/acs.analchem.0c03830

Hu D, Li T, Liang W et al (2023) Silk sericin as building blocks of bioactive materials for advanced therapeutics. J Contr Release 353:303–316. https://doi.org/10.1016/j.jconrel.2022.11.019

Huang L, Shi J, Zhou W, Zhang Q (2023) Advances in preparation and properties of regenerated silk fibroin. Insect Biochem Mol Biol 24:13153. https://doi.org/10.3390/ijms241713153

James EN, Van Doren E, Li C, Kaplan DL (2019) Silk biomaterials-mediated miRNA functionalized orthopedic devices. Tissue Eng A 25:12–23. https://doi.org/10.1089/ten.tea.2017.0455

Johnston ER, Miyagi Y, Chuah J-A et al (2018) Interplay between silk fibroin's structure and its adhesive properties. ACS Biomater Sci Eng 4:2815–2824. https://doi.org/10.1021/acsbiomaterials.8b00544

Kabir M, Hasan MK, Rafi MN et al (2024) Biowaste transformation to functional materials: structural properties, extraction methods, applications, and challenges of silk sericin. ChemistrySelect 9:e202403777. https://doi.org/10.1002/slct.202403777
Kanoujia J, Dubey R, Debgharia S et al (2023) Mini-review on analytical methods applied for analysis and characterization of sericin. Curr Anal Chem 19:119–129. https://doi.org/10.2174/1573411019666221024150235
Karageorgiou V, Kaplan D (2005) Porosity of 3D biomaterial scaffolds and osteogenesis. Biomaterials 26:5474–5491. https://doi.org/10.1016/j.biomaterials.2005.02.002
Kazarian SG, Chan KLA (2013) ATR-FTIR spectroscopic imaging: recent advances and applications to biological systems. Analyst 138:1940–1951. https://doi.org/10.1039/c3an36865c
Koczoń P, Hołaj-Krzak JT, Palani BK et al (2023) The analytical possibilities of FT-IR spectroscopy powered by vibrating molecules. Int J Mol Sci 24:1013. https://doi.org/10.3390/ijms24021013
Kopilovic B, Valente AI, Ferreira AM et al (2023) Towards the sustainable extraction and purification of non-animal proteins from biomass using alternative solvents. RSC Sustain 1:1314–1331. https://doi.org/10.1039/D3SU00062A
Kunz RI, Brancalhão RMC, Ribeiro LDFC, Natali MRM (2016) Silkworm sericin: properties and biomedical applications. BioMed Res Int 2016:1–19. https://doi.org/10.1155/2016/8175701
Lammel AS, Hu X, Park S-H et al (2010) Controlling silk fibroin particle features for drug delivery. Biomaterials 31:4583–4591. https://doi.org/10.1016/j.biomaterials.2010.02.024
Lawrence BD, Marchant JK, Pindrus MA et al (2009) Silk film biomaterials for cornea tissue engineering. Biomaterials 30:1299–1308. https://doi.org/10.1016/j.biomaterials.2008.11.018
Lee JH, Bae YS, Kim SJ et al (2018) Preparation of new natural silk non-woven fabrics by using adhesion characteristics of sericin and their characterization. Int J Biol Macromol 106:39–47. https://doi.org/10.1016/j.ijbiomac.2017.07.179
Lee OJ, Sultan MdT, Hong H et al (2020) Recent advances in fluorescent silk fibroin. Front Mater 7:50. https://doi.org/10.3389/fmats.2020.00050
Lee HG, Jang MJ, Park B-D, Um IC (2023) Structural characteristics and properties of redissolved silk sericin. Polymers 15:3405. https://doi.org/10.3390/polym15163405
Li G, Sun S (2022) Silk fibroin-based biomaterials for tissue engineering applications. Molecules 27:2757. https://doi.org/10.3390/molecules27092757
Li AB, Kluge JA, Guziewicz NA et al (2015) Silk-based stabilization of biomacromolecules. J Contr Release 219:416–430. https://doi.org/10.1016/j.jconrel.2015.09.037
Li C, Hotz B, Ling S et al (2016) Regenerated silk materials for functionalized silk orthopedic devices by mimicking natural processing. Biomaterials 110:24–33. https://doi.org/10.1016/j.biomaterials.2016.09.014
Li H, Zhu J, Chen S et al (2017) Fabrication of aqueous-based dual drug loaded silk fibroin electrospun nanofibers embedded with curcumin-loaded RSF nanospheres for drugs controlled release. RSC Adv 7:56550–56558. https://doi.org/10.1039/C7RA12394A
Li Z, Song J, Zhang J et al (2020) Topical application of silk fibroin-based hydrogel in preventing hypertrophic scars. Colloids Surf B Biointerfaces 186:110735. https://doi.org/10.1016/j.colsurfb.2019.110735
Li Y, Wei Y, Zhang G, Zhang Y (2023) Sericin from fibroin-deficient silkworms served as a promising resource for biomedicine. Polymers 15:2941. https://doi.org/10.3390/polym15132941
Liang H, Chen J, Qiu G et al (2024) Ultrasonication-induced preparation of high-mechanical-strength microneedles using stable silk fibroin. Polymers 16:3183. https://doi.org/10.3390/polym16223183
Lin D, Li M, Wang L et al (2024) Multifunctional hydrogel based on silk fibroin promotes tissue repair and regeneration. Adv Funct Materials 34:2405255. https://doi.org/10.1002/adfm.202405255
Liu X, Zhang K-Q (2014) Silk fiber—molecular formation mechanism, structure-property relationship and advanced applications. Oligomerizat Chem Biol Compd 3:69–102

Liu J, Shi L, Deng Y et al (2022) Silk sericin-based materials for biomedical applications. Biomaterials 287:121638. https://doi.org/10.1016/j.biomaterials.2022.121638

Lo CH (2020) Degumming silk by CO2 supercritical fluid and their dyeing ability with plant indigo. Int J Cloth Sci Technol 33:465–476. https://doi.org/10.1108/IJCST-06-2019-0072

Lozano-Pérez AA, Rivero HC, Pérez-Hernández MDC et al (2017) Silk fibroin nanoparticles: efficient vehicles for the natural antioxidant quercetin. Int J Pharm 518:11–19. https://doi.org/10.1016/j.ijpharm.2016.12.046

Luo Z, Zhang Q, Shi M et al (2015) Effect of pore size on the biodegradation rate of silk fibroin scaffolds. Adv Mater Sci Eng 2015:1–7. https://doi.org/10.1155/2015/315397

Ma L, Dong W, Lai E, Wang J (2024) Silk fibroin-based scaffolds for tissue engineering. Front Bioeng Biotechnol 12:1381838. https://doi.org/10.3389/fbioe.2024.1381838

Maleki H, Huesing N (2021) Silica-silk fibroin hybrid (bio)aerogels: two-step versus one-step hybridization. J Sol–gel Sci Technol 98:430–438. https://doi.org/10.1007/s10971-019-04933-4

Mandal BB, Kundu SC (2009) Cell proliferation and migration in silk fibroin 3D scaffolds. Biomaterials 30:2956–2965. https://doi.org/10.1016/j.biomaterials.2009.02.006

Mondal M, Trivedy K, Kumar SN (2007) The silk proteins, sericin and fibroin in silkworm, *B. mori* Linn., a review. Casp J Environ Sci 5:63–76

Mu X, Wang Y, Guo C et al (2020) 3D printing of silk protein structures by aqueous solvent-directed molecular assembly. Macromol Biosci 20:1900191. https://doi.org/10.1002/mabi.201900191

Narita C, Okahisa Y, Wataoka I, Yamada K (2020) Characterization of ground silk fibroin through comparison of nanofibroin and higher order structures. ACS Omega 5:22786–22792. https://doi.org/10.1021/acsomega.0c01750

Nguyen TP, Nguyen QV, Nguyen V-H et al (2019) Silk fibroin-based biomaterials for biomedical applications: a review. Polymers 11:1933. https://doi.org/10.3390/polym11121933

Nilogal P, Uppine GB, Rayaraddi R et al (2021) Conductive *in situ* reduced graphene oxide–silk fibroin bionanocomposites. ACS Omega 6:12995–13007. https://doi.org/10.1021/acsomega.1c00013

Ornell KJ, Taylor JS, Zeki J et al (2020) Local delivery of dinutuximab from lyophilized silk fibroin foams for treatment of an orthotopic neuroblastoma model. Cancer Med 9:2891–2903. https://doi.org/10.1002/cam4.2936

Partlow BP, Hanna CW, Rnjak-Kovacina J et al (2014) Highly tunable elastomeric silk biomaterials. Adv Funct Materials 24:4615–4624. https://doi.org/10.1002/adfm.201400526

Partlow BP, Tabatabai AP, Leisk GG et al (2016) Silk fibroin degradation related to rheological and mechanical properties. Macromol Biosci 16:666–675. https://doi.org/10.1002/mabi.201500370

Pereira C, Pinto TV, Santos RM, Correia N (2024) Sustainable and naturally derived wet spun fibers: a systematic literature review. Fibers 12:75. https://doi.org/10.3390/fib12090075

Pritchard EM, Normand V, Hu X et al (2014) Encapsulation of oil in silk fibroin biomaterials. J Appl Polym Sci 131:39990. https://doi.org/10.1002/app.39990

Qian Z, Sun C, Li Q et al (2024) Unravelling the antioxidant behaviour of self-assembly β-Sheet in silk fibroin. Redox Biol 76:103307. https://doi.org/10.1016/j.redox.2024.103307

Qiu W, Patil A, Hu F, Liu XY (2019) Hierarchical structure of silk materials versus mechanical performance and mesoscopic engineering principles. Small 15:1903948. https://doi.org/10.1002/smll.201903948

Rastogi S, Kandasubramanian B (2020) Processing trends of silk fibers: silk degumming, regeneration and physical functionalization. TJTI 111:1794–1810. https://doi.org/10.1080/00405000.2020.1727269

Reizabal A, Costa CM, Saiz PG et al (2021) Processing strategies to obtain highly porous silk fibroin structures with tailored microstructure and molecular characteristics and their applicability in water remediation. J Hazard Mater 403:123675. https://doi.org/10.1016/j.jhazmat.2020.123675

Reizabal A, Costa CM, Pérez-Álvarez L et al (2023) Silk fibroin as sustainable advanced material: material properties and characteristics, processing, and applications. Adv Funct Mater 33:2210764. https://doi.org/10.1002/adfm.202210764

Rising A, Harrington MJ (2023) Biological materials processing: time-tested tricks for sustainable fiber fabrication. Chem Rev 123:2155–2199. https://doi.org/10.1021/acs.chemrev.2c00465
Rockwood DN, Preda RC, Yücel T et al (2011) Materials fabrication from *B. mori* silk fibroin. Nat Protoc 6:1612–1631. https://doi.org/10.1038/nprot.2011.379
Rodriguez MJ, Dixon TA, Cohen E et al (2018) 3D freeform printing of silk fibroin. Acta Biomater 71:379–387. https://doi.org/10.1016/j.actbio.2018.02.035
Sachin B, Rao CM, Prasad CD et al (2024) Interactive design and development of an intelligent vision-driven 3D printed precision sorting mechanism for silk cocoons. Int J Interact des Manuf. https://doi.org/10.1007/s12008-024-02086-5
Sashina ES, Bochek AM, Novoselov NP, Kirichenko DA (2006) Structure and solubility of natural silk fibroin. Russ J Appl Chem 79:869–876. https://doi.org/10.1134/S1070427206060012
Schäfer S, Aavani F, Köpf M et al (2023) Silk proteins in reconstructive surgery: do they possess an inherent antibacterial activity? A systematic review. Wound Repair Regen 31:99–110. https://doi.org/10.1111/wrr.13049
Seo S-J, Das G, Shin H-S, Patra JK (2023) Silk sericin protein materials: characteristics and applications in food-sector industries. Int J Mol Sci 24:4951. https://doi.org/10.3390/ijms24054951
Sharma YC, Yadav M, Upadhyay SN (2019) Latest advances in degumming feedstock oils for large-scale biodiesel production. Biofuels Bioprod Bioref 13:174–191. https://doi.org/10.1002/bbb.1937
Silva AS, Costa EC, Reis S et al (2022) Silk sericin: a promising sustainable biomaterial for biomedical and pharmaceutical applications. Polymers 14:4931. https://doi.org/10.3390/polym14224931
Stinson JA, Palmer CR, Miller DP et al (2020) Thin silk fibroin films as a dried format for temperature stabilization of inactivated polio vaccine. Vaccine 38:1652–1660. https://doi.org/10.1016/j.vaccine.2019.12.062
Tengattini S, Orlandi G, Perteghella S et al (2020) Chromatographic profiling of silk sericin for biomedical and cosmetic use by complementary hydrophylic, reversed phase and size exclusion chromatographic methods. J Pharm Biomed Anal 186:113291. https://doi.org/10.1016/j.jpba.2020.113291
Tsioris K, Raja WK, Pritchard EM et al (2012) Fabrication of silk microneedles for controlled-release drug delivery. Adv Funct Mater 22:330–335. https://doi.org/10.1002/adfm.201102012
Vidal SEL, Tamamoto KA, Nguyen H et al (2019) 3D biomaterial matrix to support long term, full thickness, immuno-competent human skin equivalents with nervous system components. Biomaterials 198:194–203. https://doi.org/10.1016/j.biomaterials.2018.04.044
Vidya M, Rajagopal S (2021) Silk fibroin: a promising tool for wound healing and skin regeneration. Int J Polym Sci 2021:1–10. https://doi.org/10.1155/2021/9069924
Vyas SK, Shukla SR (2016) Comparative study of degumming of silk varieties by different techniques. J Text Inst 107:191–199. https://doi.org/10.1080/00405000.2015.1020670
Wang S, You Z, Ye L et al (2014) Quantitative proteomic and transcriptomic analyses of molecular mechanisms associated with low silk production in silkworm *B. mori*. J Proteome Res 13:735–751. https://doi.org/10.1021/pr4008333
Wang Y, Guo J, Zhou L et al (2018) Design, fabrication, and function of silk-based nanomaterials. Adv Funct Mater 28:1805305. https://doi.org/10.1002/adfm.201805305
Wang S, Zhu M, Zhao L et al (2019) Insulin-loaded silk fibroin microneedles as sustained release system. ACS Biomater Sci Eng 5:1887–1894. https://doi.org/10.1021/acsbiomaterials.9b00229
Wang H-Y, Wei Z-G, Zhang Y-Q (2020) Dissolution and regeneration of silk from silkworm *B. mori* in ionic liquids and its application to medical biomaterials. Int J Biol Macromol 143:594–601. https://doi.org/10.1016/j.ijbiomac.2019.12.066
Wang K, Ma Q, Zhou H-T et al (2023) Review on fabrication and application of regenerated *B. mori* silk fibroin materials. AUTEX Res J 23:164–183. https://doi.org/10.2478/aut-2021-0059
Ward OP (2019) Proteases. Comprehensive biotechnology, p 604

Wray LS, Hu X, Gallego J et al (2011) Effect of processing on silk-based biomaterials: reproducibility and biocompatibility. J Biomed Mater Res 99B:89–101. https://doi.org/10.1002/jbm.b.31875

Wu L, Leng X, Sun Y, Ren FS (2010) Analysis of the effects of pH and salt on the conformation of the sericin particles by DLS and TEM measurements. Spectrosc Spectr Anal 30:1391–1395. https://doi.org/10.3964/j.issn.1000-0593(2010)05-1391-05

Wu R, Li H, Yang Y et al (2021) Bioactive silk fibroin-based hybrid biomaterials for musculoskeletal engineering: recent progress and perspectives. ACS Appl Bio Mater 4:6630–6646. https://doi.org/10.1021/acsabm.1c00654

Xiong R, Grant AM, Ma R et al (2018) Naturally-derived biopolymer nanocomposites: interfacial design, properties and emerging applications. Mater Sci Eng R Rep 125:1–41. https://doi.org/10.1016/j.mser.2018.01.002

Yamada H, Nakao H, Takasu Y, Tsubouchi K (2001) Preparation of undegraded native molecular fibroin solution from silkworm cocoons. Mater Sci Eng C 14:41–46. https://doi.org/10.1016/S0928-4931(01)00207-7

Yang C, Yao L, Zhang L (2023) Silk sericin-based biomaterials shine in food and pharmaceutical industries. Smart Mater Med 4:447–459. https://doi.org/10.1016/j.smaim.2023.01.003

Yang C, Lin Y, Zhang X, Zhu C (2024) Study on the modification and properties of silk sericin protein/nano-titanium dioxide composite for textile applications. Fibres Text East Eur 32:68–75. https://doi.org/10.2478/ftee-2024-0008

Yeo J, Lee K, Kweon H et al (2006) Fractionation of a silk fibroin hydrolysate and its protective function of hydrogen peroxide toxicity. J Appl Polym Sci 102:772–776. https://doi.org/10.1002/app.23740

Ying S, Guan Z, Ofoegbu PC et al (2022) Green synthesis of nanoparticles: current developments and limitations. Environ Technol Innov 26:102336. https://doi.org/10.1016/j.eti.2022.102336

Yu B, Li Y, Lin Y et al (2023) Research progress of natural silk fibroin and the application for drug delivery in chemotherapies. Front Pharmacol 13:1071868. https://doi.org/10.3389/fphar.2022.1071868

Yuan Y, Nasri M, Manayi A et al (2024) Sericin coats of silk fibres, a degumming waste or future material? Mater Today Bio 29:101306. https://doi.org/10.1016/j.mtbio.2024.101306

Yukseloglu SM, Sokmen N, Canoglu S (2015) Biomaterial applications of silk fibroin electrospun nanofibres. Microelectron Eng 146:43–47. https://doi.org/10.1016/j.mee.2015.04.008

Zare A (2023) Sericin eco-friendly biomaterial as a spin finish in drawn textured PET yarn production. Res J Text Appar 28:687–698. https://doi.org/10.1108/RJTA-07-2022-0086

Zhang Y-Q (2002) Applications of natural silk protein sericin in biomaterials. Biotechnol Adv 20:91–100. https://doi.org/10.1016/S0734-9750(02)00003-4

Zhang Q, Cui L, Wang P et al (2018) Improving properties of silk sericin membranes via enzymatic oxidation with laccase and TEMPO. Biotech App Biochem 65:372–380. https://doi.org/10.1002/bab.1609

Zhao Z-L, Zhang Y-Q (2020) Greener degumming production of layered sericin peptides from a silkworm cocoon and their physicochemical characteristics and bioactivities in vitro. J Clean Prod 261:121080. https://doi.org/10.1016/j.jclepro.2020.121080

Zhao Z, Li Y, Xie M-B (2015) Silk fibroin-based nanoparticles for drug delivery. Int J Mol Sci 16:4880–4903. https://doi.org/10.3390/ijms16034880

Zhao M, Qi Z, Tao X et al (2021) Chemical, thermal, time, and enzymatic stability of silk materials with silk I structure. Int J Mol Sci 22:4136. https://doi.org/10.3390/ijms22084136

Zheng Z, Wu J, Liu M et al (2018) 3D bioprinting of self-standing silk-based bioink. Adv Healthcare Mater 7:1701026. https://doi.org/10.1002/adhm.201701026

Zhou J, Zhang B, Shi L et al (2014) Regenerated silk fibroin films with controllable nanostructure size and secondary structure for drug delivery. ACS Appl Mater Interf 6:21813–21821. https://doi.org/10.1021/am502278b

Zhu ZH, Lu X, Zhou XH (2011) Study on improving biodegradation process of regenerated silk fibroin fiber. Adv Mater Res 175–176:266–271. https://doi.org/10.4028/www.scientific.net/AMR.175-176.266

# Chapter 8
# Potential of Silkworm Pupae as a Nutritious Food Source for Humans

**Ravi Kumar Goswami, Sonal Saluja, Bronson Kumar Khangembam, Anupam V. Sharma, Trishnakhi Swargiari, Krishnakshi Misra, and Dhitri Borah**

## 8.1 Introduction

The global population is projected to grow by 34%, reaching approximately 9.6 billion people by 2050, posing a significant challenge to sustainability, nutrition, and food security (Grafton et al. 2015). Nonetheless, one of the primary sources of food is conventional livestock farming, which places enormous pressure on land, water, and other natural resources while also significantly contributing to greenhouse gas emissions. In response, researchers and policymakers have shifted their attention to more sustainable, nutritious, and affordable food sources. Edible insects are gaining popularity due to their high nutritional value, environmental sustainability, economic feasibility, and cultural acceptability in many regions of the world (Lange and Nakamura 2023). In addition, edible insects are highly nutritious, containing proteins, beneficial fats, amino acids, fatty acids, vitamins, and trace minerals (Zhou et al.

R. K. Goswami · A. V. Sharma
Department of Zoology, Hindu College, University of Delhi, Delhi, India

S. Saluja
Department of Zoology, Dyal Singh College, Karnal, Kurukshetra University, Haryana, India

B. K. Khangembam
Department of Zoology, Tripura University (A Central University), Suryamaninagar 799022, Tripura, India

T. Swargiari
Department of Zoology, Bhawanipur Anchalik College, Barpeta, Assam, India

K. Misra
Molecular Endocrinology and Reproductive Biology Research Laboratory, Department of Zoology, Rajiv Gandhi University, Itanagar, Arunachal Pradesh, India

D. Borah (✉)
Department of Zoology, Biswanath College, Gauhati University, Biswanath, Assam, India
e-mail: boradhitri@gmail.com

 
D. Kumar and M. Shahid (eds.), *Science and Technology of Silk*, Textile Science and Clothing Technology, https://doi.org/10.1007/978-3-032-16230-4_8

2022). Furthermore, raising insects for food requires less area and water, and has a lower effect on the economy and environment than livestock farming (Akande et al. 2020). Several studies showed that edible insects are increasingly being used in food products such as bread, pasta, high-energy biscuits, and snacks (Karnjanapratum et al. 2022). Moreover, edible insects are replacing animal proteins in a variety of cuisines, hence increasing their market share (Akande et al. 2020). Silkworm pupae (SWP), a byproduct of the sericulture industry, are one of the most promising choices among these insects.

SWP, both mulberry and non-mulberry varieties, have long been consumed in many Asian countries, like China, India, Thailand, Japan, and South Korea. They are valued for their rich flavor, nutritional content, and medicinal benefits (Altomare et al. 2020; Yeruva et al. 2023). Additionally, they provide an essential supply of proteins, lipids, key amino acids, polyunsaturated fatty acids (ω-3 and ω-6), vitamins, and minerals, all of which may offer health benefits. There are notable environmental benefits of using silk byproducts, including waste reduction and the development of a circular bioeconomy (Mahanta et al. 2023). Sericulture offers a viable platform for incorporating insect protein into the human diet, requiring minimal infrastructure, as it is already well-established in many parts of the world (Lisboa et al. 2024).With recent technological developments in the food processing industry, a variety of pupae-based products such as protein powders, snacks, and meat substitutes have emerged with improved palatability and acceptance across a broader customer group (Kim et al. 2017a, b and c). This book chapter explores the potential of SWP as a nutritious food source by examining its nutritional composition, health benefits, safety considerations, and environmental impact.

## 8.2 Nutritional Profile of Silkworm

### *8.2.1 Protein and Amino Acid Profile*

The pupae of silk worm are rich in protein (55 to73% dry weight) depending on species, age, and rearing conditions (Longvah et al. 2012), which surpasses most animal and plant-based protein sources. SWP has been reported to have higher protein levels compared to other edible insects, most fish, and even chicken breast and thigh meat (Jamal et al. 2020). Moreover, popular plant-based proteins in meals, such as soybean seeds, pulses, chickpeas, and lentils, provide a comparatively lower percentage (15–46.5%) of protein (Tiwari and Singh 2012; Jukanti et al. 2012; Khazaei et al. 2019; Jin et al. 2023). The proteins in SWP are easily digestible and contain all essential amino acids, which are crucial for overall human growth and metabolic processes (Longvah et al. 2012). Indeed, SWP proteins have been identified as a rich source of functional bioactive components to meet human nutritional needs in space research facilities (Altomare et al. 2020). Furthermore, silkworms and other edible insects are widely recognized as a valuable source of protein for humans

(Wendin and Nyberg 2021), making them a viable alternative to conventional animal and plant proteins.

### 8.2.2 *Lipid and Fatty Acid Profile*

In addition to protein, SWP are rich in healthy fats, accounting for 25–32% of dry weight, depending on species, developmental stage, and rearing conditions (Longvah et al. 2012; Kouřimská and Adámková 2016). The fatty acid profile of SWP oil is balanced, with saturated fatty acids (SFAs), monounsaturated fatty acids (MUFAs), and polyunsaturated fatty acids (PUFAs). Especially, α-linolenic acid (ALA) is abundant, which is beneficial in treating coronary heart disease, as well as in maintaining membrane fluidity, improving immunological function, and regulating metabolic processes in humans (Kim et al. 2017a, b and c). Dried SWP contains ω-3 essential fatty acids, such as ALA, commonly used in nutraceuticals, livestock nutrition, and cleaning agent sectors (Thirupathaiah et al. 2016; Kumar et al. 2015). Higher ALA in SWP oil, hexadecenoic acid, and palmitic acid were also reported, which is a reflection of a strong source of vitamin A due to the presence of carotenoids, as it serves as an antioxidant, as a reactive agent to scavenge oxidative activity. A significant percentage of certain unsaturated fatty acids, particularly ALA, has been documented in silkworm oil (Wang et al. 2014) in various studies (Yue et al. 2021). Fu et al. (2016) also reported higher ALA in SWP compared to sunflower oil (38.41%) and soya (54.21%). ALA is a potent ingredient in *B. mori* that has been investigated in numerous studies for its potential health benefits (Rodriguez-Ortiz et al. 2024), which include antitumor, anti-inflammatory, anti-oxidative stress, anti-obesity, neuroprotective, and intestinal microbiota regulatory properties (Yuan et al. 2022). Hence, *B. mori* pupae oil may serve as a potential component of many nutraceutical and functional foods (Rodriguez-Ortiz et al. 2024).

### 8.2.3 *Vitamins and Minerals*

SWP also contain the majority of the water and fat-soluble vitamins, such as vitamin A, B, C, D, E, and K (Paul and Dey 2014). Five tocopherol derivatives (Vitamin E forms) were reported in SWP (Hăbeanu et al. 2023a and b). Yeruva et al. (2023) reported the maximum concentration (300 ng/g) of pantothenic acid among all the B complex vitamins in the silkworm. The vitamin $B_{12}$ content of SWP was found to be equivalent to that of fish (Chatterjee et al. 2017), while other B complex vitamins (B6 and B9) were lower. Among the lipid-soluble vitamins, vitamin E was reported to be the highest, followed by vitamin $K_2$ and $K_1$, and vitamins $D_1$ and $D_2$. The fat-soluble vitamins are beneficial for bones, skin, immunity, eyes, and help with blood clotting (Sadat et al. 2022). Tangsanthatkun et al. (2022) reported that 100 g of SWP oil contains 0.785 mg of $\beta$-carotene and 9.434 mg of $\alpha$-tocopherol. Variations in α-tocopherol content of SWP may arise due to disparities in sex, maturity stage, and extraction technique, which may result in variations in the α-tocopherol content of SWP (Ravinder et al. 2016). Kumar et al. (2015) also reported the presence of vitamins A, E, and C, while Wu et al. (2021) found vitamin $B_1$, vitamin $B_2$, vitamin $B_3$, and vitamin E at high ranges in SWP. Carotenoids, such as lutein and neoxanthin, are also reported in the SWP (Kotake-Nara et al. 2002).

Major minerals, such as Fe, Zn, Mn, Mg, and Ca, are commonly reported in the majority of studies on SWP, although their levels may vary across studies due to preprocessing methods like boiling, which can cause mineral loss (Tassoni et al. 2022). The various range of minerals include Fe (2.83–4.95 mg), Ca (92.0–181 mg), Mn (1.08–2.30 mg), Zn (1.39–24.4 mg), and Mg (89–280 mg) in various studies (Köhler et al. 2019; David-Birman et al. 2019; Rodriguez-Ortega et al. 2016; Pereira et al. 2003). Few studies have shown that the amount of Na in SWP ranges from 29.6 (Pereira et al. 2003) to 363 mg/100 g (Köhler et al. 2019), while potassium ranges from 477 to 672 mg/100 g (Köhler et al. 2019; David-Birman et al. 2019). The low Na–K ratio in SWP may prove to be beneficial in reducing the likelihood of lifestyle diseases such as cerebrovascular and cardiovascular disorders (Yang et al. 2009). However, Mahanta et al. (2023) reported the higher content of phosphorus, magnesium, and calcium in SWP.

Hăbeanu et al. (2023a and b) reported that up to 25 different minerals, including calcium, potassium, iron, and zinc, can be found in SWP. Various minerals such as Ca, Mg, P, Fe, Zn, Cr, Mn, and Cu were also reported in *B. mori* (Zhou et al. 2022). Selenium may be elevated in some pupae, potentially contributing to antioxidant defense and reducing the risk of cancer. Yeruva et al.(2023) also reported high nitrogen in dried SWP (8.06%), followed by potassium (2.18%), phosphorus (1.07%), calcium (0.91%), and magnesium (0.45%). zinc (127.5 ppm), iron (52.8 ppm), copper (9.8 ppm), and manganese (11.7 ppm), while extremely low and acceptable concentrations of heavy metals, such as nickel, chromium, cadmium, arsenic, and mercury, ranging from (0.02–0.08 ppm), were also observed in the same.

## 8.3 Health Benefits of Silkworm Pupae

### *8.3.1 Antioxidant Effects and Immunomodulatory Effects*

Various peptides and polyphenols with antioxidant properties have been successfully isolated and extracted from SWP using different techniques. In vitro studies have shown that these compounds effectively neutralize DPPH (2,2-diphenyl-1-picrylhydrazyl) and ABTS (2,2'-Azino-bis (3-ethylbenzothiazoline-6-sulfonic acid)free radicals while also eliminating Reactive Oxygen Species (ROS) produced within the cell (Zhang et al. 2021; Lee et al. 2021). Research on HepG2 cells revealed that two specific peptides from SWP exhibited strong antioxidant activity by reducing ROS levels and enhancing the expression of superoxide dismutase

(SOD) and glutathione (GSH) production (Cermeno et al. 2022). Additionally, a 30% ethanolic extract of SWP showed the highest antioxidant potential in DPPH and ABTS antioxidant activity methods. The antioxidant activity of SWP varies depending on biological factors, such as gender and developmental stage, with the initial female pupation phase exhibiting higher reactive radical and ROS neutralizing activity (Lee et al. 2021). Moreover, extraction methods have a significant influence on the antioxidant potency of SWP-derived compounds. For instance, microwave-assisted extraction has been shown to enhance the total phenolic content, resulting in stronger antioxidant activity compared to conventional extraction techniques (Hu et al. 2017). Key bioactive components, including unsaturated fatty acids, peptides, and phenolic compounds, contribute to the antioxidant properties of SWP. Unlike synthetic antioxidants, which may interfere with the biochemical functions of other nutrients, naturally occurring antioxidants are typically considered more protective (Hrebień-Filisińska 2021, and Liu 2022). The promising antioxidant potential of SWP suggests its applicability in the development of functional foods and medicinal products with health-protective benefits.

SWP have been identified as a significant source of bioactive compounds with immunomodulatory ability (Table 8.1). Various innovative immunomodulatory peptides have been effectively extracted and purified from SWP, highlighting their significant role in immune regulation (Li et al. 2019). A specific peptide from SWP protein hydrolysate with a mol. wt. of 441.06 Da has been identified from SWP, which can stimulate the expression of key immunity-associated molecules, comprising IL-6, IL-12, NF-κB, Cyclin D1, and Cdks 4, indicating its potential impact on immune system modulation (Li et al. 2020). Additionally, an immunologically active polysaccharide was discovered in SWP, which was found to activate innate immune responses in RAW264 cells and small shrimps, significantly enhancing their survival rates (Ali et al. 2018). The growing body of research on silkworm pupae-derived proteins and polysaccharides suggests their promising potential for therapeutic applications in managing immune-related disorders in humans. Further studies may lead to the development of innovative immunomodulatory agents derived from SWP.

### 8.3.2 *Blood Pressure, Blood Sugar, and Blood Lipids Regulation*

SWP-derived compounds have demonstrated significant regulatory effects on blood pressure, blood sugar, and blood lipids (Table 8.1). Studies indicate that SWP protein hydrolysate effectively suppresses angiotensin-converting enzyme (ACE) activity, which is typically elevated in patients with hypertension (Cermeno et al. 2022). Moreover, a peptide isolated from the SWP protein exhibited potent ACE inhibitory activity, with a semi-inhibitory concentration of just 0.047 mg/mL (Wang et al. 2011). Spontaneously hypertensive rats showed a dose-dependent reduction in systolic blood pressure when fed with varying concentrations of SWP protein hydrolysate

**Table 8.1** Pharmacological properties and underlying mechanisms of silkworm pupae

| Therapeutic effects | Active ingredients | Animal cell/animal model used | Mechanism of action | Results assessment | References |
|---|---|---|---|---|---|
| Antioxidant | Protein hydrolysates | Hepatic HepG2 cells | Antioxidant activity was enhanced, ROS levels were reduced, superoxide dismutase expression increased, and glutathione synthesis was promoted | It may have the potential to be a natural antioxidant | Cermeño et al. (2022) |
| Immunomodulatory | Peptides | Mouse spleen cells | Upregulation of immunomodulatory factors including IL-6, IL-12, cyclin D1, and cyclin-dependent kinase 4 | Exhibited significant immunomodulatory properties and hold potential for medicinal applications | Li et al. (2020) |
| Blood pressure regulation | Peptide hydrolysates | Hypertension-prone rats | Systolic blood pressure decreased progressively with higher doses | Serve as a safe and effective antihypertensive effect, SWP peptides as a functional ingredient in antihypertensive treatment | Wang et al. (2014) |
| | Protein hydrolysates | RP- HPLC | Suppressive effect against angiotensin I-converting enzyme | Potential to produce angiotensin I-converting enzyme inhibitor medications | Wang et al. (2008) |
| Regulation of blood lipids | Peptides | 3T3-L1 pre-adipocytes | Inhibition of adipogenesis occurs when the expression of adipogenic genes and protein synthesis are suppressed, resulting in the regulation of blood lipid levels | Promising safe and effective alternatives to reducing dietary obesity | Lee et al. (2012) |
| Regulation of blood sugar levels | Soluble fibroin | 3T3-L1 adipocyte | Fibronectin enhances glucose uptake and metabolism | Significantly improves diabetic hyperglycemia | Hyun et al. (2004) |

(continued)

**Table 8.1** (continued)

| Therapeutic effects | Active ingredients | Animal cell/animal model used | Mechanism of action | Results assessment | References |
|---|---|---|---|---|---|
| Anti-cancer | Protein hydrolysates | Human gastric cancer SGC-7901 cells | Suppresses the growth of human gastric cancer cells and triggers aberrant morphological changes; facilitates apoptosis and arrests the cell cycle in the S phase; leads to ROS aggregation and mitochondrial membrane potential depolarization | Silkworm pupae have Potential as a source of anticancer agents | Li et al. (2018) |
| | Protein extracts | Breast cancer cells MCF-7 | Significant decrease in protein and nucleic acid content, reduced levels of IL-6, IL-1, and TNF-α | Promising innovative pharmacological target for breast cancer | Chukiatsiri et al. (2020) |
| | Selenium-rich amino acids | Human hepatoma cells | Exhibits significant dose-dependent inhibition of cell viability, alters cell morphology and cycle, and triggers apoptosis | Application for anticancer therapy, dietary source of essential amino acids and trace elements | Hu et al. (2005) |
| Anti-Alzheimer's disease | Silkworm pupae powder | Male Wistar rats | Significantly improved hippocampal memory impairment along with decrease in hippocampal neuron density | Promising dietary option for Alzheimer's patients | Wattanathorn et al. (2012) |
| Cardiovascular defense | Crude extract | Male New Zealand white rabbits | Histopathological analysis revealed a reduction in the size of atherosclerotic plaques | *Bombyx mori* cocoon extract shows cardioprotective and neuroprotective features | Ali and Arumugam (2011) |

for four weeks (Wang et al. 2014). These findings suggest that the in vitro and in vivo antihypertensive properties of SWP protein hydrolysate make it a promising candidate for the development of antihypertensive drugs or functional food supplements. Studies have also shown that SWP powder functions as an alpha-glucosidase inhibitor, effectively lowering postprandial blood glucose levels (Lee et al. 2012). Additionally, it enhances the metabolism of lipids and reduces fat accumulation in rats (Ryu 2014; Aznar-Cervantes et al. 2021). These metabolic studies suggest that SWP may be utilized as a natural treatment for diabetes, while also contributing to weight loss.

### *8.3.3 Anticancer Effect, Anti-Alzheimer's Disease, and Cardiovascular Protection*

SWP are reported to contain bioactive compounds with significant anticancer properties (Table 8.1). In vitro studies indicate that protein hydrolysates and amino acids derived from SWP exhibit cytotoxic effects against stomach, breast, and liver cancer cells (Hu et al. 2005; Li et al. 2018; and Chukiatsiri et al. 2020). Both SWP protein and SWP oil demonstrate anticancer activity through distinct mechanisms. SWP proteins inhibit cancer cell proliferation by disrupting the cell division cycle, triggering apoptosis through the activation of apoptotic factors, and impairing the energy metabolism of cancer cells by affecting mitochondrial function (Weixin et al. 2019). In contrast, SWP oil exerts its anticancer effects through its amino acid content, particularly in inhibiting liver cancer via the generation of ROS in cancer cells, inducing oxidative stress, and promoting apoptosis (Weixin et al. 2019). Further study, particularly in *vivo* investigations and clinical trials, is required to assess their efficacy and safety for human application.

SWP have also shown potential in mitigating neurological disorders linked to oxidative stress, including stroke and Alzheimer's disease (Pratico and Delanty 2000). In vivo studies using animal models of Alzheimer's disease suggest that SWP can improve cognitive function in Wistar rats by enhancing cholinergic activity and offering neuroprotective effects through the reduction of oxidative stress (Table 8.1). SWP contributed to decreased memory impairment and increased hippocampal neuronal density in Alzheimer's disease models (Wattanathorn et al. 2012). On the other hand, a study utilizing transgenic mice with Alzheimer's disease demonstrated that recombinant proteins expressed in SWP reduced amyloid plaque deposition in the brain, lowered malondialdehyde levels, and enhanced memory and cognitive abilities (Li et al. 2019). These findings suggest that SWP could serve as a functional food for preventing Alzheimer's disease, while also offering promising avenues for therapeutic intervention. The 30Kc6 protein from SWP effectively reduced serum levels of total triglycerides (TG), high-density lipoprotein cholesterol (HDL-C), low-density lipoprotein cholesterol (LDL-C), total cholesterol (TC), and helped mitigate aortic and liver lesions in an atherosclerotic rabbit model (Yu et al. 2013). Moreover, SWP crude extracts have been reported to improve atherosclerotic conditions, likely due to their antioxidant and lipid-lowering properties (Ali and Arumugam 2011). Another significant cardiovascular benefit of SWP is linked to its **oil** sodium salt, which has been found to inhibit platelet-derived growth factor (PDGF)-induced abnormal movement and expansion of vascular smooth muscle cells through the downregulation of ERK1/2 phosphorylation levels in these cells, which plays a critical role in vascular health (Kim et al. 2020). These results highlight the potential of SWP-derived compounds in maintaining cardiovascular health through multiple mechanisms. Further studies for their integration into novel treatment strategies may offer a natural and potentially effective approach to cardiovascular health management.

## 8.4 Processing and Preparation Techniques

### *8.4.1 Traditional Processing Methods*

Traditionally, edible insects have been prepared by roasting, curing, smoking, frying, stewing, and steaming (Fig. 8.1). Blanching is done before these methods to lower the number of foodborne microorganisms and deactivate enzymes (Marshall et al. 2016). Traditionally, insect consumption, particularly that of silkworms, is higher in tropical regions of the world at lower latitudes. Cooking enhances the sensory quality and extends the shelf life of food items by decreasing or inactivating degradative enzymes. It also enhances the absorption and biological activity of proteins in the gastrointestinal tract from a nutritional standpoint, although some nutrients may be lost due to solubilization, leakage, intra- and inter-biochemical interactions, or the formation of novel byproducts (Megido et al. 2018). Lipid oxidation, vitamin and mineral solubilization, and proteolysis are other side effects of heat treatment of foods (Juarez-Ortega et al. 2012). The simplest method to enhance the shelf life of food is drying (roasting, frying, and sun-drying). Except for a minor decrease in monounsaturated fatty acids, there were no discernible detrimental impacts on nutritional value from the heat treatment (Lautenschlager et al. 2017).

The most traditional drying method is sun-drying, which is largely used in houses due to its low energy consumption (Azzollini et al. 2016). However, given the inadequate hygienic procedures used, the dried insects may get contaminated by coming into contact with dirt and air (Megido et al. 2018). Smoking is a heating and curing method and one of the most conventional methods for preserving meat of all kinds. For insects, smoking is carried out in a dry atmosphere while a curing procedure is carried out concurrently. Heat and enzymatic activity combine to enhance protein and lipid alteration throughout the process (Tiencheu et al. 2013).

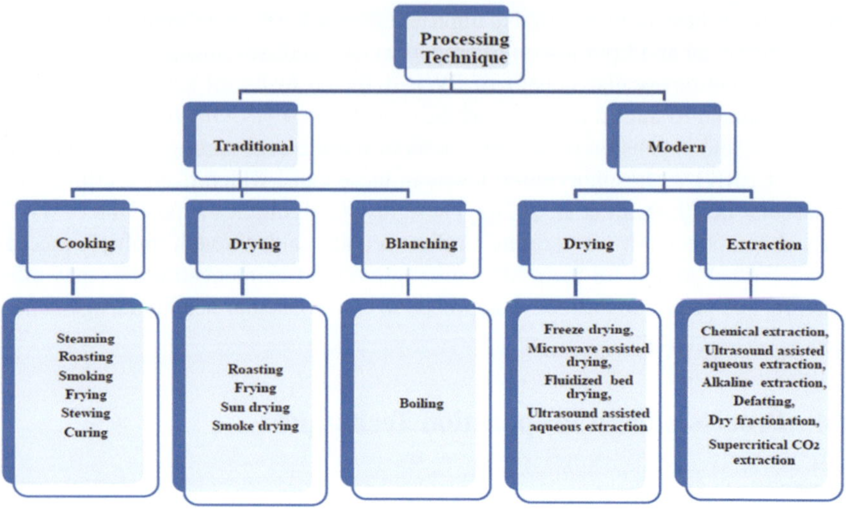


**Fig. 8.1** Different types of processing and preparation techniques (Created with Microsoft PowerPoint)

## *8.4.2 Modern Processing Methods*

Modern processing methods encompass various techniques, including drying, blanching, and extraction of bioactive molecules for diverse applications in both the food and therapeutic sectors (Fig. 8.1). The technique of blanching involves briefly submerging food in heated water, then removing it and immersing it in cold, running water to halt the heating process (Xiao et al. 2014). It is used as a pretreatment on both commercial and artisanal levels to lower bacterial populations and disable enzymes that break down proteins, which cause food poisoning and spoilage. Blanching has also been shown to lower the amount of general psychrotrophic bacteria and lactic acid bacteria (Vandeweyer et al. 2017). Slight increase in moisture (Purschke et al. 2018) after blanching is, however, reported.

Drying (sun-dry, oven-dry, or freeze-dry) remains one of the best ways for processing insects including SWP as most microorganisms stop reproducing at these temperatures. Non-traditional techniques, including oven-drying and freeze-drying, are frequently used for large-scale production. Furthermore, drying increases a product's shelf life during delivery and storage. Cooking dried insects for 30 min is an effective way to eliminate yeasts, molds, *Staphylococcus*, *Bacillus*, *Enterobacteriaceae*, and other pathogens (Grabowski and Klein 2016). Freeze-drying is a widely used technique since the extremely low temperatures and water sublimation that occurs, reduces the microbiological and oxidative deterioration significantly resulting in a high-quality final product with remarkable nutritional value and long shelf life. However, scaling up to the manufacturing level may be costly. In Europe, freeze-dried edible insects are readily accessible. Because most studies focus on physical characterization and provide little to no data on the specific conditions utilized, it might be difficult to compare the results, even for the same species (Megido et al. 2018). Oven-dried products, like freeze-dried ones, have less lipid oxidation, fewer energy costs for inputs, and higher protein solubility (Fombong et al. 2017). Microwave-assisted drying for 16 min significantly decreased the microbial count owing to a significant reduction in the amount of water in the tested samples (Lenaerts et al. 2018). After drying, both fresh and frozen larvae showed slight variations in their lipid, protein, and ash levels. Moreover, the browning index remained stable throughout four months of storage, with the only noticeable reduction in vitamin B-12 compared to freeze-drying (Lenaerts et al. 2018). Nevertheless, more research is needed to establish how to preserve the insect sensory quality after microwave drying.

### *8.4.3 Development of Silkworm-Based Food Products*

The United Nations Educational, Scientific, and Cultural Organization (UNESCO) has recognized sericulture as an important cultural resource. It also generates a number of beneficial byproducts for individuals and society (Jasmine and Mandal 2014). The byproducts that are produced at every stage of the technical value chain play a significant role in providing fresh opportunities for advancement. In order to extract minerals and active ingredients, the pupae undergo additional treatment before being used in pharmaceutical research. Sericulture's future hinges on a paradigm change that values all of the items produced by this industry in order to lessen its negative impacts on the environment and improve social and economic life. The massive volume of SWP generated is a significant byproduct of the sericulture industry which can be turned into useful products (Oduor et al. 2021). Moreover, fibromyalgia, migraine, carpal tunnel syndrome, osteoarthritis, and rheumatoid arthritis are believed to be successfully treated by silkworm extract (Olteanu et al. 2012). It has been consumed as a nutrient-dense food, particularly by individuals with polyps, diabetes, heart disease, asthma, cerebral palsy, primary trigeminal neuralgia (TN), and vocal nodules.

## 8.5 Food Safety and Quality Concerns

### *8.5.1 Allergenic Potential*

A total of 26 allergenic proteins have been identified in SWP, with arginine kinase (Bomb m 1) being the only officially recognized allergen in silkworm larvae due to its cross-reactivity with cockroach arginine kinase (Liu et al. 2009). However, no SWP allergens have been confirmed by the WHO Allergen Nomenclature Sub-committee.

Among these allergens, five belong to the lipoprotein 11 family (30-kDa proteins), three are insect cuticle proteins, and the remaining belong to various other protein families. Several SWP proteins, including heat shock protein, chitinase, tropomyosin, and triosephosphate isomerase, exhibit sequence homology with known allergens from house dust mites and crustaceans (Ivanciuc et al. 2009). Additionally, allergen expression in silkworm pupae may be influenced by factors such as gender and diet (Lamberti et al. 2019). Notably, female SWP raised on mulberry leaves showed lower allergen levels. Other allergens, such as excrement, dander, and silk, are present in silkworms, which can trigger various allergic responses, including skin reactions, respiratory issues, cardiovascular symptoms, and even anaphylactic shock (Gautreau et al. 2017; Ji et al. 2008). SWP can also cause allergic responses when eaten (Ji et al. 2008; Choi et al. 2010), suggesting that allergens are present throughout the lifecycle of silkworms. In many Asian countries, SWP are consumed raw, boiled, or fried, and have been linked to symptoms like, dizziness, skin itching, urticaria, and shock (Feng et al. 2018). While allergic reactions from SWP are comparable to those from fish and shrimp (de Gier and Verhoeckx 2018). Further research studies need to be undertaken to understand the underlying mechanisms for the allergenicity of SWP.

### *8.5.2 Safety Concerns of Silkworm Pupae*

Although SWP are considered nutrient rich, evaluating their safety is essential when considering their use as both food and medicine. The primary concerns regarding their safety revolve around potential toxicity and allergic reactions, highlighting the need for continued evaluation. Efforts to reduce food allergenicity through processing methods have gained attention, but only the 27-kDa hemolymph glycoprotein has been identified as a heat-stable allergen in SWP (Jeong et al. 2016). Hydrolysis and fermentation have shown promise in reducing allergenicity in food products and may offer similar benefits for silkworm pupae. Enzymatic hydrolysis, in particular, could be an effective strategy to modify allergenic proteins while preserving their bioactive properties. High pressure, ultrasound, and microwave treatments may further reduce allergenic components (He et al. 2021). Jiang et al. (2020) conducted a 28-day feeding bioassay on rats to assess the subacute toxicity of transgenic silkworms, revealing no adverse effects or fatalities. Additionally, no mortality or significant hematological, biochemical, or histopathological abnormalities were observed in rats fed with SWP protein for 30 days, with a safe dosage of 1.50 g/kg/d in rats (Zhou and Han 2006). Overall, the SWP are considered safe to be consumed both as food and medicine. Although most allergens in SWP are effectively minimized or eliminated by various processing methods, further research is necessary to better understand the other possible allergens in SWP proteins for their removal.

### *8.5.3 Regulatory Laws and Guidelines for Edible Insects*

The U.S. Food and Drug Administration (FDA) has demonstrated significant interest in regulating insects as a food source, as reflected in its 2016 guidelines. According to these regulations, insects, including SWP, can be used as human food, provided they meet safety and hygiene standards. The inclusion of pupae as edible insects is supported by legal frameworks that require collaboration between legislators, regulatory bodies, and law enforcement agencies to ensure compliance (Wilderspin and Halloran 2018). In Europe, recent legislative changes have facilitated the legalization of edible insect consumption, allowing for the expansion of insect-based food products. However, ensuring safety for both producers and consumers remains a key consideration. Comprehensive regulatory measures are essential to monitor the production, processing, and distribution of edible insects, including SWP, to mitigate potential health concerns and ensure food safety. To encourage the sustainable growth of SWP as a food source, clearer and more detailed regulations are necessary. Establishing well-defined guidelines would help govern the industry, ensuring that insect-based food products meet nutritional and safety standards. These restrictions would not only enhance consumer confidence but also encourage the health benefits of SWP while addressing potential risks (Belluco et al. 2017). Governments can encourage appropriate use of SWP in human diets by implementing stringent food safety standards. These initiatives will help to standardize production processes, improve quality control, and increase public acceptance of edible insects. Ultimately, a strong regulatory framework can protect consumers while promoting the potential of SWP as a nutritious, sustainable, and healthy food source.

## 8.6 Environmental and Economic Sustainability

Using insects for human consumption is one approach to mitigating the impacts of climate change and other environmental issues that the world is currently facing. The agriculture sector is responsible for around 36% of global greenhouse gas (GHG) emissions, with cattle accounting for 78% of these emissions (van Huis 2022). Since livestock contributes significantly to environmental deterioration, consumption of edible insects can be one alternative source of protein that are also known to have a minimal greenhouse gas contribution (Siddiqui et al. 2022). Lepidoptera are the most commonly consumed insect order in Africa (Hlongwane et al. 2020). They provide essential vitamins, proteins, and fats. Rural people in South Africa, Uganda, and Nigeria earn money by rearing and selling caterpillars (Hlongwane et al. 2020). In contrast to conventional farming practices in East Africa, which require arable land and water, insect rearing is rapidly gaining popularity and becoming a viable alternative (Tanga et al. 2021). The pupae, which remain after the silkworm has done reeling, are a highly nutritious byproduct that is safe to consume (Samami et al. 2019). The bioactive compounds found in SWP are beneficial to health. Hence, these pupae are utilized in pharmaceutical research after being further treated to extract minerals and active compounds (Hăbeanu et al. 2023a and b). The future of sericulture depends on a paradigm shift that values all the items produced by the industry, aiming to reduce its negative environmental impact and improve social and economic well-being (Altman and Farrell 2022).

### *8.6.1 Circular Economy: Byproducts of Silk Industry for Human Food*

The most lucrative byproduct, SWP, opened the door to potential uses in fertilizer, animal feed, cosmetics, nutrition, medicine, and other fields (Hăbeanu et al., 2023a and b). The pupae are a particularly nutrient-dense waste produced by the silkworm after reeling silk and are recommended as an edible byproduct (Lamberti et al. 2019). The chitin found in SWP skin (about 4% of dry matter basis) can be converted into a number of useful compounds, including sodium carboxymethyl chitin, chitosan, chitin sulfate, chitin nitrate, and chitin xanthate (Singh and Jayasomu 2002). In Asian countries, SWP has been used for its nutritional and therapeutic benefits for a long time. Additionally, the oil from SWP contains fatty acids with odd chain lengths that have strong anticancer properties (Priego Capote et al. 2007). However, they have mostly been developed for use in pharmacological and medical fields such as controlled medication release and wound dressing materials (Zhou et al. 2022). Furthermore, a fascinating application is as a composite bone-filling material, whose oligomers have been proposed as anticancer medications, which creates a self-hardening material for tissue regeneration in the treatment of periodontal bony abnormalities (Zhou et al. 2022). Applications for chitin and chitosan are numerous and include the food industry, cosmetics, water treatment, agriculture, biotechnology, paper, biomedicine, textile industry, and wound healing agents (Fig. 8.2).

## 8.7 Challenges and Future Prospects

To encourage entomophagy, new food processing methods are needed to create functional components and food products in whole, identifiable forms. Early adopters, usually the younger age, are the major population sector that should be addressed to shift negative attitudes against insect consumption. Products that seem different from those that are currently used in everyday meals and snacks such as such cookies and burgers are in high demand among young people. The nutritional qualities and functional qualities of certain insect-based substances make them highly promising for use in food, medicine, and cosmetic applications properties (Arfan et al. 2012). Protein isolates or concentrates, made from insects, offer strong emulsifying and

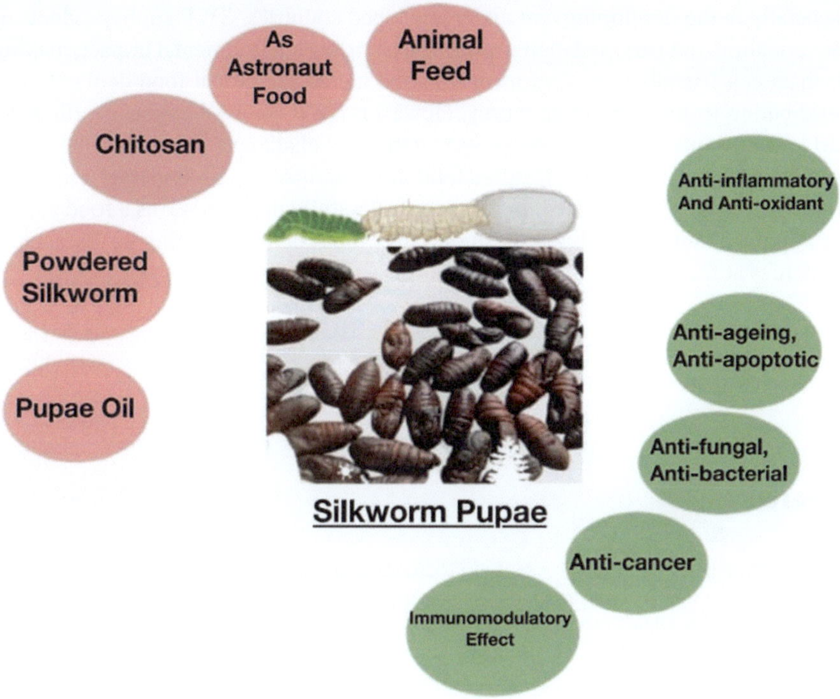


**Fig. 8.2** Multifaceted applications of silkworm pupae (Created with BioRender)

foaming qualities in addition to other techno-functional capabilities. Future studies and technical advancements ought to concentrate on these areas.

The current high cost of edible insects when compared to alternative sources of animal and vegetable proteins is a substantial financial hurdle to their commercial expansion. In order to avoid food safety issues, edible insect farmers around the world must adopt and implement appropriate hygienic procedures and best agricultural management practices (Kim et al. 2019). Therefore, standard procedure manuals, as well as hazard analyses and critical control points, are required to ensure that raw edible insects are handled, processed, preserved, stored, packed or distributed in the most secure and nutrient-dense manner possible (Raheem et al. 2019).

## 8.8 Conclusions

Silkworm pupae (SWP) are a potential nutritious food ingredient for humans. They offer a sustainable alternative to traditional animal proteins, by providing protein, fat, vital amino acids, fatty acids, bioactive molecules, vitamins, and minerals. This may be significant in the present context of growing demand for protein-rich food, especially in the developing and underdeveloped countries. SWP are byproducts of the sericulture industry, and their production has a low environmental impact, making them an eco-friendly solution for meeting the increasing global food demand, thus contributing to the circular economy. However, SWP use by humans is still at an early stage, as its acceptance by masses remains a challenge. Moreover, the large-scale industrial production of fortified foods and pharmaceuticals enriched with SWP from silkworms is also limited. In order to successfully utilize SWP as a food source for improved human health, future studies may focus on investigating its therapeutic properties at the molecular and clinical levels.

**Declaration of AI Usage** AI tools were used to assist in language editing and improving the clarity of this document. The authors have thoroughly reviewed all material, and take full responsibility for the accuracy, integrity, and originality of the final content.

## References

Akande OA, Falade OO, Badejo AA et al (2020) Assessment of mulberry silkworm pupae and African palm weevil larvae as alternative protein sources in snack fillings. Heliyon 6(4)

Ali MFZ, Yasin IA, Ohta T et al (2018) The silkrose of *Bombyx mori* effectively prevents vibriosis in penaeid prawns via activation of innate immunity. Sci Rep 8(1):8836

Ali MM, Arumugam SBA (2011) Effect of crude extract of *Bombyx mori* cocoons in hyperlipidemia and atherosclerosis. J Ayurveda Integr Med 2(2):72

Altman GH, Farrell BD (2022) Sericulture as a sustainable agroindustry. Clean Circ Bioeconomy 2:100011

Altomare AA, Baron G, Aldini G et al (2020) Silkworm pupae as a source of high-value edible proteins and bioactive peptides. Food Sci Nutr 8(6):2652–2661

Arfan M, Khan R, Rybarczyk A, Amarowicz R (2012) Antioxidant activity of mulberry fruit extracts Int. J Mol Sci 13(2):2472–2480

Aznar-Cervantes SD, Monteagudo Santesteban B, Cenis JL (2021) Products of sericulture and their hypoglycemic action evaluated using the silkworm *Bombyx mori*. Insects 12(12):1059

Azzollini D, Derossi A, Severini C (2016) Drying kinetics and hygroscopic behaviour of *Tenebrio molitor* larvae and effects on their quality. J Insects Food Feed 2(4):233–244

Belluco S, Halloran A, Ricci A (2017) New protein sources and food legislation: the case of edible insects and EU law. Food Secur 9:803–814

Cermeno M, Bascón C, Amigo-Benavent M et al (2022) Identification of antioxidant peptides from edible silkworm pupae (*Bombyx mori*) protein hydrolysates. J Funct Foods 92:105052

Chatterjee NS, Kumar KA, Ajeesh Kumar KK et al (2017) Screening water-soluble B vitamins in fish: enzymatic extraction, HILIC separation, and tandem MS. J AOAC Int 100(3):579–585

Choi GS, Shin YS, Kim JE et al (2010) Food allergy to vegetable worm (*Cordyceps sinensis*) with cross-reactivity to silkworm pupae. Allergy 65(9):1196–1197

Chukiatsiri S, Siriwong S, Thumanu K (2020) Silkworm pupae protein extracts exert anticancer effects via downregulation of IL-6, IL-1β, and TNF-α. Biomed Pharmacother 128:110278

David-Birman T, Moshe H, Lesmes U (2019) Thermal processing effects on physicochemical properties of *Bombyx mori* flour. Food Res Int 123:11–19

de Gier S, Verhoeckx K (2018) Insect (food) allergy and allergens. Mol Immunol 100:82–106

Feng Y, Chen XM, Zhao M et al (2018) Edible insects in China: utilization and prospects. Insect Sci 25(2):184–198

Fombong FT, Van Der Borght M, Vanden Broeck J (2017) Effect of freeze- and oven-drying on Ruspolia differens nutrient composition. Insects 8(3):102

Fu M, Qu Q, Yang X, Zhang X (2016) Intermittent oven drying effect on lipid oxidation and antioxidant activity in walnut. LWT Food Sci Technol 65:1126–1132

Gautreau M, Restuccia M, Senser K, Weisberg SN (2017) Familial anaphylaxis after silkworm ingestion. Prehosp Emerg Care 21(1):83–85

Grabowski NT, Klein G (2016) Microbiology of crickets and superworms after four heating treatments. Food Sci Technol Int 23(1):17–23

Grafton RQ, Daugbjerg C, Qureshi ME (2015) Towards food security by 2050. Food Secur 7:179–183

Hăbeanu M, Gheorghe A, Mihalcea T (2023a) *Bombyx mori*—sustainability and economic opportunity, particularly for Romania. Agriculture 13(6):1209

Hăbeanu M, Gheorghe A, Mihalcea T (2023b) Nutritional value of silkworm pupae (*Bombyx mori*) with emphases on fatty acids profile and their potential applications for humans and animals. Insects 14(3):254

He W, He K, Sun F et al (2021) Effect of heat, enzymatic hydrolysis and acid-alkali treatment on the allergenicity of silkworm pupa protein extract. Food Chem 343:128461

Hlongwane ZT, Slotow R, Munyai TC (2020) Nutritional composition of edible insects consumed in Africa: a systematic review. Nutrients 12(9):2786

Hrebień-Filisińska A (2021) Application of natural antioxidants in oxidative stabilization of fish oils: a mini-review. J Food Process Preserv 45(4):e15342

Hu B, Li C, Zhang Z et al (2017) Microwave-assisted extraction of silkworm pupal oil and evaluation of its fatty acid composition and antioxidant activities. Food Chem 231:348–355

Hu D, Liu Q, Cui H et al (2005) Effects of amino acids from selenium-rich silkworm pupas on human hepatoma cells. Life Sci 77(17):2098–2110

Hyun CK, Kim IY, Frost SC (2004) Soluble fibroin enhances insulin sensitivity and glucose metabolism in 3T3-L1 adipocytes. J Nutr 134(12):3257–3263

Ivanciuc O, Garcia T, Torres M et al (2009) Characteristic motifs for families of allergenic proteins. Mol Immunol 46(4):559–568

Jamal S, Jamil DM, Khidhir ZK (2020) Protein determination in some animal products from Sulaymaniyah markets using Kjeldahl procedure. J Food Dairy Sci 11(12):343–346

Jasmine S, Mandal BB (2014) Types and properties of non-mulberry silk biomaterials for tissue engineering applications. In: Silk biomaterials for tissue engineering and regenerative medicine. Woodhead, Sawston, Cambridge, pp 275–298

Jeong KY, Son M, Lee JY et al (2016) Allergenic characterization of 27-kDa glycoprotein, a novel heat-stable allergen, from silkworm pupa. J Korean Med Sci 31(1):18–24

Ji KM, Zhan ZK, Chen JJ, Liu ZG (2008) Anaphylactic shock caused by silkworm pupa consumption in China. Allergy 63(11):1407–1408

Jiang L, Wang Y, Guo H et al (2020) Toxicological evaluation of transgenic silkworms. Toxicol Res 9(6):845–853

Jin H, Yang X, Zhao H et al (2023) Genetic analysis of protein and oil content in soybean via genome-wide association study. Front Plant Sci 14:1182771

Juarez-Ortega AJA, Viesca-González FC, Barrera-García VD (2012) La recolección, venta y consumo de insectos en Toluca, México y sus alrededores. Rosa Dos Ventos: Tur Hosp 4(2):208–221

Jukanti AK, Gaur PM, Gowda CLL, Chibbar RN (2012) Nutritional quality and health benefits of chickpea (*Cicer arietinum* L.): a review. Br J Nutr 108(S1):S11–S26

Karnjanapratum S, Kaewthong P, Indriani S et al (2022) Characteristics and nutritional value of silkworm (*Bombyx mori*) pupae-fortified chicken bread spread. Sci Rep 12(1):1492

Khazaei H, Subedi M, Nickerson M et al (2019) Seed protein of lentils: current status, progress, and food applications. Foods 8(9):391

Kim HW, Setyabrata D, Lee Y et al (2017a) Effect of house cricket (*Acheta domesticus*) flour addition on physicochemical and textural properties of meat emulsion under various formulations. J Food Sci 82(12):2787–2793

Kim SK, Weaver CM, Choi MK (2017b) Proximate composition and mineral content of five edible insects consumed in Korea. CyTA J Food 15(1):143–146

Kim YJ, Chon JW, Song KY et al (2017c) Sensory profiles of protein-fortified kefir prepared using edible insects (silkworm pupae, Bombyx mori): a preliminary study. J Dairy Sci Biotechnol 35(4): 262–265

Kim TK, Yong HI, Kim YB et al (2019) Edible insects as a protein source: a review of public perception, processing technology, and research trends. Food Sci Anim Resour 39(4):521

Kim YJ, Lee KP, Lee DY et al (2020) Inhibitory effect of modified silkworm pupae oil in PDGF-BB-induced proliferation and migration of vascular smooth muscle cells. Food Sci Biotechnol 29:1091–1099

Köhler R, Kariuki L, Lambert C et al (2019) Protein, amino acid and mineral composition of some edible insects from Thailand. J Asia Pac Entomol 22(1):372–378

Kotake-Nara E, Yamamoto K, Nozawa M et al (2002) Lipid profiles and oxidative stability of silkworm pupal oil. J Oleo Sci 51(11):681–690

Kouřimská L, Adámková A (2016) Nutritional and sensory quality of edible insects. NFS j 4:22–26

Kumar D, Dev P, Kumar RV (2015) Biomedical applications of silkworm pupae proteins. In: Biomedical applications of natural proteins: an emerging era in biomedical sciences

Lamberti C, Gai F, Cirrincione S et al (2019) Investigation of the protein profile of silkworm (*Bombyx mori*) pupae reared on artificial vs. mulberry leaf diet. PeerJ 7:e6723

Lange KW, Nakamura Y (2023) Potential contribution of edible insects to sustainable consumption and production. Front Sustain 4:1112950

Lautenschlager T, Neinhuis C, Kikongo E et al (2017) Impact of preparation methods on the nutritional value of *Imbrasia epimethea*. Eur Food Res Technol 243(5):769–778

Lee JH, Jo YY, Kim SW, Kweon H (2021) Antioxidant capacity of silkworm pupa: effects of extraction conditions, variety, pupation time, and sex. Int J Ind Entomol 43(2)

Lee SH, Park D, Yang G et al (2012) Silk and silkworm pupa peptides suppress adipogenesis in preadipocytes and fat accumulation in rats. Eur J Nutr 51:1011–1019

Lenaerts S, Van Der Borght M, Callens A et al (2018) Suitability of microwave drying for *Tenebrio molitor*: nutritional quality and colour. Food Chem 254:129–136

Li X, Xie H, Chen Y et al (2018) Silkworm pupa protein hydrolysate induces apoptosis and cell cycle arrest in gastric cancer cells. Int J Mol Sci 19(4):1013

Li Z, Zhao S, Xin X et al (2019) Purification and characterization of a novel immunomodulatory hexapeptide from silkworm (*Bombyx mori*) pupa protein. J Asia Pac Entomol 22(3):633–637

Li Z, Zhao S, Xin X et al (2020) Functional analysis of a novel immunomodulatory peptide from silkworm pupa protein. Int J Pept Res Ther 26:243–249

Lisboa HM, Nascimento A, Arruda A et al (2024) Unlocking the potential of insect-based proteins: sustainable solutions for global food security. Foods 13(12):1846

Liu ZQ (2022) Why natural antioxidants are readily recognized by biological systems? 3D architecture plays a role. Food Chem 380:132143

Liu Z, Xia L, Wu Y et al (2009) Identification and characterization of an arginine kinase as a major allergen from silkworm (*Bombyx mori*) larvae. Int Arch Allergy Immunol 150(1):8–14

Longvah T, Manghtya K, Qadri SS (2012) Eri silkworm: a source of edible oil with a high content of α-linolenic acid and of significant nutritional value. J Sci Food Agric 92(9):1988–1993

Mahanta DK, Komal J, Samal I et al (2023) Nutritional aspects and dietary benefits of "Silkworms": current scenario and future outlook. Front Nutr 10:1121508

Marshall DL, Dickson JS, Nguyen NH (2016) Ensuring food safety in insect-based foods: mitigating microbiological and other foodborne hazards. Insects as sustainable food ingredients. Academic Press, Cambridge, pp 223–253

Megido RC, Poelaert C, Ernens M et al (2018) Effect of household cooking techniques on microbiological load and nutritional quality of mealworms (*Tenebrio molitor* L. 1758). Food Res Int 106:503–508

Oduor EO, Ciera L, Adolkar V (2021) Physical characterization of eri silk fibers produced in Kenya. J Nat Fibers 18(1):59–70

Olteanu M, Panaite T, Ciurescu G (2012) Effect of dietary mulberry leaves on performance and nutrient digestibility in laying hens. Indian J Anim Sci 82(8):914
Paul D, Dey S (2014) Essential amino acids, lipid profile and fat-soluble vitamins of the edible silkworm *Bombyx mori* (Lepidoptera: Bombycidae). Int J Trop Insect Sci 34:239–247
Pereira NR, Ferrarese-Filho O, Matsushita M et al (2003) Proximate composition and fatty acid profile of Bombyx mori L. chrysalis toast. J Food Compos Anal 16(4):451–457
Praticò D, Delanty N (2000) Oxidative injury in diseases of the central nervous system: focus on Alzheimer's disease. Am J Med 109(7):577–585
Priego Capote F, Jiménez JR, Granados JMM et al (2007) Identification and determination of fat-soluble vitamins and metabolites in human serum by LC-MS/MS. Rapid Commun Mass Spectrom 21(11):1745–1754
Purschke B, Brüggen H, Scheibelberger R et al (2018) Effect of pre-treatment and drying method on physico-chemical properties and dry fractionation of mealworm larvae (Tenebrio molitor L.). Eur Food Res Technol 244:269–280
Raheem D, Carrascosa C, Oluwole OB et al (2019) Traditional consumption and rearing of edible insects in Africa, Asia and Europe. Crit Rev Food Sci Nutr 59(14):2169–2188
Ravinder T, Kaki SS, Kunduru KR et al (2016) Physico-chemical characterization and oxidative stability of eri silkworm oils. Int J Mod Chem Appl Sci 3:293–300
Rodríguez-Ortega A, Pino-Moreno JM, Ángeles-Campos SC et al (2016) Valor nutritivo de larvas y pupas de gusano de seda (Bombyx mori). Rev Colomb Entomol 42(1):69–74
Rodríguez-Ortiz LM, Hincapié CA, Hincapié-Llanos GA et al (2024) Potential uses of silkworm pupae (Bombyx mori L.) in food, feed, and other industries: a systematic review. Front Insect Sci 4:1445636
Ryu SP (2014) Silkworm pupae powder ingestion increases fat metabolism in swim-trained rats. J Exerc Nutr Biochem 18(2):141
Sadat A, Biswas T, Cardoso MH et al (2022) Silkworm pupae as a future food with nutritional and medicinal benefits. Curr Opin Food Sci 44:100818
Samami R, Seidavi A, Eila N, Moarefi M, Ziaja DJ, Lis JA, Rubiu NG, Cappai MG (2019) Production performance and economic traits of silkworms (Bombyx mori L., 1758) fed with mulberry tree leaves (Morus alba, var. Ichinose) significantly differ according to hybrid lines. Livest Sci 226:133–137
Siddiqui SA, Alvi T, Sameen A et al (2022) Consumer acceptance of alternative proteins: a systematic review of protein sources and interventions to increase acceptability. Sustainability 14(22):15370
Singh KP, Jayasomu RS (2002) Bombyx mori—a review of its potential as a medicinal insect. Pharm Biol 40:28–32
Tanga CM, Egonyu JP, Beesigamukama D et al (2021) Edible insect farming as an emerging and profitable enterprise in East Africa. Curr Opin Insect Sci 48:64–71
Tangsanthatkun J, Peanparkdee M, Katekhong W et al (2022) Application of aqueous saline process to extract silkworm pupae oil (Bombyx mori): process optimization and composition analysis. Foods 11(3):291
Tassoni L, Cappellozza S, Dalle Zotte A et al (2022) Nutritional composition of Bombyx mori pupae: a systematic review. Insects 13(7):644
Thirupathaiah Y, Sivaprasad V, Bhuvaneswari E et al (2016) Extraction and characterization of Bombyx mori pupae oil. Indian J Seric 55(1–2):31–37
Tiencheu B, Womeni HM, Linder M et al (2013) Lipid changes in Rhynchophorus phoenicis during cooking and storage. Eur J Lipid Sci Technol 115(2):186–195
Tiwari BK, Singh N (2012) Pulse chemistry and technology. Royal Society of Chemistry, London
van Huis A (2022) Edible insects: challenges and prospects. Entomol Res 52(4):161–177
Vandeweyer D, Lenaerts S, Callens A, Van Campenhout L (2017) Blanching and microwave drying effects on microbial load in Tenebrio molitor. Food Control 71:311–314
Wang W, Shen S, Chen Q et al (2008) Silkworm pupae protein hydrolysates as a new source of ACE inhibitory peptides. Curr Pharm Biotechnol 9(4):307–314

Wang W, Wang N, Zhang Y (2014) Antihypertensive effects of peptide hydrolysates from silkworm pupae protein in rats. Food Nutr Sci 2014

Wang W, Wang N, Zhou Y et al (2011) Novel peptide from silkworm pupae with ACE inhibition characteristics. Eur Food Res Technol 232:29–38

Wattanathorn J, Muchimapura S, Boosel A et al (2012) Silkworm pupae protect against Alzheimer's disease. Am J Agric Biol Sci 7(3):330–336

Weixin L, Lixia M, Leiyan W et al (2019) Effects of silkworm pupa protein hydrolysates on mitochondria in gastric cancer cells. J Asia Pac Entomol 22(2):387–392

Wendin KM, Nyberg ME (2021) Factors influencing consumer perception and acceptability of insect-based foods. Curr Opin Food Sci 40:67–71

Wilderspin DE, Halloran A (2018) Effects of regulation and policy on insect consumption in the global South. Edible insects in sustainable food systems. Springer, Cham, pp 443–455

Wu X, He K, Velickovic TC et al (2021) Nutritional, functional, and allergenic properties of silkworm pupae. Food Sci Nutrition 9(8):4655–4665

Xiao HW, Bai JW, Sun DW, Gao ZJ (2014) Superheated steam impingement blanching (SSIB) in agricultural products processing—a review. J Food Eng 132:39–47

Yang Y, Tang L, Tong L, Liu H (2009) Silkworm culture as a protein source for humans in space. Adv Space Res 43(8):1236–1242

Yeruva T, Jayaram H, Aurade R et al (2023) Profiling of nutrients and bioactive compounds in pupae of silkworm Bombyx mori. Food Chem Adv 3:100382

Yu W, Ying H, Tong F et al (2013) Protective effect of silkworm protein 30Kc6 on human vascular endothelial cells damaged by oxidized LDL. PLoS ONE 8(6):e68746

Yuan Q, Xie F, Huang W et al (2022) Alpha-linolenic acid: sources, metabolism, and pharmacology. Phytother Res 36(1):164–188

Yue H, Luo RM, Qin X et al (2021) Synthesis of α-linolenic acid–rich partial glycerides from silkworm pupa oil with immobilized lipase MAS1-H108A. Food Sci Technol 42:e58221

Zhang Y, Wang J, Zhu Z et al (2021) Two novel antioxidant peptides from silkworm pupae protein hydrolysates. Eur Food Res Technol 247:343–352

Zhou J, Han D (2006) Safety evaluation of protein from Antheraea pernyi pupae. Food Chem Toxicol 44(7):1123–1130

Zhou Y, Zhou S, Duan H et al (2022) Bombyx mori pupae: a functional food with health benefits for humans. Foods 11(11):1594

# Chapter 9
# Spider Silk Structure Properties and Applications: An Introduction

**Shubham Singh, Divya Gupta, Mohit, Dhiraj Kumar, Monika Bhadauria, and Satendra Kumar Nirala**

## 9.1 Introduction

There are more than 40,000 species of spiders that have been identified worldwide so far and still it is believed that most spider species are not yet known (Perrotta et al. 2025). Every spider spin at least one type of silk and sometime more than one, majorly depend on the spider gland present in their abdomen. These spider silks have different characteristics depending on their applications, which includes, making the strands of the nest, the connection radius of the main strands or the prey capturing (Wang et al. 2025a, b). Spider silk, which is also well known as one of the hardest natural substances produced by any living organism, is made up of proteins, due to which it is a biocompatible and environment friendly material. Other unique feature

S. Singh
Department of Zoology, Government E. Raghvendra Rao P.G. Science College, Bilaspur, Chhattisgarh, India
e-mail: shubhamsinghggu@rediffmail.com

D. Gupta
Amity Institute of Biotechnology, Amity University, Gwalior, Madhya Pradesh, India

Mohit · D. Kumar
Centre for Skill Development and Seri-Technology, Department of Zoology, Hansraj College, University of Delhi, Delhi, India
e-mail: dhirajindia01@gmail.com

M. Bhadauria
Toxicology and Pharmacology Lab, Department of Zoology, Guru Ghasidas Vishwavidyalaya, Bilaspur, Chhattisgarh, India
e-mail: monikabhadauria@rediffmail.com

S. K. Nirala (✉)
Laboratory of Natural Products, Department of Rural Technology and Social Development, Guru Ghasidas Vishwavidyalaya, Bilaspur, Chhattisgarh, India
e-mail: satendrak75@rediffmail.com



D. Kumar and M. Shahid (eds.), *Science and Technology of Silk*, Textile Science and Clothing Technology, https://doi.org/10.1007/978-3-032-16230-4_9

related with spider silk is its mechanical properties, its excellent fatigue behaviour, biocompatibility, and low biodegradation rate (Jiang et al. 2025). Different properties of silk depend on its type such as low density, high expandability, and its high tensile strength, e.g. MaSp1 has very high tensile strength, and flagelliform silks have high expandability. These unique characteristics of spider silk have attracted the focus of researchers for industrial and medicinal purposes (Hong et al. 2025).

Spider silk is a remarkable natural fibre produced by spiders, renowned for their unique combination of strength, elasticity, and lightness. It is composed of primarily structural proteins called as spidroins, which provide its extraordinary mechanical properties. The silk is produced within the specialized gland present in the spider abdomen and is extruded through spinnerets, forming fine but incredibly strong threads (Heim et al. 2010; Landreh et al. 2024). Each spider species can produce multiple types of silk and each type serves different function in their survival and ecological roles. The most common uses of spider silk are in web building. Many species of spider, particularly orb-weaving spiders, construct intricate, geometrically precise webs designed to capture prey. These webs represent the incredible tensile strength of spider silk, which is frequently associated to steel and Kevlar in terms of strength to weight ratio (Ruan and Li 2024). Some spider silk thread particularly dragline silk is strong enough to withstand significant force while remaining highly flexible. This allows the web to absorb the impact of flying insects without breaking and helps spiders to capture their prey (Han et al. 2023; Zou et al. 2024).

Beyond web building, spiders use their silk for various other survival strategies. Some species, like the bolas spider, use a single strand of sticky silk to ensnare moths, mimicking pheromones to attract their prey. Others like trapdoor spiders line their burrows with silk, creating hidden entrances that help them ambush unsuspecting insects (Wu et al. 2024). Apart from that jumping spiders use a silk safety line when leaping between surfaces, ensuring they can recover from missed jumps (Zhao et al. 2024).

Spider silk also plays an important role in reproduction. Female spiders weave protective eggs sac from silk to safeguard their offsprings from predators and environmental related issues. In some species, male uses silk to wrap nuptial gifts, offering them to potential mates as part of courtship rituals (Land et al. 2024). Silk importance in reproduction ensures the continuation of spider population and contributes to the stability of various ecosystem where spiders play a vital role as natural pest controllers. Another fascinating use of spider silk in nature is ballooning, which is a dispersal method employed by some spiderlings and even adult spiders. By releasing silk threads into the air, spiders can catch wing current and travels a long distance. This behaviour aids in population distribution, enabling spiders to colonize new areas and maintain genetic diversity across different habitats (Sethy and Ahi 2022).

Due to its exceptional properties, spider silk has attracted significant interest from researchers to engineers. Researchers are exploring ways to replicate its characteristics for various applications, including biomedical uses such as sutures, artificial ligaments, and tissue engineering (Li et al. 2022a; b). Apart from that its combination of strength and lightweight nature have made it an ideal material for military and industrial applications such as bulletproof clothing's and biodegradable fishing lines.

However, large production of natural spider silk remains challenging due to the difficulties of farming spiders, leading to efforts in synthetic biology to engineer spider silk proteins using bacteria, yeast, and even genetically modified plants (Whittall et al. 2021).

## 9.2 Historical Perspective of Studying Spider Silk

The study of spider silk has a long and fascinating history. Ancient civilizations recognized the unique properties of spider silk and utilized it in various practical applications, but it was not until the modern era that its structure and mechanical properties became a subject of detailed scientific study.

### 9.2.1 *Early Uses and Observations*

The earliest recorded application of spider silk dates back thousands of years. In ancient Greece and Rome, physicians reportedly used spider silk for wound dressing due to its fine, adhesive quality and potential antibacterial properties. Indigenous communities were known to use spider silk to create fishing nets, taking advantage of its strength and durability (Greco et al. 2021). During the eighteenth and nineteenth centuries, European naturalists became increasingly fascinated by spiders and their silk-producing abilities. French scientist Rene-Antoine Ferchault de Reaumur performed one of the first systematic studies of spider silk, documenting how different species produced various types of silk for different purposes. He also explored the possibilities of using spider silk in textiles, which led to early experiments in harvesting silk for fabric production (Hollsten 2025).

### 9.2.2 *Nineteenth and Twentieth Century Research*

With the advancement in microscopy and textile sciences, researchers have begun to understand spider silks' composition and structure more deeply. In the late nineteenth century, attempts had been made to commercialize spider silk as a luxury fabric. A notable example occurred in the early 1900s when Madagascar weavers successfully created garments from the silk of golden orb-weaver spiders. However, large-scale production proved difficult due to the territorial and cannibalistic nature of spiders, making them unsuitable for farming like silkworms (Brunetta and Craig 2010). In the early twentieth century, scientists focused on the mechanical properties of spider silk. They have discovered that different types of silk exhibited varying levels of elasticity and tensile strength, with dragline silk being particularly strong and resilient. The studies about silk proteins and their molecular structures gained momentum, laying the foundation for modern biomaterials research with spider silk (Lu et al. 2024).

### 9.2.3 Modern Scientific Advancement

By the mid to late twentieth century, technological advancements enabled deeper analysis of spider silk at the molecular level. Researchers discovered that silk proteins, which is also known as spidroins, played a key role in the material's unique mechanical properties. The rise of molecular biology and genetic engineering in the late twentieth and early twenty-first century opened new possibilities for replicating spider silk synthetically (Khan et al. 2023). In the current scenario, spider silk research continues to be a thriving field, with applications spanning medicine, engineering, and material science. Efforts to produce artificial spider silk using genetically modified bacteria, yeast, and even goats have shown promising results, bringing scientists closer to its full potential (Gupta and Rastogi 2023). Apart from this, genetic engineering is also applied to produce spider silk from bacteria, yeast, which could revolutionize material science and biomedical applications. Researchers are also exploring methods to replicate the exceptional strength and flexibility of spider silk synthetically, aiming for scalable production and diverse industrial uses.

## 9.3 Biological Basis of Silk Production

Spiders produce silk through specialized glands located in their abdomen, which secrete liquid silk proteins called spidroins. These proteins undergo a phase transition, solidifying as they pass through the spinnerets, an external structure that controls the extrusion and manipulation of silk threads. Unlike silkworms, which produce a single type of silk, spiders can generate multiple silk types (Table 9.1), each adapted for different purposes such as web construction, prey capture, egg case protection, and locomotion (Guessous et al. 2024).

Spidroins, the fundamental protein of spider silk, exhibit a distinctive molecular structure that lends exceptional strength and elasticity. Their composition primarily consists of amino acids like glycine, alanine, and serine, arranged in repetitive patterns. These sequences form crystalline regions through hydrogen bonding, folded into beta-pleated sheets that stack to create the silk's remarkable nano-crystalline structure. This unique molecular arrangement imparts spider silk with outstanding properties, including high tensile strength, elasticity, and toughness, combined with low weight and biocompatibility (Eisoldt et al. 2011; Rajput et al. 2022).

**Table 9.1** Types of spider silk and their function

| Type of spider silk | Function | Reference |
|---|---|---|
| Dragline | Used for web frame, radii, and lifelines (safety lines)<br>Extremely strong and tough, comparable to steel in tensile strength | Malay et al. (2022) |
| Capture spiral | Forms the sticky spirals of orb webs to trap prey<br>Highly elastic, allowing it to stretch without breaking | Baumgart et al. (2022) |
| Viscid | Used to ensure and immobilize prey, allowing the spider to wrap its prey and secure it for feeding<br>Enables spiders to adhere to surfaces, facilitate locomotion, web attachment, and even egg sac anchorage | Alicea-Serrano et al. (2021) |
| Aciniform | Used in wrapping and immobilizing prey<br>Strong and slightly elastic, aiding prey capture | Wijerathna et al. (2025) |
| Tubuliform | Used to create protective egg sacs<br>Thick and though, providing a strong barrier against predators | Yarger et al. (2018) |
| Piriform | Used to attach silk threads to surfaces (silk anchors)<br>Extremely adhesive, ensuring strong attachment points for webs and draglines | Greco et al. (2020) |
| Aggregate | Coats flagelliform silk with glue-like droplets to enhance prey capture<br>Sticky and adhesive, increasing the effectiveness of capture spirals | Rising and Johansson (2015) |

## 9.4 Characteristic of Spider Silk

Spider silk is an amazing biomaterial with unique mechanical properties that surpass many synthetic fibres. It combines strength, elasticity, and toughness in ways that make it one of nature's most efficient materials (Li et al. 2022a; b).

- **Tensile strength**: Spider silk, particularly dragline silk, possesses a tensile strength 3–5 times that of ordinary steel while being significantly lighter (Li et al. 2021; Wang et al. 2025b).
- **Elasticity:** Some types of spider silk, such as flagelliform silk, can stretch up to five times their original length without breaking. This flexibility allows it to absorb high-energy impacts, such as those from flying prey (Eisoldt et al. 2011; Hopfe et al. 2024).
- **Toughness:** Spider silk can absorb more energy before breaking than most synthetic materials. Its toughness surpasses that of Kevlar, making it ideal for absorbing shock without snapping (Wang et al. 2025a; b).

## 9.5 Application of Spider Silk

The biocompatibility, strength, and biodegradability of spider silk make it highly promising in various medical fields (Fig. 9.1).

### 9.5.1 *Surgical Sutures*

The fine, strong, and flexible nature of spider silk enables sutures that minimize tissue damage and dissolve naturally over time, thereby reducing the need for removal (Branković et al. 2024; Kumar et al. 2023). The antimicrobial properties of spider silk prevent microbial infection at the site of injury, making it a suitable material for use as a suture (Franco et al. 2019).

### 9.5.2 *Artificial Ligaments and Tendons*

Due to its unique combination of elasticity and toughness, spider silk has potential for use in making artificial ligaments and tendons, aiding in reconstructive surgery (Dong et al. 2020; Gellynck et al. 2008). In a study, the growth of chondrocyte cells on spider cocoon silk resulted in the development of articular cartilage (Straub et al. 2024). It has also been reported that micropatterns can facilitate signal transduction and thus differentiation by providing physical and mechanical induction on MSCs without

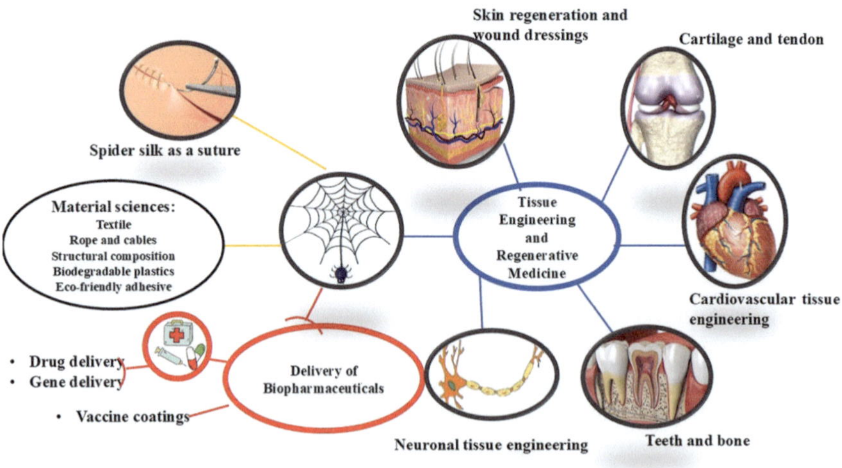


**Fig. 9.1** Spider silk applications in various medical fields

the need for growth factors like transforming growth factors in culture medium (Hernando et al. 2021).

### *9.5.3 Drug Delivery System*

Drug delivery system reduces toxic side effects caused by any xenobiotics and improving cell absorption and bioavailability and also improving cell absorption and confirms to deliver medicine at the specific target site or organ system (Rathour et al. 2024; Singh et al. 2024a, b). Spider silk's ability to encapsulate and release substances gradually makes it an ideal material for drug delivery, particularly for targeted therapies and wound healing applications (Branković et al. 2024; Singh et al. 2024a, b). Spider silk, which is a hydrophilic biological material, having all required characteristics for drug delivery. Several factors, like pH, type, and concentration of ions, are also effective in the characterization of synthetic spider silk for drug delivery (Wang et al. 2025a; Kundapur et al. 2023). These features make spider silk an appropriate delivery agent in drug delivery.

### *9.5.4 Bone and Teeth Engineering*

Despite remodelling ability, bone healing faces more problems from non-uniform breakages and vast defects. Therefore, bone engineering using osteoconductive agents, biocompatible materials, and suitable cell types remains an area of ongoing research (Chen et al. 2024). The capacity to hold bones and teeth with its stiffness, spider silk may be a replacement option for fractures in bones and teeth. Biomineralization is utilized to produce materials that play a crucial role in forming hard tissues, such as bone and teeth (Neubauer et al. 2021).

### *9.5.5 Material Science Application*

Spider silk's excellent mechanical characteristics have inspired its use in various industrial and commercial materials such as:

#### **9.5.5.1 High-Performance Textiles**

Future fabrics made from synthetic spider silk could be lightweight, durable, and resistant to wear, with potential applications in sportswear, military gear, and protective clothing (Wu and Chen 2022; Khan and Jawaid 2022). Wang et al. (2025a; b) have reported a multifunctional, wearable fabric that provides neutron and electromagnetic interference shielding, energy harvesting, and thermal management, confirming its use in both protection and comfort in harsh conditions such as high cold conditions.

#### 9.5.5.2 Ropes and Cables

The high tensile strength of spider silk could lead to ultra-strong yet lightweight ropes and cables for aerospace, climbing, and construction industries (Li et al. 2022a; b; Jiang et al. 2025). Jiang et al. (2025) prepared nanofiber yarns using spider silk and reported an increase in their strengthening properties.

#### 9.5.5.3 Structural Composites

When integrated into composites, spider silk fibres could improve materials used in the aircraft, automotive industries, and even bulletproof vests (Lentz et al. 2022). Liu et al. (2024) prepared hydrogels using spider proteins and reported increased tensile strength, fracture strain, toughness, and fracture energy.

### *9.5.6 Biodegradable Plastics*

Researchers are developing spider silk-inspired polymers that could replace petroleum-based plastics, thereby reducing environmental waste (Chang et al. 2023). Kedir et al. (2024) have prepared an active biodegradable polymer from spider silk by mixing it with chitosan. Xia (2024) has suggested using chitosan-silk fibroin as a strong, tough, and biodegradable alternative to plastics for packaging materials.

### *9.5.7 Eco-Friendly Adhesives*

The sticky nature of certain spider silks could lead to bio-based adhesives that are strong yet biodegradable (Liu et al 2021; Li et al. 2022a, b). Feng et al. (2025) have prepared a spider silk-inspired bio-based adhesive that exhibits excellent properties of flame retardancy and full recyclability.

## 9.6 Future Perspective and Challenges

From this chapter, the advantage of biomaterials is currently applied in all fields of biological and medical sciences as they possess extraordinary mechanical properties and are made up of proteins, which have led scientists to add efficient motifs to them to promote their effectiveness. The advantage of this biomaterial is not limited to this and also expands to promoting differentiation of stem cells by addition of specific signalling and adhesion motifs, making it permanent implants like bone through reduction of silk biodegradability, and producing safe cosmetics, thread lifts, hydrogel form as a lip filler used in beauty surgeries, and burn healing products. For achieving all these benefits from spider silk proteins, we still face several challenges in modification process, more research is required to maintain a balance and safety on the properties of spider silk-based materials, such as required experimental conditions involved in chemical modifications and changes or loss of partial intrinsic characteristics of the original proteins. Silk may be utilized to develop an organ-on-chip in drug discovery and toxicity testing. Spider silk can be tested for use in sprays and eye drops or local injection to repair tissue damage. Another possible use of spider silk could be to store sensitive substances, such as enzymes, vaccines, and certain drugs, for an extended period and to serve as a container for their safe transfer. All these potential advantages of using spider silk require extensive research in future to deliver suitable and safe spider silk-based biomaterials for use in drug delivery, industrial production, and regenerative medicine.

## 9.7 Conclusion

Spider silk is a remarkable natural product that combines excellent strength, elasticity, toughness, and biocompatibility, making it a model for advanced biomimetic applications. Its unparalleled combination of strength, elasticity, and biocompatibility has already influenced fields ranging from medicine to material sciences, offering solutions that were once thought impossible. Historically admired and scientifically decoded, spider silk has inspired ground-breaking advancement in biomedical applications such as sutures, artificial ligament, drug delivery, and even in tissue engineering, along with its industrial applications such as textile, biodegradable plastics, and adhesive. Despite several challenges in large-scale production and functional customization, synthetic biology is steadily advancing the ability to replicate spider silk using several vectors such as bacteria, yeast, and transgenic system. All these efforts make it a sustainable alternative to synthetic material science. Last but not the least, spider silk not only showcases natures advance engineering excellence but also serves as a blueprint for creating innovative, eco-friendly technologies that bridge the gap between biology and modern science.

## References

Alicea-Serrano AM, Onyak A, Dhinojwala A, Blackledge TA (2021) Robust performance of spider viscid silk on hairy and smooth insect substrates. Integr Comp Biol 61(4):1432–1439

Arndt T, Greco G, Schmuck B, Bunz J et al (2022) Engineered spider silk proteins for biomimetic spinning of fibers with toughness equal to dragline silks. Adv Funct Mater 32(23):2200986
Bakhshandeh B, Nateghi SS, Gazani MM et al (2021) A review on advances in the applications of spider silk in biomedical issues. Int J Biol Macromol 192:258–271
Baumgart L, Schaa EM, Menzel F et al (2022) Change of mechanical characteristics in spider silk capture threads after contact with prey. Acta Biomater 153:355–363
Branković M, Zivic F, Grujovic N, Stojadinovic I, Milenkovic S, Kotorcevic N (2024) Review of spider silk applications in biomedical and tissue engineering. Biomimetics 9(3):169
Brunetta L, Craig CL (2010) Spider silk: evolution and 400 million years of spinning, waiting, snagging, and mating. Yale University Press
Chang Z, Shen Y, Xue J et al (2023) Fabrication of spider silk-inspired bio-based polymeric materials under dynamic nanoconfinement as high-strong, ultra-tough, and multifunctional plastic substitutes. Chem Eng J 457:140984
Chen W, Liu K, Liao X et al (2024) Harmonizing thickness and permeability in bone tissue engineering: a novel silk fibroin membrane inspired by spider silk dynamics. Adv Mater 36(13):2310697
Dong Q, Cai J, Wang H et al (2020) Artificial ligament made from silk protein/Laponite hybrid fibers. Acta Biomater 106:102–113
Eisoldt L, Smith A, Scheibel T (2011) Decoding the secrets of spider silk. Mater Today 14(3):80–86
Feng JF, Wang T, Xu C et al (2025) Spider silk-inspired bio-based adhesive: robust adhesion strength, exceptional flame retardancy, and full recyclability. ACS Mater Lett 7(6):2031–2040
Franco AR, Fernandes EM, Rodrigues MT, Rodrigues FJ, Gomes ME, Leonor IB, Reis RL (2019) Antimicrobial coating of spider silk to prevent bacterial attachment on silk surgical sutures. Acta Biomater 99:236–246
Gellynck K, Verdonk PC, Van Nimmen E (2008) Silkworm and spider silk scaffolds for chondrocyte support. J Mater Sci Mater Med 19:3399–3409
Greco G, Mastellari V, Holland C et al (2021) Comparing modern and classical perspectives on spider silks and webs. Perspect Sci 29(2):133–156
Greco G, Wolff JO, Pugno NM (2020) Strong and tough silk for resilient attachment discs: the mechanical properties of piriform silk in the spider Cupiennius salei (Keyserling, 1877). Front Mater 7:138
Guessous G, Blake L, Bui A (2024) Disentangling the web: an interdisciplinary review on the potential and feasibility of spider silk bioproduction. ACS Biomater Sci Eng 10(9):5412–5438
Gupta V, Rastogi S (2023) Engineering spider genes for high-tensile silk and patenting challenges. J Adv Zool 44
Han SI, Alicea-Serrano AM, Blackledge TA (2023) Anchor threads can double the insect flight energy absorbed by spider orb webs. J Exp Biol 226(2):jeb245123
Heim M, Römer L, Scheibel T (2010) Hierarchical structures made of proteins. The complex architecture of spider webs and their constituent silk proteins. Chem Soc Rev 39(1):156–164
Hernando A, Saputri DHA, Tan MI et al (2021) Directing the chondrogenic differentiation of human Wharton's jelly mesenchymal stem cells using spider silk-based micropattern. In: AIP conference proceedings, vol 2346, No. 1. AIP Publishing
Hollsten L (2025) Wild spiders in fragile knowledge networks: spiders in medicine, natural history, and silk production in eighteenth-century Europe. Brill's series in the history of the environment 109
Hong YQ, Zhang XR, Wu LH et al (2025) Analysis of spider silk in loading-unloading cycles using Raman spectroscopy based on molecular bioinformatics of spidrion. Polymer 317:127910
Hopfe C, Ospina-Jara B, Schulze T et al (2024) Impact of environmental factors on spider silk properties. Curr Biol 34(1):56–67
Jiang Y, Han Y, Gong D (2025) Mimicking the hierarchical structure of spider silk: pseudoprotein nanofiber yarns with unprecedented toughness. Small 21(17):2412432
Kedir WM, Geletu AK, Weldegirum GS (2024) Spider web-reinforced chitosan/starch biopolymer for active biodegradable food packaging. Appl Food Res 4(2):100526

Khan A, Jawaid M (2022) Novel spider silk fiber for high-performance textiles and eco-friendly armor applications. J Nat Fibers 19(5):1879–1884
Khan AQ, Shafiq M, Li J et al (2023) Recent developments in artificial spider silk and functional gel fibers. SmartMat 4(6):e1189
Kumar D, Singh S, Kundapur RR et al (2023) Introduction and history of insect biotechnology. Introduction to insect biotechnology. Springer International Publishing, Cham, pp 1–15
Kundapur R, Kumar D, Singh S et al (2023) Insect metagenomics. Introduction to insect biotechnology. Springer International Publishing, Cham, pp 81–91
Land H, Christenson TE, Uhl G (2024) Sperm storage organs change with female age and mating history in the Golden Silk spider Trichonephila clavipes (Araneae: Araneidae). J Zool 323(1):19–28
Landreh M, Osterholz H, Chen G et al (2024) Liquid-liquid crystalline phase separation of spider silk proteins. Commun Chem 7(1):260
Lentz S, Trossmann VT, Borkner CB et al (2022) Structure-property relationship based on the amino acid composition of recombinant spider silk proteins for potential biomedical applications. ACS Appl Mater Interfaces 14(28):31751–31766
Li J, Chen Y, Zhang F et al (2022) Spider silk-inspired high-performance soybean meal-based adhesive reinforced by greenly produced chitosan-functionalized boron nitride nanosheets. Chem Eng J 438:135442
Li J, Li S, Huang J et al (2022) Spider silk-inspired artificial fibers. Adv Sci 9(5):2103965
Li X, Qi X, Cai YM (2021) Customized flagelliform spidroins form spider silk-like fibers at pH 8.0 with outstanding tensile strength. ACS Biomater Sci Eng 8(1):119–127
Liu Z, Chu Y, Wu Y et al (2024) Spider silk inspired strong yet tough composite hydrogels. Compos Sci Technol 252:110613
Liu X, Lianxin S, Xizi W, Bing D, Yupeng C, Shutao W (2021) Recent progress of spider-silk-inspired adhesive materials. ACS Mater Lett 3(10):1453–1467
Lu W, Kaplan DL, Buehler MJ (2024) Generative modeling, design, and analysis of spider silk protein sequences for enhanced mechanical properties. Adv Funct Mater 34(11):2311324
Malay AD, Craig HC, Chen J et al (2022) Complexity of spider dragline silk. Biomacromol 23(5):1827–1840
Neubauer VJ, Döbl A, Scheibel T (2021) Silk-based materials for hard tissue engineering. Materials 14(3):674
Perrotta BG, Kidd KA, Marcarelli AM et al (2025) Effects of chronic metal exposure and metamorphosis on the microbiomes of larval and adult insects and riparian spiders through the aquatic-riparian food web. Environ Pollut 371:125867
Rajput A, Upma SSK et al (2022) Advanced polymeric materials for aerospace applications. Aerosp Polym Mater 1:117–136
Rathour A, Gupte SS, Gupta D et al (2024) Modulatory potential of Bacopa monnieri against aflatoxin B1 induced biochemical, molecular and histological alterations in rats. Toxicol Res 13(2):tfae060
Rising A, Johansson J (2015) Toward spinning artificial spider silk. Nat Chem Biol 11(5):309–315
Ruan H, Li D (2024) Band gap characteristics of bionic acoustic metamaterials based on spider web. Eng Struct 308:118003
Sethy TR, Ahi J (2022) Spider silk and the silk of egg sacs with its astonishing concealed attributes: a review. J Nat Fibers 19(15):11492–11506
Singh S, Nirala SK, Kumar D et al (2024) Physicochemical characterization and in vitro biological activity of silver nanoparticles of propolis extract. BioNanoScience 14(3):2181–2196
Singh S, Nirala SK, Bhadauria M (2024) Comparative role of acetaminophen, carbon tetrachloride and thioacetamide in development of fibrosis in rats. Toxicol Res 13(1):p.tfad114
Strauß S, Diemer M, Bucan V et al (2024) Spider silk enhanced tissue engineering of cartilage tissue: Approach of a novel bioreactor model using adipose derived stromal cells. J Appl Biomater Funct Mater 22:22808000241226656

Wang B, Li H, Chen Y, Chen Z, Li P, Zhang X, Lin X (2025) Preparation and characterization of a iRGD-modified recombinant spider silk particles for antitumor polypeptide drug delivery into cancer cells. BMC Biotechnol 25(1):1–15

Wang M, Du Z, Tong A et al (2025) Silkworm silk can become a high-strength and super-toughness spider-silk-like fiber: multipathway engineering strategies and applications. Small Struct 2400639

Whittall DR, Baker KV, Breitling R et al (2021) Host systems for the production of recombinant spider silk. Trends Biotechnol 39(6):560–573

Wijerathna T, Wolff JO, Schneider JM (2025) Structure, properties, and functional diversity of spider aciniform silk. J Arachnol 52(3):266–273

Wu D, Li X, Zhang Y et al (2024) Novel biomimetic "Spider Web" robust, super-contractile liquid crystal elastomer active yarn soft actuator. Adv Sci 11(17):2400557

Wu X, Chen L (2022) New textile material—spider silk fiber. In: Journal of physics: conference series, vol 2256, No. 1. IOP Publishing, p 012003

Xia C (2024) Engineering chitosan-silk fibroin laminates for use as strong, tough, and biodegradable alternatives to plastic packaging

Yarger JL, Cherry BR, Van Der Vaart A (2018) Uncovering the structure–function relationship in spider silk. Nat Rev Mater 3(3):1–11

Zhao Y, Ming J, Cai S et al (2024) One-step fabrication of polylactic acid (PLA) nanofibrous membranes with spider-web-like structure for high-efficiency PM0. 3 capture. J Hazard Mater 465:133232

Zou J, Chen X, Song B et al (2024) Bionic spider web flexible strain sensor based on CF-L and machine learning. ACS Appl Mater Interfaces 16(18):23761–23770

# Chapter 10
# Unravelling the Mysteries of Spider Silk: A Path to Biomedical Breakthroughs

**Abhijit Mukherjee and Asmita Pramanik**

## 10.1 Introduction

Spiders are some of the oldest creatures on Earth, having been around for hundreds of millions of years. They have the ability to survive in any environment. They belong to the phylum Arthropod and possess a unique talent for spinning silk, one of nature's remarkable materials (Platnick 2020). With over 70,000 species, spiders have mastered the art of silk-making, using it for everything from catching prey and building nests to reproduction and survival. They are nature's ultimate silk weavers, showcasing remarkable skill in creating and using this incredible material (Wu and Chen 2022). Their silk is incredibly essential to their survival, helping them catch prey and escape predators.

It is found that Darwin's Bark Spider (*Caerostris darwini*), (Fig. 10.1) a member of the family *Araneidae* and native to Madagascar, produces the strongest Dragline silk whose origin is the MA gland. Almost all 41,000 + spider species worldwide spin dragline silk using their major ampullate silk glands (Brunetta and Craig 2010). According to the researchers, the extremely tough silk's evolution is to support the large orb webs it creates by spinning single threads that can span up to 25 m long, enough to stretch across a narrow river. Spider silk is difficult to harvest, but ways have been found to use the webs of orb-weaving spiders (*Araneidae*) for centuries. From fishing to treating wounds, these delicate yet incredibly strong webs have served as natural tools, valued for their durability and healing properties (Eisoldt et al. 2011).

A. Mukherjee (✉) · A. Pramanik
Department of Fashion Technology, National Institute of Fashion Technology, Kolkata, West Bengal, India
e-mail: abhijit.mukherjee@nift.ac.in

A. Pramanik
e-mail: asmita.pramanik1@nift.ac.in



D. Kumar and M. Shahid (eds.), *Science and Technology of Silk*, Textile Science and Clothing Technology, https://doi.org/10.1007/978-3-032-16230-4_10

**Fig. 10.1** Darwin's bark spider [Reprinted from (Wu and Chen 2022)]

Darwin's spider silk (Fig. 10.2) is an incredibly tough silk, with an average tensile strength of 350 $MJ/m^3$. Few strands even reach 520 $MJ/m^3$. That makes it the strongest silk ever discovered—more than 10 times tougher than Kevlar (a strong, heat-resistant synthetic fibre). Naturally, it is one of the most incredible materials, attracting more and more scientists because of its strength, flexibility, and durability. Since it is made from natural proteins, it is a perfect example of how nature creates strong yet lightweight materials (Gu et al. 2020). What makes spider silk so fascinating is its amazing natural properties. It is incredibly strong and tough, conducts heat well and has unique abilities, such as contracting when wet and twisting in response to certain forces (Blackledge 2012). Unlike most materials that are either strong or flexible, spider silk is both tough and stretchy, resilient and lightweight (Belbéoch et al. 2021). These qualities have led scientists to explore their applications, from textiles to exceptional medical treatments like artificial ligaments and tendons. These strong and tough fibres also promise applications in areas such as cables, construction materials, ordinary clothing (civilian apparel) and fibre-reinforced composites designed for energy management (absorption and conservation) (Saravanan 2006).

It is a perfect example of how nature inspires innovation (Blackledge 2012). There are very limited ways of naturally collecting spider silk. Hence, the process of creating recombinant spider silk protein has begun (Brankovi et al. 2024).

From this review, the construction, features, and application of spider silk, mainly in the biomedical field, can be easily understood. It also highlights advancements in creating artificial fibres with similar strength and toughness. These efforts use recombinant spidroins, non-spidroin proteins, polypeptides, and synthetic polymers. Additionally, we address challenges in the field and discuss the outlook for future research.

**Fig. 10.2** Spider silk [Reprinted from (Wu and Chen 2022)]

## 10.2 Types of Spider Silk

The large and diverse superfamily *Araneoidea* is created by the female orb web-creating spiders, which produce seven distinct types of silk, each created by specialised silk glands (Fig. 10.3). Among these,

- Major ampullate (MA) silk stands out for its exceptional mechanical properties, combining remarkable strength with high extensibility. This silk makes the framework and radii of orb webs and protects almost all spiders.

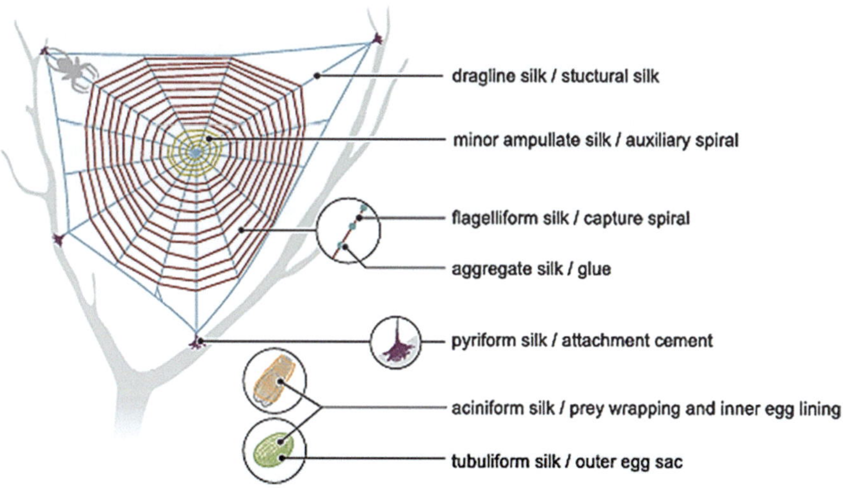


**Fig. 10.3** Schematic diagram for different types of spider silk and their role in spider webs [Reprinted from (Ramezaniaghdam et al. 2022)]

- Minor ampullate (MiA) silk, on the other hand, plays a different role. Orb web spiders use it as a temporary spiral during web construction, while cobweb spiders employ it to wrap their prey. Every silk is uniquely extruded in such a way that caters to the particular needs of the lifestyle and environment of that spider.
- The spirals of orb webs are captured by the Flagelliform silk, forming stretchy axial threads. Though it is weaker (less than half) than MA silk, it is about seven times more elastic, making it perfect for catching prey. To enhance its function, this silk is coated with a sticky, glue-like aggregate that helps trap anything that flies into the web.
- Pyriform silk acts like a natural adhesive, anchoring draglines and web frames to surfaces. When the web is constructed, it even binds the threads of the silks together.
- Aciniform silk, which is about as strong as MA silk but much stiffer, serves several roles, such as wrapping prey, forming the outer covering of egg capsules, and sometimes even decorating the orb webs.
- Tubuliform (or cylindriform) silk is specially designed to protect spider eggs, forming the tough outer shell of egg sacs (Eisoldt et al. 2011).

- Major ampullate spidroin 1, also known as MaSp1, and Major ampullate spidroin 2, better known as MaSp2, are encoded by multiple similar genes, which scientists have sequenced for species such as (a) Western black widow (*Latrodectus hesperus*) and (b) Golden orb-weaver (*Nephila clavipes*).

These gene sequences serve as models to estimate the protein composition of MA silk in other spider species. Interestingly, MaSp2-like genes have only been found in orb-web spiders, suggesting that MaSp2 likely evolved from a progenitor form of MaSp1 specific to this group. Both MaSp1 and MaSp2 have regions with highly repetitive amino acid sequences, as well as conserved non-repetitive regions (comprising approx. 100 amino acids) at their ends (C-terminal and N-terminal domains) (Wen et al. 2023). The N-terminal domain indicates the initiation of a protein or polypeptide, which is terminated by an amino acid having a free amine group (−NH2), while the C-terminal domain refers to the end of an amino acid chain (protein or polypeptide), which is terminated by a free carboxyl group (−COOH). The intermolecular disulphide bonds help stabilise these protein structures under the direct influence of the cysteine residues. These terminal domains play a significant role in assisting proteins in folding into their specific secondary structures, a factor that is key to silk's remarkable properties.

### *10.2.1 MA Silk*

MA silk, or dragline silk, is the most researched as it is extremely strong and tough. It forms the main framework and spokes of the spider's web, called the orb web, which makes it crucial for stability and survival (Salehi et al. 2020). This silk is

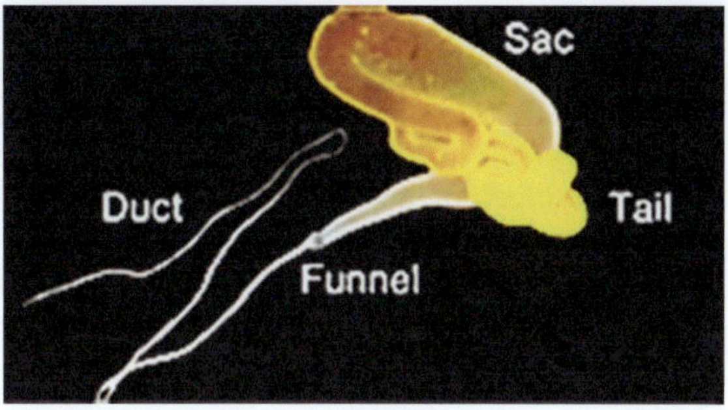


**Fig. 10.4** Four regions of the MA gland [Reprinted from (Gomes and Salgueiro 2022)]

three times stronger than aramid fibres, five times stronger than steel, and twice as flexible as nylon. Therefore, spider silk is considered to be nature's most impressive structural material (Gu et al. 2020).

MA silk is a complex and fascinating material composed of several layers. On the outside, it has a lipid-rich coating and a glycoprotein-rich skin that covers a fibrous inner and outer core. Within the major ampullate (MA) gland, we distinguish four regions (Fig. 10.4) (Brankovi et al. 2024), which are as follows:

- Tail zone
- Lumen (bag)
- Fibre alignment channel
- Output

Two key proteins, namely, MaSp1, which is virtually proline-free, and MaSp2, the proline-rich major ampullate spidroins, make up the core (Salehi et al. 2020). The organised regions of the proteins correspond to the crystalline areas, whereas the less-organised areas are the non-crystalline areas, or amorphous areas (Gomes and Salgueiro 2022). Major ampullate silk is incredibly strong, comparable to Kevlar, and also has some flexibility. In all known orb-weaving spiders, this silk has unique properties from a mix of two key proteins, MaSp1 and MaSp2. However, the exact balance varies among these proteins (Brooks et al. 2008).

The two silk proteins are built from four key amino acid sequences that work together to create their extraordinary properties, such as follows:

- **Strength**: The **(GA)n** (Glycine-Alanine repeats; where *n* represents any number) motif forms rigid $\beta$-sheet structures, imparting the silk its impressive strength.
- **Elasticity**: The stackable $\beta$-turns created by the **GPGXX** motif [Glycine-Proline-Glycine-X-X (where X represents a variable amino acid)] make silk stretchy and flexible.
- **Stability**: The **GGX** [Glycine-Glycine-X (where X can be different amino acids)] regions form glycine II helices, which help stabilise the silk fibre by interacting with each other (Brankovi' et al. 2024).

These repeating sequences give spider silk its unique combination of strength, flexibility, and durability (Bratzel and Buehler 2012). The pattern of alternating polyalanine and glycine regions comprises crystalline and amorphous regions. In major ampullate silk, the structural proteins MaSp1 and MaSp2 arrange themselves into tight $\beta$-sheet crystallites compared to more flexible non-crystalline regions, forming structures like 310 helices, $\beta$-turns or, $\beta$-spirals determined by their amino acid composition (Wen et al. 2023). This allows the proteins to interlock, almost like a biological zipper, with their molecular side chains fitting together perfectly. Hydrogen bonds and van der Waals forces further strengthen this structure, making it highly water-resistant, a crucial feature for spiders relying on silk for survival (Brooks et al. 2008).

### *10.2.2 Orb Web*

Orb webs are the most studied webs, consisting of seven varieties of silk, as observed in the *Araneus diadematus* species of the European garden spider (Römer and Scheibel 2008).

The female orb-weaving spiders spin their webs all over the world. The sizes range from just a few centimetres to several metres across. Scientists recently discovered spiders that can spin single threads stretching an astonishing 25 m across rivers in Madagascar. Since evolution, these webs have absorbed the impact of large prey despite the silk being incredibly thin. This has taken place over millions of years. At first glance, every thread in an orb web might look similar, though the webs are made from different varieties of silk, each with its own special purpose (Eisoldt et al. 2011).

The strong, slightly stiff silk that constructs the frame and spokes of an orb web gives it structure and support. This silk comes from the special gland called the MA gland in the spider's body, which comprises two types of proteins, hence the name MA silk (Blackledge 2012). The spiders also use this same silk as a safety line, keeping it always within reach for a quick escape from predators. Because they constantly trail it behind them, it is often called 'dragline silk' (Römer and Scheibel 2008). The framework and radii of orb webs are made from this silk, which acts as safety lines for the spiders as they navigate their surroundings. The dragline fibres can be harvested from the orb webs or by milking (silking forcefully) spiders. Flagelliform silk is responsible for forming the capture spiral in the orb web, serving as a clever trap for flying prey (Eisoldt et al. 2011). The design of the orb webs is to catch flying insects, which imposes two unique demands on the silk. First, silk must withstand the high-speed impact of insects crashing into the web. Second, it must be sticky enough to keep them from escaping once caught. Over time, these challenges have shaped the evolution of spider silk, making it incredibly tough and highly adhesive (Blackledge 2012). Flag silk is incredibly stretchy—able to stretch up to 300% and is designed to absorb the impact of prey without breaking. Imagine a honeybee weighing about 120 mg and flying at a speed of 3.1 m per second, crashing into a web. That tiny insect hits with a surprising amount of force—around 0.55 millijoules of energy. Yet, despite being just one to five micrometres thick, flag silk can easily handle that impact. This impressive flexibility and strength make spider webs so effective at catching prey, sometimes even creatures larger than the spider species.

In addition to MA and Flagelliform silk, two other varieties of silk are used by the orb-weaving spiders to build their webs. One comes from the minor ampullate gland and creates a temporary spiral that helps shape and stabilise the web before adding the sticky capture spiral. Another essential component is a unique silk 'glue' made in the pyriform gland. This acts like a natural cement, securing the web's frame to surfaces like trees, bushes, or walls and forming strong connections between different silk threads, keeping the web sturdy and functional. It was recently found that wet glue consists of small peptides that function as metal chelators. The metal ions in spider silk help prevent bacteria and other microbes from growing. This could be important for keeping the web intact and protecting developing spider eggs (Römer and Scheibel 2008).

While orb webs may help with courtship, temperature regulation, and even predator defence, their main job is catching flying insects. This process happens in three key steps: first, the web intercepts the insect, then absorbs the impact to stop it, and finally, it holds onto the prey until the spider can subdue it. The effectiveness of each step depends on the web's design and the unique properties of its silk (Blackledge 2012).

## 10.3 Mechanical Properties of the Silk

Spider silk is amazingly strong and flexible (Wu and Chen 2022), and its secret lies in tiny nano-crystals hidden within its structure. These microscopic crystals are perfectly arranged and evenly spread throughout the silk, reinforcing it like nature's high-tech fibre. This clever design gives spider silk incredible toughness, elasticity, and durability (as shown in Fig. 10.5), rendering it one of nature's most impressive materials (Gu et al. 2020).

Spider silk is well known for being incredibly strong with a high modulus value and toughness. The dragline silk that spiders use—often for their webs or lifelines—is not only stronger but also more stretchable than silkworm silk. In fact, it is even tougher than many synthetic fibres, such as nylon (6,6), Kevlar, or carbon fibres. All the varieties of spider silks have a variety of functions and mechanical properties (Fig. 10.6), and their synergism is crucial for a spider's survival (Li et al. 2022).

### *10.3.1 Strength and Durability*

Despite being incredibly lightweight, spider silk is stronger than steel of the same thickness. Its tensile strength of around 1.1 GPa surpasses high-performance nylon

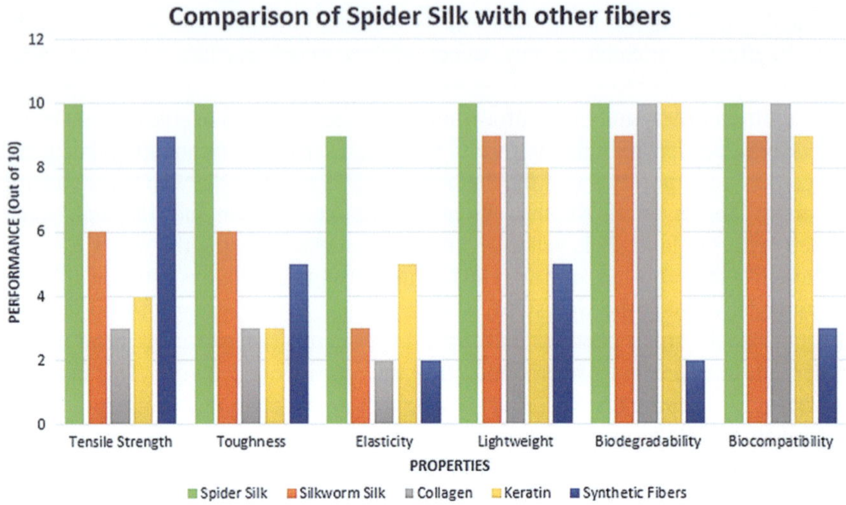


**Fig. 10.5** A comparison of spider silk's mechanical properties with those of other fibres

(0.9 GPa) and industrial polyester (0.5 GPa). This means it can endure heavy loads and external impacts far better than most man-made fibres.

## *10.3.2 Heat and Cold Resistance*

Unlike many natural fibres, spider silk holds up remarkably well under extreme temperatures. It stays stable up to 200 °C and only begins to yellow at 300 °C, whereas silk degrades at 140 °C. On the other end of the spectrum, spider silk remains flexible and elastic even at − 40 °C, while materials like wool, silk, and synthetic fibres become brittle and lose their stretch.

## *10.3.3 Lightweight and Smooth*

Spider silk is not just strong—it is also incredibly light and smooth. With a density of 1.2–1.35 g/cm$^3$, it feels soft to the touch and glides effortlessly against the skin. In fact, its friction coefficient is lower than that of silk, making it one of the most comfortable natural fibres.

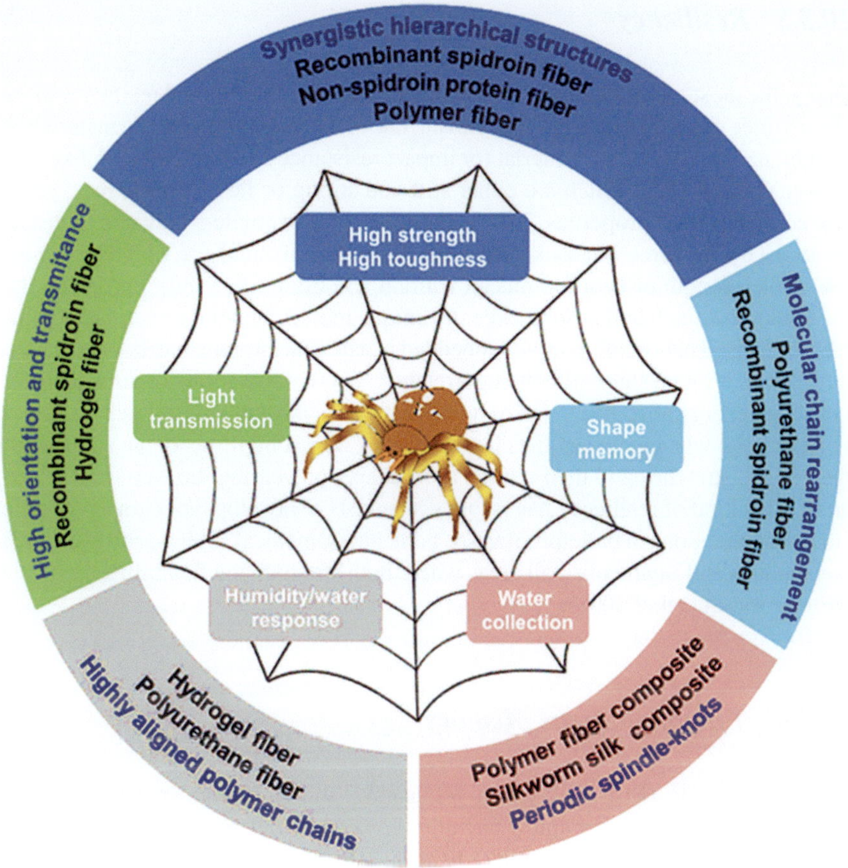


**Fig. 10.6** Properties of spider silk and the artificial fibres inspired by them [Reprinted from (Li et al. 2022)]

## *10.3.4 Eco-Friendly and Biodegradable*

What makes spider silk even more remarkable is its sustainability. It is a natural protein fibre that resists water, dilute acids and alkalis, but it is biodegradable, meaning it decomposes naturally without harming the environment. Unlike synthetic materials, which contribute to pollution, spider silk offers a green alternative for future textiles and materials (Wu and Chen 2022).

### 10.3.5 Resiliency

One of the most remarkable qualities of spider silk is its exceptional resiliency. It can absorb three times more energy than Kevlar, one of the toughest synthetic materials by weight, making it an ideal material for impact resistance and durability. Unlike most man-made materials, which are either stiff and strong or flexible and weak, spider silk combines both properties. Take carbon fibre, for example—they can withstand up to 4 GPa of force but lack elasticity. They snap instantly if stretched beyond 1% of their length or bent too much (Vollrath and Edmonds 1989). Spider silk has the ability to stretch and absorb force. Its unique molecular structure—made of tiny, well-organised nanoscale crystals embedded in a flexible protein matrix—allows it to return to its original shape after being stretched. This combination of toughness, elasticity, and energy absorption makes it one of nature's most efficient shock absorbers. For example, when an insect flies into a spider's web at high speed, the silk absorbs the impact, stretches and then returns to its shape, preventing the web from tearing apart. This kind of resilience has inspired scientists to develop bioengineered spider silk for applications in bulletproof vests, parachutes, medical sutures, and even artificial tendons and ligaments—all areas where both strength and flexibility are crucial (Römer and Scheibel 2008).

### 10.3.6 Torsional Shape Memory

Major Ampullate (MA) silk has a fascinating ability to 'remember' its shape, preventing a spider from spinning uncontrollably while it dangles from a silk thread. This built-in torsional shape memory helps the spider stay stable during its descent. The more remarkable part is that the MA silk naturally returns to its original position without any extra stimulus. Unlike other materials that might keep spinning or swinging after being twisted, MA silk barely oscillates thanks to its high damping ability, which absorbs movement and stabilises quickly (Blamires et al. 2016). This ensures a smooth and controlled descent, helping spiders easily navigate their environment.

### 10.3.7 Super Contraction

Spider silk has a fascinating ability to shrink and tighten when it comes into contact with water, a phenomenon known as super-contraction. This built-in feature helps keep a spider's web firm and stable, even as it faces changes in humidity and weather (Römer and Scheibel 2008). Essentially, when the silk absorbs moisture, it pulls itself together, preventing the web from sagging or losing shape. Scientists believe this happens because silk proteins are arranged in a way that allows the fibres to naturally adjust and remain strong over time (Gu et al. 2020).

## 10.4 Comparison Between Natural Silk and Spider Silk

People often compare spider silk to silkworm silk—especially the silk made from *Bombyx mori*. Unlike spider silk, which is challenging to harvest, silkworm silk is produced in a well-established, cost-effective process that has remained largely unchanged for centuries. It is collected by unravelling the cocoons of silkworm pupae, making large-scale production relatively easy (Hakimi et al. 2007). In comparison, spider silk is much harder to obtain. It can only be collected manually by carefully drawing the silk thread from a restrained spider's silk gland. However, this method applies only to MA silk, the type used for web draglines, which is both time-consuming and expensive. Additionally, farming spiders is not a viable option because most species are cannibalistic, making large-scale production difficult and costly.

The differences between spider and silkworm silk go beyond just production methods. At a molecular level, silkworm silk is made of sericin proteins, which are not present in spider silk (Lewis 1992). The key structural proteins that create silk fibres differ as well. In silkworm silk, these proteins called fibroins—consist of both light and heavy chains, whereas spider silk proteins, known as spidroins, follow a different structural pattern. These molecular differences translate into distinct mechanical properties (Huet al. 2006). Silkworm silk is generally weaker and less elastic than spider silk. Interestingly, silkworm silk can be spun to be either strong or stretchy, but not both at the same time.

In contrast, spider silk naturally combines both strength and flexibility, making it one of the toughest natural fibres. While spinning conditions influence the final properties of both types of silk, the real difference lies in the proteins that make up the fibres (Römer and Scheibel 2008). Spider silk is biocompatible. The main component of spider silk, a protein in nature, is non-toxic, making it biocompatible (Gu et al. 2020). Because of spider silk's impressive properties, scientists have long sought ways to produce it artificially through genetic engineering. By replicating the silk-producing genes and inserting them into bacteria, yeast, or other host organisms, researchers aim to create spider silk without the need for actual spiders, paving the way for its large-scale use in various applications (Fig. 10.6) (Römer and Scheibel 2008).

## 10.5 The Spider Silk Structure

Spider silk is made from complex protein structures composed of secondary, tertiary, and even quaternary structures, which are discussed below.

### 10.5.1 Primary Structure

Spider silk is mostly composed of proteins that are rich in nonpolar, water-repelling amino acids, such as glycine and alanine, with little to no tryptophan. Compared to typical cellular enzymes, the amino acid makeup of silk proteins is quite unusual (Heim et al. 2010). One of the standout features of these proteins (amino acids) is their highly repetitive protein sequences (Fig. 10.7), particularly in the larger section of the protein architecture, which is crucial in giving spider silk its special mix of strength and flexibility (Lindberg 2024).

Spider silk proteins are fascinatingly repetitive, with these sequences accounting for often more than 90% of the proteins. These repetitive segments, each made of 10 to 50 building blocks of amino acids, can show up more than 100 times in just one protein chain. Each repeat has a specific useful role, conferring the spider silk its incredible mechanical properties (Koh et al. 2015).

Major Ampullate (MA) and Flagelliform silks, for example, include up to four main types of repeating motifs:

- **(GA)n/(A)n [Repeating Sequence of Glycine and Alanine]**: Forms α-helices in solution and β-sheet structures in fibres.
- **GPGGX/GPGQQ [Repeating Glycine-Proline-Glycine-X Motifs]**: It likely adopts a rubber-like amorphous structure or possibly a 31-helical structure.
- **GGX [Repeating Unit Where Glycine-Glycine Is Followed by a Variable Amino Acid] (Where X Can Be Alanine, Serine or Tyrosine):** Similar to GPGGX, it may form amorphous or helical structures.
- **Spacer Sequences**: These include charged amino acids, adding variety and flexibility to the silk's composition (Greving et al. 2020).

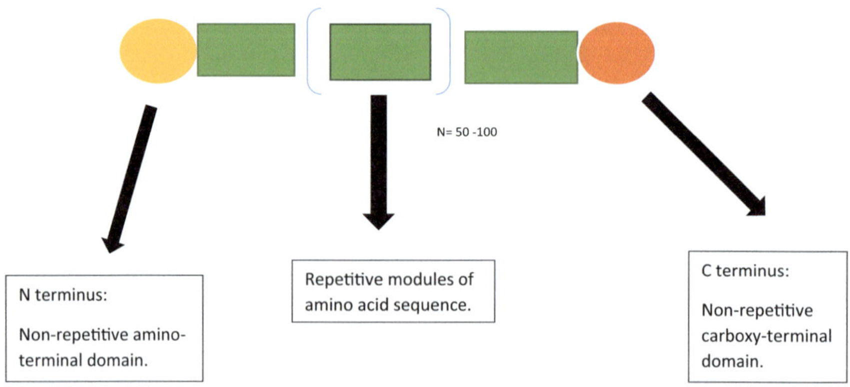


**Fig. 10.7** Illustrated image of a spidroin composed of up to a hundred repetitive modules (green) and two terminal domains (yellow and orange). The components are not proportional to scale [Reprinted and modified from (Eisoldt et al. 2011)]

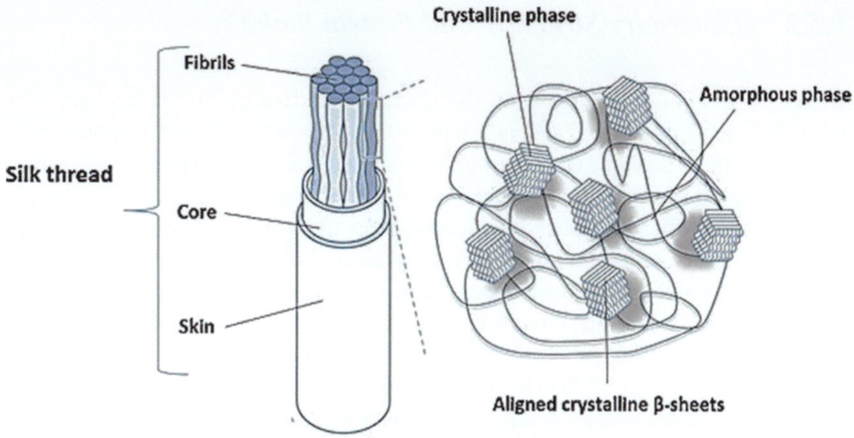


**Fig. 10.8** Structural organisation of spider silk thread [Reprinted from (Brankovi´ et al. 2024)]

Spider silk has a semi-crystalline microstructure (Fig. 10.8) and a fascinatingly repetitive protein architecture, with these sequences often comprising more than 90% of amino acid chains. These repetitive stretches of proteins, typically 10–50 units long, may appear more than 100 times in a single protein chain. Each repeated sequence plays specific functional roles, contributing to the incredible mechanical properties of the silk (Qiu et al. 2019).

Flagelliform silk, rich in GPGGX and GGX motifs, tends to form β-turn structures that stack into a right-handed *β*-spiral helix.

In addition to these repetitive regions, silk proteins also have non-repetitive terminal domains at their ends, consisting of about 100 amino acids. These terminal regions play a crucial role in assembling silk proteins into fibres. Unlike the repetitive core, these domains have well-defined structures. They can form disulphide bonds (-S–S-) under certain conditions, stabilising the protein dimers and multimers necessary for silk formation (Gosline et al. 1999).

Interestingly, despite the wide variety of spider species, the terminal segment of silk protein chains is surprisingly similar. For instance, the full-length segment (sequence) of Flagelliform silk was first identified in *Nephila clavipes* (golden orb-weaver), while the entire sequence of MA silk was only recently reported for the black widow spider. Unlike Flagelliform silk genes, major Ampullate silk genes do not contain introns, comprising exceptionally large exons of over 9000 base pairs. This might result from gene duplication during evolution and may help maintain high silk production levels throughout a spider's life (Sun et al. 2020).

Spider silk proteins exhibit a distinctive pattern of hydrophobicity, with alternating hydrophilic (MaSp2) and hydrophobic (MaSp1) regions—in the core architecture. This amphiphilic structure resembles surfactants or biological membranes and is thought to play a crucial role in the spinning process (Tokareva et al. 2014). These properties allow silk proteins to phase-separate during spinning and may even help form micelle-like intermediate structures during thread assembly.

### *10.5.2 Quaternary Structure and Protein Stability*

The secretion of silk proteins from the spider's glands starts in an aqueous solution with little to no defined secondary or tertiary structure (Jin and Kaplan 2003). However, as these proteins travel through the spinning duct, the repetitive sequences in their core domains allow numerous weak interactions to form between neighbouring protein chains. These interactions help the silk develop more complex secondary, tertiary, and even quaternary structures.

Using X-ray diffraction, researchers have discovered that the final structure of major Ampullate (MA) silk threads contains regions of high electron density interspersed with regions of low electron density. The dense regions are said to include crystalline sub-structures rich in β-sheets, which give them incredible strength. Meanwhile, the less-dense areas are amorphous, with loosely organised structures, such as hydrogels. The silk stretches and demonstrate elastic properties due to these amorphous regions (Dicko et al. 2004).

The proportion of crystalline and hydrogel-like regions differ across silk types, indicating the specific roles of the silk proteins. For instance, MA silk is highly concentrated in $\beta$-sheet crystalline structures, making it strong and durable, which is essential for building the web frame. In contrast, Flagelliform silk, which needs to be highly flexible, is made almost entirely of amorphous, hydrogel-like regions. The unique properties of spider silk are explored to understand how its structure correlates with its extraordinary properties and to apply them in various fields (Xu et al. 2014).

## 10.6 Production of Spider Silk

Spiders are cold-blooded creatures (poikilotherms), so they spin silk with elevated strength and toughness (Liu et al. 2023). The temperature of the environment affects the mechanical and structural properties, which, in turn, also affects the web-spinning.

### *10.6.1 Natural Spinning Process*

The spider silk development process happens in the following four main stages inside the MA gland:

- **Tail Zone**: This is where the silk proteins are made and released.
- **Sac/Ampulla (Storage Area)**: The proteins are stored here until they are needed for silk production (Andersson et al. 2013).

As the silk fibre develops, additional compounds are incorporated to form its outer layer. Within the ampulla, the protein concentration can reach an impressive 50%

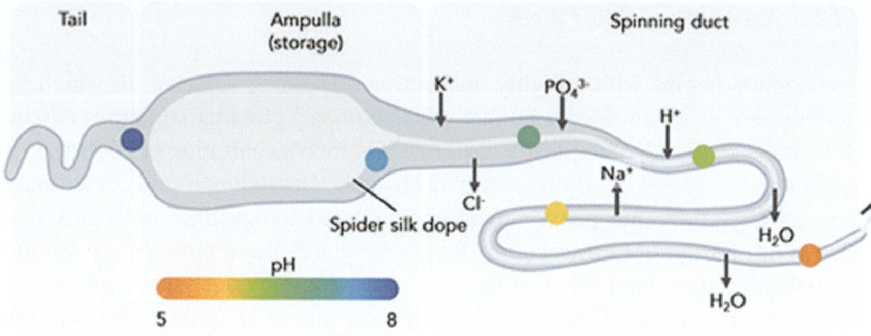


**Fig. 10.9** Spinning of silk in the spider [Reprinted from (Whittall et al. 2021)]

(w/v), which is unusually high, as most proteins would clump irreversibly under such conditions. However, spider silk proteins, known as spidroins, avoid this by assembling into micelle-like structures. This organisation is guided by its unique amphiphilic nature and specialised non-repetitive terminal regions, which help regulate the formation of these molecular structures. The stored silk proteins, known as the spinning dope (as shown in Fig. 10.9), exist in a liquid-crystalline state, ensuring a smooth and controlled transition from liquid to solid silk when spun.

- **Fibre Alignment Channel**: Proteins are carefully arranged to give silk remarkable strength and flexibility.

As the liquid silk moves through an **S-shaped duct**, it is gradually stretched, helping the silk proteins align in an orderly fashion. This natural pre-alignment allows the molecules to flow smoothly and organise even further along the way. As the duct narrows, the increasing pressure and friction trigger the formation of ***β*-sheet crystals**, which are key to the silk's impressive strength.

- **Spinning Outlet**: As the liquid silk comes into contact with the outside air, it solidifies into a strong, durable fibre (Liu et al. 2023).

In the final stage of silk formation, the liquid silk is gently pulled through a tiny opening called the '**spigot**', where the tension increases even more. This pulling action nudges the silk proteins into their final structure, exposing their water-repelling regions. At the same time, the spider's body actively removes excess water, helping the proteins solidify. This natural process transforms liquid silk into a remarkably strong and flexible fibre, making spider silk one of the toughest natural materials (Andersson et al. 2013).

As the silk proteins move through the spinning process, their terminal regions shift, rearranging the core structures within their micelle-like formations. This, combined with the mechanical forces at play, helps separate the proteins from the liquid solution, allowing them to organise and solidify into the final silk fibre. This finely tuned process enables spiders to produce silk with incredible properties, making it one of nature's most fascinating materials.

#### 10.6.1.1 Assembly of Silk Fibres

Silk production begins with a highly concentrated spinning solution, in which the silk proteins are in an unfolded state rather than compact globular structures (Rising and Johansson 2015). To maintain this extreme protein concentration of 50% (w/v) in the silk gland, the spider relies on several mechanisms, including liquid crystallinity, surface glycosylation, and phase separation triggered by specific molecules such as polyols or phospholipid surfactants. Inside the gland, these proteins remain in a random, flexible state, but as they move through the spinning duct, they rapidly self-assemble into an organised, water-resistant fibre (Heim et al. 2009). This transformation is carefully regulated, requiring precise protein folding and accurate control of environmental conditions such as pH, ion concentration, and water content. This ensures that the spider silk solidifies with its remarkable strength and durability. Spider silk formation is a delicate and highly controlled process (Kronqvist et al. 2017).

#### 10.6.1.2 Phase Transition in the Spinning Duct

The formation of spider silk is a delicate and highly controlled process. The silk proteins begin to stretch, align and organise precisely when they start travelling through the spinning duct. The hydrophobic (water-repelling) regions, mainly composed of polyalanine sequences, naturally come together, guided by both the proteins' structure and the forces within the duct (Rising and Johansson 2015). As these regions align, they begin to form strong $\beta$-sheet structures, held together by countless hydrogen bonds. This intricate process transforms a liquid protein solution into solid silk, giving it remarkable strength, flexibility, and durability—qualities that produce spider silk as one of the most extraordinary materials in nature.

The spider silk is formed when silk proteins are precisely aligned as they move through the spinning duct (Pinto et al. 2016). These proteins, either as single molecules or linked by disulphide (-S–S-) bonds, exist in a highly concentrated state, exhibiting a liquid-crystalline-like behaviour. This ordered state helps neighbouring molecules form strong interactions, such as van der Waals forces and hydrogen bonds, which are essential for silk's strength. As silk proteins lose water, they undergo structural changes, forming a fibre (Rising and Johansson 2015).

Another perspective suggests that the proteins first assemble into tiny micelles—small spherical structures around 100–200 nm in diameter —due to their unique water-attracting and water-repelling properties. These micelles then cluster into larger globules, which are then stretched and elongated by the forces within the spinning duct, ultimately transforming into silk fibres. This careful orchestration of molecular interactions and mechanical forces enables spiders to produce one of nature's strongest and most versatile materials (Römer and Scheibel 2008).

### *10.6.2 Recombinant Production*

Advancements in biotechnology have enabled the production of spider silk proteins to be more efficient and scalable, using bacteria, plants, animal cells, yeast, and even genetically modified animal sources (Vendrely and Scheibel 2007). This method offers a more stable and cost-effective approach than harvesting silk directly from spiders. This biotechnological approach not only enables large-scale production but also opens the doors to customising silk properties for medical, industrial, and textile applications (Guessous et al. 2024).

Scientists use different host organisms to produce spider silk proteins, each with its own benefits and challenges. Some of the most common hosts include bacteria such as *E. coli*, yeast species such as *Pichia pastoris* and even mammalian cells such as hamster kidney cells. Insects like *Spodoptera Frugiperda* have also been tested. Among these, *E. coli* is a favourite because it grows quickly, is easy to work with, and keeps production costs low (Makirita 2020). Advancements in biotechnology have helped researchers modify and enhance synthetic spider silk. However, several hurdles remain. One of the biggest challenges is producing spider silk proteins at their natural size (Kronqvist et al. 2017). Genetically modified mammals and insects can produce larger, more easily purified proteins, but maintaining these organisms is expensive, and their silk output is often lower.

Alternatively, while *E. coli* is cheap and efficient; its inability to produce full-length silk is its major disadvantage. Despite significant progress, the large-scale production of synthetic spider silk is an ongoing endeavour.

Unlike silkworms, which can be easily farmed, most spiders are cannibalistic, making spider farming impractical (Edlund 2016; Cardoso et al. 2024). Collecting silk from webs is not a great solution either—it is slow, inefficient, and only works for certain types of silk, like dragline silk. Because of these limitations, scientists have turned to biotechnology as the best method for the mass production of spider silk (Glissovic and Volrath 2010).

The process begins by taking silk-producing genes from spiders and inserting them into hosts such as bacteria, yeast, and mammalian cells that can function as tiny silk factories (Fig. 10.10). *E. coli* is the most common choice because it grows fast and can start producing silk proteins within a few days (Collins et al. 2013). However, spider DNA does not naturally function well in bacteria, so researchers tweak its genetic code to make it more compatible with *E. coli*, thereby improving production efficiency (Bhattacharyya et al. 2021).

Spider silk proteins have a repetitive structure comprising smaller amino acid sequences (such as polyalanine). While these sequences vary across different kinds of spider genera and their unique output types, they follow a predictable pattern within each specific silk type. By analysing these patterns, scientists can create synthetic versions that closely mimic real spider silk—or even design new silk materials with improved properties. So far, researchers have successfully engineered silk proteins based on well-known spidroins, such as MaSp1 and MaSp2. However, a significant challenge remains that artificial silk proteins are often much smaller than their

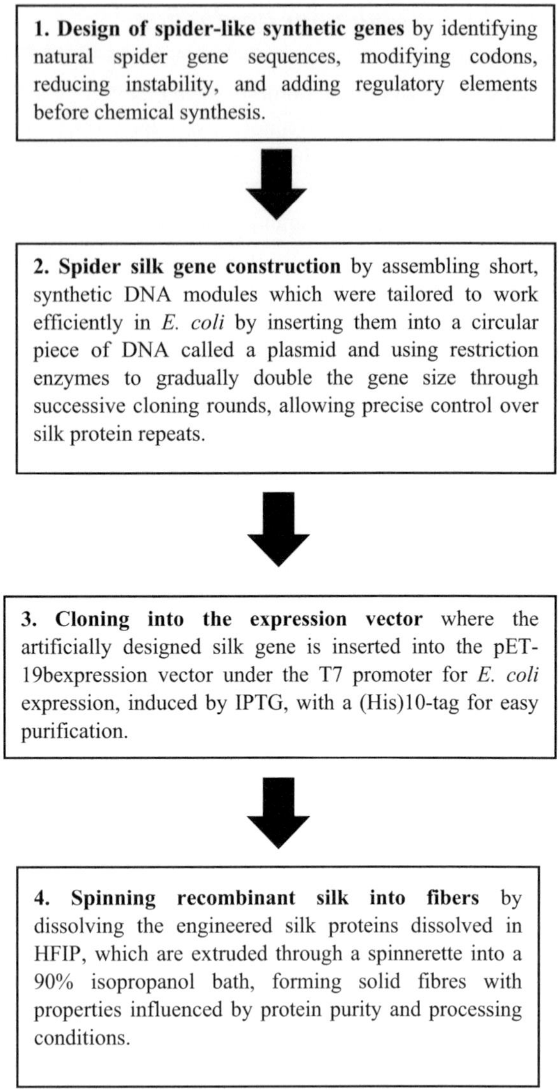


**Fig. 10.10** Flowchart showing the recombinant silk production process [Reproduced from (Teulé et al. 2009)]

natural counterparts. While real spider silk proteins are massive (300–350 kDa), most lab-produced versions are only 30–110 kDa (Guessous et al. 2024). This size difference makes it harder to replicate silk's strength and flexibility, especially since *E. coli* struggles to produce larger proteins. Scientists are exploring alternative host organisms to address this issue (Brown et al. 2015).

Another problem is the solubility. Unlike natural spidroins, most lab-made spider silk proteins dissolve poorly in water. Instead of forming smooth, continuous silk strands, they tend to clump together into tiny fibres. While these artificial fibres share some structural similarities with real silk, they lack the precise molecular alignment needed for full strength and flexibility. Finding ways to better control protein assembly is one of the biggest challenges researchers are working to solve in the quest for fully synthetic spider silk (Eisoldt et al. 2011).

### *10.6.3 Transgenic Goat: The Synthesis of Spider Dragline Silk*

Research has found that synthesising spider silk proteins in the mammary gland extracts of transgenic goats can be advantageous. Goats are small animals that produce large quantities of protein-rich milk. It also produces less waste that can be disposed of easily. The herd size can be easily scaled up with standard goat breeding as the spider genes are heritable (Decker Jr. 2018). Hence, scientists are targeting the goat alpha-S2 casein gene for replacement with spider silk-coding genes.

In an experiment, an artificially synthesised dragline spider silk fragment was transferred into Albas cashmere goats to obtain transgenic goats expressing the spider silk protein. The results showed that the wool from transgenic goats had greater tensile strength and elongation in the foreleg and hind leg than that of non-transgenic goats of the same breed (Fang et al. 2025a, b). Synthesising spider silk in goat milk can increase spider silk's potential for use in medical and industrial applications (Sabikhi 2007). Also, the large quantities of spider silk produced can be used for in-depth analysis of the silk, even more so (Decker Jr. 2018).

An American company, Nexia, has already created a special breed of genetically modified goats, known as BELE (Breed Early, Lactate Early), that grow quickly and produce spider silk proteins in their milk. These proteins, which weigh between 60 and 140 kilo Daltons, can be extracted and used to make strong, flexible fibres. Nexia has even partnered with the U.S. Army to explore spider silk for products such as bulletproof vests and surgical threads (Sabikhi 2007).

## 10.7 Applications of Spider Silk in Various Fields

The diversified spectrum of spider silk applications is discussed below.

### *10.7.1 Biomedical Field*

Spider silk is a renewable, biocompatible fibre that originates from natural resources. For centuries, the extraordinary mechanical performances of spider silk have been harnessed in various biomedical uses (Brankovi´et al. 2024). Some of them are discussed below.

#### 10.7.1.1 Artificial Blood Vessel Formation

Spider silk has the ability to control bleeding and promote natural wound healing. The development of a new vascular graft by researchers using this material is awaited. In recent times, the inability to maintain proper vascular resistance has been the main drawback of artificially created blood vessels (Shehata et al., 2019). Hence, finding alternatives for enhancing the physical properties of fabricated blood vessels is the need of the hour. Spider silk is a perfect alternative as it is strong, flexible, and well biocompatible (Bittencourt 2016).

For instance, by using spider silk as the supporting matrix, researchers created a new tissue-engineered vascular structure (Dastagir et al. 2020). C2C12 and ST1.6R cells were seeded on opposite sides of the spider silk scaffold and grown in a bioreactor under pulsatile flow, ultimately developing into a vascular prosthesis (Cordelle and Mantello 2020). The main process is shown in Fig. 10.11. According to tests, artificial blood vessels are highly biocompatible and exhibit mechanical properties that closely resemble those of natural blood vessels. They can also induce cell adhesion, differentiation, and proliferation, offering promising applications in tissue engineering (Wang et al. 2020).

#### 10.7.1.2 Tissue Regeneration

In tissue engineering, choosing the right biomaterial is crucial (Khan and Tanaka 2017). It needs to provide the right physical, chemical, and biological signals to support cell movement, attachment, and development—ultimately helping to build functional tissues (Bakhshandeh et al. 2017). Another important factor is how quickly the material breaks down; it must degrade at a rate that matches new tissue growth and allows the body to form its own extracellular matrix (ECM).

Cell adhesion is a key process in the formation and maintenance of tissues. As silk-like polymers (SLPs) can be genetically modified, scientists can enhance their biological function by introducing bioactive components, such as fibronectin (Kapoor and Kundu 2016). Fibronectin is a versatile glycoprotein that plays a crucial role in cell behaviour by interacting with integrin receptors in the ECM (Halper 2021). Type II (FNII), among the three major forms of fibronectin, is particularly known for its role in improving cell attachment and collagen binding due to its presence in matrix metalloproteinase-2 (MMP-2) and MMP-9 proteins.

In a study, researchers combined spider dragline silk protein (6mer—six repeats of the protein chain in bioengineered spider silk) with the FNII gene sequence to develop a new silk-based biopolymer (Pereira 2020). This hybrid material was not only safe for human cells, but it also significantly outperformed other functionalised 6mer-based spider silk proteins in production efficiency, yielding 10 times more biomaterial. Beyond its impressive production rate, this chimeric silk material exhibited mechanical properties comparable to those of well-known biopolymers such as polylactic acid (PLA), fibroin–collagen blends, and recombinant spider silk films (Reizabal et al. 2023). When tested in a lab setting, it showed no toxic effects

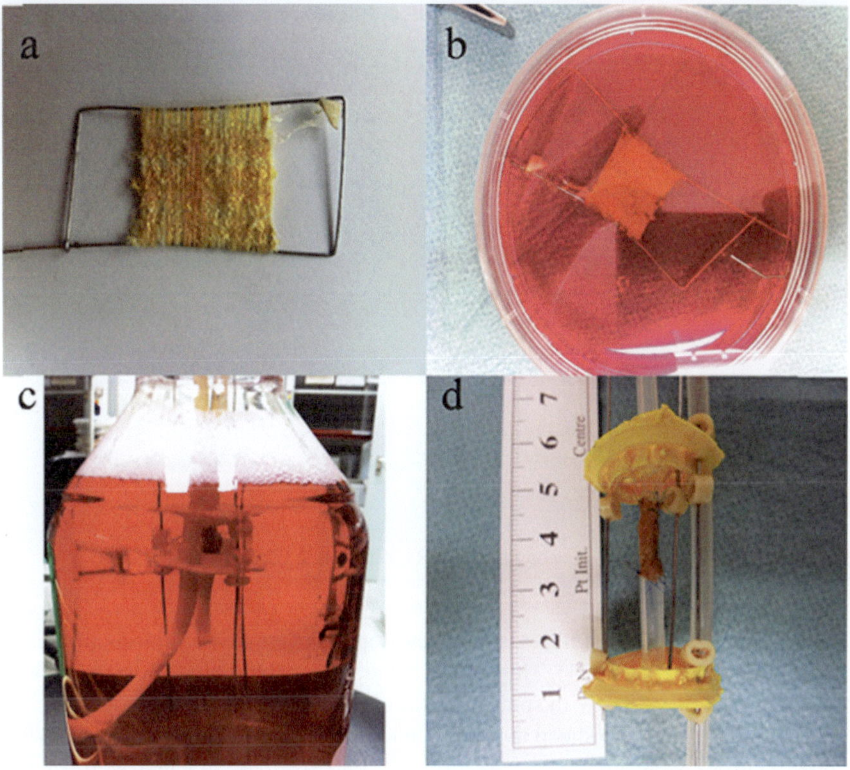


**Fig. 10.11** Process of artificial blood vessel formation—**a** Spider silk scaffold, **b** C2C12 and ST1.6R being cultured on both sides of the scaffolds, **c** Induction of constructed blood vessels by pulsatile blood flow in the bioreactor, **d** Tissue-engineered vascular graft after 3 weeks of induction. Copyright 2020, Elsevier [Reprinted from (Gu et al. 2020)]

on human skin fibroblasts and even enhanced cell adhesion, making it a promising option for tissue engineering (Brankovi' et al. 2024).

#### 10.7.1.3 Peripheral Nerve Regeneration

One promising way to help nerves heal after injury is to use nerve guidance conduits (NGCs) composed of biodegradable, biocompatible scaffold materials. These act like tiny bridges, guiding nerve fibres (Fig. 10.12) to reconnect across damaged areas (Mankavi et al. 2023). However, most current NGCs can support nerve growth over only short distances—generally up to 30 mm—and work best for small-diameter nerves (Hansson et al. 2021).

To improve this, scientists have developed NGCs with unique internal structures, such as guiding fibres, microchannels, or growth-factor-filled compartments. These features help steer nerve re-growth in the right direction. In one study, researchers

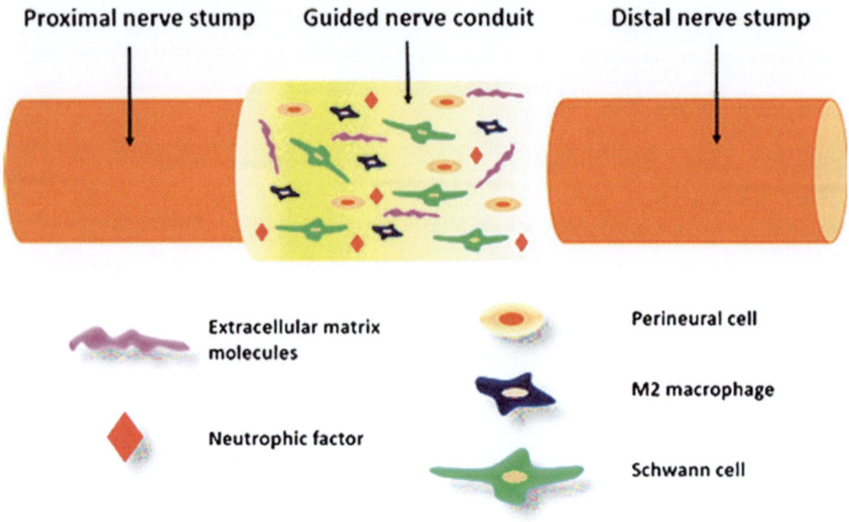


**Fig. 10.12** Peripheral nerve tissue engineering [Reprinted from (Brankovi´ et al. 2024)]

tested whether spider silk fibres could help repair larger nerve injuries in animal models (Kornfeld et al. 2021). They used vein grafts—natural blood vessels stripped of cells—and filled them with spider silk fibres from *Nephila clavipes*. These grafts were then implanted into sheep to repair a 6 cm nerve gap in the tibial nerve. When human neurons were grown on silk fibres in the lab, they attached well, moved along the fibres, and formed ganglion-like clusters, indicating healthy nerve development (Roloff et al. 2014). The implanted scaffolds were observed to repair and myelinate nerve fibres between 6 and 10 months, indicating that the nerve fibres were growing and forming protective coverings (Xue et al. 2021). The presence of Schwann cells—essential for nerve repair—confirmed that the healing process was progressing. Scientists also identified Ranvier's nodes, which are key structures for efficiently transmitting nerve signals. Most importantly, electrical tests showed that the repaired nerves functioned just as well as those treated with traditional nerve grafts. This suggests that spider silk-based scaffolds could offer a powerful new approach to repairing large nerve injuries, improving patient' chances of recovery (Yonesi et al. 2021).

#### 10.7.1.4 Heart Muscle Regeneration

A study investigated the interaction between primary rat cardiac cells and films composed of eADF4 (16), a type of spider silk protein (Petzold et al. 2017). The results showed that cardiomyocytes—the muscle cells responsible for heart contractions—formed strong connections with the positively charged spider silk surface. Interestingly, other cell types, such as fibroblasts, endothelial cells, and smooth muscle cells, did not attach as well, suggesting that the material has a natural selectivity for heart muscle cells (Han et al. 2019). To fabricate the films, glass cover slips were immersed in a solution of spider silk dripping in formic acid. To evaluate whether these films could be useful for heart tissue engineering, the behaviour of cell types, including myocardiocytes, endothelial cells, fibroblasts, and myocytes, was studied over short (3-h) and long (48-h) time periods. Specific markers were used to identify different cell types, including sarcomeric-actinin and troponin I for myocardiocytes, collagen 1 for fibroblasts, smooth muscle actin for smooth muscle cells, and VE-cadherin for endothelial cells (Cho et al. 2019). According to the results, other cell types did not bind well, to the spider silk film, cardiomyocytes did (Esser et al. 2021). Unlike fibronectin, which can sometimes cause an excessive growth response (hypertrophy), eADF4 (16) films did not have this effect, although they did support controlled cardiomyocyte hypertrophy when needed.

One of the most promising findings was that cardiomyocytes grown on the spider silk films expressed connexin 43, a protein that helps synchronise electrical signals between cells (Hou et al. 2018). Even more importantly, the heart cells could contract in a coordinated manner, with no signs of irregular beating (arrhythmia), similar to those grown on fibronectin (Tasiopoulos 2016). This suggests that spider silk-based materials could be a promising tool for heart tissue engineering, providing a stable and selective surface for heart muscle cell growth and function (Song et al. 2020).

#### 10.7.1.5 Bone and Cartilage Tissue Regeneration

Bone is a remarkable material—it is both strong and flexible due to the combination of minerals and proteins. The hard part is mostly calcium phosphate, while the softer, more elastic part is made of collagen. Scientists have been trying to create better bone implants using biodegradable materials, and collagen has been a top choice (Patino et al. 2002). The problem is that it is not very strong on its own, especially outside the body, and tends to break down over time (Ferreira et al. 2012).

That is where spider silk comes in—it is incredibly strong, flexible, and surprisingly compatible with the human body (Neubauer and Scheibel 2020). Researchers like Hardy and Yang have found that lab-engineered spider silk proteins can help minerals, such as calcium phosphate, build up on their surface—a process called biomineralisation (Delaquilla et al. 2020). When human stem cells were placed on these silk-based materials, they produced more of an enzyme called alkaline phosphatase, which plays a key role in bone formation (Correira et al. 2012). Spider silk naturally attracts calcium ions, helping the minerals stick and grow (Fang et al. 2025). Taking this idea further, Gomes and his team combined spider silk (*MaSp1* from the golden orb-weaver) with bone sialoprotein (BSP), a protein that helps bone cells attach and grow. The result was impressive—after just 6 h at body temperature, the silk-based material was already covered in calcium phosphate, mimicking real bone (Fig. 10.13). Not only did this speed up the mineralisation process, but it also helped human stem cells stick to the surface and develop into bone cells more efficiently. Other scientists have tried adding special proteins to spider silk to make it even better for bone growth. Plowright's team, for example, fused silk with a silica-binding peptide (R5), making the material even better at attracting minerals (Plowright 2017). Dinjaski's team took a similar approach, using a peptide called VTK to help build hydroxyapatite—a key component of bone (Dinjaski et al. 2017). They found that attaching VTK to one end of the silk made a huge difference, creating more bonding sites for minerals and significantly boosting bone growth (Fig. 10.14 and 10.15). Even better, the modification did not change the silk's physical properties, but it made the material three times more effective at encouraging bone development.

But spider silk is not just good for bones—it might also help rebuild cartilage. Morgan's team found that blending spider silk with silkworm silk created a material that encouraged early-stage bone cells to mature. The silk's unique structure helped guide cell growth, and built-in protein sequences (RGD domains) made it easier for cells to attach and function properly.

Meanwhile, Scheller's team explored how spider silk interacts with chondrocytes, the cells that produce cartilage (Tuchinda 2023). They created a special spider silk–elastin protein using genetically modified tobacco and potato plants and then tested it with chondrocytes. The results were promising—cells grew faster and multiplied more readily on silk than on regular plastic surfaces (Zhou et al. 2022).

Gellynck's research went even further, and he tested various types of spider silk, such as egg-sac silk and dragline silk, as scaffolds for cartilage repair (Gellynck et al. 2008). After 6 weeks, human cartilage cells started forming new tissue inside the silk's porous structure. The smaller pores (100–200 μm) were especially effective, helping cells settle in and produce key cartilage-building proteins, such as collagen type II and aggrecan.

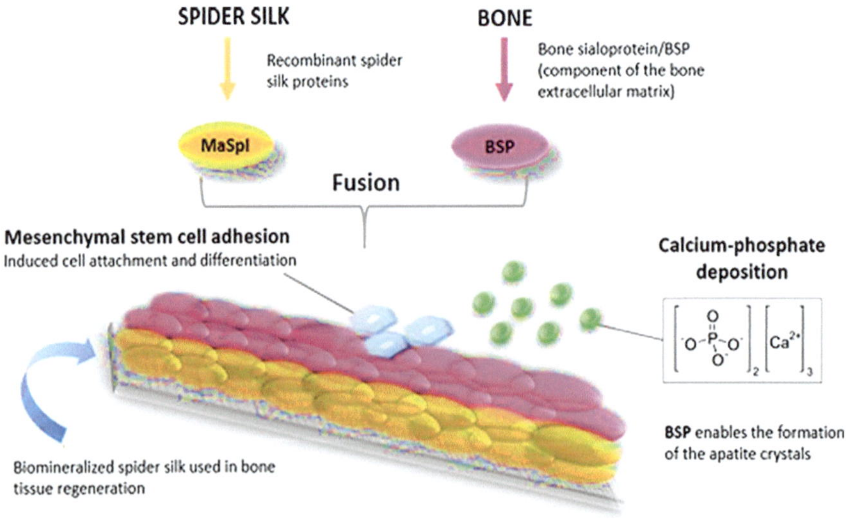


**Fig. 10.13** Use of spider silk protein (MA spidroin protein) in fabricating scaffolds for bone tissue regeneration [Reprinted from (Brankovi´ et al. 2024)]

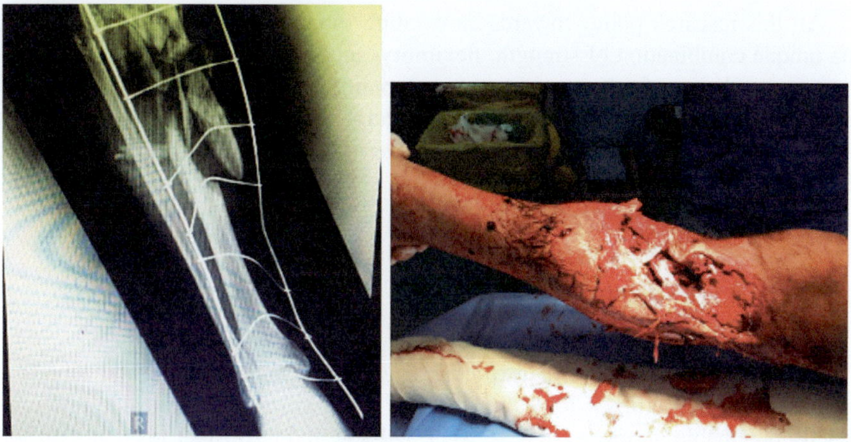

(a) (b)

**Fig. 10.14** Open multi-fragmented fracture of the lower leg. Bone graft was applied, **a** a radiography image of the fracture and, **b** an actual image of the complex fracture before the surgical procedure [Reprinted from (Brankovi´ et al. 2024)]

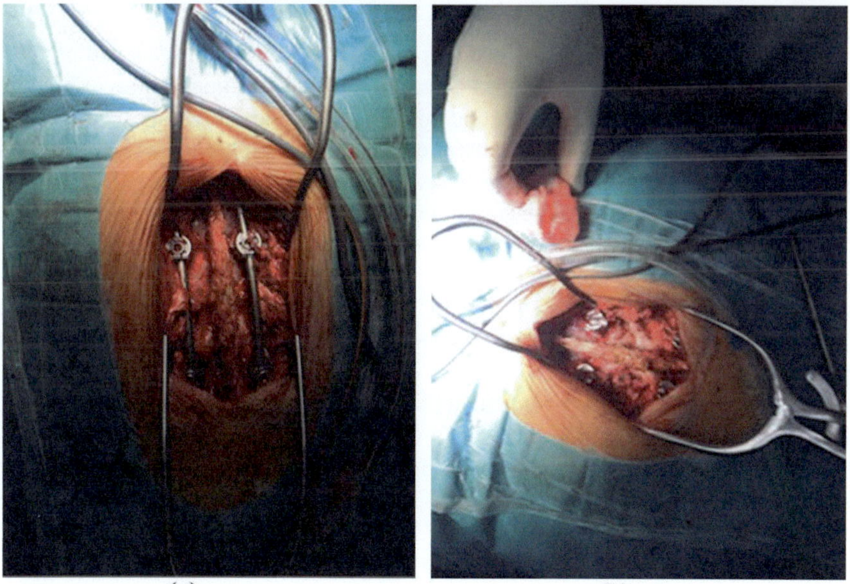

(a) (b)

**Fig. 10.15** An intraoperative image after stabilisation of spinal fracture and positioning of the transpedicular screws, where bone grafts are applied for better final fusion—**a** and **b** show the two views of the stabilisation method, with prepared bone graft [Reprinted from (Brankovi' et al. 2024)]

All this research points towards an exciting future for spider silk in medicine. Its unique combination of strength, flexibility, and biocompatibility could one day be a game-changer for bone and cartilage repair. Scientists are still fine-tuning the process, but the idea of using something as natural as spider silk to heal human bones and joints is looking more promising than ever (Gu et al. 2020; Salehi et al. 2020; Brankovi' et al. 2024).

### 10.7.2 Textile Field

Natural spider silk is used in textiles because of its unique molecular structure and excellent properties. Clothes made from spider silk are soft, wrinkle resistant, highly breathable, more resistant to wear and tear and ultraviolet rays, and do not yellow easily. As science and technology continue to advance, the incredible properties of spider silk fibre have drawn increasing interest from the textile industry. Researchers continually explore new ways to develop and incorporate this remarkable material into fashion and clothing (Wu and Chen 2022). The yield of spider silk is very low; hence, clothing from spider silk is always considered a luxury (Gu et al. 2020).

Electro-spun silk scaffolds can be incorporated with a diverse array of compounds, making them useful for drug delivery and embedding functional particles. These compounds have specific properties that might not be stable or effective on their own, but when supported by fibroin fibres, they can function in different environments and over time (Sill and Raccum 2008). In the textile industry, this technology has exciting applications. For example, some compounds that react under UV light do not work well in regular fabrics or lose their activity quickly. However, when embedded in electro-spun silk mats, their functionality is preserved. One well-known example is the use of titanium dioxide ($TiO_2$) and silicon dioxide ($SiO_2$) in textiles, which not only protects fabrics from UV damage but also helps break down wine and coffee stains (Oksuz et al. 2021).

Researchers have even developed self-cleaning fabrics using electro-spun silk. They have developed composites made from silkworm fibroin, PVA, and $TiO_2$ that decompose stains upon exposure to light. Three fluorescent dyes—rhodamine, acridine orange, and fluorescent sodium—have been successfully incorporated into electro-spun spider silk (Maltseva et al. 2022). The resulting silk glowed under UV light in shades ranging from deep red to bright yellow-green, making it suitable for glowing fabric strips or symbols (Belbéoch et al. 2021). A further step was taken by embedding cerium oxide nanoparticles into electro-spun spider silk mats. When exposed to UV light, the mats emitted a green glow and showed changes in electrical conductivity with humidity levels. This opens the door to intelligent textiles that can respond to environmental conditions, offering opportunities for futuristic clothing and smart fabric technology (Gu et al. 2020).

## 10.8 Challenges Faced by Spider Silk

Harvesting natural spider silk is a slow and costly process. One way to obtain small amounts is to forcibly extract silk from spiders or collect it from egg sacs (Hsia et al. 2011). However, this method is highly inefficient. For instance, producing just 3.4 m of textile from the silk of *Nephila madagascariensis* spiders requires around a million spiders and a team of over 70 workers, at an estimated cost of more than $500,000 (GS 2012). Additionally, almost all spiders exhibit territorial and cannibalistic behaviours, which makes large-scale farming challenging.

Due to these limitations, researchers have turned to recombinant technology to produce spider silk artificially. Many host organisms, including bacteria, yeast, plants, insects, mammalian cells, and even genetically modified animals, have been explored. However, the challenge of producing spider silk on a large scale persists, as it is difficult to find the right host organism to produce it efficiently. This issue not only limits industrial-scale production but also slows down efforts to develop new or improved versions of spider silk. In comparison to silkworms, which have undergone domestication for millennia, most spiders are territorial, carnivorous, and even cannibalistic—making spider farming nearly impossible.

To work around this, scientists have turned to synthetic biology, testing various organisms as hosts for producing spider silk proteins (spidroins) (Whittall et al. 2021). The first attempts used baker's yeast (*Saccharomyces cerevisiae*) and *Escherichia coli*, both of which are easy to genetically modify and inexpensive to grow at an industrial scale. In fact, *E. coli* has become a widely used host for the production of recombinant proteins. However, early experiments faced significant challenges—initially, only small amounts of spider silk proteins were produced, and the largest spidroin successfully expressed was just 43 kDa (kilodaltons) in size.

A major reason for this difficulty lies in the unique structure of spidroins. These proteins contain highly repetitive amino acid sequences, which can cause problems during DNA replication and disrupt transcription and translation (Heim et al. 2010). Additionally, studies have shown that in spider silk glands, levels of certain transfer RNAs (tRNAs) are increased before protein synthesis begins to meet the high demand for amino acids, such as alanine and glycine—something that standard production systems struggle to replicate (Poddar et al. 2020).

Hence, one of the most significant challenges is that natural spider silk proteins are extremely long and contain highly repetitive sequences, which pose an obstacle to their full-length production. To overcome this, scientists have developed chimeric (hybrid) versions of these proteins, which can be more easily expressed and later spun into fibres using different techniques (Ramezaniaghdam et al. 2022).

## 10.9 Conclusion

Spider Silk can be considered an excellent natural fibre, combining strength, elasticity, light weight, biocompatibility, and toughness. The complex structure of the silk is made of seven different silk types from seven different glands, which creates super-functional orb webs. Silk from silkworms could be of high strength or elasticity. In contrast, spider silk combines both properties, making it the most sought-after material and a field of intense scientific research. The biomedical field is the most experimental field of research in terms of the application of spider silk, as it is seen that spider silk has immense uses, namely:

- Wound dressings
- Tissue engineering
- Peripheral nerve regeneration
- Drug delivery systems.

Its biocompatibility with the human body and its ability to support cell growth make it ideal for biomedical applications. It also finds its uses in lightweight textiles and self-cleaning textiles using electro-spun silk.

Despite its immense qualities, spider silk creates the biggest drawback in its production. The major challenge is to commercially produce spider silk on a large scale. Unlike silkworm silk, which can be easily harvested on a large scale from the cocoons by increasing the number of cocoons, producing spider silk directly from spiders is incredibly difficult because they tend to eat each other when placed too close together. That is why scientists have focused on creating artificial versions of natural spider silk that mimic its structure. Although many different methods have been tried, we are still far from matching the mechanical performance of fabricated fibre with that of real spider dragline silk. This gap exists because replicating the intricate, hierarchical structure of spider silk is incredibly challenging. The challenge lies in spinning artificial fibres by mimicking the ampulla of the spider.

In addition, the types of silk spider products vary, so they must be sorted and selected when a specific variety of silk is needed. Also, compared with the more popular silkworm silk, spider silk's internal structure and macro properties vary significantly, which are still not thoroughly researched in detail. Hence, there should be a greater understanding of the molecular-level structure of spider silk and its supramolecular states. Moreover, the formation process of spider silk needs to be further explored to better understand its applicability across various areas and to make it more commercially viable and cost-effective.

**Declaration** The authors acknowledge the use of AI tools during the preparation of this manuscript for the purposes of language editing and grammar checking. All content was subsequently reviewed, verified, and edited by the authors to ensure accuracy, originality, and compliance with ethical standards. The authors take full responsibility for the final content of the manuscript.

## References

Agnarsson I, Kuntner M, Blackledge TA (2010) Bioprospecting finds the toughest biological material: extraordinary silk from a giant riverine orb spider. PLoS ONE 5(9):e11234

Andersson M, Holm L, Ridderstrale Y, Johansson J, Rising A (2013) Morphology and composition of the spider major ampullate gland and dragline silk. Biomacromol 14(8):2945–2952

Bakhshandeh B, Zarrintaj P, Oftadeh MO, Keramati F, Fouladiha H, Sohrabi-Jahromi S, Ziraksaz Z (2017) Tissue engineering; strategies, tissues, and biomaterials. Biotechnol Genet Eng Rev 33(2):144–172

Belbéoch C, Lejeune J, Vroman P, Salaün F (2021) Silkworm and spider silk electrospinning: a review. Environ Chem Lett 19:1737–1763

Bhattacharyya G, Oliveira P, Krishnaji ST, Chen D, Hinman M, Bell B, Jones JA (2021) Large scale production of synthetic spider silk proteins in Escherichia coli. Protein Expr Purif 183:105839

Blackledge TA (2012) Spider silk: a brief review and prospectus on research linking biomechanics and ecology in draglines and orb webs. J Arachnol 40(1):1–12

Blamires SJ, Tseng YH, Wu CL, Toft S, Raubenheimer D, Tso IM (2016) Spider web and silk performance landscapes across nutrient space. Scientific reports 6(1):26383

Blamires SJ, Blackledge TA, Tso IM (2017) Physicochemical property variation in spider silk: ecology, evolution, and synthetic production. Annu Rev Entomol 62(1):443–460

Branković M, Zivic F, Grujovic N, Stojadinovic I, Milenkovic S, Kotorcevic N (2024) Review of spider silk applications in biomedical and tissue engineering. Biomimetics 9(3):169

Bratzel G, Buehler MJ (2012) Sequence-structure correlations in silk: poly-Ala repeat of N. clavipes MaSp1 is naturally optimized at a critical length scale. J Mech Behav Biomed Mater 7:30–40

Brooks AE, Nelson SR, Jones JA, Koenig C, Hinman M, Stricker S, Lewis RV (2008) Distinct contributions of model MaSp1 and MaSp2 like peptides to the mechanical properties of synthetic major ampullate silk fibers as revealed in silico. Nanotechnol Sci Appl 9–16

Brown CP, Whaite AD, MacLeod JM, Macdonald J, Rosei F (2015) With great structure comes great functionality: understanding and emulating spider silk. J Mater Res 30(1):108–120

Brunetta L, Craig CL (2010) Spider silk: evolution and 400 million years of spinning, waiting, snagging, and mating. Yale University Press, New Haven

Cardoso P, Pekar S, Birkhofer K, Chuang A, Fukushima CS, Hebets EA, Mammola S (2024) Ecosystem services provided by spiders

Chen R (2012) Bacterial expression systems for recombinant protein production: E. coli and beyond. Biotechnol Adv 30(5):1102–1107

Cho DW, Kim BS, Jang J, Gao G, Han W, Singh NK (2019) 3D bioprinting: modeling in vitro tissues and organs using tissue-specific bioinks. Springer Nature, Cham

Collins T, Azevedo-Silva J, Da Costa A, Branca F, Machado R, Casal M (2013) Batch production of a silk-elastin-like protein in E. coli BL21 (DE3): key parameters for optimisation. Microb Cell Factories 12:1–16

Cordelle J, Mantero S (2020) Insight on the endothelialization of small silk-based tissue-engineered vascular grafts. Int J Artif Organs 43(10):631–644

Correia C, Bhumiratana S, Yan LP, Oliveira AL, Gimble JM, Rockwood D, Vunjak-Novakovic G (2012) Development of silk-based scaffolds for tissue engineering of bone from human adipose-derived stem cells. Actabiomaterialia 8(7):2483–2492

Dastagir K, Dastagir N, Limbourg A, Reimers K, Strauß S, Vogt PM (2020) In vitro construction of artificial blood vessels using spider silk as a supporting matrix. J Mech Behav Biomed Mater 101:103436

de C Bittencourt DM (2016) Spider silks and their biotechnological applications. In: Short views on insect genomics and proteomics: insect proteomics, vol 2, pp 211–227

De Oliveira DH, Gowda V, Sparrman T, Gustafsson L, SanchesPires R, Riekel C, Hedhammar M (2024) Structural conversion of the spidroin C-terminal domain during assembly of spider silk fibers. Nat Commun 15(1):4670

DeCambra W (2024) The morphology and biochemical structure of spider egg cases: potential arachnid-based silk nanosealants in medicine. Master's thesis, Arizona State University
Decker RE (2018) Production and biocompatibility of spider silk proteins in goat milk. Doctoral dissertation, Utah State University
Dellaquila A, Greco G, Campodoni E, Mazzocchi M, Mazzolai B, Tampieri A, Sandri M (2020) Optimized production of a high-performance hybrid biomaterial: biomineralized spider silk for bone tissue engineering. J Appl Polym Sci 137(22):48739
Dicko C, Knight D, Kenney JM, Vollrath F (2004) Secondary structures and conformational changes in flagelliform, cylindrical, major, and minor ampullate silk proteins temperature and concentration effects. Biomacromol 5(6):2105–2115
Ding Y, Li W, Zhang F, Liu Z, ZanjanizadehEzazi N, Liu D, Santos HA (2019) Electrospun fibrous architectures for drug delivery, tissue engineering and cancer therapy. Adv Funct Mater 29(2):1802852
Dinjaski N, Plowright R, Zhou S, Belton DJ, Perry CC, Kaplan DL (2017) Osteoinductive recombinant silk fusion proteins for bone regeneration. Actabiomaterialia 49:127–139
dos Santos-Pinto JRA, Garcia AMC, Arcuri HA, Esteves FG, Salles HC, Lubec G, Palma MS (2016) Silkomics: insight into the silk spinning process of spiders. J Proteome Res 15(4):1179–1193
Edlund A (2016) Synthetic spider silk sustainability verification by techno-economic and life cycle analysis. Utah State University
Eisoldt L, Smith A, Scheibel T (2011) Decoding the secrets of spider silk. Mater Today 14(3):80–86
Eisoldt L, Thamm C, Scheibel T (2012) The role of terminal domains during storage and assembly of spider silk proteins. Biopolymers 97(6):355–361
Esser TU, Trossmann VT, Lentz S, Engel FB, Scheibel T (2021) Designing of spider silk proteins for human induced pluripotent stem cell-based cardiac tissue engineering. Materials Today Bio 11:100114
Fang J, Meng FH, Wang F, Wang HT, Zhou HM (2025) Production of transgenic goat encoding spider dragline silk. Pakistan J Zool 58(2):527–534
Fang X, Zeng Q, Zhang Y, Hou H, Yang D, Xu K, Meng T (2025) A robust spider-silk-like calcium alginate fiber with biomineralized rough spindle-knots for water collection. Int J Biol Macromol 305(1):141011
Feng P, Zhao R, Tang W, Yang F, Tian H, Peng S, Shuai C (2023) Structural and functional adaptive artificial bone: materials, fabrications, and properties. Adv Funct Mater 33(23):2214726
Ferreira AM, Gentile P, Chiono V, Ciardelli G (2012) Collagen for bone tissue regeneration. Actabiomaterialia 8(9):3191–3200
Gellynck K, Verdonk PC, Van Nimmen E, Almqvist KF, Gheysens T, Schoukens G, Verbruggen G (2008) Silkworm and spider silk scaffolds for chondrocyte support. J Mater Sci Mater Med 19:3399–3409
Glišovic A, Vollrath F (2010) Mulberry silk, spider dragline and recombinant silks. In: Industrial applications of natural fibres: structure, properties and technical applications. Wiley, Hoboken, NJ, p 237
Gomes V, Salgueiro SP (2022) From small to large-scale: a review of recombinant spider silk and collagen bioproduction. Discov Mater 2(1):3
Gosline JM, Guerette PA, Ortlepp CS, Savage KN (1999) The mechanical design of spider silks: from fibroin sequence to mechanical function. J Exp Biol 202(23):3295–3303
Greving I, Terry AE, Holland C, Boulet-Audet M, Grillo I, Vollrath F, Dicko C (2020) Structural diversity of native major ampullate, minor ampullate, cylindriform, and flagelliform silk proteins in solution. Biomacromol 21(8):3387–3393
Gu Y, Yu L, Mou J, Wu D, Zhou P, Xu M (2020) Mechanical properties and application analysis of spider silk bionic material. e-Polymers 20(1):443–457
Guessous G, Blake L, Bui A, Woo Y, Manzanarez G (2024) Disentangling the web: an interdisciplinary review on the potential and feasibility of spider silk bioproduction. ACS Biomater Sci Eng 10(9):5412–5438

Hakimi O, Knight DP, Vollrath F, Vadgama P (2007) Spider and mulberry silkworm silks as compatible biomaterials. Compos Part B: Eng 38(3):324–337

Halper J (2021) Basic components of connective tissues and extracellular matrix: fibronectin, fibrinogen, laminin, elastin, fibrillins, fibulins, matrilins, tenascins and thrombospondins. In: Progress in heritable soft connective tissue diseases, pp 105–126

Han C, Luo X, Zou D, Li J, Zhang K, Yang P, Huang N (2019) Nature-inspired extracellular matrix coating produced by micro-patterned smooth muscle and endothelial cells endows cardiovascular materials with better biocompatibility. Biomater Sci 7(7):2686–2701

Hansson ML, Chatterjee U, Francis J, Arndt T, Broman C, Johansson J, Rising A (2021) Artificial spider silk supports and guides neurite extension in vitro. FASEB J 35(11):e21896

Heim M, Keerl D, Scheibel T (2009) Spider silk: from soluble protein to extraordinary fiber. AngewandteChemie Int Ed 48(20):3584–3596

Heim M, Römer L, Scheibel T (2010) Hierarchical structures made of proteins. The complex architecture of spider webs and their constituent silk proteins. Chem Soc Rev 39(1):156–164

Hou J, Xie Y, Ji A, Cao A, Fang Y, Shi E (2018) Carbon-nanotube-wrapped spider silks for directed cardiomyocyte growth and electrophysiological detection. ACS Appl Mater Interfaces 10(8):6793–6798

Hsia Y, Gnesa E, Jeffery F, Tang S, Vierra C (2011) Spider silk composites and applications. In: Metal, ceramic and polymeric composites for various uses 2, pp 303–324

Hu X, Vasanthavada K, Kohler K, McNary S, Moore AMF, Vierra CA (2006) Molecular mechanisms of spider silk. Cell Mol Life Sci CMLS 63:1986–1999

Jin HJ, Kaplan DL (2003) Mechanism of silk processing in insects and spiders. Nature 424(6952):1057–1061

Kapoor S, Kundu SC (2016) Silk protein-based hydrogels: promising advanced materials for biomedical applications. Actabiomaterialia 31:17–32

Khan F, Tanaka M (2017) Designing smart biomaterials for tissue engineering. Int J Mol Sci 19(1):17

Kiseleva AP, Krivoshapkin PV, Krivoshapkina EF (2020) Recent advances in development of functional spider silk-based hybrid materials. Front Chem 8:554

Koh LD, Cheng Y, Teng CP, Khin YW, Loh XJ, Tee SY, Han MY (2015) Structures, mechanical properties and applications of silk fibroin materials. Prog Polym Sci 46:86–110

Koop F, Strauß S, Peck CT, Aper T, Wilhelmi M, Hartmann C, Bucan V (2022) Preliminary application of native Nephilaedulis spider silk and fibrin implant causes granulomatous foreign body reaction in vivo in rat's spinal cord. PLoS ONE 17(3):e0264486

Kornfeld T, Nessler J, Helmer C, Hannemann R, Waldmann KH, Peck CT, Radtke C (2021) Spider silk nerve graft promotes axonal regeneration on long distance nerve defect in a sheep model. Biomaterials 271:120692

Kronqvist N, Sarr M, Lindqvist A, Nordling K, Otikovs M, Venturi L, Johansson J (2017) Efficient protein production inspired by how spiders make silk. Nat Commun 8(1):15504

Lewis RV (1992) Spider silk: the unraveling of a mystery. Acc Chem Res 25(9):392–398

Lefèvre T, Auger M (2016) Spider silk as a blueprint for greener materials: a review. Int Mater Rev 61(2):127-153

Li J, Li S, Huang J, Khan AQ, An B, Zhou X, Zhu M (2022) Spider silk-inspired artificial fibers. Adv Sci 9(5):2103965

Lindberg E (2024) Decoding spider silk: machine learning approaches to uncovering property determinants

Liu S, Cui Z, Liu Z, Zhao W, Zhou X (2023) Research progress on spider-inspired tough fibers. Chin J Chem 41(23):3401–3418

Makirita WE (2020) Development of bio-pesticide for management of spodopterafrugiperda (je smith) and other lepidoptera pests of maize in Tanzania. Doctoral dissertation, NM-AIST

Maltseva ES, Nikolaeva VO, Savin AM, Dobryakov MY, Koshel EI, Krivoshapkin PV, Krivoshapkina EF (2022) Fluorescent hybrid material based on natural spider silk and carbon dots for bioapplication. ACS Biomater Sci Eng 8(8):3310–3319

Mankavi F, Ibrahim R, Wang H (2023) Advances in biomimetic nerve guidance conduits for peripheral nerve regeneration. Nanomaterials 13(18):2528
Neubauer VJ, Scheibel T (2020) Spider silk fusion proteins for controlled collagen binding and biomineralization. ACS Biomater Sci Eng 6(10):5599–5608
Öksüz KE, Özkaya NK, İnan ZDŞ, Özer A (2021) Novel natural spider silk embedded electrospun nanofiber mats for wound healing. Mater Today Commun 26:101942
Patino MG, Neiders ME, Andreana S, Noble B, Cohen RE (2002) Collagen as an implantable material in medicine and dentistry. J Oral Implant 28(5):220–225
Pereira AMMB (2020) Insights into the mechanism of antimicrobial activity and its exploitation for developing new therapies against clinically relevant bacteria. Doctoral dissertation, Universidade do Minho (Portugal)
Pérez-Rigueiro J, Elices M, Plaza GR, Guinea GV (2021) Basic principles in the design of spider silk fibers. Molecules 26(6):1794
Petzold J, Aigner TB, Touska F, Zimmermann K, Scheibel T, Engel FB (2017) Surface features of recombinant spider silk protein eADF4 (κ16)-made materials are well-suited for cardiac tissue engineering. Adv Funct Mater 27(36):1701427
Platnick NI (2020) Spiders of the world: a natural history. Princeton University Press, Princeton
Plowright R (2017) Experimental studies of silk and silk composites for biomedical application. Nottingham Trent University (United Kingdom)
Poddar H, Breitling R, Takano E (2020) Towards engineering and production of artificial spider silk using tools of synthetic biology. Eng Biol 4(1):1–6
Qiu W, Patil A, Hu F, Liu XY (2019) Hierarchical structure of silk materials versus mechanical performance and mesoscopic engineering principles. Small 15(51):1903948
Ramezaniaghdam M, Nahdi ND, Reski R (2022) Recombinant spider silk: promises and bottlenecks. Front Bioeng Biotechnol 10:835637
Reizabal A, Costa CM, Pérez-Álvarez L, Vilas-Vilela JL, Lanceros-Méndez S (2023) The new silk road: silk fibroin blends and composites for next generation functional and multifunctional materials design. Polym Rev 63(4):1014–1077
Rising A, Johansson J (2015) Toward spinning artificial spider silk. Nat Chem Biol 11(5):309–315
Roloff F, Strauß S, Vogt PM, Bicker G, Radtke C (2014) Spider silk as guiding biomaterial for human model neurons. BioMed Res Int 2014(1):906819
Römer L, Scheibel T (2008) The elaborate structure of spider silk: structure and function of a natural high performance fiber. Prion 2(4):154–161
Sabikhi L (2007) Designer milk. Adv Food Nutr Res 53:161–198
Salehi S, Koeck K, Scheibel T (2020) Spider silk for tissue engineering applications. Molecules 25(3):737
Saravanan D (2006) Spider silk-structure, properties and spinning. J Text Appar Technol Manag 5(1):1–20
Shehata N, Hassounah I, Sobolciak P, Krupa I, Lewis R, Kandas I (2019) Spider silk fibers: synthesis, characterization, and related biomedical applications. Materials for biomedical engineering. Elsevier, Amsterdam, pp 289–307
Silk GS Golden spider silk
Sill TJ, Von Recum HA (2008) Electrospinning: applications in drug delivery and tissue engineering. Biomaterials 29(13):1989–2006
Song Y, Wang H, Yue F, Lv Q, Cai B, Dong N, Wang L (2020) Silk-based biomaterials for cardiac tissue engineering. Adv Healthc Mater 9(23):2000735
Spiess K, Lammel A, Scheibel T (2010) Recombinant spider silk proteins for applications in biomaterials. Macromol Biosci 10(9):998–1007
Sun J, Su J, Ma C, Göstl R, Herrmann A, Liu K, Zhang H (2020) Fabrication and mechanical properties of engineered protein-based adhesives and fibers. Adv Mater 32(6):1906360
Suthan R, Jayakumar V, Madhu S (2018) Evaluation of mechanical properties of kevlar fibre epoxy composites: an experimental study. Int J Veh Struct Syst 10(6):389–394
Tasiopoulos CP (2016) Partial spider silk as scaffold for tissue engineering the aortic valve

Teulé F, Cooper AR, Furin WA, Bittencourt D, Rech EL, Brooks A, Lewis RV (2009) A protocol for the production of recombinant spider silk-like proteins for artificial fiber spinning. Nat Protoc 4(3):341–355

Tokareva O, Jacobsen M, Buehler M, Wong J, Kaplan DL (2014) Structure–function–property–design interplay in biopolymers: spider silk. Actabiomaterialia 10(4):1612–1626

Tuchinda P (2023) 3D Printed silk scaffold for bone and cartilage regeneration. Master's thesis, Tufts University

Vendrely C, Scheibel T (2007) Biotechnological production of spider-silk proteins enables new applications. Macromol Biosci 7(4):401–409

Vollrath F, Edmonds DT (1989) Modulation of the mechanical properties of spider silk by coating with water. Nature 340(6231):305–307

Wang D, Xu Y, Li Q, Turng LS (2020) Artificial small-diameter blood vessels: materials, fabrication, surface modification, mechanical properties, and bioactive functionalities. J Mater Chem B 8(9):1801–1822

Wen R, Wang S, Wang K, Yang D, Zan X, Meng Q (2023) Complete gene sequence and mechanical property of the fourth type of major ampullate silk protein. Actabiomaterialia 155:282–291

Whittall DR, Baker KV, Breitling R, Takano E (2021) Host systems for the production of recombinant spider silk. Trends Biotechnol 39(6):560–573

Wu X, Chen L (2022) New textile material—spider silk fiber. J Phys: Conf Ser 2256(1):012003

Xu G, Gong L, Yang Z, Liu XY (2014) What makes spider silk fibers so strong? From molecular-crystallite network to hierarchical network structures. Soft Matter 10(13):2116–2123

Xue W, Shi W, Kong Y, Kuss M, Duan B (2021) Anisotropic scaffolds for peripheral nerve and spinal cord regeneration. Bioact Mater 6(11):4141–4160

Yonesi M, Garcia-Nieto M, Guinea GV, Panetsos F, Perez-Rigueiro J, Gonzalez-Nieto D (2021) Silk fibroin: an ancient material for repairing the injured nervous system. Pharmaceutics 13(3):429

Zhou Z, Cui J, Wu S, Geng Z, Su J (2022) Silk fibroin-based biomaterials for cartilage/osteochondral repair. Theranostics 12(11):5103

# Chapter 11
# Sustainability Aspects of Silk Production: Environmental Concerns and Solutions

**Siam Sarker, Shah Md. Maruf Hasan, Rafaj Hossain Rakin, Esmat Jahan Mitu, Md. Kamrul Hassan Chowdhury, and Mahmuda Akter**

## 11.1 Introduction

Silk is a natural protein fiber produced by the silkworm Bombyx mori and valued for its optical brightness, the soft feel of its texture, and its strong mechanical qualities. The compound filament structure comprises fibroin, a naturally hydrophobic protein, and sericin, a readily hydrolyzable hydrophilic glycoprotein. Conventional degumming techniques facilitate the extraction of sericin, yielding glossy, high-tensile yarns (ElleSilk 2023). The modern application of sericin in biomedical research involves the ability to modify and adjust the properties, composition, and physical structure of the material into a variety of polymer modifications, such as crosslinking, copolymerization, and blending, all of which can lead to hydrogel, thin film, sponges, and engineered fibers (Dhilip Kumar and Abrahamse 2022; Qi et al. 2017; Shitole et al. 2020). At the same time, the industry's focus on developing sustainable and biodegradable materials for medical and biomedical applications has brought silk materials into the limelight due to their established biocompatibility, biodegradability, and mechanical properties (Barcelos et al. 2020). Recent developments in biotechnology have expanded the commercial potential of silk fibroin by leveraging additional bioproduction options, thereby making silk a multifaceted platform with significant commercial and biomedical promise (Altman and Farrell 2022). Based on these properties, scientists have also placed spider silk as a paradigm of next-generation fibers.

Spider silk, with its exceptional strength, elasticity, biodegradability, and biocompatibility, has a range of potential applications, including protective apparel, regenerative medicine, and artificial muscles (Vollrath and Edmonds 1989; Vollrath and Porter 2006). The possibility of producing large quantities of recombinant spider silk proteins has become available with the field of genetic engineering (Hauptmann

S. Sarker · S. Md. Maruf Hasan · R. H. Rakin · E. J. Mitu · Md. K. H. Chowdhury · M. Akter (✉)
Department of Apparel Engineering, Bangladesh University of Textiles, Dhaka, Bangladesh
e-mail: mahmuda.akter.mitu88@gmail.com



D. Kumar and M. Shahid (eds.), *Science and Technology of Silk*, Textile Science and Clothing Technology, https://doi.org/10.1007/978-3-032-16230-4_11

et al. 2013; Humenik et al. 2014; Lang et al. 2013; Rammensee et al. 2008; Xu et al. 2012). New fiber fabrication technologies, such as electrospinning, wet spinning, self-assembly, and microfluidics, are utilized to produce regenerated protein-based fibers. However, regenerated silk could not acquire sufficient mechanical properties as those of its natural partner due to insufficient knowledge of the spinning process in terms of factors affecting the spinning process, such as the morphology of the protein, spinning conditions, pH (Sukigara et al. 2003), metal ions (Zhu et al. 2014), coagulation baths (Oroudjev et al. 2002), post-treatment (Garay et al. 2014), and solvents (Liu et al. 2017; Pawcenis et al. 2014). Measuring the environmental impact of silk is a complex task due to the wide range of life-cycle phases. The stages identified in ISO 14067:2018 are cradle-to-grave, cradle-to-gate, and gate-to-gate analyses for estimating greenhouse gases (Can et al. 2024). Organic sericulture minimizes its environmental impact through the natural control of pests, integrated pest management, non-toxic dyeing, and the recycling of fibers. In addition, it supports millions of farmers and artisans, promotes women's empowerment, and enhances competitiveness in the global market (Baker et al. 2020; Ekka and Bais 2023; Sharma et al. 2024).

The primary objective of this chapter is to review the environmental impacts of conventional silk manufacturing. This chapter also covers sustainable sericulture methods, biodiversity, and technical developments in this area. We examined innovative alternatives to traditional silk, as well as the latest eco-friendly techniques for silk finishing and coloring. With an emphasis on the most recent important research, this chapter provides a comprehensive overview of the use of the circular economy in sericulture. The chapter also provides a summary of the ways in which the silk industry collaborates and develops policy, highlighting the potential and challenges for sustainable silk in the international market.

## 11.2 Environmental Impact of Traditional Production of Silk

### *11.2.1 Pesticide and Chemical Contamination of the Land and Aquatic Environment*

In sericulture, chemicals are utilized at various stages of production. During mulberry cultivation, farmers apply pesticides, fungicides, herbicides, and synthetic fertilizers to improve plant growth and yield. In silkworm rearing, disinfectants such as sodium hypochlorite, lime, formaldehyde, and naphthalene are used to prevent the spread of diseases (Hazarika 2025). Although mulberry trees require fewer chemical inputs than most other fiber crops, sericulture can still lead to environmental contamination. The chemicals used in farming can persist in the soil, reduce beneficial microbes, lower soil fertility, and, in some cases, render the soil acidic or saline (Fajfer and Łochyńska 2022). Silkworm waste, which is sometimes used as an organic fertilizer, may still contain chemical residues and can release greenhouse gases such as methane and nitrous oxide if not composted properly (DS et al. 2024). Water pollution can also occur when nitrates, phosphates, and other farm chemicals wash into rivers, ponds, or lakes. This runoff can lead to eutrophication, which causes excessive algae growth, reduces oxygen levels, and harms aquatic life. Moreover, experimental studies have shown that the insecticide chlorfenapyr causes very high mortality in silkworm larvae, even at low concentrations. In the larvae that survive, growth is slower, cocoons are lighter with thinner shells, and overall silk production decreases. The severity of these effects increases with higher insecticide doses (Vasanth et al. 2025). To mitigate these negative effects, Integrated Pest Management (IPM) is recommended, as it combines several strategies, including the use of disease-resistant mulberry varieties, adherence to proper cultural practices, application of biological control methods, and the use of natural plant-based pesticides, such as neem, along with the judicious use of chemical insecticides. Research indicates that neem-based products are generally non-toxic to silkworms, help maintain soil and water health, support biodiversity, and reduce the need for synthetic pesticides (CM et al. 2023). However, the use of IPM methods is still limited in many developing countries, where studies have reported huge adoption gaps in biological controls and improved chemical alternatives in some regions (Sakthivel et al. 2012). To make sericulture more sustainable, it is important to educate farmers, enforce proper regulations, and improve waste management systems to reduce reliance on harmful chemicals and protect the environment in the long term.

### *11.2.2 Carbon Footprint in Silk Processing*

The carbon footprint of silk processing encompasses all greenhouse gas emissions generated throughout the silk's life cycle (Liu et al. 2022)—from mulberry cultivation to silk production and product use. During sericulture, mulberry trees absorb atmospheric carbon dioxide through the process of photosynthesis. Research indicates that the amount of carbon sequestered by these trees exceeds the emissions produced during cultivation, leading to a net negative carbon footprint at this stage (Li et al. 2020). For example, China's mulberry fields collectively fix about 64.40 million tons of $CO_2$ annually, demonstrating the significant role of sericulture in carbon sequestration (Fig. 11.1) (Liu et al. 2022). After being cultivated, silk undergoes several processing steps, including degumming, spinning, weaving, dyeing, and finishing (Thamizharasu et al. 2025). Unlike the cultivation of mulberry plants, which can result in net negative emissions, this sequence of manufacturing places significant energy requirements and high carbon emissions.

A comparative analysis by WRAP showed that silk has a carbon footprint of approximately 25,425 kg of $CO_2$ equivalent per tonne of silk fibre, encompassing cocoon production through to end-of-life projections, excluding the mitigation effect of the mulberry trees. This is equivalent to, or even greater than, the footprint of some synthetic fibers (Thomas et al. 2012). The post-production stage of silk clothes also adds to the environmental load. Drying, washing, and ironing generate about 0.78

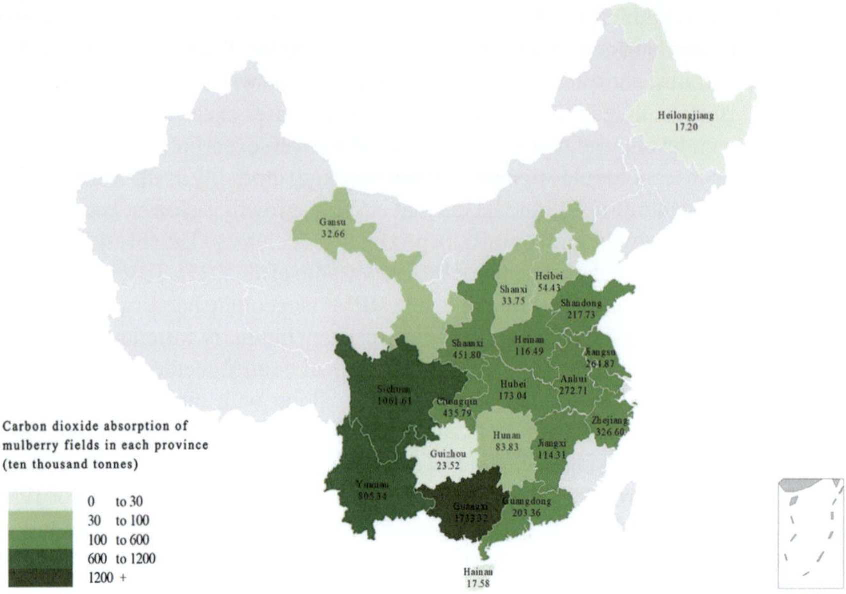


**Fig. 11.1** Carbon dioxide absorption of mulberry fields of each province in China in 2017 (Liu et al. 2022). Copyright©2022 Shuyi Liu et al., published by Sciendo, (under the Creative Commons Attribution-3.0 License)

kg of $CO_2$ emissions per kilogram of garments washed (Liu et al. 2022). Since most studies on the carbon footprint of silk vary widely due to methodological differences, scope, and data availability, it is necessary to achieve uniformity in methods and reporting to ensure an accurate measure of the carbon footprint of silk products and to mitigate their environmental consequences.

## 11.3 Sustainable Sericulture Practices

### *11.3.1 Organic Farming Techniques of Silk*

Sustainable practices are increasingly incorporated into organic silkworm and mulberry production, and may be referred to as peace silkworm or Ahimsa silk. The use of organic manures and biofertilizers, such as vermicompost, vermiwash, poultry manure, Dalweed, oil cakes, farmyard manure, and press mud, is one of the biofertilizers that are proven to improve mulberry growth, leaf production, and nutrient content, consequently stimulating excellent silkworm health, larval weight, high-quality cocoons, and high silk production. As a comparative analysis suggests, vermicompost and vermiwash can provide plants with essential nutrients, plant hormones, and beneficial microbes. Coupled with biofertilizers and oil cakes, resistance to pests and diseases, improved soil fertility, and increased cocoon production are some of the factors that make the use of these organic inputs both cost-effective and environmentally friendly substitutes for traditional chemical fertilizers in sericulture (Singh and Kapila 2021). In conventional reeling, cocoons are boiled in hot water to preserve their fiber structure, whereas in organic silk harvesting, moths are allowed to emerge on their own. The former produces longer and more consistent fibers and has a very specific texture, but the latter gives shorter fibers and does not compromise on animal welfare (Kind 2025). Due to reduced water usage and lower energy consumption, the process relies on natural soaps, hot-water baths, and sunlight to dry its products. The end textile product is environmentally friendly, skin-friendly, and often colored with vegetable dyes. In addition, these practices contribute to the preservation of soil, biodiversity, and equitable trade, making organic silk production a sustainable alternative to traditional silk production (ElleSilk 2023).

### 11.3.2 Integrated Pest Management and Biodiversity Conservation

Sericulture Integrated Pest Management (IPM) involves the incorporation of host-plant resistance control plans, culturally sustainable activities, biological controls, and the mindful use of chemicals to limit pests while preserving biodiversity. As empirical studies show, breeding mulberry varieties that are resistant to the main sericultural pests coupled with cultural practices, pruning, optimum spacing, spatial orientation, suitable nitrogen use, destroying weeds, and strict sanitation methods, form the first line of defense against insect population and subsequent secondary infestation frequently caused by indiscriminate insecticidal application (Singh and Saratchandra 2002). In complement to this arsenal, measures to control insect populations using bio-control processes operating within an ecosystem are routinely employed, such as the conservation and augmentation of parasitoids, predators, and microbial pathogens. The use of natural enemies, including Trichogramma spp., Acerophagus papayae, ladybird beetles, and fungal and bacterial organisms, helps decrease the use of synthetic pesticides and strengthens ecological soundness (Kishore et al. 2024). The presence of these biodiversity-rich elements not only inhibits proliferation of the pests but also builds ecosystem services, such as pollination, soil stabilization, and habitat complexity, by growing a diversity of mulberry species, conserving the old strains of silkworms and taking the agroforestry systems and minimal use of agroecological chemicals into consideration, thereby protecting the genetic and species diversity within the sericultural landscape (Attri et al. 2024). Altogether, IPM in sericulture complements pest control, enhances yield, and conserves biodiversity, aligning sericultural regimes with the goal of wider conservation.

### *11.3.3 Water and Soil Conservation in Sericulture Farms*

The long-term sustainability of sericulture relies on water and soil conservation measures, as mulberry growth processes require a steady source of moisture and nutrients to produce high-quality and uniform leaves, which in turn determine silk quality and, consequently, cocoon production. One important intervention is the introduction of precision irrigation, including the use of drip irrigation (Fig. 11.2a), to improve the efficiency of water use and moisture delivery at the desired location in the root zone. When used together with mulching through either plastics or organic materials, irrigation systems using drip techniques maintain moisture in the soils, reduce water losses through evaporation, improve vegetable growth by reducing weed growth, and increase nutrient uptake, which results in developing healthier and more productive mulberry plants (Ranjitha and Chandrashekhar 2021). Solar-powered drip irrigation platforms (Fig. 11.2b) provide a cost-efficient and environmentally friendly solution for resource-starved, rainfed environments, ensuring consistent irrigation during droughts, higher overall productivity, and reduced reliance on monsoon rainfall (Rajendran et al. 2025).

Moreover, integrating drip fertigation with reduced fertilizer doses supports nutrient-use efficiency without compromising leaf quality or yield, which is particularly important for environmentally responsible sericulture practices (Mahesh et al. 2022). In addition to water conservation, mulching is beneficial for enhancing the initial growth of the mulberry tree, as it stabilizes soil temperature, improves the

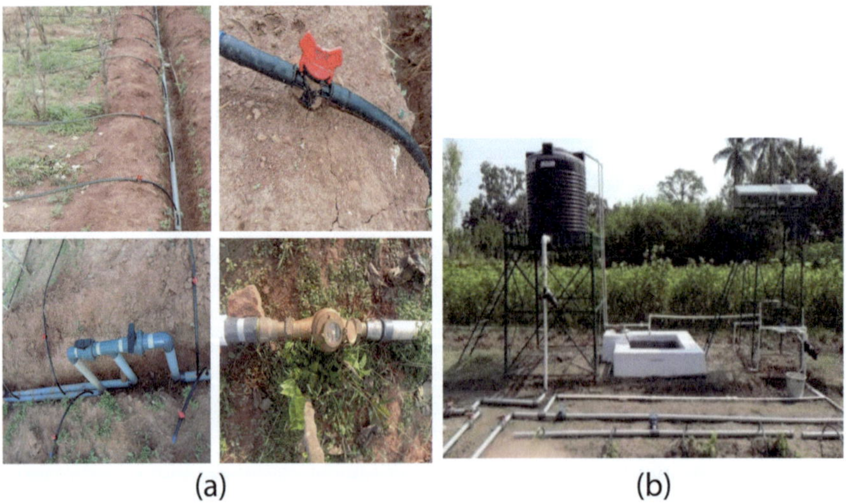


**Fig. 11.2** Drip Irrigation Setup in the crop field **a** surface drip irrigation system (Ranjitha and Chandrashekhar 2021) Copyright©2021 Ranjhita et al., published by University of Agricultural Sciences, **b** Solar Drip irrigation system (Rajendran et al. 2025) Copyright©2025 Rajendran et al., published by International Journal of Environment, Agriculture and Biotechnology, (under the Creative Commons Attribution-4.0 License)

**Table 11.1** Performance outcomes of water-efficient technologies in sericulture

| Conservation practice | Outcome/Impact | Numerical value(s) | Reference |
|---|---|---|---|
| Drip irrigation + Black plastic mulch | Mulberry leaf yield | ≈ 51 t/ha/year | Ranjitha and Chandrashekhar (2021) |
| | Leaf moisture | ~ 66% | |
| | Water productivity increase | > 7.4% | |
| | Cost–Benefit Ratio | 1: 2 | |
| Drip fertigation (75% RDF) | Leaf yield increase | + 26% | Rajendran et al. (2025) |
| | Soluble protein increase | + 56% | |
| | Sugar content increase | + 28% | |
| | Water savings | –24% | |
| | Water Use Efficiency (WUE) increase | + 61% | |
| | Nutrient Use Efficiency (NUE) increase | + 63% | |
| Solar-Powered Drip Irrigation Kit (SDIK) | Leaf yield increase | + 37% | Mahesh et al. (2022) |
| | Water savings | –24% | |
| | Water productivity increase | + 44% | |
| Mulching (organic and plastic) | Root–shoot ratio, seedling vigor, germination rate, plant height, and early growth all improved | | (Wani et al. 2018) |

root-to-shoot ratio, and promotes strong seedling growth, as confirmed by field verification in various mulch situations (Wani et al. 2018). Together, these measures of conservation not only streamline resource utilization and help ensure cost-effective practices but also enhance climate resilience in sericulture ecosystems. A summary of key impacts from these practices is presented in Table 11.1, highlighting their individual contributions to productivity and sustainability.

## *11.3.4 Technological Innovations in Sustainable Sericulture*

Some of the major technological improvements made in the area of sustainable sericulture include significant enhancements in productivity, resource sustainability, and environmental sustainability, particularly in silk waste valorization, silkworm rearing, and cocoon processing, as well as digital transformation. In the process of raising silkworms, a climate-controlled system featuring Internet-of-Things sensors and automation enables precise control of temperature and humidity levels, thereby reducing silkworm mortality and increasing the number of cocoons produced (Elizabeth 2025; Fatimaa et al. 2024). During the post-cocoon phase, technical innovations such as solar-powered reeling devices can serve as an energy-efficient and labor-saving replacement for manual reeling, thereby aiding the sari industry by providing high-quality filament silk for use in delicate saris (Textile and Ltd. 2025). Other emerging possibilities for utilizing silk waste include the processing of defective cocoons, pupae, and gheecha/baikal yarn, which can lead to increased income generation and a sustainable, multi-raw source of silk-blend saris (Textile and Ltd. 2025). Coupled with such an endeavor, drying of cocoons can also be accomplished feasibly using solar convection dryers to attain such sustainability goals through reduced operational costs and emissions levels, in a similar drying efficiency as that of ovens with energy amounting at around 0.75 kWh/kg cocoons (Singh 2011). Digitization is also undergoing further refinement, with research into mobile applications to provide real-time alerts against pests, computer-vision-enhanced environmental controls, and image-derived disease identification still in the early stages of development, outside of specific styles such as Banarasi, Kanchipuram, or Bhagalpuri (Elizabeth 2025). New traceability platforms designed around blockchain are being studied as a tool of increasing transparency and authenticity in the silk value chain (Sharma and Kalra 2021). These tools are not necessarily culturally specific to sari products, but they do form a solid technological system that serves as the foundation of sustainable production and, subsequently, the certification of silk. Without this, the sustainable production and marketing of saris would simply not be possible. As shown in Table 11.2, emerging technologies such as IoT-based precision sericulture and nanotechnology integration are playing a crucial role in enhancing silk quality and production efficiency.

## 11.4 Ecological Technologies of Finishing and Dyeing of Silk

The wet-chemical processing of textiles, including preparation, dyeing or printing, and finishing, is essential for improving fabric appearance, color, and functionality (Samanta et al. 2016). However, this process generates significant environmental pollution through toxic wastewater and air emissions (Jajpura et al. 2004). Silk, used widely in both textile and non-textile sectors, consists mainly of two proteins: fibroin (core) and sericin (outer layer) (Zhu et al. 2022). Degumming removes sericin to enhance silk's texture and shine. Eco-friendly methods like steam, citric acid, and enzymes such as papain offer effective results with less damage and pollution (Amirian et al. 2023; Zhu et al. 2022). While natural dyeing is improving, sustainable finishing still needs further research (Zhu et al. 2022).

**Table 11.2** Emerging technological innovations in sericulture and their impacts (Elizabeth 2025)

| Innovation area | Key technologies/practices | Potential impact |
|---|---|---|
| Precision sericulture | IoT sensors, automation, data-driven analytics | Minimizes silkworm mortality, enhances cocoon quality, optimizes resource use |
| Genetic engineering | Selective breeding, CRISPR, transgenic strain development | Improves silk strength and yield, adds disease resistance, enables custom traits |
| Sustainable practices | Organic mulberry farming, closed-loop water systems, waste reuse | Lowers ecological footprint, supports biodiversity, promotes circular economy |
| Advanced processing | Automated reeling, eco-friendly dyeing, innovative finishing | Produces premium silk, reduces environmental harm, enables multifunctional textiles |
| Blockchain integration | Digital supply chain tracking, secure ledgers, transaction transparency | Ensures traceability, encourages ethical sourcing, builds consumer confidence |
| Nanotechnology in silk | Embedding nanoparticles, silk–nanomaterial hybrids | Enhances performance (e.g. strength, conductivity), opens biomedical/smart uses |
| Controlled environment | Indoor rearing, vertical farming, CEA (Controlled Environment Agriculture) | Enables year-round output, reduces space and water needs, supports urban farming |

### *11.4.1 Chemical-Free Silk Processing Methods*

As environmental issues escalate and people demand more environmentally friendly products, the need to develop chemical-free methods of processing silks also grows. These eco-friendly, sustainable methods will minimize the use of harmful chemicals, reduce environmental pollution, and preserve the inherent attributes of silk, thereby protecting both suppliers and customers. One of the most important steps in silk processing is degumming, which involves the removal of the sericin layer —a thick, sticky protein coating surrounding the twin filamentous fibroin threads. This can occur through the cleavage of the peptide bonds, which hold sericin to fibroin to yield silk fibers, which are tridimensional, soft, lustrous, and mechanically refined. The extent to which sericin is removed imparts a conclusive influence on the end physical and mechanical properties of the consequent silk fibers. The percent material loss can be calculated based on the following standard formula:

$$\text{Degumming Loss (\%)} = \left(\frac{W_1 - W_2}{W_1}\right) \times 100$$

where the weight before degumming is $W1$ and the weight after the degumming procedure is $W2$.

Various degumming techniques can be employed, depending on the desired degumming characteristics, and each technique has its own set of advantages and impacts on the silk structure (Haggag et al. 2007). Water degumming is one of the most cost-effective and commonly used processes. In this technique, silk is treated with deionized water under hot conditions (115–120°C). Even though up to 96% of sericin can be removed during water degumming, the process is often limited by the possible degradation of fibroin and the failure of full sericin extraction, which depends on high temperatures and a lack of control. To overcome these drawbacks, degumming of soap is often employed (Gulrajanid et al. 1992). Typically, soap degumming is carried out in a mildly alkaline solution under high-temperature conditions for an extended period, using soaps such as Marseille soap, sodium laurate, or sodium stearate. Relative to water degumming, soap degumming is often milder and capable of retaining the softness and lustre of the silk, but unlike water, it can be degraded by hard water, which combines with calcium and magnesium to form insoluble salts. The problem can be alleviated by adding surfactants that increase the efficacy of soap and protect the quality of the fiber (Bray 1928; Tsunokaye 1932).

In addition to these conventional methods, enzymatic degumming offers a more environmentally friendly and selective alternative. Utilizing enzymes such as papain, trypsin, and bacterial proteases, this method operates under mild pH (7–9) and temperature conditions, thereby minimizing damage to the fibroin structure. However, its gentle nature may result in incomplete removal of hydrophobic impurities (Shahid et al. 2016), (Chopra and Gulrajani 1994). To further improve efficiency and precision, ultrasonic degumming has been introduced, which utilizes acoustic cavitation to remove sericin without damaging the inner fibroin layer. This technique is particularly effective when combined with chemical agents such as alkalis, acids, or enzymes, offering a balance of speed and safety (Haggag et al. 2007). An advanced alternative is microwave-assisted degumming, which uses microwave energy to enhance the action of various degumming agents, including hydrochloric acid, sodium carbonate, soaps, and enzymes. Studies have demonstrated the effectiveness of this method, all while maintaining the desired properties of silk fibroin. Overall, the choice of degumming technique is guided by the intended application and required characteristics of the final silk product, as each method offers a unique combination of efficiency, cost, and impact on fiber quality (Haggag et al. 2007; Mahmoodi et al. 2010; Mahmoodi et al. 2010). A comparative overview of common silk degumming methods, including their reagents, mechanisms, advantages, and limitations, is presented in Table 11.3.

### *11.4.2 Reduction in Water, Chemical, and Energy Consumption in Sustainable Dyeing*

Textile dyeing is a crucial process in the fabric finishing industry, utilizing natural or synthetic dyes to impart color and enhance the value of textiles (Khattab et al. 2020).

**Table 11.3** Comparison of different silk degumming methods

| Method | Reagents | Mechanism | Advantages | Disadvantages | References |
|---|---|---|---|---|---|
| Water degumming | Deionized water | High-temperature boiling (115–120°C) removes sericin | Cost-effective, removes up to 96% sericin | Risk of fibroin degradation, incomplete removal | Gulrajanid et al. (1992) |
| Soap degumming | Marseille soap, sodium salts | Alkaline hydrolysis converts sericin into soluble salts | Enhances luster, mild on fibers | Water quality affects efficiency, possible fibroin damage | Bray (1928), Tsunokaye (1932) |
| Enzyme degumming | Trypsin, papain, bacterial proteases | Enzymatic hydrolysis of sericin peptide bonds | Mild conditions, minimal fibroin damage | Ineffective on hydrophobic impurities, requires controlled pH/temp | Shahid et al. (2016), Chopra and Gulrajani (1994) |
| Ultrasonic degumming | Acids, alkalis, enzymes | Acoustic cavitation removes sericin | Fast, uniform removal | Requires additional reagents | Vyas and Shukla (2016), Wang et al. (2019) |
| Microwave degumming | HCl, $Na_2CO_3$, enzymes, soap, sodium silicate | Microwave energy enhances reagent efficiency | Fast, high efficiency, preserves fibroin | Requires power level optimization | Haggag et al. (2007), Mahmoodi et al. (2010) |

The process depends on factors such as material type, fabric construction, and chemical compatibility (Chequer et al. 2013; Perkins 1991). However, it relies heavily on freshwater, making the textile sector the third-largest global water consumer after agriculture and energy (learners 2014; Nandhakumar et al. 2012). Since only a small fraction of the Earth's water is available as surface freshwater (Peter 1993), this dependence raises serious environmental concerns. Moreover, untreated dye effluents containing toxic chemicals increase biological oxygen demand and damage ecosystems (Mahmud and Kaiser 2020). To address these issues, the silk industry is adopting sustainable dyeing technologies that conserve water and minimize environmental harm. Techniques such as enzymatic scouring, natural dyeing, and plant-based functional finishes reduce reliance on harmful synthetic chemicals. Natural substances like neem and aloe vera are used for antimicrobial treatments, plant-based molecules provide ultraviolet protection, and biopolymer-derived agents offer flame resistance. These eco-conscious treatments help reduce chemical pollution and align with global objectives for green production and sustainable supply chains (Lara et al. 2022; Samanta et al. 2016). To further minimize environmental impact, biological coloration methods have been introduced that bypass conventional wet processing. One innovative approach involves feeding silkworms mulberry leaves that have been infused with non-toxic dye solutions. As the silkworms consume the dyed leaves, pigments embed into the fibroin layer of the silk filament, producing naturally colored cocoons. This technique eliminates the need for dye baths, significantly reducing water, energy, and chemical use (Nisal et al. 2014; Sangappa et al. 2007; Tansil et al. 2011a, b). Experimental trials using Direct Acid Fast Red (Fig. 11.3) demonstrated uniform coloration with good color strength and minimal ecological footprint (Karthik et al. 2015). Additionally, natural dyes extracted from plants such as Diospyros malabarica and Ceriops decandra have been successfully applied to silk fabrics using environmentally safe mordants like aluminum sulfate and ferrous sulfate. These methods achieve acceptable color strength and fastness when clean water sources are used. Trace metal analysis confirms that the dyed silk fabrics comply with Oeko-Tex safety limits, making them suitable for eco-friendly textile applications (Mridha et al. 2021). Supporting this trend, recent research has demonstrated the effective dyeing of silk and cotton with natural dyes from Acacia catechu, Alkanna tinctoria, and Curcuma longa through exhaustion dyeing, without the use of synthetic mordants. This process resulted in higher dye uptake in silk, with some dyes acting as natural mordants to enhance the fixation of the dye. Environmental assessments of dye effluent, based on pH, total dissolved solids, and chemical oxygen demand, confirmed biodegradability (Jose et al. 2017). Microwave-assisted dyeing using Alkanna tinctoria extract and natural bio-mordants also improved color strength and fastness, eliminating the need for synthetic chemicals. These findings suggest that combining natural dyes with low-impact technologies presents a viable and sustainable alternative to conventional silk dyeing (Adeel et al. 2022). Overall, these developments represent significant progress toward sustainable silk dyeing, as they reduce dependency on hazardous inputs, preserve water resources, and promote environmentally responsible production.

### *11.4.3 Advantages of Silk Recycling and Waste Reduction*

The silk industry generates substantial waste across various production stages, including silkworm rearing, cocoon harvesting, unwinding, spinning, weaving, dyeing, and finishing. Only about 20% of cocoons are classified as first-grade and can be completely reeled with no waste, while the remaining second and third-grade cocoons are only partially processed in silk factories. Including waste generated during spinning and weaving, the total waste across the silk industry amounts to about 55% (Sashina and Yakovleva 2023). This high level of production waste not only contributes to environmental degradation but also results in economic inefficiencies by underutilizing valuable raw materials. Recycling and reducing silk waste offer notable advantages by transforming discarded material into useful products and reducing the demand for virgin resources. Mechanical recycling, which involves processing spinnable silk waste into yarn, contributes to the production of textiles with up to 50 percent material recovery. Although certain types of unspinnable waste

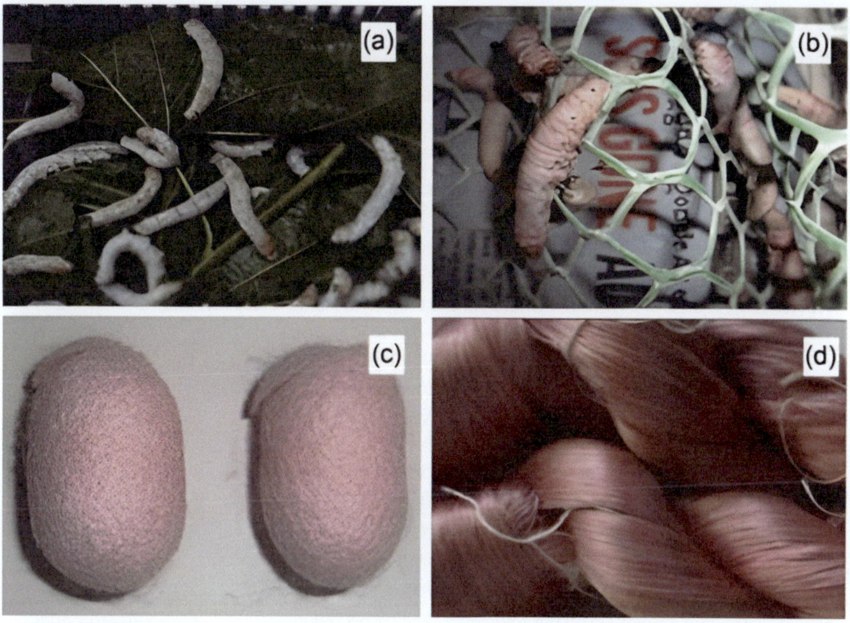


**Fig. 11.3** **a** Fifth instar *Bombyx mori* larvae feeding on treated mulberry leaves with a Direct Acid Fast Red dye solution; **b** Active silkworm larvae spinning cocoons on a patterned mesh; **c** Color cocoon shells which remain after the larvae are removed; **d** Stands of Direct Acid Fast Red silk threads after the degumming process. Reprint with permission from (Karthik et al. 2015). Copyright@2015 Springer

are more challenging to recycle, advancements in fiber processing and spinning technology continue to expand the scope of recovery (Sashina and Yakovleva 2023). Furthermore, the conversion of silk waste into nonwoven fabrics provides additional value, especially in the medical sector, due to silk's intrinsic hypoallergenic and antibacterial properties. These nonwovens are being explored for wound dressings, tissue engineering, and high-efficiency filtration materials (Zhang et al. 2019). Chemical recycling of waste silk through fibroin extraction enables the production of regenerated silk-based materials, including nanofibers, hydrogels, and composite films. These materials are widely used in biomedical applications, wearable electronics, and high-performance textiles due to silk fibroin's excellent biocompatibility, mechanical strength, and versatility. These new methods, which have been developed to transform waste silk into high-value materials, are discussed in Table 11.4.

Additionally, waste silk has found alternative uses beyond textiles, including as poultry feed, where studies have shown that silk powder supplements can improve egg production and alter nutritional content favorably (Kim et al. 2002). It is also being considered for bioenergy production, although incineration releases high levels of carbon dioxide and is less preferred from a sustainability standpoint (Leal Filho et al. 2019; Zamani et al. 2015). In design and fashion, the creative reuse of silk

**Table 11.4** Silk fibroin-based materials and their applications

| Material type | Production method | Applications | References |
|---|---|---|---|
| Porous scaffold | Freeze-drying | Filtration, soundproofing, controlled drug release | Tandon et al. (2020) |
| Silk fibroin film | Drying silk protein | Corneal applications, filtration, adsorption | Puerta et al. (2020) |
| Silk fibroin/PU composite film | Blending silk fibroin with polyurethane (PU) | High strain resistance (1254.3%) and toughness (54.9 $MJ \cdot m^{-3}$) for enhanced mechanical performance | Zhang et al. (2020) |
| Silk nanofibers | Electrospinning of silk selvedges | Waterproof and breathable materials, drug delivery carriers, battery separators | Lee et al. (2020) |
| Regenerated fibroin fibers | Dry–wet spinning method | Surgical sutures, artificial ligaments, nerve conduits | Zhang and Pan (2019) |
| Extruded silk filaments | Direct extrusion and drafting on rollers with temperature-controlled baths | Improved fiber cross-sections for varied applications | Mollahosseini et al. (2019) |
| Silk fibroin hydrogels | Processing of different fibroin fibers | Biomedical applications, including cell culture scaffolds and drug delivery | Bexiga et al. (2017) |

by-products, such as silk straw and weaving selvedges, has led to their incorporation into decorative and commercial textile items, supporting waste valorization and local craftsmanship (Giacomin et al. 2017). Despite challenges such as solvent toxicity in chemical recycling and limited infrastructure for in-house processing, the continued development of environmentally friendly technologies and closed-loop systems is making silk recycling increasingly feasible and impactful (Nisal et al. 2014; Rammensee et al. 2008). Overall, silk recycling and waste reduction not only mitigate environmental harm but also open up pathways for sustainable innovation, value creation, and resource efficiency across textile and allied industries.

## 11.5 Innovative Alternatives to Traditional Silk

As the fashion world comes to terms with its ecological footprint, sustainability has emerged as a prime priority. Leading the way in this revolution is the creation of sustainable materials, which are necessary for the manufacturing of garments and textiles. These new materials are designed to minimize waste, reduce pollution, and lower the overall environmental cost associated with conventional fashion production. The focus is now on creating textiles that are not only practical, durable, and aesthetically pleasing but also biodegradable. This shift towards sustainable textiles is transforming the fashion business. Developing alternatives to traditional silk that are biodegradable while offering similar performance characteristics is crucial for addressing the challenges faced by sustainable fashion today.

### *11.5.1 Bio-Silk Development: Improvement of Spider Silk and Silk-Equivalent Materials*

Spider silk is a part of the very exceptional biomaterials of the current century, which also boasts surprisingly high mechanical properties in terms of strength, elasticity, and toughness. Unlike silkworm silk, which is synthesized by one type of gland, spiders have a multifunctional range of glands, which includes about 37,000 different species that utilize different types of silk to perform specific biological functions. The two model systems, Aranea diadematus and Nephila clavipes, which possess seven distinct silk glands, have been well characterized due to the structural and chemical variations that play out (Tokareva et al. 2014). Such diversity makes spider silk a particularly good natural example to use as the basis of creating and improving bio-silk and other comparable made-made fibers.

Recent studies have helped scientists to understand the pathway of native production, to guide the artificial synthesis plans. The biogenesis begins in specialized abdominal glands in which the fibroin proteins (spidroins) are secreted as micelles by columnar epithelial cells to avoid premature aggregation. These micelles are aligned, elongated, and crystallized into β-sheets through migration of these micelles through an S-shaped duct. Active ion replacement and increasing dehydration in the duct continue to bring the fibers to be formed together (Eisoldt et al. 2011). At the late stages of silk formation, the aggregate nanostructures are extruded as thread like nanofibrils, 20 nm in diameter that assemble into the fiber core (Nentwig 2012; Römer and Scheibel 2008). This core is then covered by glycoproteins and lipids. The spider then rapidly retracts the posterior legs to set the mechanical properties of the emerging filament into an optimal configuration before it is deployed (He et al. 2008). Such a naturally existing spinning mechanism was a key paradigm for the modern silk-generating technologies.

The mechanical properties of spider silk are a myriad of characteristics that make it so very appealing as a biomaterial. A notable characteristic is supercontraction in which dragline silk of Nephila clavipes can contract as far as 47% in damp environments, producing contraction forces that can be comparable to 22% of the energy needed to break the fiber (Bell et al. 2002). The repetitive cycles of humidity-influenced contracting and relaxing lead to a better resistance to fatigue and opposition to sagging (Blackledge et al. 2009) (Boutry and Blackledge 2010). In addition to this, spider silk shows a unique ability known as the electrocapillary winch effect, where the adherent water on the surfaces acts as micro-tension controllers, allowing them to extend very effectively and recoil fast absorbing kinetic energy that is gained during prey capture (Neukirch et al. 2017). The combination of these qualities, namely high strength coupled with elasticity, hygroscopic responsiveness, and energy absorbing capacity makes spider silk far superior to most naturally occurring and synthetic fibres. The different properties of spider silk threads are explained in Table 11.5 contrasting with silk threads of *Bombyx mori.*

Therefore, the ongoing research focuses on recreating or even improving these functions via bioengineering, synthetic biology, and nanotechnology. The goal is to develop bio-silk and silk-like fibers that maintain the adaptive properties of spider silk and still allow a huge quantity production to textile markets as well as biomedical devices and engineering. One of the ideas for developing biosilk is illustrated in Fig. 11.4 (Ryszard et al. 2012).

### *11.5.2 Artificial and In Vitro Silk Developments*

Silk fibroin possesses a number of prominent qualities/characteristics, such as good biocompatibility, controllable biodegradability, self-assembly, and demonstrated high mechanical strength, which make silk fibroin irreplaceable in biomedical research and industrial use. Such natural properties are further manipulated by controlled chemical manipulation expanding its application in tissue engineering, enzyme entrapment, wound dressing, vascular prostheses, and structural implants (Acharya et al. 2008; Dal Pra et al. 2005; Meinel et al. 2005; Murphy and Kaplan 2009). Recent developments in bioengineered silk include BNES, a sericin-like protein that is secreted by the bacterium Bacillus sp. strain NE, without genetic engineering and was isolated in a petroleum-polluted environment (Kamoun et al. 2019). Unlike transgenic expression systems of spider silk, BNES is produced within a wild-type organism and thus carries the advantage of high secretion rates, low cost of production, ease of recovery, and environmental sustainability of large-scale manufacture, especially on a bacterial scale of sericin production (Fahnestock et al. 2000). The traditional sericin usually has to become an industrial waste, now however it can be retrieved and reused again, also returning both economic and ecological benefits (El-Fakharany et al. 2020). Existing studies indicate that BNES also possess a strong biologic activity, a DPPH and ABTS antioxidant potential, anticancer activity against carcinoma cell lines and antimicrobial competence. SDS-PAGE analysis reveals the presence of fibroin entities (30, 35, 400 kDa) and sericin entities (6.5, 16, 66, 95 kDa), indicating a structural resemblance to native silk proteins (Fahnestock et al. 2000). The potential rebirth of silk fibroin in regenerative medicine has garnered its due attention in recent literature, in addition to the likelihood that it has already been extensively studied in comparison to alternatives such as poly (lactic acid) (PLA), poly (glycolic acid) (PGA), and collagen. Such substitutes have the limitations of low mechanical strength and faster degradation. Silk fibroin-based scaffolds, on the other

**Table 11.5** Comparison of silk properties: *Bombyx mori* (Silkworm) versus *Aranea diadematus* (Spider)

| Characteristic | Silkworm (*Bombyx mori*) | Spider (*Aranea diadematus*) | References |
|---|---|---|---|
| Molecular weight (kDa) | Fibroin: 360 (heavy chain), 25 (light chain) Sericin: 80–309 | Fibroin (spidroin): Up to 350 | Ayoub et al. (2007), Garb et al. (2010) |
| Composition (Main Amino Acids) | Fibroin: Alanine, glycine, serine, tyrosine Sericin: Serine, glycine, glutamic acid, aspartic acid, threonine, tyrosine | Dragline silk: Glycine, alanine, proline, glutamic acid, serine | Altman et al. (2003), Numata and Kaplan (2010), Andersen (1970) |
| Structure of the Fiber | Core of fibrils' structure: Bipartite | Core of fibrils' structure: Simple | Du et al. (2011) |
| Thread diameter (μm) | 10–20 | 3–5 | Du et al. (2011) |
| Ultimate tensile strength | Natural fiber: 500 MPa Degummed: 240–700 MPa | Dragline silk: 1.1–1.2 GPa Spiral silk: 0.3 GPa | Pérez-Rigueiro et al. (2000), Zhao et al. (2007), Du et al. (2011) |
| Strain at break (%) | Natural fiber: 15–19 Degummed: 4–16 | Dragline silk: 20–27 Spiral silk: 270 | Pérez-Rigueiro et al. (2000; Zhao et al. (2007), Du et al. (2011) |
| Stiffness (GPa) | Natural fiber: 7 | Dragline silk: 10 Spiral silk: 0.003 | Heim et al. (2009), Shao and Vollrath (2002), Du et al. (2011); Pérez-Rigueiro et al. (2000) |
| Toughness ($MJ.m^{-3}$) | 70 | Dragline silk: 180 Spiral silk: 150 | Heim et al. (2009), Shao and Vollrath (2002), Du et al. (2011), Pérez-Rigueiro et al. (2000) |
| Density ($g.cm^{-3}$) | 1.3 | Dragline silk: 1.3 Spiral silk: 1.3 | Römer and Scheibel (2008) |

hand, provide temporary structural support to enhance tissue formation (Cooper et al. 2005; Lu et al. 2005; Solheim et al. 2000). In the case of anterior cruciate ligament (ACL) reconstruction, the characteristics of silk fibroin, strong tensile strength, good biocompatibility, a long time duration of degradation, and the low level of contamination make it a very good candidate (Greenwald et al. 1994; Inouye et al. 1998; Sofia et al. 2001; Vepari and Kaplan 2007). Other newer technologies are polymer blends, such as with poly (L-lactic acid-co-glycolic acid) (PLGA) (Sahoo et al. 2010a, b),

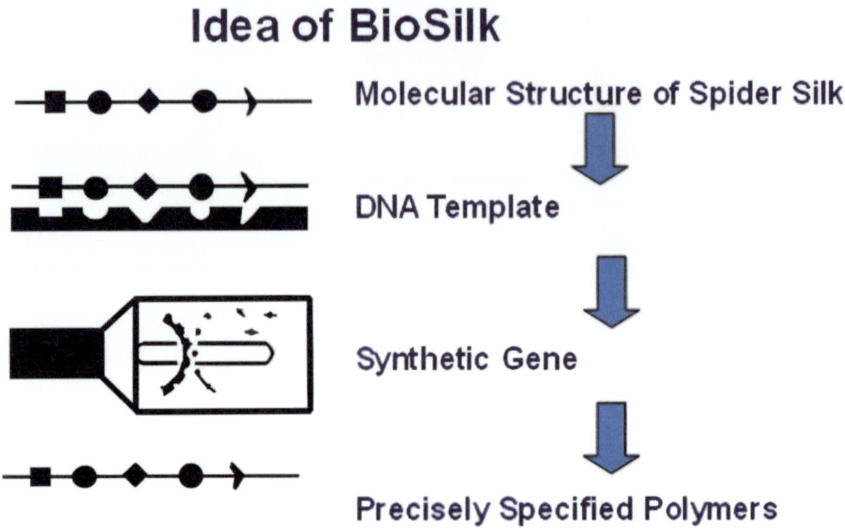


**Fig. 11.4** Idea to synthesize bio silk. Reprint with permission from (Ryszard et al. 2012). Copyright@2012 Taylor and Francis

peptide nanofiber meshes (Chen et al. 2012), and textile-based scaffolds (Altman et al. 2002); clinical translation of these advances is not yet complete.

### *11.5.3 Biodegradable and Non-toxic Alternative Silk*

The synthesis of biodegradable polymers through chemical treatment, microorganisms, and enzymes has been deemed an eco-friendly alternative to synthetic materials. They are synthesized to degrade after a certain time frame, disintegrating into easily disposable by-products. They are made from several renewable sources, including food waste, agricultural residues, starch, and cellulose. Bioplastics made from renewable resources are cheaper than those made from microorganisms (Mukherjee et al. 2023; Samir et al. 2022). Polymers are characterized as natural (polysaccharides, proteins), synthetic (amides, esters, ethers, urethanes), and hybrid synthetic biopolymers. They are affected by the rate of biodegradation, which is influenced by composition, morphology, molecular structure, and molecular weight (Mukherjee et al. 2023; Samir et al. 2022). They are widely applied in cosmetics, coatings, wound dressings, gene delivery, nanotechnology, medical implants, prosthetics, cryopreservation, and sanitation products, such as surgical sutures (Mukherjee et al. 2023; Samir et al. 2022). Recent advances in silk-based and bioengineered materials hold great promise for medical, textile, and sustainable applications. Silk proteins. Bacteria of the Bacillus sp. strain NE, an oil-degrading microorganism, have strong antiviral activity. These proteins, which consist of fibroin and sericin, are very effective against herpes simplex virus type 1 (HSV1), adenovirus type 7 (ADV7), and hepatitis C virus (HCV). They have antiviral effects that include inhibiting viral entry and replication, as well as decreasing the production of reactive oxygen species (ROS) in infected cells, which can result in enhanced cell defenses. Additionally, the affordability and simplicity of proteins make them appealing targets for future antiviral treatments (El-Fakharany et al. 2022). Likewise, the electrospraying of natural and synthetic porphyrins, such as hemin and iron-containing porphyrin complexes, into polyhydroxybutyrate (PHB) fibers by electrospinning has been shown to significantly enhance the antimicrobial effect of the fibers, as well as increase mechanical strength (a 125 percent improvement). The additives further adjusted the structure of the fibers, resulting in significant enhancements of electrical conductivity and viscosity qualities that are crucial in high-quality fiber production. The results of the conducted study support the fact that enriched PHB fiber can be used in biomedical applications (Tyubaeva et al. 2022). Vegan leathers made from bacterial cellulose (BC) are gaining traction as alternatives to animal leathers in the sustainable fashion sector. BC was waterproofed with essential oils of tea tree (Melaleuca alternifolia) and the waxy layer of a carnauba palm (Copernicia prunifera) and treated with extracts of an onion skin (Allium cepa), pomegranate (Punica granatum), and eucalyptus (Eucalyptus globulus). The end product demonstrated resilience in terms of tensile strength, flexibility, and water resistance, making it a potent and environmentally friendly alternative. While further optimization is needed for waterproofing, BC-based vegan leather is a promising, eco-friendly, and ethical solution to reduce environmental damage caused by conventional leather production (da Silva Junior et al. 2022). Together, these innovations in bacterial silk, enhanced polymer fibers, and sustainable vegan leather represent significant progress toward developing scalable, cost-effective, and sustainable materials for use in the healthcare, textile, and fashion industries.

## 11.6 Sericulture and the Circular Economy

A circular economy is an economic model that aims to minimize waste and promote sustainable use of natural resources. Where the traditional linear economy consists of take, make, dispose, the circular economy includes reducing, reusing, refurbishing, repairing, and recycling. The circular economy supports sustainable agriculture, and sericulture is a prime example of sustainable agricultural practices. Mulberry silkworm breeding is a sustainable practice because of its organic life cycle, optimized waste management and versatile product output.

### 11.6.1 Value Addition to Sericulture By-Products Through Waste Management-Upcycling

Silk production involves excessive wastage in the production process, characterized by the cocooning, spinning, weaving, and dyeing processes, where the input material loses an estimated 55 percent of its value as waste. This environmental impact requires the formulation of strategies that promote sustainable waste management (Sashina and Yakovleva 2023). With recent innovations, however, it is evident that such by-products can be repurposed, thereby transforming waste into resources and generating economic opportunities. The organic excretions of silkworms, which are the byproducts of feeding on mulberry leaves, can be utilized as a biofertilizer that enriches the soil and promotes sustainable farming practices. Likewise, the cocoon's outer layers, which do not provide sufficiently good quality silk to be used in textile making, can be utilized as animal food or mushroom growing media. The silkworm pupae that are not used in the reeling process contain a lot of protein and nitrogen (making them usable as animal feed or as natural fertilizers). Moreover, the mulberry leaves that are not consumed could be used to make green tea (containing antioxidants), traditional medicine, or animal food. The silkworm exuviae, including the outer skins shed as they grow, have unexploited potential in biomedicine, cosmetics, and biodegradable wrapping. In addition, silk gum (sericin), the natural glue used to attach silk fibers, which is removed during the degumming process, can be converted into natural adhesives, biodegradable films, and wound dressings. By implementing sustainable practices, the silk industry can significantly reduce its ecological footprint and create new opportunities for innovation and value addition (Sharma et al. 2022).

Vermicomposting is another sustainable method for managing waste in sericulture. Such waste has been described by the presence of a large majority of organic matter, which, by the operation of the earthworm, is readily converted into humus of a highly nutritious value. It takes place in sets of stages. Sericulture waste is first collected in the bed or on the floor, and then it is protected from rain and direct sunlight. A mixture of sericulture wastes, cow dung, and water is thereafter prepared to activate the degradation process; this is continued with more organic material until the pile reaches a height of 30–40 cm. In this step, earthworms are added to accelerate the decomposition of organic matter. To ensure optimal conditions, the compost is covered with a wet jute sack, and regular watering, aeration, and tunnel formation are applied to maintain good moisture and airflow. The nutrient-rich vermicompost can then be harvested when the compost has matured and used in gardens or on farms. Its quality needs to be monitored at all times. Utilization of vermicomposting in sericulture has high benefits as the farmers are able to minimize waste production, environmental conservation, generate organic, high-quality fertilizer, and long-term profitability (Ganesan et al. 2022).

### 11.6.2 Product Life Cycle Analysis of Silk Products

Life Cycle Analysis (LCA) is a systematic method for evaluating the environmental impacts of a product, material, process, or activity throughout its entire life cycle. A Life Cycle Analysis of silk products describes the environmental impact of silk-based products, such as clothing, accessories, and home textiles, across their entire life span (Vollrath et al. 2013). All silk products undergo five major stages during their Life Cycle Assessment, which include the acquisition of cocoons, industrial processing, distribution, consumer use, and end-of-life. The evaluation begins with cocoon procurement, in which silkworms fed on mulberry leaves build up cocoon strings, which are later harvested by farmers. The ensuing cocoons are the primary raw material of silk manufacture. The next step is the industrial stage, which involves processing the cocoons into a fiber via spinning, weaving, and sewing. The silk can also be processed further to add dyeing, prints, or combine it with other fibers to add strength, texture or even color. Then comes the distribution stage, which involves packaging and transporting the finished silk products to retailers and consumers. This step is vital because transportation and packaging have noticeable environmental effects. Silk is quite popular among consumers due to its elegance, comfort, and durability. It is used in clothing, soft furnishings, home furnishings, accessories, underwear, insect-proof clothing, and various sports. Silk products require regular washing and proper care during this period. The final end-of-life stage examines what happens to the silk products after they are no longer in use. There may be discarding, reusing, or recycling. While the world seems to be embracing sustainability, the importance of recycling and proper disposal is on the rise. Both stages thus have to impact the total environmental implications of silk, emphasizing the need to practice sustainability throughout its entire production life cycle (Liu et al. 2022). In the context of the sericulture industry, sustainability is further reinforced through a circular economy approach (Fig. 11.5), where waste management is integrated into the core production process. The sustainability chain begins with the growth of white mulberry, along with the rearing of silkworms, and the mechanism then proceeds to manufacture cocoons and other raw materials into finished goods, such as cosmetics and biomaterials. Notably, all-natural waste produced during these processes can be converted into highly valuable resources, including organic fertilizer, pellets, and biogas. This re-use of by-products by recycling them back into the manufacturing system can be used to cultivate other saturing mulberry trees, creating a sustainable loop and reducing waste production, and making the system more resource-efficient and sustainable.

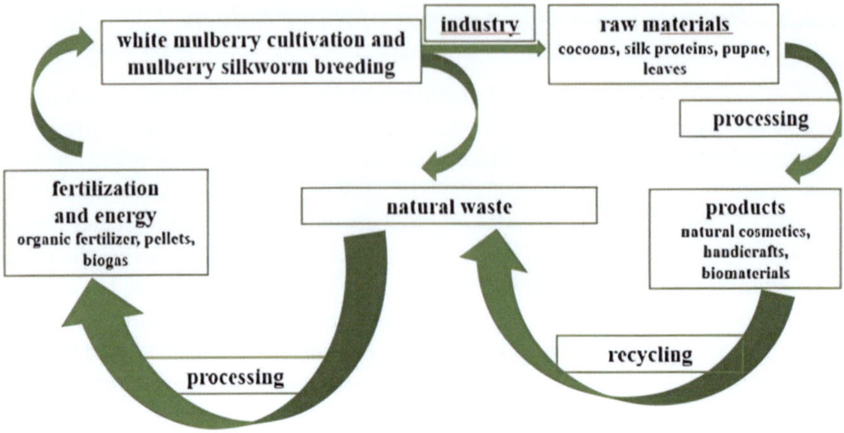


**Fig. 11.5** Utilizing mulberry silkworm waste and byproducts for sustainability in a circular economy (Grześkowiak et al. 2022). Copyright©2022 Grześkowiak et al., published by Lublin University of Technology

### *11.6.3 Importance and Procedure for Reuse and Recycling of Silken Fabrics*

The silk is typically preferred due to its soft texture, high breathability, and aesthetic value. However, traditional silk manufacturing has some shortcomings, including ethical issues, excessive water usage, and the use of highly toxic substances (Ghonmode 2020; Gregory and Gregory 2002). As a result, the reusing and recycling of silk fabrics have drawn academic and industrial attention as a more sustainable alternative. Recycled silk can help reduce the demand for raw primary materials by transforming unwanted garments, fabric scraps, and unused cocoons into reclaimed yarns and new textile products (Hassan et al. 2025). Such an environment-based practice minimizes the amount of water and energy used, limits the quantity of textile waste that ends up in landfills, and reduces greenhouse gas emissions, thereby lowering the overall amount of greenhouse gases produced by the fashion industry. The concept of using and upcycling waste silk fabric is a key component of a circular economy model, where the focus is on product reuse and maximizing life cycles.

One viable way to extend the life cycles of textiles is to encourage the use of second-hand clothes, clothes swapping, and renting, as well as shopping at thrift shops (Gray et al. 2022). The speed of such repurposing is now achieved online via online marketplaces and peer-to-peer exchange networks (Korvenranta 2023). The reprocessing of the silk fabric is not only environmentally beneficial in terms of how carefully the fabric can be cleaned and handled so that the material can be reengineered into accessories, home textiles, or handcrafted jewelry but also economically as even the most damaged pieces can be broken down without leaving behind any toxic residue, and thus ecological damage is minimal (Das et al. 2025). Additionally, recycled silk can be used to produce innovative new fabrics, supporting sustainability

**Table 11.6** Recycling methods of silk fabrics

| Recycling method | Description |
|---|---|
| Bioenergy conversion | Waste silk (WS) can be burned for bioenergy, reducing landfill waste but releasing harmful $CO_2$ (Leal Filho et al. 2019; Zamani et al. 2015). Alternatively, protein-rich WS powder can be used as poultry feed, improving egg production and lowering cholesterol. WS also holds potential for sustainable biofuel production (Kim et al. 2002) |
| Mechanical recycling | WS can be turned into fiber assemblies for yarns, fabrics, or composite reinforcements. WS can be blended with pineapple leaf fibers to create stronger textiles (Hazarika et al. 2018). Furthermore, because of its lightweight it is combined with materials like high-density polyethylene (HDPE) (Koçak et al. 2008), polypropylene (PP) (Govindaraju and Jagannathan 2018), and vinyl ester (VE) resin (Ravindra Rama and Rai 2011) |
| Non-woven conversion | Used silk garments are recycled by recovering fibers, needling them into strong non-woven fabric, and winding for storage and transport (Sashina and Yakovleva 2023) |

in the fashion industry. By choosing recycled silk, both consumers and brands can contribute to eco-conscious fashion without compromising on style or quality. Well-known fashion brands such as EILEEN FISHER, Patagonia, Reformation, and Stella McCartney have already demonstrated that fashion and sustainability can successfully go hand in hand. Some processes exist for recycling silk fabrics, which are described in Table 11.6.

### *11.6.4 Socio-Economic Impact of Sustainable Sericulture*

Sustainable sericulture is a significant source of high social and economic benefits, particularly for rural and tribal communities (Mushtaq et al. 2023). The activities of silkworm breeding to produce silk are also described as profitable and flexible, which help enhance the living standards in underdeveloped areas. Its key strength is that it creates jobs: sericulture is also labor-intensive, generating employment throughout the entire value chain, including the cultivation of mulberry trees, the rearing of silkworms, the processing of silk, and the marketing of the products (Sharma and Kapoor 2020). Through this, it increases the incomes of families, offering a more reliable and consistent income compared to seasonal farming, and is therefore particularly useful to low-income families. Environmentally, a mulberry plantation is beneficial in maintaining ecological balance and improving air quality. Sericulture is a lucrative sector that is expected to yield quick economic returns. Furthermore, in addition to financial results, it fosters community participation and social integration, while also empowering women and educating rural youth by providing readily available opportunities in entrepreneurship and self-employment (Sharma and Kapoor 2020; Ssemugenze et al. 2021).

## 11.7 Collaboration and Policymaking in the Silk Industry

The sustainability of the silk industry is not only enabled by technological innovations but also by policy initiatives and collaborative partnerships. Involvement by multiple stakeholders, such as governments, private sector organizations, research centers, and consumers, can enable enhanced adoption of green silk production. Policies and certification enable responsible production, while partnerships enable the development of green supply chains. This section discusses key industry certifications and partnership structures that enable green silk production.

### *11.7.1 Industry Certification: International Initiatives on Sustainable Silk*

Several global certification programs have been developed to make the silk industry more sustainable with a focus on environmental conservation and fair treatment of labor. The OEKO-TEX Standard 100 is one of the most commonly used frameworks and standards because it guarantees that the given textile product does not contain harmful substances, such as heavy metals, formaldehyde, and dyes that will trigger an allergic response. In this certification, the health of human beings is preserved through the evaluation of whether the silk textiles can be used and come into contact with the skin, (Association 2025; Silk 2025). The Bluesign certification system also promotes the sustainability of silk product manufacturing by integrating the use of safe chemicals, promoting water and energy conservation, and reducing pollution in production (AG 2025). In addition to such measures, the Fair-Trade certification protects the silk farmers and workers by ensuring fair payment, a good working environment, the absence of child labor, and the provision of equal opportunities. At the same time, it promotes responsible, environment-friendly agronomy by using fewer chemical inputs in mulberry cultivation (International 2025). Cradle to Cradle Certified has also been used to test some silk yarns, including Slow Silk Yarns. This standard evaluates the entire life cycle of the product, encompassing material safety, recycling, renewable energy consumption, water consumption, and social fairness (Giuseppe and S.P.A. 2024). All these certification mechanisms promote an accountable and ecological silk industry as a whole and thus they ensure confidence to consumers that products are made available in a responsible and ecologically conscious manner.

### *11.7.2 Partnership Models and Practices to Build Sustainable Silk Value Chains*

The development of a sustainable value chain in the production of silk requires collaboration of government, private industry, NGOs, research institutions, and farmer cooperatives. The existence of the public–private partnerships (PPPs) is now observed as an indispensable framework, whereby the government provides infrastructure, subsidies, and policy support, whereas the private sector provides capital, technology, as well as markets, especially in case of organic sericulture and eco-friendly practices like cluster level shared processing unit (Alimjonovna 2025; Government of India 2007). In this context, the Silk Samagra scheme (2021–26) of India consolidates training, R&D, seed systems, certification, and market development, domestically coordinated by the central and state governments of India (Press Information Bureau 2025). The other stakeholders (NGOs and research institutions) provide capacity building programs and technological development, such as eco dyeing, closed loop processing, and peace silk procedures. Multistakeholder organizations, such as the Sustainable Apparel Coalition (SAC), offer common solutions, like the Higg Index, to evaluate and improve their environmental and social performance across the entire textile supply chain, including silk goods (Coalition 2025). The local farmer cooperatives and associations enable joint training, fair trade, and better access to inputs and markets, and a stronger negotiation position among silk producers (Meena et al. 2023; Taufique and Hoque 2021).

## 11.8 Challenges and Future Prospects of Sustainable Silk in the Global Market

Sericulture, a subsidiary industry of agriculture, presents opportunities for rural economic development. The importation of cheap cloth, outdated manufacturing methods, low-quality seeds, a lack of bivoltine seed production, and farmers' ignorance regarding agricultural diseases are some of its problems. The low cost of production, recurring droughts, fluctuating prices, limited market coverage, long distances to markets, inadequate means of transportation and storage, limited market awareness, and lack of financial resources are some of the problems that impede its growth. The silk industry, a key player on the international stage, is expected to expand despite these challenges due to a favorable climate and increased female involvement. The cost of the end product is high due to the labor-intensive and costly process of producing silk. Due to ethical concerns, alternative methods of producing silk are necessary, as traditional silk-making involves the killing of silkworms. The cost of cocoons and silk yarn production is determined by supply and demand conditions, climate, and fluctuations in raw material prices, all of which impact profit. Demand for silk is increasing as customers invest more in designer clothing and demand organic and sustainable fabrics. Sericulture is promoted by nations through incentives, research funding, and subsidies. Meanwhile, online shopping websites offer a huge market for silk clothing and accessories. The silk industry will continue to grow as new developments in green silk manufacturing and advanced applications emerge. Some of them include creating alternatives that are cruelty-free to address ethical issues, combining silk with nanotechnology to improve function and longevity, and helping to promote organic sericulture and small-scale farming. The consumer's need for luxury, sustainability, and innovation is driving the silk market to expand. The market will continue to grow as long as governments promote silk production and companies seek new uses for it. To help companies capitalize on new trends and remain competitive in the evolving silk industry, personalized market insights provide comprehensive market research.

## 11.9 Conclusion

This chapter describes the environmental concerns of traditional and contemporary sericulture processes, as well as the sustainability issues associated with silk production, with a focus on the environmental implications of traditional sericulture practices. There are environmental concerns associated with the production of silk fibers, including the use of freshwater, solid waste, and harmful chemicals, as well as wastewater and air and water pollution. Silkworm waste can be a sustainable alternative to synthetic fertilizers, but contamination may occur. Composting promotes carbon sequestration and reduces reliance on synthetic fertilizers. Large textile manufacturers are attempting to overcome these challenges by switching to eco-friendly silk fibers, as awareness of the problem is growing worldwide. Conventional sericulture that takes up land, uses water, is pesticide-based, and carbon-emitting is studied. Reductions on these effects are considered, such as synthetic and bio-silk substitutes, eco-friendly processing technologies, improved water and waste management, the industry efforts towards adopting the principles of circular economy, and technical advancements and policy making side by side, create a sustainable system for silk production. This chapter outlines how and why silk, as an industry, must change if it has to confront environmental issues but hold economic and cultural significance. It discusses bio-silk manufacturing, artificial silk, biodegradable alternatives, and the concept of a circular economy. It also explores waste management, product life cycle analysis, and the socioeconomic impacts of sericulture sustainably.

**Declaration of generative AI and AI-assisted technologies in the writing process** The authors used QuillBot and Grammarly to correct grammatical errors and enhance the language quality while preparing our text. The only purpose of these technologies was to improve language clarity and readability. The authors are the original creators of all the main ideas, findings, and writings. The authors accept full responsibility for the accuracy and integrity of the final text, which we have carefully examined.

## References


Acharya C, Kumar V, Sen R, Kundu SC (2008) Performance evaluation of a silk protein-based matrix for the enzymatic conversion of tyrosine to l-dopa. Biotechnol J 3(2):226–233. https://doi.org/10.1002/biot.200700120

Adeel S, Liaqat S, Hussaan M, Mia R, Ahmed B, Wafa H (2022) Environmental friendly bio-dyeing of silk using alkanna tinctoria based alkannin natural dye. Ind Crops Prod 186:115301. https://doi.org/10.1016/j.indcrop.2022.115301

AG BT (2025) Bluesign | sustainable solutions for a safer, cleaner textile industry. https://www.bluesign.com/en. Accessed 18 Aug 2025

Alimjonovna MN (2025) Green innovation strategies for advancing the silk industry: problems and prospects. Mod Am J Bus Econ Entrep 1(3):456–462

Altman GH, Diaz F, Jakuba C, Calabro T, Horan RL, Chen J, Lu H, Richmond J, Kaplan DL (2003) Silk-based biomaterials. Biomaterials 24(3):401–416. https://doi.org/10.1016/S0142-9612(02)00353-8

Altman GH, Farrell BD (2022) Sericulture as a sustainable agroindustry. Clean Circ Bioecon 2:100011. https://doi.org/10.1016/j.clcb.2022.100011

Altman GH, Horan RL, Lu HH, Moreau J, Martin I, Richmond JC, Kaplan DL (2002) Silk matrix for tissue engineered anterior cruciate ligaments. Biomaterials 23(20):4131–4141. https://doi.org/10.1016/S0142-9612(02)00156-4

Amirian J, Wychowaniec JK, Amel Zendehdel E, Sharma G, Brangule A, Bandere D (2023) Versatile potential of photo-cross-linkable silk fibroin: roadmap from chemical processing toward regenerative medicine and biofabrication applications. Biomacromolecules 24(7):2957–2981. https://doi.org/10.1021/acs.biomac.3c00098

Andersen SO (1970) Amino acid composition of spider silks. Comp Biochem Physiol 35(3):705–711. https://doi.org/10.1016/0010-406X(70)90988-6

Association OT (2025) Oeko-tex standard 100. https://www.oeko-tex.com/en/our-standards/oeko-tex-standard-100. Accessed 18 Aug 2025

Attri K, Vakayil S, Shruthi G, Parmar S, Harika KR, Teja KSS, Panigrahi CK (2024) Integrative approaches in sericulture from traditional practices to modern technologies sericulture and its contribution to biodiversity conservation. J Adv Biol Biotechnol 27(8):480–490

Ayoub NA, Garb JE, Tinghitella RM, Collin MA, Hayashi CY (2007) Blueprint for a high performance biomaterial: full-length spider dragline silk genes. PLoS ONE 2(6):e514. https://doi.org/10.1371/journal.pone.0000514

Baker BP, Green TA, Loker AJ (2020) Biological control and integrated pest management in organic and conventional systems. Biol Control 140:104095. https://doi.org/10.1016/j.biocontrol.2019.104095

Barcelos SM, Salvador R, Guedes MD, de Francisco AC (2020) Opportunities for improving the environmental profile of silk cocoon production under brazilian conditions. Sustainability 12(8). https://doi.org/10.3390/su12083214

Bell FI, McEwen IJ, Viney C (2002) Supercontraction stress in wet spider dragline. Nature 416(6876):37–37. https://doi.org/10.1038/416037a

Bexiga NM, Bloise AC, de Moraes MA, Converti A, Beppu MM, Polakiewicz B (2017) Production and characterization of fibroin hydrogel using waste silk fibers. Fibers Polym 18:57–63. https://doi.org/10.1007/s12221-017-6805-8

Blackledge TA, Boutry C, Wong S-C, Baji A, Dhinojwala A, Sahni V, Agnarsson I (2009) How super is supercontraction? persistent versus cyclic responses to humidity in spider dragline silk. J Exp Biol 212(13):1981–1989. https://doi.org/10.1242/jeb.028944

Boutry C, Blackledge TA (2010) Evolution of supercontraction in spider silk: structure–function relationship from tarantulas to orb-weavers. J Exp Biol 213(20):3505–3514. https://doi.org/10.1242/jeb.046110

Bray C (1928) Boiling off silk: methods. Am Dyestuff Rep 17:743–747

Can LIU, Qizheng LI, Shuyi LIU, Ying Z, Laili W (2024) Discussion on key issues of carbon footprint quantification of silk products. Ind Text 75:251–258. https://doi.org/10.35530/IT.075.03.202345

Chen K, Sahoo S, He P, Ng KS, Toh SL, Goh JC (2012) A hybrid silk/rada-based fibrous scaffold with triple hierarchy for ligament regeneration. Tissue Eng Part A 18(13–14):1399–1409. https://doi.org/10.1089/ten.tea.2011.0376

Chequer FMD, de Oliveira GAR, Ferraz ERA, Cardoso JC, Zanoni MVB, de Oliveira DP (2013) Textile dyes: dyeing process and environmental impact. In: Eco-friendly textile dyeing and finishing ed.. IntechOpen

Chopra S, Gulrajani M (1994) Comparative evaluation of the various methods of degumming silk

CM RS, VL, U AK, K LP, B MK (2023) The role of neem in integrated pest management: a sustainable approach for mulberry ecosystem protection. J Entomol Zool Stud 11(6):130–134. https://doi.org/10.22271/j.ento.2023.v11.i6b.9449

Coalition SA (2025) Sustainable apparel coalition/higg product tools. https://product.higg.org/page/sustainable-apparel-coalition. Accessed 18 Aug 2025

Cooper JA, Lu HH, Ko FK, Freeman JW, Laurencin CT (2005) Fiber-based tissue-engineered scaffold for ligament replacement: design considerations and in vitro evaluation. Biomaterials 26(13):1523–1532. https://doi.org/10.1016/j.biomaterials.2004.05.014

da Silva Junior CJG, de Amorim JDP, de Medeiros ADLM, de Holanda Cavalcanti AKL, do Nascimento HA, Henrique MA, do Nascimento Maranhão LJC, Vinhas GM, de Oliveira Souto Silva KK, de Santana Costa AF (2022) Design of a naturally dyed and waterproof biotechnological leather from reconstituted cellulose. J Funct Biomater 13(2):49. https://doi.org/10.3390/jfb13020049

Dal Pra I, Freddi G, Minic J, Chiarini A, Armato U (2005) De novo engineering of reticular connective tissue in vivo by silk fibroin nonwoven materials. Biomaterials 26(14):1987–1999. https://doi.org/10.1016/j.biomaterials.2004.06.036

Das AK, Hossain MF, Khan BU, Rahman MM, Asad M, Akter M (2025) Circular economy: a sustainable model for waste reduction and wealth creation in the textile supply chain. SPE Polym 6(1):e10171. https://doi.org/10.1002/pls2.10171

Dhilip Kumar SS, Abrahamse H (2022) Sericin-based nanomaterials and their applications in drug delivery. In: Bio-based nanomaterials: synthesis protocols, mechanisms and applications ed., pp 211–229. Elsevier

Ck DS, Gowthami B, Durga VPN (2024) Valuating the ecological footprint of sericulture: a comparative analysis with other textile industries. Int J Agric Ext Soc Dev 7(8):657–661. https://doi.org/10.33545/26180723.2024.v7.i8i.1618

Du N, Yang Z, Liu XY, Li Y, Xu HY (2011) Structural origin of the strain-hardening of spider silk. Adv Funct Mater 21(4):772–778. https://doi.org/10.1002/adfm.201001397

Eisoldt L, Smith A, Scheibel T (2011) Decoding the secrets of spider silk. Mater Today 14(3):80–86. https://doi.org/10.1016/S1369-7021(11)70057-8

Ekka V, Bais D (2023) Contribution of sericulture in the socio-economic transformation of the workers. Asian J Appl Sci Technol 07:217–224. https://doi.org/10.38177/ajast.2023.7223

El-Fakharany EM, Abu-Elreesh GM, Kamoun EA, Zaki S, Abd-EL-Haleem DA (2020) In vitro assessment of the bioactivities of sericin protein extracted from a bacterial silk-like biopolymer. RSC Adv 10(9):5098–5107. https://doi.org/10.1039/C9RA09419A

El-Fakharany EM, Abu-Serie MM, Habashy NH, El-Deeb NM, Abu-Elreesh GM, Zaki S, Abd-EL-Haleem D (2022) Inhibitory effects of bacterial silk-like biopolymer on herpes simplex virus type 1, adenovirus type 7 and hepatitis c virus infection. J Funct Biomater 13(1):17. https://doi.org/10.3390/jfb13010017

Elizabeth (2025) Future innovations in sericulture. https://www.pandasilk.com/future-innovations-in-sericulture/. Accessed 20 Aug 2025

ElleSilk (2023) Sustainable silk production practices. https://www.ellesilk.com/blog/sustainable-silk-production-practices. Accessed 20 Aug 2025

Fahnestock SR, Yao Z, Bedzyk LA (2000) Microbial production of spider silk proteins. Rev Mol Biotechnol 74(2):105–119. https://doi.org/10.1016/S1389-0352(00)00008-8
Fajfer D, Łochyńska M (2022) Impact of ecological plant protection products on mortality and cocoon shell ratio of mulberry silkworms (bombyx mori l.)–pilot studies. J Plant Prot Res 281–286
Fatimaa S, Isharb AK, Jeeva P, Devi HD, Rajeswari SU, Kumari B (2024) Recent innovations in sericulture: a comprehensive review of advancements in silk production and quality enhancement. Uttar Pradesh J Zool 45(23):75–83
Ganesan S, Padmapriya G, Debbarma A, De Zoysa ASA, Tharudini JH (2022) Eco-friendly approach for sericulture waste management and nutrient enhancement using vermicomposting technology to improve socio-economic status of karnataka: a review. Asian J Microbiol Biotechnol Environ Sci 24(2):357–360
Garay L, Nembri A, Oro A, Fassina V, Saez C, Chiarello A, Climas Pereira N, Pessini G, Munhoz R, Fernandez M (2014) New technique to produce large amount of flat silk by biospinning. Agric Sci 05:1483–1490. https://doi.org/10.4236/as.2014.514159
Garb JE, Ayoub NA, Hayashi CY (2010) Untangling spider silk evolution with spidroin terminal domains. BMC Evol Biol 10:1–16. https://doi.org/10.1186/1471-2148-10-243
Ghonmode S (2020) Study on health issues faced by workers of sericulture industry and safety measures at work place. Vidyabharati Int Interdiscip Res J
Giacomin AM, Garcia JB Jr, Zonatti WF, Silva-Santos MC, Laktim MC, Baruque-Ramos J (2017) Brazilian silk production: economic and sustainability aspects. Procedia Eng 200:89–95. https://doi.org/10.1016/j.proeng.2017.07.014
Giuseppe B, SPA F (2024) Slow silk yarns – cradle to cradle certified® gold. https://c2ccertified.org/products/registry/slow-silk-yarns. Accessed 23 Aug 2025
Government of India MoT, Office of the Textile Commissioner (2007) Sericulture and silk industry (chapter 12, xi plan). https://www.txcindia.gov.in/html/XIplanch12.pdf. Accessed 23 Aug 2025
Govindaraju R, Jagannathan S (2018) Optimization of mechanical properties of silk fiber-reinforced polypropylene composite using box behnken experimental design. J Ind Text 47(5):602–621. https://doi.org/10.1177/1528083716667257
Gray S, Druckman A, Sadhukhan J, James K (2022) Reducing the environmental impact of clothing: an exploration of the potential of alternative business models. Sustainability 14(10):6292. https://doi.org/10.3390/su14106292
Greenwald D, Shumway S, Albear P, Gottlieb L (1994) Mechanical comparison of 10 suture materials before and after in vivo incubation. J Surg Res 56(4):372–377. https://doi.org/10.1006/jsre.1994.1058
Gregory S, Gregory J (2002) Agronomic practices and social constraints in water management: the case of sericultur. Small Enterp Dev Manag Ext J (SEDME) 29(2):65–76. https://doi.org/10.1177/0970846420020207
Grześkowiak J, Łochyńska M, Frankowski J (2022) Sericulture in terms of sustainable development in agriculture. Probl Ekoroz 17(2):210–217. https://doi.org/10.35784/pe.2022.2.23
Gulrajanid M, Sethi S, Gupta S (1992) Some studies in degumming of silk with organic acids. J Soc Dyers Colour 108(2):79–86. https://doi.org/10.1111/j.1478-4408.1992.tb01420.x
Haggag K, El-Sayed H, Allam O (2007) Degumming of silk using microwave-assisted treatments. J Nat Fibers 4(3):1–22
Hassan R, Acerbi F, Terzi S, Rosa P (2025) Circular economy solutions for sustainability in textile and silk sector. In: Circular business models in the apparel industry, vol 5–32. Springer
Hauptmann V, Weichert N, Menzel M, Knoch D, Paege N, Scheller J, Spohn U, Conrad U, Gils M (2013) Native-sized spider silk proteins synthesized in planta via intein-based multimerization. Transgenic Res 22(2):369–377. https://doi.org/10.1007/s11248-012-9655-6
Hazarika M (2025) Pollution and its impact on sericulture industry. J Exp Zool 20:33–37
Hazarika P, Hazarika D, Kalita B, Gogoi N, Jose S, Basu G (2018) Development of apparels from silk waste and pineapple leaf fiber. J Nat Fibers 15(3):416–424. https://doi.org/10.1080/15440478.2017.1333071

He J-H, Liu Y, Xu L, Yu J-Y, Sun G (2008) Biomimic fabrication of electrospun nanofibers with high-throughput. Chaos, Solitons Frac 37(3):643–651. https://doi.org/10.1016/j.chaos.2007.11.028
Heim M, Keerl D, Scheibel T (2009) Spider silk: From soluble protein to extraordinary fiber. Angew Chem Int Ed 48(20):3584–3596. https://doi.org/10.1002/anie.200803341
Humenik M, Magdeburg M, Scheibel T (2014) Influence of repeat numbers on self-assembly rates of repetitive recombinant spider silk proteins. J Struct Biol 186. https://doi.org/10.1016/j.jsb.2014.03.010
Inouye K, Kurokawa M, Nishikawa S, Tsukada M (1998) Use of bombyx mori silk fibroin as a substratum for cultivation of animal cells. J Biochem Biophys Meth 37(3):159–164. https://doi.org/10.1016/S0165-022X(98)00024-4
International F (2025) What is fairtrade? https://www.fairtrade.net/en/why-fairtrade/what-we-do/what-is-fairtrade.html. Accessed 23 Aug 2025
Jajpura L, Khandual A, Pai R (2004) Effluent treatment in textile industries. Text Mag 4(45):34
Jose S, Gurumallesh Prabu H, Ammayappan L (2017) Eco-friendly dyeing of silk and cotton textiles using combination of three natural colorants. J Nat Fibers 14(1):40–49. https://doi.org/10.1080/15440478.2015.1137530
Kamoun EA, Abu-Elreesh GM, El-Fakharany EM, Abd-El-Haleem D (2019) A novel bacterial polymeric silk-like protein from a petroleum origin bacillus sp. Strain ne: Isolation and characterization. J Polym Environ 27:1629–1641. https://doi.org/10.1007/s10924-019-01459-2
Karthik T, Rathinamoorthy R, Ganesan P (2015) Sustainable luxury natural fibers—production, properties, and prospects. In: Handbook of sustainable luxury textiles and fashion: vol1, pp 59–98
Khattab TA, Abdelrahman MS, Rehan M (2020) Textile dyeing industry: environmental impacts and remediation. Environ Sci Pollut Res 27(4):3803–3818. https://doi.org/10.1007/s11356-019-07137-z
Kim D, Kim Y, Kim K, Chin Y, Ueda M (2002) Recycled silk wastes as feed ingredient for poultry. SEN-I GAKKAISHI 58(3):106–110
Kind E (2025) Why choose organic peace silk. Ethical Kind. https://ethicalkind.com/blogs/news/why-choose-organic-peace-silk. Accessed 23 Aug 2025
Kishore S, Khajuria M, Shalini K, Saini A, Srivastav S, Manideep K, Chethan M, Morabad P (2024) Advancing sustainable sericulture: a review on biological control agents in managing pests and diseases of mulberry and silkworms. J Exp Agric Int 46(9):1139–1146
Koçak D, Taşdemir M, Usta İ, Merdan N, Akalin M (2008) Mechanical, thermal, and microstructure analysis of silk-and cotton-waste-fiber-reinforced high-density polyethylene composites. Polym Plast Technol Eng 47(5):502–507. https://doi.org/10.1080/03602550801977919
Korvenranta L (2023) Fashion peer-to-peer platforms: An exploratory user study. Master's thesis, Aalto University, Espoo. https://aaltodoc.aalto.fi/items/e3a45581-53f0-4910-ad60-5e5b5be95d08
Lang G, Jokisch S, Scheibel T (2013) Air filter devices including nonwoven meshes of electrospun recombinant spider silk proteins. J vis Exp. https://doi.org/10.3791/50492
Lara L, Cabral I, Cunha J (2022) Ecological approaches to textile dyeing: A review. Sustainability 14(14):8353. https://doi.org/10.3390/su14148353
Leal Filho W, Ellams D, Han S, Tyler D, Boiten VJ, Paço A, Moora H, Balogun A-L (2019) A review of the socio-economic advantages of textile recycling. J Clean Prod 218:10–20. https://doi.org/10.1016/j.jclepro.2019.01.210
learners T (2014) Water consumption in textile industry. Accessed 20 Aug 2025.
Lee H, Park SJ, Lee M-e, Choi K-m, Choi HY, Hasegawa Y, Kim M, Kim KB (2020) Fabrication of nanofibers using fibroin regenerated by recycling waste silk selvage. Polym Bull 77(7):3853–3862. https://doi.org/10.1007/s00289-020-03113-7
Li Y, Wang Y, He Q, Yang Y (2020) Calculation and evaluation of carbon footprint in mulberry production: a case of haining in china. Int J Environ Res Public Health 17(4):1339. https://doi.org/10.3390/ijerph17041339

Liu S, Liu H, Meng Y, Li Q, Wang L (2022) Review of carbon emission and carbon neutrality in the life cycle of silk products. Fibres Text East Eur 30(2)
Liu y, Qu J, Li M (2017) Accelerated vascularization of silk fibroin scaffolds through immobilized basic fibroblast growth factor (bfgf). DEStech Trans Eng Technol Res. https://doi.org/10.12783/dtetr/apetc2017/11439
Lu HH, Cooper JA Jr, Manuel S, Freeman JW, Attawia MA, Ko FK, Laurencin CT (2005) Anterior cruciate ligament regeneration using braided biodegradable scaffolds: In vitro optimization studies. Biomaterials 26(23):4805–4816. https://doi.org/10.1016/j.biomaterials.2004.11.050
Mahesh R, Anil P, Debashish C, Sivaprasad V (2022) Improved mulberry productivity and resource efficiency through low-cost drip fertigation. Arch Agron Soil Sci 68(6):749–763. https://doi.org/10.1080/03650340.2020.1852552
Mahmoodi NM, Moghimi F, Arami M, Mazaheri F (2010) Silk degumming using microwave irradiation as an environmentally friendly surface modification method. Fibers Polym 11:234–240. https://doi.org/10.1007/s12221-010-0234-2
Mahmud I, Kaiser S (2020) Recent progress in waterless textile dyeing. J Text Sci Eng 10(6):1–3
Meena B, Tyagi S, Sharma P (2023) Sericulture development in india with special reference to public-private partnership. In: Essentials of economic entomology, vol 5, pp 131–137. AkiNik Publications
Meinel L, Fajardo R, Hofmann S, Langer R, Chen J, Snyder B, Vunjak-Novakovic G, Kaplan D (2005) Silk implants for the healing of critical size bone defects. Bone 37(5):688–698. https://doi.org/10.1016/j.bone.2005.06.010
Mollahosseini H, Fashandi H, Khoddami A, Zarrebini M, Nikukar H (2019) Recycling of waste silk fibers towards silk fibroin fibers with different structures through wet spinning technique. J Clean Prod 236:117653. https://doi.org/10.1016/j.jclepro.2019.117653
Mridha A, Mukhopadhyay A, Das D, Sarkar B, Roy Maulik S (2021) Sustainable dyeing of silk with natural dyes using natural sources of water. J Inst Eng (India) Ser E 102:61–74. https://doi.org/10.1007/s40034-020-00195-w
Mukherjee C, Varghese D, Krishna J, Boominathan T, Rakeshkumar R, Dineshkumar S, Rao CB, Sivaramakrishna A (2023) Recent advances in biodegradable polymers–properties, applications and future prospects. Eur Polym J 192:112068. https://doi.org/10.1016/j.eurpolymj.2023.112068
Murphy AR, Kaplan DL (2009) Biomedical applications of chemically-modified silk fibroin. J Mater Chem 19(36):6443–6450. https://doi.org/10.1039/B905802H
Mushtaq R, Qadiri B, Lone FA, Raja TA, Singh H, Ahmed P, Sharma R (2023) Role of sericulture in achieving sustainable development goals. Problemy Ekorozwoju 18(1):199–206
Nandhakumar R, Kaviyarasu R, Kalidass M (2012) Dyeing of fabrics without water: a review. Indian Text J 2:37–39
Nentwig W (2012) Ecophysiology of spiders. Springer Science and Business Media
Neukirch S, Antkowiak A, Rollard C, Vollrath F (2017) Le treuil élastocapillaire dans les toiles d'araignées
Nisal A, Trivedy K, Mohammad H, Panneri S, Sen Gupta S, Lele A, Manchala R, Kumar NS, Gadgil M, Khandelwal H (2014) Uptake of azo dyes into silk glands for production of colored silk cocoons using a green feeding approach. ACS Sustain Chem Eng 2(2):312–317. https://doi.org/10.1021/sc400355k
Numata K, Kaplan DL (2010) Silk-based delivery systems of bioactive molecules. Adv Drug Deliv Rev 62(15):1497–1508. https://doi.org/10.1016/j.addr.2010.03.009
Oroudjev E, Soares J, Arcidiacono S, Thompson JB, Fossey S, Hansma H (2002) Segmented nanofibers of spider dragline silk: Atomic force microscopy and single-molecule force spectroscopy (vol 99, pg 6460, 2002). Proc Natl Acad Sci USA 99:9606–9606. https://doi.org/10.1073/pnas.132282899
Pawcenis D, Koperska M, Milczarek J, Łojewski T (2014) Size exclusion chromatography for analyses of fibroin in silk: optimization of sampling and separation conditions. Appl Phys A 114. https://doi.org/10.1007/s00339-013-8143-z

Pérez-Rigueiro J, Viney C, Llorca J, Elices M (2000) Mechanical properties of single-brin silkworm silk. J Appl Polym Sci 75(10):1270–1277

Perkins WS (1991) A review of textile dyeing processes. In: Text chem colorist ed., vol 23

Peter HG (1993) Water and conflict: fresh water resources and international security. Int Security 18(1):79–112

Press Information Bureau GoI (2025) The magic of indian silk (press release, prid 2120877). https://www.pib.gov.in/PressReleasePage.aspx?PRID=2120877. Accessed 23 Aug 2025

Puerta M, Peresin MS, Restrepo-Osorio A (2020) Effects of chemical post-treatments on structural and physicochemical properties of silk fibroin films obtained from silk fibrous waste. Front Bioeng Biotechnol 8:523949. https://doi.org/10.3389/fbioe.2020.523949

Qi Y, Wang H, Wei K, Yang Y, Zheng R-Y, Kim IS, Zhang K-Q (2017) A review of structure construction of silk fibroin biomaterials from single structures to multi-level structures. Int J Mol Sci 18(3). https://doi.org/10.3390/ijms18030237

Rajendran M, Rahman T, Trivedy K, Subramanium GD, Babu C (2025) Solar drip irrigation kit (sdik) sustaining the mulberry productivity under rainfed sericulture. Int J Environ Agric Biotechnol 10(1):590472. https://doi.org/10.22161/ijeab.101.5

Rammensee S, Slotta U, Scheibel T, Bausch A (2008) Assembly mechanism of recombinant spider silk proteins. Proc Natl Acad Sci USA 105:6590–6595. https://doi.org/10.1073/pnas.0709246105

Ranjitha H, Chandrashekhar S (2021) Enhancing water productivity in mulberry through different levels of drip irrigation and mulching. Mysore J Agric Sci 55(3):264–264

Ravindra Rama S, Rai S (2011) Performance analysis of waste silk fabric-reinforced vinyl ester resin laminates. J Compos Mater 45(23):2475–2480. https://doi.org/10.1177/0021998311401097

Römer L, Scheibel T (2008) The elaborate structure of spider silk: Structure and function of a natural high performance fiber. Prion 2(4):154–161. https://doi.org/10.4161/pri.2.4.7490

Ryszard MK, Maria MT, Malgorzata M, Jorge BB (2012) Future of natural fibers, their coexistence and competition with man-made fibers in 21st century. Mol Cryst Liq Cryst 556(1):200–222. https://doi.org/10.1080/15421406.2011.635962

Sahoo S, Lok Toh S, Hong Goh JC (2010a) Plga nanofiber-coated silk microfibrous scaffold for connective tissue engineering. J Biomed Mater Res B Appl Biomater 95(1):19–28. https://doi.org/10.1002/jbm.b.31678

Sahoo S, Toh SL, Goh JC (2010b) A bfgf-releasing silk/plga-based biohybrid scaffold for ligament/tendon tissue engineering using mesenchymal progenitor cells. Biomaterials 31(11):2990–2998. https://doi.org/10.1016/j.biomaterials.2010.01.004

Sakthivel N, Kumaresan P, Qadri, Jayachandran R, Balakrishna R. (2012) Adoption of integrated pest management practices in sericulture–a case study in tamil nadu (vol 5)

Samanta KK, Basak S, Chattopadhyay S (2016) Sustainable dyeing and finishing of textiles using natural ingredients and water-free technologies. In: (Ed.), Textiles and clothing sustainability: sustainable textile chemical processes, pp 99–131. Springer

Samir A, Ashour FH, Hakim AA, Bassyouni M (2022) Recent advances in biodegradable polymers for sustainable applications. npj Mater Degrad 6(1):68. https://doi.org/10.1038/s41529-022-00277-7

Sangappa S, Dandin S, Trivedy K, Kumar S (2007) Coloured cocoons to coloured silk. Indian Silk 22–24

Sashina ES, Yakovleva OI (2023) The current state and prospects of recycling silk industry waste into nonwoven materials. Fibers 11(6):56. https://doi.org/10.3390/fib11060056

Shahid M, Mohammad F, Chen G, Tang R-C, Xing T (2016) Enzymatic processing of natural fibres: white biotechnology for sustainable development. Green Chem 18(8):2256–2281. https://doi.org/10.1039/C6GC00201C

Shao Z, Vollrath F (2002) Surprising strength of silkworm silk. Nature 418(6899):741–741. https://doi.org/10.1038/418741a

Sharma A, Kalra M (2021) A blockchain based approach for improving transparency and traceability in silk production and marketing. J Phys Conf Ser 1998(1):012013. https://doi.org/10.1088/1742-6596/1998/1/012013
Sharma K, Kapoor B (2020) Sericulture as a profit-based industry—a review. Indian J Pure Appl Biosci 8(4):550–562. https://doi.org/10.18782/2582-2845.8210
Sharma M, Tellili N, Kacem I, Rouissi T (2024) Microbial biopolymers: From production to environmental applications—a review. Appl Sci 14(12). https://doi.org/10.3390/app14125081
Sharma V, Rattan M, Chauhan S (2022) Potential use of sericultural by products: a review. Pharma Innov 1154–1158
Shitole M, Dugam S, Tade R, Nangare S (2020) Pharmaceutical applications of silk sericin. Ann Pharm Fr 78(6):469–486. https://doi.org/10.1016/j.pharma.2020.06.005
Silk J (2025) Our silk. https://junesilk.com/our-silk. Accessed 23 Aug 2025
Singh PL (2011) Silk cocoon drying in forced convection type solar dryer. Appl Energy 88(5):1720–1726. https://doi.org/10.1016/j.apenergy.2010.11.016
Singh R, Saratchandra B (2002) An integrated approach in the pest management in sericulture. Int J Ind Entomol Biomater 5(2):141–151
Singh T, Kapila R (2021) Role of organic production system in improving sericulture. J Text Assoc 82:214–217
Sofia S, McCarthy MB, Gronowicz G, Kaplan DL (2001) Functionalized silk-based biomaterials for bone formation. J Biomed Mater Res 54(1):139–148. 10.1002/1097-4636(200101)54:1%3C139::AID-JBM17%3E3.0.CO;2-7
Solheim E, Sudmann B, Bang G, Sudmann E (2000) Biocompatibility and effect on osteogenesis of poly (ortho ester) compared to poly (dl-lactic acid). J Biomed Mater Res 49(2):257–263. 10.1002/(SICI)1097-4636(200002)49:2%3C257::AID-JBM15%3E3.0.CO;2-5
Ssemugenze B, Esimu J, Nagasha J, Masiga CW (2021) Sericulture: agro-based industry for sustainable socio-economic development: a review. Int J Sci Res Publ 11(9):474–482. https://doi.org/10.29322/IJSRP.11.09.2021.p11756
Sukigara S, Gandhi M, Ayutsede J, Micklus M, Ko F (2003) Regeneration of bombyx mori silk by electrospinning-part 1: processing parameters and geometric properties. Polymer 44:5721–5727. https://doi.org/10.1016/S0032-3861(03)00532-9
Tandon S, Kandasubramanian B, Ibrahim SM (2020) Silk-based composite scaffolds for tissue engineering applications. Ind Eng Chem Res 59(40):17593–17611. https://doi.org/10.1021/acs.iecr.0c02195
Tansil NC, Li Y, Koh LD, Peng TC, Win KY, Liu XY, Han M-Y (2011a) The use of molecular fluorescent markers to monitor absorption and distribution of xenobiotics in a silkworm model. Biomaterials 32(36):9576–9583. https://doi.org/10.1016/j.biomaterials.2011.08.081
Tansil NC, Li Y, Teng CP, Zhang S, Win KY, Chen X, Liu XY, Han MY (2011b) Intrinsically colored and luminescent silk. Adv Mater 23(12):1463. https://doi.org/10.1002/adma.201003860
Taufique M, Hoque MA (2021) Current scenario of sericulture production in india: a spatio-temporal analysis. Int Res J Educ Technol 2(4):12–23
Textile JS, Ltd. HDC (2025) Sericulture. https://www.jharcraft.in/sericulture/. Accessed 20 Aug 2025.
Thamizharasu T, Srikanth P, Madhanram G, Vengateshkumar M, Maheswari S (2025) Silk production systems, making of silk fabric: Industrial aspects. In: Silk marketing dynamics: Scientific insights and future prospects for silk consumers, pp 107–128. SR edu Publications, [Place of publication if available]
Thomas B, Fishwick M, Joyce J, Van Santen A (2012) A carbon footprint for uk clothing and opportunities for savings. Banbury, UK, WRAP
Tokareva O, Jacobsen M, Buehler M, Wong J, Kaplan DL (2014) Structure–function–property–design interplay in biopolymers: spider silk. Acta Biomater 10(4):1612–1626. https://doi.org/10.1016/j.actbio.2013.08.020
Tsunokaye AR (1932) Degumming action of soap on raw silk. J Soc Dyers Colour 48(6):164–167. https://doi.org/10.1111/j.1478-4408.1932.tb01705.x

Tyubaeva P, Varyan I, Krivandin A, Shatalova O, Karpova S, Lobanov A, Olkhov A, Popov A (2022) The comparison of advanced electrospun materials based on poly (-3-hydroxybutyrate) with natural and synthetic additives. J Funct Biomater 13(1):23. https://doi.org/10.3390/jfb13010023

Vasanth V, Senguttuvan K, Priyadharsini P, Gracy RG, Ponnuvel KM, Jayakanthan M (2025) Impact of the insecticide chlorfenapyr on rearing efficiency of silkworms, bombyx mori. Entomol News 132(3):325–334, 310. https://doi.org/10.3157/021.132.0306

Vepari C, Kaplan DL (2007) Silk as a biomaterial. Prog Polym Sci 32(8–9):991–1007

Vollrath F, Carter R, Rajesh G, Thalwitz G, Astudillo MF (2013) Life cycle analysis of cumulative energy demand on sericulture in karnataka, India. Padua, Italy 12

Vollrath F, Edmonds DT (1989) Modulation of the mechanical properties of spider silk by coating with water. Nature 340:305–307. https://doi.org/10.1038/340305a0

Vollrath F, Porter D (2006) Spider silk as archetypal protein elastomer. Soft Matter 2(5):377–385. https://doi.org/10.1039/B600098N

Vyas SK, Shukla SR (2016) Comparative study of degumming of silk varieties by different techniques. J Text Inst 107(2):191–199. https://doi.org/10.1080/00405000.2015.1020670

Wang W, Pan Y, Gong K, Zhou Q, Zhang T, Li Q (2019) A comparative study of ultrasonic degumming of silk sericin using citric acid, sodium carbonate and papain. Color Technol 135(3):195–201. https://doi.org/10.1111/cote.12392

Wani M, Mir M, Mehraj S, Rather D-R, Ganie N, Sahaf K, Hussain A, Baqual M (2018) Effect of different types of mulches on the germination and seedling growth of mulberry (morus sp.). https://doi.org/10.13140/RG.2.2.32360.37129

Xu L, Rainey J, Meng Q, Liu X-Q (2012) Recombinant minimalist spider wrapping silk proteins capable of native-like fiber formation. PLoS ONE 7:e50227. https://doi.org/10.1371/journal.pone.0050227

Zamani B, Svanström M, Peters G, Rydberg T (2015) A carbon footprint of textile recycling: a case study in Sweden. J Ind Ecol 19(4):676–687. https://doi.org/10.1111/jiec.12208

Zhang C, Xia L, Deng B, Li C, Wang Y, Li R, Dai F, Liu X, Xu W (2020) Fabrication of a high-toughness polyurethane/fibroin composite without interfacial treatment and its toughening mechanism. ACS Appl Mater Interfaces 12(22):25409–25418. https://doi.org/10.1021/acsami.0c03936

Zhang W, Yang Z-Y, Cheng X-W, Tang R-C, Qiao Y-F (2019) Adsorption, antibacterial and antioxidant properties of tannic acid on silk fiber. Polymers 11(6):970. https://doi.org/10.3390/polym11060970

Zhang X, Pan Z (2019) Microstructure transitions and dry-wet spinnability of silk fibroin protein from waste silk quilt. Polymers 11(10):1622. https://doi.org/10.3390/polym11101622

Zhao H-P, Feng X-Q, Shi H-J (2007) Variability in mechanical properties of bombyx mori silk. Mater Sci Eng C 27(4):675–683. https://doi.org/10.1016/j.msec.2006.06.031

Zhu J, Zhang Y, Shao H, Hu X (2014) Effects of environment parameters on sol-gel transition and dry-spinnability of regenerated silk fibroin aqueous solution. Fibers Polym 15:540–546. https://doi.org/10.1007/s12221-014-0540-1

Zhu L, Lin J, Pei L, Luo Y, Li D, Huang Z (2022) Recent advances in environmentally friendly and green degumming processes of silk for textile and non-textile applications. Polymers 14(4):659. https://doi.org/10.3390/polym14040659

# Chapter 12
# Beyond the Loom: A Sustainable Prospect for Silk Manufacturing

**Veena S. More, R. Srinivasan, P. Darshan, Aishwarya Lomte, R. Janavi, and B. Kiran Kumar**

## 12.1 Introduction

Since ancient times, people around the world have been fascinated by silk, a beautiful and unique fiber (Dandin 2025). Sericulture involves the rearing of silkworms for silk production, which requires cultivating crops to provide them with nourishment. The four primary types of silkworms used in this process are Eri, Mulberry, Muga, and Tasar, each producing different varieties of natural silk. Silk from Bombyx mori cocoons, primarily composed of fibroin, has been utilized in the textile industry for thousands of years due to its exceptional mechanical features and biocompatibility. This adaptable protein can be extracted and processed for various applications, including biomaterials for medical implants and photonic devices such as sensors and lasers (Pugina and Caiut 2021). Some moth species in the bombycid and saturnid families produce silk from their larvae, with the domesticated mulberry silkworm, Bombyx mori, being the most common for industrial silk production. Additionally, various saturnid species are known for producing wild silk in their natural habitats (Czaplicki et al. 2021).

### *12.1.1 Overview of Silk's History and Historical Weaving Cultural Importance*

The commerce of silk, which dates back to the early era of human history, was a significant activity that facilitated the first wave of globalization. It connected Europe and Asia via the "Silk Road," a sophisticated, technologically advanced enterprise. Currently, the production of silk accounts for over 0.2% of the global textile output,

V. S. More (✉) · R. Srinivasan · P. Darshan · A. Lomte · R. Janavi · B. Kiran Kumar
Department of Biotechnology, Sapthagiri College of Engineering, Bangalore, India
e-mail: veenasmore@sapthagiri.edu.in



D. Kumar and M. Shahid (eds.), *Science and Technology of Silk*, Textile Science and Clothing Technology, https://doi.org/10.1007/978-3-032-16230-4_12

which is distributed across approximately 60 countries. According to Giora et al. (2022) and Pavithra et al. (2024), India, China, Uzbekistan, Vietnam, Thailand, and Brazil are the most significant producers of silk. From protein molecules to webs and cocoons, silk is one of nature's most fascinating materials in terms of its strength, biological function, and toughness in relation to mechanical and biological qualities. The Silk Road Programme was launched by UNESCO in 1988 to give a forum for showcasing and recording the remarkable and revolutionary effects of the silk trade through the cultural exchange that took place as traders, starting with a diplomatic mission from China to western Asia in 139 BCE, traded silk from China throughout Asia, Europe, and North Africa. Throughout Southeast Asia, the production of silk and its many applications expanded in the early years of the new century.

Each cocoon contains the bave, a long silk filament ranging from 700 to 1500 m. This bave is divided into sericin, an external glue, and fibroin, an internal structural protein (Bucciarelli and Motta 2022). The silkworm's salivary glands produce a fiber for cocoon-spinning. This fiber is made up of brins (70–80% of cocoon weight) and sericin (20–25% of brins), with < 3% natural contaminants (Reizabal et al. 2023), and represents the silkworm's life cycle, Fig. 12.1, adopted from Peng et al. (2019).

Sericulture in India produces four kinds of silk fibers: mulberry, Tasar, Eri, and Muga. The silkworm secretes sericin and fibroin proteins to produce silk fibers, with sericin accounting for 25–30% of the mulberry cocoon's weight (Jaiswal et al.

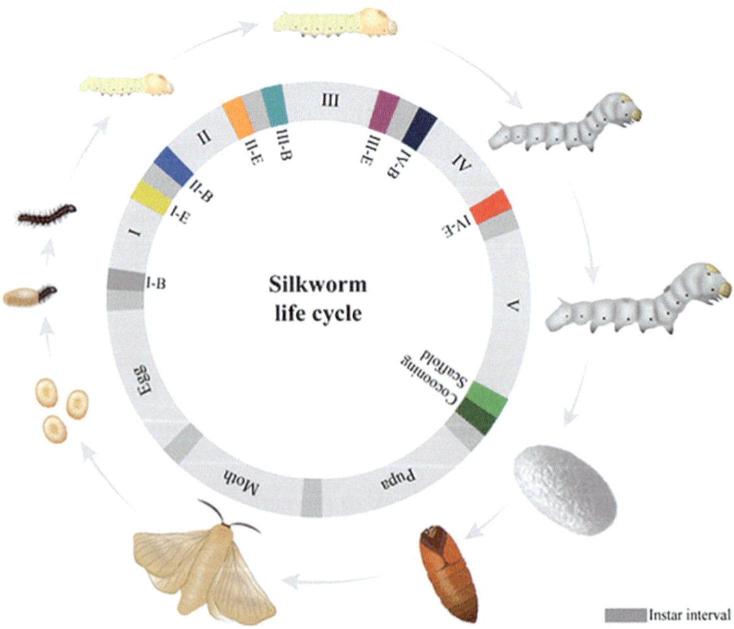


**Fig. 12.1** Exemplifies the silkworm's life cycle: egg, five larval instars (I-V, with B/E for beginning/end), pupal, and moth stages. Silk is spun at the start and end of each instar, except the fifth. The 5th instar includes a scaffold stage, followed by cocooning (Peng et al. 2019)

2021). Popular silk textiles, such as organza, silk satin, georgette, and chiffon, originate from warmer Asian countries, including Thailand, Vietnam, Korea, India, and China. Major global producers of silk textiles include China, India, Uzbekistan, Brazil, Iran, and Thailand (Nayak et al. 2023). The Silk Road has been a popular destination for scholars to uncover the achievements of one of the most significant movement corridors in human history (Punsara Amarasinghe 2023). China, a leading silk producer, has a rich sericulture history spanning over 5,000 years and an annual production of 200.7 thousand boxes, accounting for 80.1% of the global supply (Bakhtiyar et al. 2023).

### 12.1.2 *Silk's Role in Trade and Global Economies Throughout History*

Sericulture offers sustainable rural employment, promotes economic development, and improves livelihoods in developing countries by reducing poverty and reducing migration to urban areas. Policymakers and planners have taken notice of this industry and recognized it as one of the best opportunities for the socio-economic expansion of an economy rooted in agriculture because of its nature, which is primarily composed of rural on- and off-farm activities and has a significant amount of job-creation potential (Dasari and Venkataramana 2023). Trade fuels economic and societal progress, with routes evolving from the Silk Road to modern airways and seaways that shape the global economy (Dangre and Gaurav 2024). Silk, a globally popular fiber, is crucial for poverty alleviation and rural development due to its widespread use in sectors such as medicine and textiles (Lu et al. 2025). India, the second-largest silk producer after China, generated 34,903 tonnes in 2021. The sericulture sector in India offers farmers varied employment and modest investment options, contributing minimally to the global textile market at less than 0.2% (Dasari et al. 2024).

Silk, prized for its exceptional quality, is now used in the production of luxury textiles. Despite fiber competition and rising prices, raw silk production continues to attract investors (Popescu 2018). India's silk industry is primarily based in Karnataka, Tamil Nadu, Andhra Pradesh, Jammu & Kashmir, and West Bengal, with non-mulberry silk being produced in Chhattisgarh, Jharkhand, Odisha, and the north-eastern states (Anitha 2011). Silk, valued for its warmth, softness, and eco-friendliness, is a widely used natural fiber, and its growing production highlights the need for greater awareness and accessibility. Figure 12.2 shows the global silk production (in metric tons), illustrating the overall trends and variations in silk output across different years.

India's development hinges on its 48.18% female population, especially rural women, who are vital in agriculture, particularly in goat, chicken, apiculture, and sericulture. Sericulture, an intensive industry, provides rural women with employment and income opportunities in states such as Tamil Nadu, Himachal Pradesh, West Bengal, Karnataka, Assam, and Jammu and Kashmir (Bukhari et al. 2019).

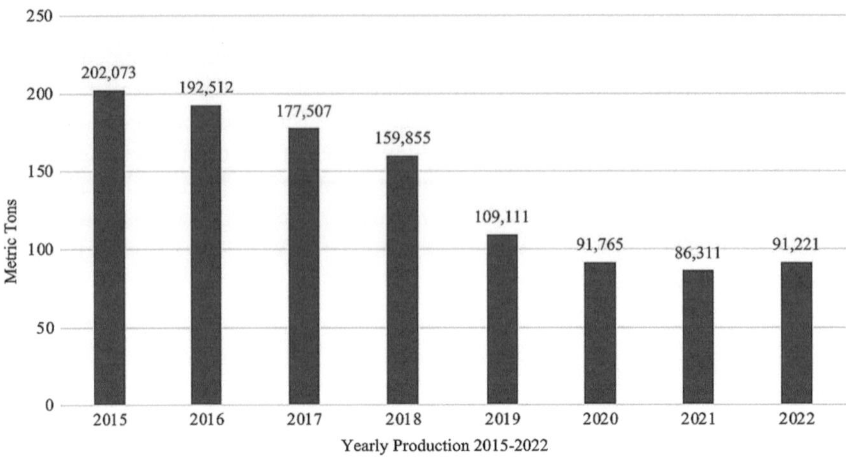


**Fig. 12.2** Global silk production (Metric Tons). Adopted from: Hassan et al. (2025)

Government intervention and training support enhance sericulture's income, crop diversification, land use efficiency, and sustainability, benefiting millions of rural families and marginalized groups. Sericulture is crucial to India's economy and silk production (Reddy and Parasuramudu 2024). Silk, a luxury fabric, is produced using silk worms, which are a sustainable and eco-friendly source. Handloom weavers produce silk with a minimal energy footprint, ensuring it meets the most stringent criteria for sustainable fabric production (Karthik et al. 2015).

The sericulture industry's ability to add value and diversify the methods of valorizing the main product and byproducts will determine its economic potential (Fig. 12.3). The industry begins with the production of mulberries (Morus sp.) and runs downstream with the rearing of silkworms to produce silk threads. However, because most importing countries lack reliable information on finished silk goods, it is difficult to estimate the precise worldwide value. Together, China and India produce approximately 95% of the world's silk, making China the leading producer. According to Hassan et al. (2023), the image depicted the economic prospects and viewpoint of sericulture.

## 12.2 Sustainable Practices in Silk Production

The market for natural fibers, such as silk, is growing, but more study is needed to figure out how to move the silk sector from a rectilinear to a circular economy. By pinpointing areas of interest, the study Hassan et al. (2025) attempts to bridge this gap and helps silk firms use circular processes to achieve sustainability. Beyond environmental issues such as poor recycling, water overuse, and chemical production, the sector also faces social sustainability challenges. These include child labor,

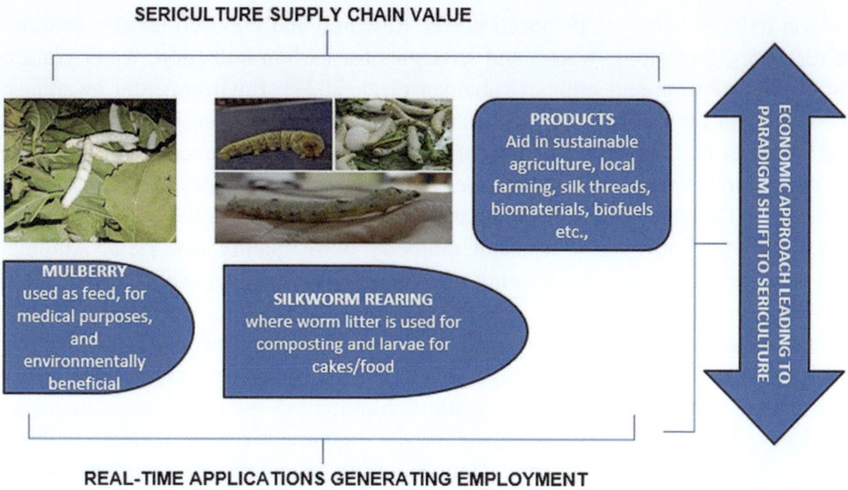


**Fig. 12.3** Economic opportunities and perspective of sericulture

low wages, unsafe conditions, and human rights violations, which impact overall ethical practices (Sahu and Nranjan 2025; Agrawal et al. 2021; Jia et al. 2020). Business survival, from the short to the long term, hinges on sustainability. Companies must integrate sustainable development into their operations to maintain market competitiveness (Prado et al. 2025). Green business practices are a growing trend, particularly in the West, in response to global social and environmental issues such as inequality, climate change, and habitat loss. The private sector's adoption of eco-friendly practices can alleviate these problems, which are most severe in developing nations (Chungyalpa and Rosing 2025). Environmental sustainability aims to minimize its impact by conserving natural resources, reducing waste, and using fewer resources. The textile industry is actively pursuing sustainability (Rani et al. 2025). All suppliers and sub-suppliers are accountable for ensuring the sustainability of the textile supply chain (Wilhelm et al. 2016; Fares et al. 2024).

### *12.2.1 Growing Consumer Awareness of Environmental and Ethical Concerns Leading to the Rise of Sustainable and Ethical Consumption*

Sustainability is a core business strategy that impacts environmental, social, and financial aspects. Stakeholders must understand the relationship between sustainable practices and brand loyalty (Agu et al. 2024). Ethical fashion is driven by concerns over pollution, microplastics, and the rise of conscious consumers. Environmental concerns and self-expression drive ethical fashion purchases through functional, social, and emotional values (Upadhyay and Kamble 2025). Sustainable fashion reduces harm via balanced needs. Research analyzes eco-textiles in terms of resources, emissions, waste, and strategies for carbon reduction, water conservation, traceability, and ethics (Rahaman et al. 2024). Environmental awareness fuels demand for eco-friendly goods, driving green innovation. Businesses adapt, and research explores the impacts of sustainability, with the concept extending beyond simple "green" practices to encompass responsible development (Yadav et al. 2024). Biotechnology and automation modernize sericulture. Genetically modified silkworms and disease-resistant mulberry boost quality and yield. Automation enhances efficiency, maintaining silk's competitiveness against synthetics amid growing demand.

Modern sericulture has transitioned from manual labor to technology, enhancing silk quality and productivity through biotechnology, including genetic engineering and disease-resistant mulberry cultivation. Automation reduces labor, enhancing efficiency. These advancements maintain silk's competitiveness against synthetic fibers amidst rising global demand. Genetic improvement, through methods such as marker-assisted selection and genetic engineering, targets fiber strength, length, and biochemical composition. This accelerates breeding, enhances disease resistance, and modifies silk fibroin, boosting quality and quantity in sericulture (E et al. 2024).

### 12.2.2 Defining Sustainable Silk Production: Balancing Economic, Environmental, and Social Factors

Sustainable silk production necessitates a delicate balancing act between economic viability, environmental responsibility, and social fairness. Its objective is to minimize the undesirable effects that the manufacturing of silk has on the atmosphere and its people while preserving long-term economic stability and respectable working conditions. According to Wu et al. (2025a, b), this entails conserving water and energy, employing fewer chemicals, promoting ethical work practices, and safeguarding the welfare of silkworms.

The multibillion-dollar textile industry, which produces a vast array of items, has a profound impact on the global market. Silk, despite its small scale, requires high energy, water, and chemicals, resulting in pollution. Sustainable silk requires identifying pollution sources and optimizing processes. Life cycle costing (LCC) and life cycle assessment (LCA) help quantify environmental and financial burdens (Wu et al. 2025a, b). The global urgency for sustainable economies intensified during the pandemic, driven by issues of inequality, resource depletion, and climate change (Ghisellini et al. 2024). The 21st Century Maritime Silk Road emphasizes sustainable logistics, green modes, and practices to transform trade, enhance connectivity, and minimize environmental impact. This fosters economic growth, attracts investment, and aligns with global sustainability goals, promoting responsible resource management and innovation (Wang et al. 2024).

## 12.3 Sustainable Solutions for the Silk Industry

The capacity of a living organism to fulfill its requirements without posing threats to nature and its diversity, in an attempt to preserve it for the future, is referred to as sustainability, which involves social, environmental, and economic concerns. The silk industry, a crucial component of India's economy, generates substantial revenue and employment opportunities. With mulberry silk accounting for 90% of global production, ISRO and the Central Silk Board are expanding cultivation into wastelands to boost India's national output (Ricciardi et al., 2020).

Figure 12.4 illustrates the silk manufacturing process, its associated natural silk-derivative industries, key inputs and outputs, and the interconnections among them. Water management is a critical element of sustainability. Drip irrigation and rainwater harvesting are two efficient irrigation methods that minimize waste and conserve this precious resource. Fewer chemical pesticides are also needed when integrated pest management (IPM) methods are employed (Rauf and Nazli 2024). While contemporary sustainable sericulture employs advanced technologies such as genetic engineering and mechanical systems to improve efficiency, it causes ecological problems. Traditional sericulture, on the other hand, focuses on organic agriculture and agroforestry, which enhance biodiversity. Both approaches integrate innovation and tradition with the aim of sustainability (Attri et al. 2024).

### *12.3.1 Organic Sericulture*

Organic farming, including sericulture, prioritizes natural methods by utilizing farm-generated fertilizers, green manure, and crop rotation over mineral fertilizers; employing mechanical weed control instead of herbicides; and using natural agents and resistant varieties to replace synthetic pesticides, thereby fostering soil health. A sustainable and environmentally responsible method of producing silk, organic sericulture prioritizes the well-being of the silkworms, the environment, and ultimately, the quality of the finished product. Cultivating mulberry trees, which serve as the primary food source for silkworms, is crucial to this technique. Mulberry leaves are grown in organic systems without the use of fertilizers, synthetic pesticides, or genetically modified organisms (Baciu et al. 2023). Rather, to preserve soil fertility and plant health, farmers use crop rotation, green manuring, and organic composts (Alemayehu et al. 2020). In addition to maintaining ecological equilibrium, this ensures that the leaves the silkworms eat are free of chemical residues that can compromise the well-being of the silkworms and the quality of their cocoons (Xu 2025).

Organic sericulture integrates mulberry leaf production, employing "integrated plant protection." The use of integrated plant protection(IPP) techniques in mulberry growing is a fundamental aspect of organic sericulture. IPP minimizes the need for

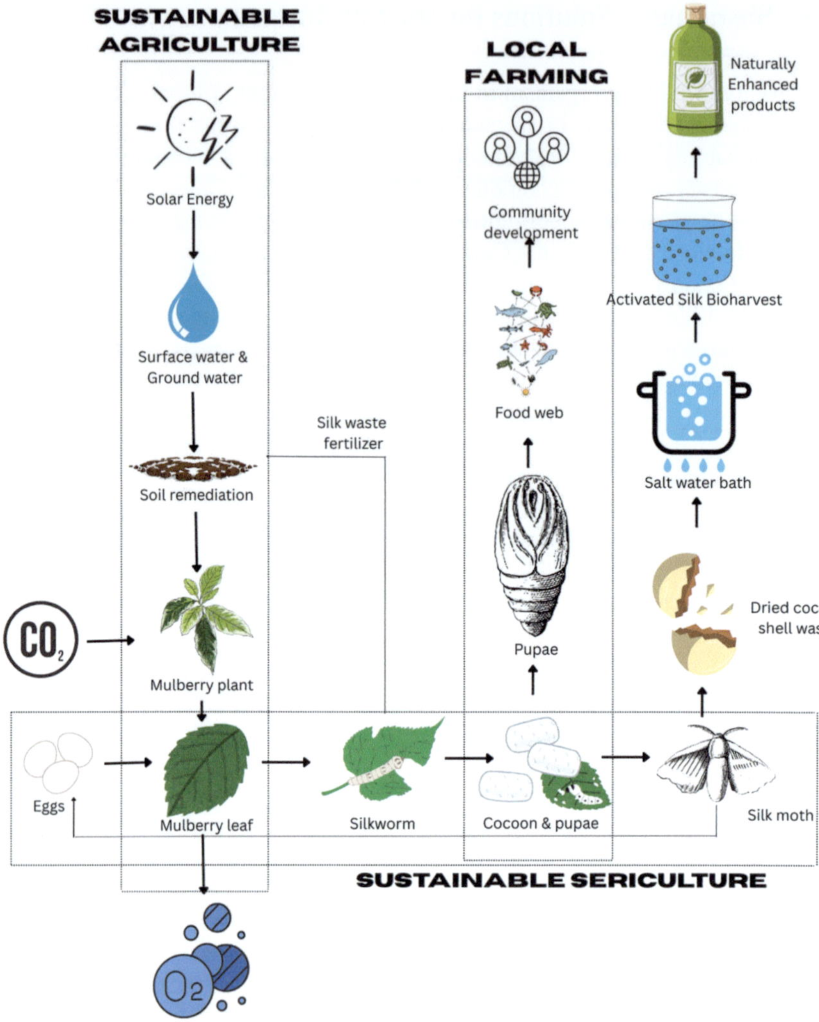


**Fig. 12.4** Schematic representation of the Silk Manufacturing process, connected and resulting natural silk-derivatives industries, key inputs and outputs, and related interdependencies. Adopted from: Altman and Farrell (2022a, b)

chemical pesticides by controlling pests and diseases using a combination of biological, cultural, mechanical, and botanical techniques (Dandin and Kumari 2021). Threats are managed organically through the use of strategies such as intercropping to confuse pests, employing neem-based biopesticides, introducing beneficial insects, and pruning at the right time (Rameshbhai 2021). Organic sericulture enhances soil health and biodiversity by including these environmentally friendly protection strategies. It also helps the larger objectives of climate-resilient agriculture and sustainable rural livelihoods (Diyaolu and Folarin 2024). Due to its comprehensive approach, organic sericulture serves as a model for producing silk in an environmentally conscious manner (Xu 2025).

The organic production of mulberry plants, the sole source of food for the silkworm, Bombyx mori, is the first step in the environmentally friendly process of organic sericulture (Attri et al. 2024). In contrast to traditional techniques, organic sericulture rigorously forbids the use of synthetic fertilizers, chemical pesticides, and genetically modified organisms (GMOs) throughout the whole production process (Dinesh et al. 2025). The primary objective is to establish a balanced agro-ecosystem that fosters robust mulberry growth, enhances leaf production, and promotes disease-resistant conditions for silkworm rearing (Metaxas 2013). This is accomplished through the use of organic soil amendments, such as farmyard manure (FYM), vermicompost, green manures, and biofertilizers, including Azospirillum, Azotobacter, and phosphate-solubilizing bacteria (PSB), which naturally enhance soil fertility and microbial activity in the cultivation of mulberry leaves. (Bamdad et al. 2022).

The implementation of integrated plant protection (IPP) strategies, which emphasize the prevention, monitoring, and control of pests and diseases using environmentally friendly and scientifically sound methodologies, is a key element of this organic framework. IPP integrates several pest management strategies in a way that is both compatible and synergistic, rather than depending on chemical treatments (Zhou et al. 2024). These include mechanical techniques, such as hand-picking pests and using light and sticky traps; biological control agents, including parasitoids, predators, and microbial pesticides (like Bacillus thuringiensis); and the use of botanicals, like neem oil or plant-based insect repellents, which are authorized by organic certification standards (Adhikari 2022). Cultural practices include timely pruning, intercropping, and sanitation. Pest control is made to be both efficient and to disturb the agro-ecological balance as little as possible by threshold-based treatments and routine field monitoring (Djeddour et al. 2021).

Organic sericulture offers a comprehensive approach that enhances the sustainability, productivity, and financial viability of silk farming by integrating the production of mulberry leaves with IPP (Islam et al. 2023). This approach preserves the health of silkworms and the quality of their cocoons while lowering the environmental impact of sericulture. It also contributes to biodiversity preservation, better soil and water quality, and the long-term sustainability of agricultural systems (Sharma 2024). In line with international initiatives in climate-smart agriculture and organic certification procedures, the convergence of integrated plant protection and organic farming practices represents a paradigm shift toward more sustainable and ethical forms of rural economy (Garwi et al. 2024).

Figure 12.5 illustrates the useful byproducts of the sericulture process. This supports various areas of provision through a sustainable index of the circular economy. Mulberry sericulture offers high-earning family employment in India's favorable climate. Embracing organic farming, it utilizes a diverse range of organic

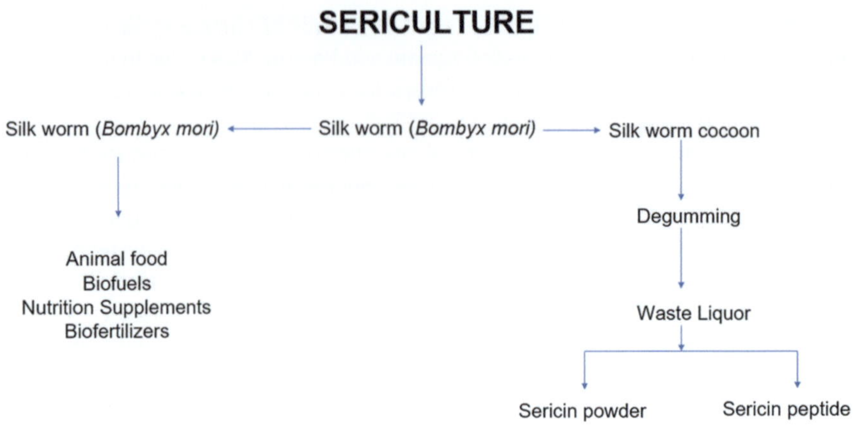


**Fig. 12.5** Byproducts of the sericulture process

manures and biofertilizers, ensuring a healthy environment and optimal crop productivity. Organic silkworm breeding enhances cocoon quality, commanding premium market prices, fostering sustainable, high-value agriculture (Kumar et al. 2012).

#### 12.3.1.1 Benefits of Organic Silk: Reduced Environmental Impact, Improved Quality

Water, vital for life, faces contamination, disrupting ecosystems, and causing disease. Bio-based water treatments are urgently needed. Unlike synthetics, they offer efficient purification and biodegrade harmlessly, preserving ecological balance and safeguarding living organisms (Gore et al. 2019). Silk's biodegradability and biocompatibility make it a promising biomaterial for implantable devices and optical sensors. Its optical transparency aids biomedical sensing, while its use in electronics, like organic thin-film transistors, improves crystal quality. Carbonized silk sensors and transient electronic devices are also actively researched, highlighting silk's versatility in diverse technological applications (Lee et al. 2022).

Mulberry leaves serve as the primary food source for silkworms, making them essential for sericulture. Figure 12.6 highlights the various applications of mulberry leaves in industries such as sericulture, animal feed, pharmaceuticals, and nutraceuticals.

Silk protein, primarily composed of fibroin and sericin, is advantageous for water treatment due to its biodegradability, which prevents environmental contamination (Sun et al. 2024). Fibroin typically constitutes a fibrous protein thread having a disulfide linkage. The consecutive adhesive layers of sericin form an envelope structure that aids in producing the cocoon. While Sericin is a globular protein, accounting for approximately 25 to 30% of silk proteins. Sericin possesses 18 amino acids, among which most have hydroxyl, carboxyl, and amino groups (Silva et al. 2022).

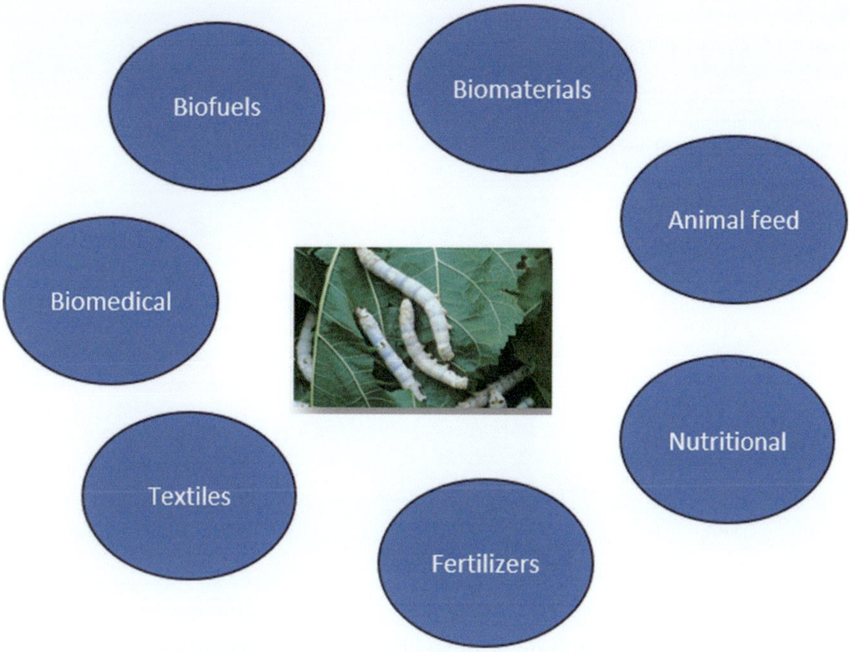


**Fig. 12.6** Applications of mulberry leaves

Its mechanical strength ensures durability and stability in composite materials. Versatility in processing enables the production of various forms, including particles and membranes. It is chemically compatible, enabling toxin adsorption and surface modification, with a high surface area and binding sites. Figure 12.7 presents a schematic diagram of sericin sources, extraction methods, properties, and biomedical and pharmaceutical applications. Silkworm spinning yields silk protein with extremely low carbon emissions, offering a sustainable source for functional materials (Sun et al. 2024).

Pupae contain protein that contains numerous essential amino acids, which help them meet widely recognized dietary requirements. A mere 105 g (dry weight) of silkworm pupae can provide a human with up to 58 g of protein, the adult recommended daily dietary allowance for protein. Additionally, n-3 fatty acids, which are essential for healthy metabolism and have been linked to improvements in cardiovascular, cancer, and inflammatory conditions as well as specific brain and cognitive processes, are abundant in silkworm pupae. Additionally, silkworm pupae can be used to make medicinal oil. Furthermore, pupae possess natural inhibitors of 1-deoxynojirimycin (DNJ), a potent alpha-glucosidase inhibitor also found in the leaves of their hosts, which has been shown to lower blood glucose levels (Altman and Farrell 2022a, b).

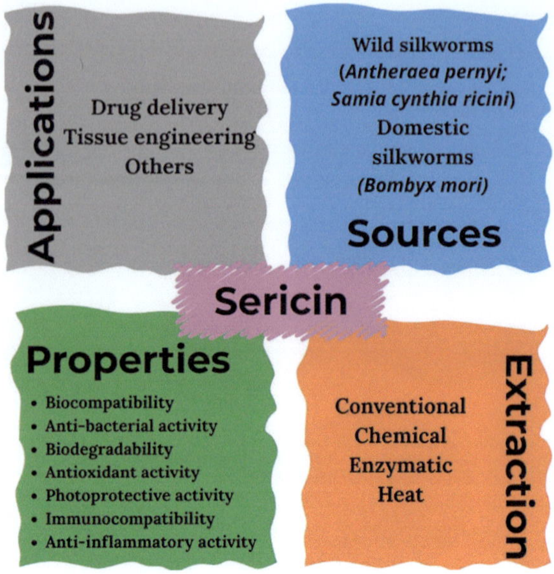


**Fig. 12.7** Schematic diagram of sericin sources, extraction methods, properties, and biomedical and pharmaceutical applications (Silva et al. 2022)

#### 12.3.1.2 Challenges and Opportunities for Organic Sericulture

Silkworms, being cold-blooded, mirror ambient temperatures. Slight air changes impact their behavior, development, survival, and reproduction. Consequently, climate and weather strongly influence silkworm rearing (Senapati et al. 2024). Several biological and commercial features, like the silkworm's expression of heat shock proteins, may be affected by changes in temperature. A high temperature during the chawki stage may bar later stages from developing toward post-cocoon parameters. Climatic factors, primarily temperature and humidity, can impact the development of silkworms, including cocoon production and egg production by the female *Bombyx mori,* in terms of both quantity and quality. The effect of temperature on silkworm growth can lead to several changes, including differences in cocoon quality, productivity, and survival rates. The quantity of cocoon yielded per box can decline by 15–30 kg if worms are not provided with the optimal temperature condition, which may also lead to a lower quality of up to 14% (Sujatha et al. 2024).

Optimal cocoons thrive between 22 and 27 °C; higher temperatures degrade quality. Cocoon and silk quality, crucial for sericulture's economy, are impacted by temperature, humidity, ventilation, genetics, and seasonal environmental factors (Senapati et al.). Organic yields vary, sometimes matching or exceeding conventional yields, often due to nutrient deficiencies, weeds, rotations, or management inexperience. Yield drivers are complex and location-specific (Ram et al. 2022). Organic farming thrives on trust-based connections, ensuring food security across the supply chain. While a leading alternative to conventional agriculture, alongside others like permaculture, its success hinges on assurance, regional growth, and ecological, low-cost practices, demonstrating leadership within the organic movement (Ram et al.

2022). Contrary to the mid to third quarter of the twentieth century, the government today invests in the development of these firms and is less concerned with their promotion. Again, repeated governments have been unable to understand the need to develop industries which offer ready markets to individuals who would otherwise be involved in mulberry cultivation and silkworm rearing. Since industries themselves are not even in a position to purchase their produce, a few who can produce at such scales still get miserably low wages (Asinyo et al. 2025).

## *12.3.2 Water Conservation Strategies*

Water conservation involves reducing water loss, use, or waste through efficiency measures, behavioral changes, or improved technologies. Water efficiency, a key conservation strategy, lowers demand. Effective measures require evaluating the impacts on other resources, such as energy and chemicals, considering both cost and value (Kumari and Singh 2016). Water is primarily used for irrigation during the early phases of sericulture, specifically during the mulberry cultivation period. Silk suppliers may employ manual irrigation, natural irrigation (rainfed), or a combination of both. Natural irrigation methods appear to be the most effective sustainable input, as they consume less water. Alternatively, water-treated, closed-system irrigation systems can also be considered as a sustainable alternative. Water bodies may be eutrophicated and acidified due to agricultural runoff. It is essential to consider the chemical inputs of water in sericulture. To enhance mulberry yield, quality, and protection, chemical formulations are utilized at various stages of sericulture (Altman and Farrell 2022a, b).

### 12.3.2.1 Drought-Resistant Mulberry Varieties

Mulberry, a Moraceae genus with roughly 36 species, thrives globally in diverse climates, from tropical to temperate (Manzoor and Qayoom 2024). Mulberry-based agroforestry reduces agricultural environmental impact. Sericulture uses mulberry leaves, cutting chemical use. Mulberry trees like Morus alba and Morus nigra grow fast, resist urban conditions, and survive semi-arid regions. A study found high mulberry plantations enhance air quality by lowering $PM_{2.5}$ and $NO_2$ (Rohela et al. 2025). Sericulture, especially with mulberry farming, offers environmental benefits like carbon sequestration, air purification, and soil/water conservation. A single hectare of mulberry can absorb a ton of $CO_2$ daily and release 0.73 tons of oxygen, supporting a thousand people's breathing needs. Mulberry afforestation improves soil carbon, water retention, erosion control, nutrient recycling, and microflora preservation. Mulberry planting reduces flood runoff by 10–20% and its deep root system aids in water and soil conservation. (Manzoor and Qayoom 2024).

In Nutthapornnitchakul's study, four proposed Thai mulberry cultivars (*Sakhonnakorn, Sakhonnakorn 85, Burirum 60,* and *Srisaket 84*) and one general drought-resistant cultivar (SRCM9809-34) were tested for proline content and Pyrroline-5-carboxylate reductase (P5CR) gene expression levels in drought stress conditions. The findings indicated that SRCM9809-34, the drought-resistant variety, contained the least amount of proline, followed by *Sakhonnakorn 85 and Burirum 60*, while *Srisaket 84* contained the most proline. Based on this study, *Sakhonnakorn* 85 had the highest potential for drought tolerance among the four varieties of Thai mulberry; cross-pollination with SRCM9809-34 can enhance this potential (Nutthapornnitchakul et al. 2024).

Genotypes were categorized into low, medium, and high frequency classes based on evaluation results. All of the evaluated germplasm genotypes showed significant genetic variability. On the basis of yield components, morphological, anatomical, and physiological characteristics of drought resistance, 23 genotypes were ultimately identified. The leaf yield of these genotypes under conditions of natural stress was also assessed using check cultivars. The two genotypes with the highest yields among the 23 were ACC-142 and ACC-109, which were regarded as genotypes resistant to drought (Susheelamma et al. 1990).

### *12.3.3 Integrated Pest Management (IPM):*

Implementing eco-friendly crop protection methods is crucial for sustainable agriculture. However, farmers' adoption of these approaches is hindered by high costs due to inadequate coordination within integrated pest management systems (IPM) (Han et al. 2024). (IPM) aims to reduce agriculture's environmental impact by combining diverse strategies like chemical and biological control methods, focusing on pest monitoring, prevention, and data-driven decision-making (Reitz 2009, Erler et al. 2022; Ahmed et al. 2024). Biopesticides, derived from natural sources like bacteria, plants, and minerals, are effective for pest control (Chowdhury et al. 2024). Through pest monitoring, promotion of natural enemies, and minimizing the application of pesticides using biological control, integrated pest management (IPM) enhances sericulture. The environmentally friendly practice enhances silk production quality, reduces its adverse impacts on the environment, and enables rural agriculture communities to stay financially sustainable. To prevent bioconcentration and biomagnification, the ideal pesticide should be highly selective toward target species, have less toxicity to non-target organisms, be effective in small doses, and break down easily in the environment. Biopesticide use is increasing due to their benefits outweighing drawbacks. They are categorized as microbial-based, plant-incorporated (using transgenes like BT toxins), or biochemically based (from natural compounds like plant essential oils) (Hezakiel et al. 2024).

#### 12.3.3.1 Waste Management and Resource Recovery:

Waste management is vital for the bioeconomy, as it helps keep the environment clean and promotes the use of recycled products, leading to positive economic impacts (Vambol et al. 2023). The rising levels of waste due to urbanization, industrialization, and population growth have severe impacts on the environment, health, and economy (Durak 2023). The harvest and processing of raw silk and silkworm cocoons generate enormous amounts of waste, including unwound cocoons, silkworm cultivation residue, cocoon unwinding, silk spinning, and weaving byproducts. Only a small portion of cocoons are of first-grade quality and can be fully unwound without loss, while the rest are of lower grades and partially processed. Overall, a significant proportion of material is wasted across all stages of the silk industry, with silk waste typically classified according to its origin, such as cocoon, winding, unwinding, spinning, dyeing, weaving, and finishing.

A multitude of fibers of different thicknesses, lengths, and linear densities usually comprise natural silk's unspinnable waste. When selecting a method for producing fiber webs, one of the key considerations is the length of the fibers. Various nonwoven processes can be employed for processing staple fibers with lengths ranging from 6 to 100 mm. Short fibers with lengths between 6 and 25 mm are generally utilized in making paper (e.g., cotton–silk paper). Intermediate-length fibers may be managed on carding machines with the web then bonded, but 100 mm length fibers are suitable for the aerodynamic process of nonwoven production.

The main constituents of short unspinnable silk waste, fibroin and sericin, are recovered in the form of part of the formulated recycling techniques (*US20090133814A1—Producing Method for Non Woven Silk Fabric—Google Patents*, 2005). Dust, dirt, and pupal residues are eliminated from the waste before extraction, and waste from cocoon is also treated with cocoon cutting machines. The recovered sericin powder is additionally applied in cotton fiber sizing which makes blended cotton–silk yarn preparation easier. Hydrogels, which are then used in the spinning processes of the fibers, are another method by which sericin is obtained from waste material (Bexiga et al. 2017).

#### 12.3.3.2 Composting of Mulberry Leaves and Silkworm Litter

Silkworm waste in sericulture, especially in mulberry farms, may contain pathogens that affect silkworms when mulberries are consumed. Proper treatment of silkworm waste is crucial for assessing environmental impact and promoting ecological responsibility (Shen et al. 2024). Therefore, it is essential to use appropriate composting technology to convert waste from sericulture farms into valuable compost (Kambai et al. 2020). Composting, the process of converting organic waste into valuable agricultural resources, is one potential solution for addressing this waste. The use of organic waste to produce quality agricultural fertilizers relies on a diverse range of microbial communities (Shen et al. 2024). Silkworm rearing generates mulberry leaves and dung, producing silkworm litter that serves as valuable organic manure.

Only a portion of the leaves is consumed and transformed into litter, while the remainder is digested or unused. The litter yield is higher on irrigated land compared to rainfed land, and its content of nitrogen, phosphorus, and potassium makes it particularly beneficial for mulberry cultivation (Shanmugam et al. 2024).

Organic manure, lime, phosphates, sulfates, nitrogen, and potash (a potassium-based fertilizer) are all found in mulberry fertilizers. As a result of the use of artificial nitrogenous fertilizers, nitrous oxide is most frequently released into the atmosphere, volatilized into ammonia, and leached into streams as nitrates. Similar to the impact of nitrogen, phosphorus released into streams by fertilizers containing phosphate can be harmful. As a result, silkworm waste is being utilized as a sustainable natural fertilizer substitute. Methane and nitrous oxide can be released during the decomposition of silkworm litter or waste. Other techniques, such as composting, which is a cleaner alternative to landfilling, can sequester carbon and help achieve a crucial sericulture goal by recycling the mulberry growth cycle and reducing the use of artificial fertilizers (Altman and Farrell 2022a, b).

## 12.4 Ethical Considerations in Silk Production

Recent surveys indicate a rise in customer awareness of ethical purchasing, suggesting that the current era prioritizes ethics. The UK has published the Ethical Consumer Markets Report since 1999, assessing consumers' ethical spending. Ethical consumerism involves allocating funds based on issues like social justice, human rights, the environment, or animal welfare (Ozkavruk Adanir et al. 2024). Disease varies in impact on silkworm mortality across nations, accounting for 10% to 47% of deaths due to environmental and behavioral factors. Improving disease control could enhance silkworm welfare and reduce animal deaths, potentially lowering costs and boosting silk demand in the long term. Boiling cocoons to kill silkworms before they become adult silk moths is a common method. This process ends their lives as pupae during the transformation from larvae (Rowe 2021). Despite the rarer occurrence of pests in silkworm farms, dermestid beetles and uzi flies are common. The four most common illnesses are pebrine, muscardine, grasserie, and flacherie. These sicknesses, which can be viral, bacterial, or fungal, can all be lethal within a few days or over ten. A lack of irregular molting, appetite, or other strange behavior could be signs of the illness. Thick disinfectant treatments are used to treat most infections, which may or may not affect silkworm health (Nayak 2006).

### *12.4.1 Concerns About the Traditional Method of Silk Production*

The silk production process reflects the rich cultural heritage of producing regions, rooted in centuries-old techniques. Craftsmen's expertise is crucial for the laborious and time-consuming process, beginning with selecting premium cocoons based on size and shape. Heating cocoons melts the sericin glue, allowing artisans to carefully unwind silk threads (Gowrisankar et al. 2023). Asia, where silk manufacturing dominates, produces most of the world's silk, including non-mulberry silk. Despite this, silk accounts for just 0.2% of all natural fibers produced globally (Popescu et al. 2024). Traditional silkworm farming efficiently recycles waste for various products, utilizing materials obtained during breeding. Products are entirely organic. The bug's short month-long lifecycle minimally disrupts farm tasks. The stages include eggs, caterpillars, pupae, and butterflies. Caterpillars feed on mulberry leaves. Adhering to hygiene prevents infections. Banana leaves are used for bedding in eco-friendly methods, reducing waste due to their fast decomposition (Grześkowiak et al. 2022).

#### 12.4.1.1 Peace Silk Production and Promoting Animal Welfare in Silkworm Rearing Practices

Ahimsa silk, also called "Peace silk," is produced sustainably by letting adult moths appear live from mulberry silk cocoons before spinning into yarn. The fashion industry may favor ahimsa silk products due to growing global concern for animal welfare (Sannapapamma and Naik 2015). While ahimsa silk is breathable and easy to care for, it can have an irregular appearance, gritty texture, and limited antibacterial properties. To treat the silk fabric effectively, it needs to be degummed first due to surface contaminants and sericin. Various degumming methods, such as enzyme, soap, alkali, or synthetic detergents, can be used during finishing processes (Aishwariya 2020). To avoid pupae death in the course of silk processing, peace silk production involves allowing silkworms to emerge from their cocoons naturally (Adanir et al., 2022). This method not only adheres to animal welfare protocols but also serves the interests of an increasingly ethically aware consumer base. As consumers progressively become conscious of the ethical implications of traditional silk production, the creation of peace silk, which emphasizes animal welfare in sericulture practice, has become popular. Peace silk or "Ahimsa" silk promotes a more humane way of sericulture by giving silkworms the opportunity to complete their life cycle before being harvested. Advances in biotechnology and eco-friendly processes that augment silk production without harming silkworms are aiding this transition. Increased silk production and improved disease control are facilitated by advances in sericulture technology that can reduce the use of disruptive approaches (Kalpana et al. 2024; Fatima et al. 2024). Although promoting sustainable activities such as organic agriculture, technologies such as mechanized technology and climate-controlling rearing systems make the process more efficient. Through the creation of new markets and the enhancement of socio-economic status of silk producers, the promotion of peace silk can consolidate rural communities (Fatima et al. 2024). In addition to that, employing silkworm pupae as a protein source for animal feeding presents an opportunity to utilize its byproduct during the production of silk, thus ensuring sustainability even more (G. et al. 2024).

Although peace silk is a giant leap forward for moral sericulture, there are also problems balancing traditional practices with modern customer demands and technical advancements. In an effort to meet these shifting demands while ensuring animal welfare, the silk industry has to constantly come up with new concepts. The Material Innovation Initiative promotes silk substitutes that could reduce total silk production. Affordable and adaptable substitutes may replace silk in many applications like tire and medical equipment manufacturing. However, it is uncertain if they will significantly impact retail silk fashion. These substitutes serve as alternatives to non-animal textiles with environmental benefits, advocating for replacing animal silk (Rowe 2021).

## *12.4.2 Social and Labor Issues*

The sericulture industry, through various activities such as mulberry production, silkworm rearing, and silk fabric production, benefits both rural and urban communities. Approximately 1.1 billion individuals work in agriculture, accounting for 60% of the global population, whereas worldwide unemployment rates were around 5.4% in 2019 and 5.42% in 2021. With predictions indicating the need to feed 9.7 billion individuals by 2050, agriculture must be diversified for improvement (Ssemugenze et al. 2021). Breeding silkworms was a family enterprise. Production for household consumption was the household mode of production. The commercial's sole purpose was to supplement household income. The Soufli households earned seasonal income from cocoon production, and labor was structured within the background of a specific division of labor, by age and sex. In the mature period of silkworms, men were engaged in commerce, mulberry tree cultivation, and cutting and transferring of mulberry leaves. The household tasks of leaf chopping, feeding, and harvesting cocoons were associated with the woman's domestic maintenance and management, which is a feature of "female patience," or a careful and complex process. Children were also involved in the activity (Macha-Bizoumi et al. 2025).

### 12.4.2.1 Advancing Fair Practices in the Silk Industry

In India, Farmer-Producer Organizations (FPOs) empower farmers and boost agriculture. FPOs benefit sericulture by enhancing bargaining power with silk reelers and processors, securing better prices. They also help members access financial services, such as credit and insurance, through collective negotiation. By eliminating middlemen, FPOs connect sericulture growers directly to markets, reducing costs and boosting profits. Furthermore, they open new markets for silk products, thereby elevating the standard of living for rural communities dependent on sericulture (Nayak 2022).

Government programs supporting Sericulture Farmer-Producer Organizations are vital to boost output, enhance living standards, and promote sustainable practices. The National Sericulture Policy aims to enhance silk quality, leverage new technologies, and strengthen the value chain to improve productivity in India. This initiative has led to the establishment of sericulture development centers, elevating production, and sericulture area in multiple states (Likhi 2020). The Silk Mark Organization of India (SMOI) focuses on enhancing silk quality and promoting its branding. The Integrated Sericulture Development Project (ISDP) focuses on holistic development, offering marketing assistance and technological support (Gunashekhar et al. 2024). The Karnataka State Sericulture Department supports sericulture FPOs with infrastructure, training, and subsidies, resulting in over 31 FPOs and a significant boost in revenue. The PMKSY supports sericulture by improving water supply for mulberry cultivation and enhancing water efficiency through conservation and micro-irrigation. Over 50,000 hectares benefit from improved irrigation infrastructure, fostering industry growth, and sustainability (Balkrishna et al. 2022).

Many communities face unemployment and vulnerability to shocks. The pandemic has hindered women's employment globally, with millions of jobs lost during the period. Despite some progress in social protection, women lack equal access to economic resources. Laws aim to ensure women's rights to inheritance, access to financial services, and control over natural resources. Women's empowerment involves challenging gender norms in various social institutions. Women play a vital role in agriculture, with sericulture being a predominantly female-driven field (Table 12.1). It is crucial to assess whether increased participation leads to greater empowerment, including control over resources, decision making, and public expression. This study focused on measuring the empowerment of women in sericulture production through the Group Empowerment Index (GEI) (Madhu et al. 2023).

**Table 12.1** Share of women in different activities of silk production

| Accomplishments | Men | Women | Total |
|---|---|---|---|
| Establishment of mulberry plantation (man-days/acre) | 24.02 (48.62) | 25.38 (51.38) | 49.40 (100.00) |
| Mulberry maintenance (man-days/acre/crop) | 8.66 (43.00) | 11.48 (57.00) | 20.14 (100.00) |
| Silk cocoon production (man-days/batch) | 12.38 (39.98) | 18.59 (60.02) | 30.97 (100.00) |
| Raw silk production (man-days/year) | 706.19 (37.70) | 1167.75 (62.30) | 1872.00 (100.00) |

*Note* Figures in parentheses are the percentage of total (Madhu et al. 2023)

## 12.5 Technological Innovations for Sustainable Sericulture

Recent technical advancements in the sericulture sector aim to boost productivity and quality (Table 12.2). Innovations such as the genetic modification of silkworms and disease-resistant mulberry cultivars have led to increased silk production. Automation and robotics have streamlined the raising process, cutting costs, and enhancing production. Precision agricultural methods, including drone technology and IoT monitoring systems, have further improved mulberry production and silkworm rearing. One of the best examples of how technology has changed conventional methods is the advent of automated silk reeling devices. By replacing hand-reeling techniques, these devices ensure silk thread homogeneity and significantly reduce wastage. Consequently, silk's total quality has improved, enhancing its competitiveness in the global market (Mana et al. 2023; Valiyev et al. 2024).

Sericulture has made significant progress, with the integration of knowledge on drone technology, particularly in the development of smart fertilization systems. Drones are employed to approximate fertilization requirements, track farming performance, and evaluate the effects of fertilizers and pesticides (Farid et al. 2022). With temperature, humidity, and other allied factors controlled by IoT, the host plant and silkworm industries will be promoted. Remote sensing is emerging as a valuable tool for identifying favorable planting sites. Through constant monitoring of host plants and silkworms, image processing-based automatic sericulture can help prevent diseases and pest infestations by regulating climatic conditions (Saikia and Saikia 2023). Modern silk reeling technology has evolved in response to technological advancements and the need for increased productivity and efficiency in the silk manufacturing process. The process can now be mechanized with mechanical reeling machines, permitting faster, and more reliable filament extraction. Silk production involves the process of reeling, which entails unwinding silk threads from cocoons. Reeling technology has evolved over time in various ways to enhance the efficiency, sustainability, and quality of silk manufacturing (Gowrisankar et al. 2023).

**Table 12.2** Technological innovations in sericulture

| Effect | Explanation | Technological innovations |
|---|---|---|
| Enhanced resilience and productivity | Creation of silkworms that are more resistant to illness and produce more silk | Silkworms with genetic modifications |
| Increased stability of the crop | Mulberry plants that have been genetically modified to resist harsh weather | Climate-adaptable mulberry types |
| Increased effectiveness in raising | Using the Internet of Things (IoT) to track silkworm rearing conditions in real time | Monitoring systems based on IoT |
| Better utilization of resources and output | Application of data-driven methodologies to enhance mulberry agriculture | Techniques for precision agriculture |

## 12.5.1 Biotechnological Innovations

Micrococoons, a thousand times smaller than silkworms, consist of a resilient silk nano-fibril shell protecting a liquid core cargo. Indigenous silk feed stocks from Bombyx mori larvae's fifth instar silk glands are used to produce these micrococoons (Laity et al. 2015). Inspired by silkworm cocoons, micrococoons can be a powerful vehicle for delivering genes, proteins, vaccines, micromolecules, and beneficial medications. Silk nanostructures produce hydrogels, fibers, films, coatings, and matrices. They are both structurally and functionally effective. Silk nanoparticles are utilized in systemic applications like drug delivery (Hota et al. 2012; Solomun et al. 2020). Natural silk is synthetically spun through a microfluidic process to create biomaterials known as "micro cocoons." These micron-scale forms have potential applications in biotechnology and medicine due to their tendency to aggregate and suitability for long-term storage (Sharma et al. 2022a, b). The creation of the silkworm genome database, known as Silkbase (http://silkbase.ab.a.u-tokyo.ac.jp/cgi-bin/index.cgi), was aided by global research on silkworm genetics. This is now a crucial component of studying the silkworm genome to advance the sericulture sector. By employing linkage mapping to identify genetic markers, the exploration of the silkworm genome has significantly transformed the sericulture sector. This led to the identification of crucial areas within the silkworm genome that influence specific phenotypes, thereby altering the breeding process (Alam et al. 2022).

### 12.5.1.1 Genetic Improvement of Mulberry Varieties and Development of Disease-Resistant Silkworms

Mulberries are versatile; they serve as excellent animal feed and can be processed into various products, such as wine, ice cream, and jam. Additionally, mulberry leaves are an important component of livestock diets. Furthermore, mulberries possess medicinal properties, including antibacterial and anti-diabetic effects (Dutta et al. 2023). Through selective breeding, disease-resistant types have been developed. Precision agriculture utilizes technology to optimize irrigation, thereby increasing resource efficiency. Sustainable practices like organic cultivation and agroforestry enhance soil health and reduce environmental impact.

Recent advancements in genetic engineering have expedited the breeding program for mulberries. To directly regenerate plants from explants and introduce desired genes into the plant genome, efficient techniques have been developed using *Agrobacterium tumefaciens* and particle bombardment (Manjunath et al. 2023). Biotechnology has revolutionized silkworm breeding and disease control by enhancing genetic traits like silk production, environmental adaptability, and disease resistance. CRISPR/Cas9 enables precise editing of the genome for desired improvements. Disease management includes using microbial agents and vaccines against common silkworm illnesses, enhancing population health and productivity (Ibitoye et al. 2024).

## 12.5.2 Automation and Mechanization

Genetically modified silkworms are utilized in Indian sericulture to improve sericulture and quality by addressing the lack of natural relatives of Bombyx mori. Mechanization, like the Farmer's Friendly Mulberry Plant Cutter, is also being investigated to improve sericulture processes (Rafiq et al. 2022). The quality of the output is certain to be enhanced by artificial intelligence (AI), the Internet of Things (IoT), and image processing systems, among others (Saikia and Saikia 2023). Pesticide application is an important aspect of pest management. To ensure the success of pest control operations, an effective pesticide application method and suitable equipment are necessary. The objective encompasses not only pest control but also the maintenance of crop health, weed management, and fertilizer application (Ali et al. 2025). The art of using mechanical devices to enhance the production and storage of crops with reduced labor and increased efficiency is referred to as mechanization. Two land and two industrial activities constitute sericulture. Whereas rearing silkworms and cultivating food plants are land-based activities, industrial processes, including silk spinning, weaving, knitting, drying, and other operations, produce silk fabric. Most operations, such as rearing silkworms, cultivating food plants, and egg production, are manually performed in India; however, in countries like China and Japan, all steps involved in the process are highly mechanized. The yield of leaves and cocoons, the price of cocoons, the price of food plantation cultivation, and rearing costs all influence the profitability of sericulture (Marak et al. 2024). Weaving is the process of creating all or a portion of a cloth by interlacing the warp and weft threads in a certain way. Three basic textile weaves may be employed to create fabrics: satin, twill, and tabby. The simplest of those weaves, tabby, is built on a unit of two ends and two picks, where each end goes under and over one pick. A deeper look at silk fabrics, supported by digital microscope images, enables us to model silk strands while accounting for their radiometric and geometric characteristics. Combining this data with mathematical models that depict weaving processes and picture components is essential for creating three-dimensional (3D) representations of ancient silk fabrics that incorporate their interior structure. On each subsequent selection, the binding points are positioned over one end. The sequence of interlacements must be planned or determined throughout the design, production, and analysis of woven textiles in order to derive mathematical models. Identify textiles which is represented with a sole matrix and those that need a multiple matrix representation by utilizing a strategy for modeling yarns as matrices with various levels. The foundation for creating the 3D models in the Virtual Loom will be this method of depicting the textiles (Gaitán et al. 2019).

#### 12.5.2.1 Improving Efficiency and Reducing Labor Costs in Sericulture

Mechanization and automation in the silk industry have increased productivity, reduced labor costs, and simplified silkworm rearing through the use of automated feeding and monitoring systems. Reeling devices have also been mechanized for better silk extraction. Robotics and AI enhance production, quality control, and environmental monitoring. Automation ensures consistent, high-quality silk, while data analytics and IoT offer real-time insights for optimized operations (Ibitoye et al. 2024). Modern mechanized procedures, such as cocoon unwinding and silk reeling, surpass traditional methods, enhancing efficiency while maintaining silk quality. Advanced analytical technologies, such as spectroscopy and chromatography, ensure precise quality control to meet global standards (Attri et al. 2024). Mechanization and automation can increase silk production efficiency and lower costs but may be limited by high investment and skilled labor requirements, especially for smaller producers. Concerns about environmental impact and sustainability must also be addressed. Finding a balance between benefits and challenges is key in integrating technology in sericulture (Ibitoye et al. 2024).

New technologies in sericulture have enhanced sustainability, production, and efficiency by improving mulberry cultivation, silkworm rearing, and silk production techniques. These advancements have boosted economic efficiency, reduced environmental impact, and enhanced silk yield and quality. The integration of robots and AI in silk production processes has optimized resource utilization by reducing the need for human intervention. AI-driven robotic systems now handle intricate silk operations like twisting and weaving with unparalleled precision and speed (Attri et al. 2024).

### 12.5.3 Precision Agriculture

Precision agricultural techniques, like soil testing and irrigation scheduling, have enhanced mulberry growth in sericulture. Sericulturists can boost output sustainably by utilizing data-driven methods to optimize inputs and increase yields. For example, precision irrigation systems enable mulberry trees to receive water and nutrients directly, conserving water and enhancing productivity (Sudhakar et al. 2019). Drones are emerging as crucial tools for enhancing precision agriculture in the face of growing challenges in modern agriculture, including food security, sustainable resource use, and climate change. The efficiency of drones in various crops, ranging from viticulture to cereal farming, is demonstrated by certain case studies (Guebsi et al. 2024). In agriculture and livestock farming, a vast array of data from diverse sources like sensor and weather data, government statistics, imagery, and maps is generated. Precision agriculture and livestock farming can harness these statistics for real-time operations, process improvement, policymaking, and predictive insights. Integrating and analyzing varied data sources can enhance predictability and yield additional information. Challenges arise in consolidating data due to the need to locate data across different sources and deal with incompatible formats (Zeginis et al. 2024).

#### 12.5.3.1 Monitoring and Controlling Environmental Factors in Sericulture to Optimize Resource Use and Improve Productivity

The goal of a sericultural development plan in developing countries is to enhance land productivity, cut costs, and improve input efficiency while safeguarding the environment and human health. This aims to establish a sustainable soil–plant-environment system to combat overexploitation and soil degradation. Encouraging eco-friendly farming by adjusting current systems and reviving soil nutrients can promote sericultural farming. Minimizing environmental impacts also fosters well-being beyond the market economy, providing a higher quality of life in a beautiful, healthy environment (Mushtaq et al. 2022). Temperature variations significantly impact the growth and development of silkworms (Bombyx mori L.). Silkworms require a specific temperature range, with optimal growth occurring between 25 and 26 °C. But the fluctuations that are either warmer (20–21 °C) or cooler (28–29 °C) impede growth. Heat shocks increase mortality rates, accelerate larval stages, reduce weight, hinder food intake, leading to inefficiency, hinder cocoon creation, and stall transformation stages such as pupae and imagoes (Rahmathulla 2012). Whereas worms that have attained the adult stage are maintained at 23–24 °C with relative humidity of approximately 70–75%, immature worms are maintained at 26–28 °C with relative humidity of approximately 80–85% (Zhang 2018).

## 12.6 Consumer Awareness and Market Trends

Technological advancements in sericulture have expanded the diversity of silk products to meet evolving customer demands and market trends. The sector now offers a wide range of products, including traditional silk textiles and modern silk-based materials for various technical applications, enhancing market presence and global competitiveness (de Kluiver and Neethling 2025). The sericulture industry generates regular revenue on an equitable basis, benefiting both the community and individuals, regardless of gender, religion, or faith. This creates numerous opportunities for future social, economic, and political independence (Geetha and Indira 2011).

### *12.6.1 Role of Certification and Labeling: Importance of Certifications (E.G., Organic, Fair Trade) in Promoting Sustainable Silk*

Certification and labeling in the silk sector are crucial for confirming authenticity, sustainability, ethical manufacturing, and quality. Silk Mark (India) and OEKO-TEX Standard 100 certifications help verify genuine silk products and ensure they are free from harmful chemicals, addressing concerns regarding fake silk and mixed textiles in global markets (Ibrahim 2022). Certification programs like GOTS enable silk producers to set fair prices and protect customers from deceptive transactions, while promoting eco-friendly practices such as organic mulberry cultivation and non-toxic dyeing to minimize environmental impact (Samanta et al. 2023). Certifications such as fair trade, organic, and eco-friendly labels are crucial in the textile industry for sustainable silk production. They ensure ethical consumerism and support social justice by upholding moral labor standards in silk-producing communities, guaranteeing ecologically conscious practices (Mishra et al. 2023). Certifications not only promote environmental sustainability but also ensure fair labor standards and pay for silk workers, which is vital in regions like China, India, and Thailand, where labor-intensive workforces often consist of women and underprivileged individuals (Karthiga 2017, Le et al. 2025).

Consumers are increasingly seeking transparent and ethically sourced products, prompting silk producers to adopt sustainability certifications. This shift in consumer behavior fosters trust and encourages the industry to embrace more equitable and eco-friendly practices, ensuring long-term sustainability (Zhao et al. 2023, Swamy 2024). Certifications promote sustainable silk by ensuring adherence to environmental and ethical standards, promoting fair trade, and ensuring cruelty-free practices. They protect workers, ecosystems, and empower consumers. The demand for certified silk is expected to grow as sustainability drives the textile industry toward responsible practices. Investing in certified silk facilitates a sustainable and just future (Garg et al. 2025; Wang et al. 2025).

#### 12.6.1.1 Consumer Education and Awareness Campaigns

Consumer education campaigns are crucial for promoting ethical and sustainable silk production by informing consumers about the social, economic, and environmental impacts of their purchasing decisions (Doyle et al. 2023). Many buyers are unaware of the challenges associated with silk manufacturing, including environmental issues, unethical labor practices, and the widespread presence of counterfeit silk on the market. Through educational campaigns, businesses, NGOs, and industry leaders can increase consumer understanding and enable them to make informed decisions that support ethical labor practices, fair trade, and sustainable silk production (Conrad et al. 2022, Chakravarty et al. 2024). To reduce ecological impact, these efforts encourage consumers to choose silk that is sourced responsibly. Fair trade initiatives educate on how WFTO and Fair Trade Certified silk ensure fair labor, safe conditions, and equal opportunities for workers, urging support for ethical companies prioritizing worker welfare and community development through responsible silk purchases (Lee et al. 2021, Dutta 2024). Consumer education is crucial in promoting cruelty-free silk options, such as Ahimsa Silk. This silk is harvested without harming silkworms, allowing them to complete their life cycle naturally. Ethical businesses emphasize humane silk practices through various educational initiatives, targeting eco-conscious customers seeking sustainable and animal-friendly fabrics. They encourage designers to integrate ethical fashion in their creations (Bandyopadhyay et al. 2024). Counterfeit silk is another issue that consumer education campaigns aim to address. Unknowingly, many people who buy silk end up with mixed or synthetic textiles that are misrepresented as genuine silk. To assist customers in identifying real silk items, recognizing certification markings, and differentiating between real and imitation silk, organizations like the Silk Mark Organization of India (SMOI) run awareness campaigns. These programs safeguard customers and real silk producers from market fraud by educating purchasers about quality assurance through interactive demonstrations, silk-testing techniques, and online guidelines (Elumalai et al. 2021, Gunashekhar et al. 2024).

Consumer education initiatives utilize digital marketing, influencer collaborations, and community engagement to amplify their impact. Blogs, social media sites, and video material are effective means of educating people about sustainable silk and connecting with environmentally concerned shoppers worldwide. When it comes to promoting sustainable silk manufacturers and educating their followers about ethical buying practices, fashion influencers and ethical lifestyle advocates play a crucial role. Furthermore, fair trade and non-governmental groups organize exhibits, panel discussions, and events to interact with customers directly and foster a greater awareness of sustainable silk practices (Murzyn-Kupisz et al. 2021; Harper et al. 2025).

## 12.7 Conclusion

Silk remains a globally valued natural fiber with noteworthy importance. While the silk industry is facing many environmental and ethical challenges due to the traditional silk rearing mechanism. This chapter focuses on the promotion of sustainable methods such as peace silk production, precision agriculture, genetic improvement of mulberry varieties through biotechnological innovations, composting the silk litter, and so on, which further aligns with evolving consumer demands, ethical concerns, waste management, water conservation, carbon footprint, global awareness, and social and labor issues are addressed. The market trends are enhanced by the role of certification and labeling, consumer awareness will enable the silk industry to mitigate its impact, ensuring a further responsible and sustainable future for this luxurious textile.

## 12.8 Declaration

The authors acknowledge the use of AI tools during the preparation of this manuscript for the purposes of language editing and grammar checking. All content was subsequently reviewed, verified, and edited by the authors to ensure accuracy, originality, and compliance with ethical standards. The authors take full responsibility for the final content of the manuscript.

## References

Adhikari U (2022) Insect pest management: mechanical and physical techniques. Rev Food Agricult (RFNA) 3(1):48–53

Agrawal TK, Kumar V, Pal R, Wang L, Chen Y (2021) Blockchain-based framework for supply chain traceability: a case example of textile and clothing industry. Comput Ind Eng 154:107130

Agu EE, Iyelolu TV, Idemudia C, Ijomah TI (2024) Exploring the relationship between sustainable business practices and increased brand loyalty. Int J Manag Entrepreneur Res 6(8):2463–2475

Ahmed B, Javed S (2024) Comparative effects of entomopathogenic nematodes, biopesticides, and plant extracts against agricultural pest meloidogyne incognita (Tylenchida: Heteroderidae). Pak J Nematol 42(2). https://doi.org/10.17582/journal.pjn/2024/42.2.171.179

Ahmed S, Repon MR, Pranta AD, Haji A, Al-Humaidi JY, Rahman MM (2025) Performance analysis of functional properties of silk fabric enhanced via in situ grafting modification. Polym Eng Sci

Aishwariya S (2020) Interesting dimensions of vegan/ahimsa silk. Indian Tex J 108:112

Alam K, Raviraj VS, Chowdhury T, Bhuimali A, Ghosh P, Saha S (2022) Application of biotechnology in sericulture: progress, scope and prospect. The Nucleus 65(1):129–150

Alemayehu G, Shibabaw A, Adgo E, Asch F, Freyer B (2020) Crop rotation and organic matter application restore soil health and productivity of degraded highland crop farms in northwest Ethiopia. Cogent Food Agricult 6(1):1831124

Ali AMESH, Ahmed AAYM, Ali ZHRK, Mahmoud FMTM, Kouna FRAZ, Ahmed MANI, Pesticide spraying machines

Altman GH, Farrell BD (2022a) Sericulture as a sustainable agroindustry. Clean Circul Bioecon 2:100011

Altman GH, Farrell BD (2022b) Sericulture as a sustainable agroindustry. Clean Circul Bioecon 2:100011. https://doi.org/10.1016/j.clcb.2022.100011

Anitha R (2011) Indian silk industry in the global scenario. Int J Multidisciplin Manag Stud 1(3):100–110

Asinyo BK, Anokye D, Tawiah B, Godomey BS, Gbadegbe RS (2025) Challenges and prospects of sericulture in Ghana. Ghana J Sci Technol Develop 10(2):116–134

Attri K, Vakayil S, Shruthi GH, Parmar S, Harika KR, Teja KSS, Panigrahi CK (2024) Integrative approaches in sericulture from traditional practices to modern technologies sericulture and its contribution to biodiversity conservation. J Adv Biol Biotechnol 27(8):480–490

Baciu ED, Baci GM, Moise AR, Dezmirean DS (2023) A status review on the importance of mulberry (Morus spp.) and prospects towards its cultivation in a controlled environment. Horticulturae 9(4):444

Bakhtiyar N, Narzulla OR, Murodkhuja A, Khurshida PF (2023) History and development prospects of silk farming in Uzbekistan. E3S web of Conferen 376:02005–02005. https://doi.org/10.1051/e3sconf/202337602005

Balkrishna A, Rana A, Sharma RS, Sharma G, Arya V (2022) Indian farmer and government initiatives: policies, gaps and way forward. Indian J Ecol 49(3):1036–1043

Bamdad H, Papari S, Lazarovits G, Berruti F (2022) Soil amendments for sustainable agriculture: microbial organic fertilizers. Soil Use Manag 38(1):94–120

Bandyopadhyay PK, Sengupta D, Reddy J, Roy J (2024) Fashion design and carbon footprint–a critical examination. In: The Routledge companion to the future of management research. Routledge, pp 279–292

Bexiga NM, Bloise AC, de Moraes MA, Converti A, Beppu MM, Polakiewicz B (2017) Production and characterization of fibroin hydrogel using waste silk fibers. Fibers Polym 18:57–63
Bucciarelli A, Motta A (2022) Use of Bombyx mori silk fibroin in tissue engineering: From cocoons to medical devices, challenges, and future perspectives. Biomaterials Advances 139:212982
Bukhari R, Kour H, Aziz A (2019) Women and the Indian sericulture industry. Int J Curr Microbiol App Sci 8(05):2319–7706
Chakravarty D, Satpathy I, Patnaik BCM (2024) Embellishing fashion with sustainable goals: challenges in fashion supply chain. In: Illustrating digital innovations towards intelligent fashion: leveraging information system engineering and digital twins for efficient design of next-generation fashion. Springer Nature Switzerland, Cham, pp 179–202
Chowdhury SK, Banerjee M, Basnett D, Mazumdar T (2024) Natural pesticides for pest control in agricultural crops: an alternative and eco-friendly method. Plant Sci. Today 11:433–450
Chungyalpa W, von Rosing M (2025) Understanding business sustainability: the what, why, and how of sustainable business practices. In: The sustainability handbook, vol 1. Elsevier, pp 579–600
Conrad V (2022) Ethical and sustainable manifestations of "made in Nepal" an investigation of the global value chain of the production of hemp apparel (Master's thesis, University of New South Wales (Australia))
Czaplicki Z, Gliścińska E, Machnowski W (2021) Natural silk–an unusual fibre: origin, processing and world production. Fibr Tex East Europe 29(5(149)):22–28
Dandin SB, Kumari V (2021) Mulberry (Morus sps.) cultivation for sustainable sericulture. In: Mulberry. CRC Press, pp 188–207
Dandin SB (2025) Silk-a global textile. GAMANAM: Glob Adv Multidisciplin Appl Next-Gen Mod Technol 1(1):32–37
Dangre G (2024) From ancient paths to modern highways: mapping the evolution of world trade routes and their crucial role in today's global supply chains gaurav dangre. Indian Sci J Res Eng Manag 08(05):1–5. https://doi.org/10.55041/ijsrem33981
Dasari J, Venkataraman M, Nandini H (2024) Revealed comparative advantage analysis of trade of Indian silk and silk products. J Crop Weed 20(1):49–55
Dasari JR, Venkataramana MN (2023) Performance of global and Indian silk industry: an economic analysis. J Pharm Innov 1206–1210
Dinesh A, Naidu SA, Priya MK (2025) Organic farming: principles and practices. OrangeBooks Publication
Diyaolu CO, Folarin IO (2024) The role of biodiversity in agricultural resilience: protecting ecosystem services for sustainable food production. [60] Jain N, Kohli P (2024) Beyond Borders. DeepMisti Publication
Djeddour D, Pratt C, Constantine K, Rwomushana I, Day R (2021) The Asian Citrus greening disease (Huanglongbing): evidence note on invasiveness and potential economic impacts for East Africa
do Prado GF, de Souza JT, Piekarski CM (2025) Sustainable and innovative: how can open innovation enhance sustainability practices?. Sustainability 17(2):454
Doyle ML (2023) Conveying material impact: exploring effective communication strategies around the ecological, social, and cultural dimensions of fashion materials to support shifts in perception and behavior of businesses and end users. The American University of Paris (France)
Durak H (2023) Comprehensive assessment of thermochemical processes for sustainable waste management and resource recovery. Processes 11(7):2092
Dutta H, Bhattacharya S, Sawarkar A, Pradhan A, Raman RB, Panigrahi KK, Dutta KBS (2023) High yielding mulberry production through controlled pollination for enhanced vegetative growth and early sprouting suitable for tropical agroclimatic regions. The Pharma Innov 12(3):4485–4492
Dutta M (2024) Complexities of applying certifications in the Indian handicraft industry. In: The Routledge handbook of craft and sustainability in India. Routledge India, pp 281–299
E VM, JS R, B K, Mani B, GH S, S A, Reddy NC, Chellem S, S KK (2024) Genetic engineering: a tool for enhanced quality silk production. Int J Adv Biochem Res 8(8S):1208–1211. https://doi.org/10.33545/26174693.2024.v8.i8sp.2011

Elumalai D, Raj PM, Ramamoorthy R, Mohan C, Poovizhiraja B (2021) Introduction to non-mulberry silkworms. CRC Press
Erler S, Eckert JH, Steinert M, Alkassab AT (2022) Impact of microorganisms and entomopathogenic nematodes used for plant protection on solitary and social bee pollinators: host range, specificity, pathogenicity, toxicity, and effects of experimental parameters. Environ Pollut 302:119051. https://doi.org/10.1016/j.envpol.2022.119051
Fares N, Lloret J, Kumar V, De Leeuw S, Barnes L (2024) Optimisation of multi-tier supply chain distribution networks with corporate social responsibility concerns in fast-fashion retail. Corp Soc Responsib Environ Manag 31(1):311–330
Farid M, Djufry F, Abdul Hakim Y, Muhammad Fuad A, Musa Y, Nasruddin M, Aqil AFA, Iswoyo H, Jamil MH, Pati S (2022) Integrated corn cultivation technology based on morphology, drone imaging, and participatory plant breeding. Sabrao J Breed Genet 54(2):267–279. https://doi.org/10.54910/sabrao2022.54.2.5
Fatima S, Ishar A, Kumar G, Ajrawat A, Saurabh S, Jeeva PS, Devi HD, Rajeswari SU, Kumari B (2024) Recent innovations in sericulture: a comprehensive review of advancements in silk production and quality enhancement. Uttar Pradesh J Zool 45(23):75–83. https://doi.org/10.56557/upjoz/2024/v45i234688
Gaitán M, Alba E, León A, Pérez M, Sevilla J, Portalés C (2019) Towards the preservation and dissemination of historical silk weaving techniques in the digital era. Heritage 2(3):1892–1911
Garwi J, Masengu R, Chiwaridzo OT (2024) Sustainable practices for agriculture and marketing convergence. IGI Global
Geetha GS, Indira R (2011) Silkworm rearing by rural women in Karnataka: a path to empowerment. Indian J Gend Stud 18(1):89–102
Ghisellini P, Quinto I, Passaro R, Ulgiati S (2024) Exploring environmental and social performances of circular start-ups: an orientation and certification assessment. Bus Strateg Environ 33(4):3222–3241
Giora D, Marchetti G, Cappellozza S, Assirelli A, Saviane A, Sartori L, Marinello F (2022) Bibliometric analysis of trends in mulberry and silkworm research on the production of silk and its by-products. Insects 13(7):568
Gore PM, Naebe M, Wang X, Kandasubramanian B (2019) Progress in silk materials for integrated water treatments: fabrication, modification and applications. Chem Eng J 374:437–470
Gowrisankar R, Dasari S, Anas M (2023) Silk reeling techniques: exploring traditional and advanced methods. J Exp Agricult Int 45(9):91–99
Grześkowiak J, Łochyńska M, Frankowski J (2022) Sericulture in terms of sustainable development in agriculture. Problem Ekorozwoju 17(2):210–217
Guebsi R, Mami S, Chokmani K (2024) Drones in precision agriculture: a comprehensive review of applications, technologies, and challenges. Drones 8(11):686
Gunashekhar H, Patil D, Kiran NR, Prem M, Damodhara GN, Manjunatha B, Priyanka BN, Harishkumar J (2024) Sericulture in Karnataka: revitalizing through farmer producer organizations. Int J Agricult Environ Biotechnol
Han P, Rodriguez-Saona C, Zalucki MP, Liu SS, Desneux N (2024) A theoretical framework to improve the adoption of green integrated pest management tactics. Commun Biol 7(1):337
Harper Ph C, Alahdal T, Agha RG (2025) Wearing the keffiyeh: solidarity, resistance, activism
Hassan R, Acerbi F, Terzi S, Rosa P (2025) Transitioning the silk industry towards circularity: a thematic analysis of sustainable value chain practices. Sustain Product Consumpt 56:94–111
Hassan R, Terzi S, Rosa P (2023) Silk industry value chain and circular economy
Hazarika S, Saikia B Sericulture: a notable approach for encouraging women to lead sustainable lives
Hezakiel HE, Thampi M, Rebello S, Sheikhmoideen JM (2024) Biopesticides: a green approach towards agricultural pests. Appl Biochem Biotechnol 196(8):5533–5562
Hota MK, Bera MK, Kundu B, Kundu SC, Maiti CK (2012) A natural silk fibroin protein-based transparent bio-memristor. Adv Func Mater 22(21):4493–4499

Ibitoye OS, Ayanniyi OA, Ayeni OH, Adekoya OO, Aremu EA, Muritala DS, Wealth AS, Murtala MO (2024) Technological innovations and environmental sustainability in Nigerian sericulture: pathways to industry revival. J Appl Sci Environ Manag 28(9):2737–2743

Ibrahim F (2022) The eco-legalization and quality requirements for textiles, state of the art and current developments. J Meas Sci Appl JMSA 2:97–120

Islam MS, Ahmed F, Rahman MM (2023) Efficacy of vegetables intercropping in mulberry on sericulture productivity and economy. Euro J Agricult Food Sci 5(6):9–18

Jaiswal KK, Banerjee I, Mayookha VP (2021) Recent trends in the development and diversification of sericulture natural products for innovative and sustainable applications. Bioresource Technol Rep 13:100614

Jia F, Yin S, Chen L, Chen X (2020) The circular economy in the textile and apparel industry: a systematic literature review. J Clean Prod 259:120728

Kalpana R, Mithilasri M, Bharathi KM, Sabarish M, KT P (2024) Bridging tradition with innovation: revolutionzing sericulture through biotechnological advancements. Int J Zool Appl Biosci 9(4):78–82. https://doi.org/10.55126/ijzab.2024.v09.i04.013

Kambai C, Olatunji RO, Francis MJ (2020) Assessment of compost quality of sericulture waste on the growth of Anogeissus leiocarpus using two composting methods. Open J Environ Res (ISSN: 2734–2085) 1(1):pp.32–46

Karthiga S (2017) Production and marketing management of ilkal and kanchipuram pure silk saree–a comparative analysis (Doctoral dissertation, University of Agricultural Sciences, Dharwad).

Karthik T, Rathinamoorthy R, Ganesan P (2015) Sustainable luxury natural fibers—production, properties, and prospects. Handbook of Sustainable Luxury Textiles and Fash 1:59–98

de Kluiver J, Neethling T (2025) The digital silk road: implications for Africa's digital governance and transformation. In: Digitalisation and public policy in Africa: GovTech and CivicTech Innovations. Springer Nature Switzerland, Cham, pp 335–353

Kumar CP, Naika R, Prasad VM, Bhaskar RN, Narayanaswamy B (2012) Knowledge of sericulturists on organic farming practices in chikballapur District. J Agricult Vet Sci 1(3):1–4

Kumari M, Singh J (2016) Water conservation: strategies and solutions. Int J Adv Res Rev 1(4):75–79

Laity PR, Gilks SE, Holland C (2015) Rheological behaviour of native silk feedstocks. Polymer 67:28–39

Le THT, Singyabuth S, Nguyen TTT, Tran TTT (2025) Quang Nam silk, Vietnam: material culture and storytelling in the process of cultural revitalization. Arts Educa 42

Lee C, Kim S, Cho YH (2022) Silk and paper: progress and prospects in green photonics and electronics. Adv Sustain Syst 6(2):2000216

Lee E (2021) Female consumers' purchase intention of fair trade textile and clothing handicrafts: the roles of a geographical indication and fair trade knowledge in a brand equity model (Doctoral dissertation, University of Missouri-Columbia).

Likhi A (2020) Strengthening market access for small and marginal farmers through farmer producer companies in India: issues and way forward. MANAGE Knowl Series 1

Macha-Bizoumi N, Tranta A, Rigopoulou E (2025) Reflections on women's labour in museum narratives: the case of silk workers. J Textile Eng Fashion Technol 11(1):21–28

Madhu DM, Sinha DK, Singh KM, Kumari T, Singh RP, Kumari S, Ahmad N (2023) Achieving sustainable development goal by empowering women through mulberry silk production in Karnataka, India. Agric Assoc Text Chem Crit Rev J 10:1–6

Mana PW, Wang-Bara B, Mvondo VYE, Bourou S, Palaï O (2023) Evaluation of the agronomic and technological performance of three new cotton varieties in the cotton zone of Cameroon. Acta Botanica Plantae 2:28–39

Manjunath KK, Harshitha BS, Anusha NM, Bhargavi HA (2023) Breeding for mulberry improvement: a review

Manzoor S, Qayoom K (2024) Environmental Importance of Mulberry: a review. J Exp Agric Int 46(8):95–105

Marak M, Th. Aruna S, Dipankar S (2024) Mechanization in sericulture: an overview. J Adv Biol Biotechnol 27(8):820–825
Metaxas AM (2013) Chestnut (Castanea spp.) cultivar evaluation for commercial chestnut production in Hamilton County, Tennessee (Master's thesis, The University of Tennessee at Chattanooga)
Mishra D, Singh AP (2023) Organic food certification: overcoming barriers and enhancing sustainability in the agricultural industry. Jamshedpur Res Rev 78
Murzyn-Kupisz M, Hołuj D (2021) Fashion design education and sustainability: towards an equilibrium between craftsmanship and artistic and business skills ? Educ Sci 11(9):531
Mushtaq R, Raja TA, Fayaz A, Singh H, Ahmed P, Mir MR, Yadav R (2022) Resource use efficiency of temperate silk cocoons in North-Western Himalayan region of Kashmir valley. Jammu and Kashmir, India
Nayak R, Jajpura L, Khandual A (2023) Traditional fibres for fashion and textiles: associated problems and future sustainable fibres. In: Sustainable fibres for fashion and textile manufacturing. Woodhead Publishing, pp 3–25
Nayak P (2006) Influence of antibiotic on feed conversion efficiency of mulberry silkworm (Bombyx mori L.). Animal Biol 56(1):13–22
Nayak AK (2022) Case studies of FPOs in India 2019–2020. NABARD research study–25. National Bank for Agriculture and Rural Development, Mumbai
Nutthapornnitchakul S, Sonjaroon W, Putthisawong N, Thumthuan N, Tasanasuwan P, Jantasuriyarat C (2024) Effect of drought stress on proline gene expression, enzyme activity, and physiological responses in Thai Mulberry (Morus spp.). HAYATI J Biosci 31(3):559–571
Ozkavruk Adanir E, İleri B, Can F, Ulasli B (2022) An ethical approach to sericulture: production of peace silk in Hatay/Turkey. TEXTILE 22(1):1–11. https://doi.org/10.1080/14759756.2022.2036072
Ozkavruk Adanir E, İleri B, Can F, Ulasli B (2024) An ethical approach to sericulture: production of peace silk in Hatay/Turkey. TEXTILE 22(1):20–30
Pavithra A, Rajashekar J, Saikrishna K, Shreya B, Pramod Kumar M, Thanuja A, Hariharan S (2024) The evolution of the sericulture industry in India: from tradition to modernity. J Sci Res Rep 30(9):282–291
Peng Z, Yang X, Liu C, Dong Z, Wang F, Wang X, Hu W, Zhang X, Zhao P, Xia Q (2019) Structural and mechanical properties of silk from different instars of Bombyx mori. Biomacromol 20(3):1203–1216
Popescu A, Şerban V, Ciocan HN (2024) New trends in the global silk production in the period 2011–2022. Sci Pap Series Manag Econ Eng Agricult Rural Develop 24(1)
Popescu A (2018) Considerations upon the trends in the world silk trade
Punsara A (2023) Revival of the Silk Road legacy. Examining how China uses history to legitimize its ‹‹Belt and Road Initiative››. Historia i Świat 12. https://doi.org/10.34739/his.2023.12.23.
Pugina RS, Caiut JMA (2021) Silk: history, obtaining, structure and properties of an old material in the development of new technologies. Int J Adv Med Biotechnol IJAMB 4(1):27–45
Rafiq I, Salim D, Bhat A, Bhat SA, Buhroo ZI, Nagoo SA (2022) Emerging technologies for sericulture development
Rahaman MT, Pranta AD, Repon MR, Ahmed MS, Islam T (2024) Green production and consumption of textiles and apparel: importance, fabrication, challenges and future prospects. J Open Innov Technol Mark Complex 100280
Rahmathulla VK (2012) Management of climatic factors for successful silkworm (Bombyx mori L.) crop and higher silk production: a review. Psyche: J Entomol 2012(1):121234
Ram D, Saini PK, Bhati J (2022) Chapter-2 organic farming in india: opportunities challenges and perspectives. Res Rev 21:21
Rameshbhai CA (2021) Management of insect-pests through intercropping and evaluation of insecticides against earias vittella (fabricius) and Helicoverpa armigera (hubner) hardwick in okra (Doctoral dissertation, ANAND AGRICULTURAL UNIVERSITY)
Rani N, Verma M, Ruchi N (2025) Sustainable fibre in the textile industry. Vigyan Varta 6(2):94–100

Rauf S, Nazli Z (2024) Organic horticulture techniques for eco-friendly agriculture. Trends Biotech Plant Sci 2(1):20–24

Reddy YP, Parasuramudu M (2024) Role of sericulture in the Indian economy. J Res Appl Sci Biotechnol 3(4):107–116. https://doi.org/10.55544/jrasb.3.4.13

Reitz SR (2009) Biology and ecology of the western flower thrips (thysanoptera: thripidae): the making of a pest. Florida Entomol 92(1):7–13. https://doi.org/10.1653/024.092.0102

Reizabal A, Costa CM, Pérez-Álvarez L, Vilas-Vilela JL, Lanceros-Méndez S (2023) Silk fibroin as sustainable advanced material: material properties and characteristics, processing, and applications. Adv Func Mater 33(3):2210764

Ricciardi L, Karatas S, Chiarelli DD, Rulli MC (2020) Environmental sustainability of increasing silk demand in India. In: EGU general assembly conference abstracts, p 19986

Rohela GK, Saini P, Syam S, Roy P, Gonal B, Aziz D, Ayoub OB, Maheswari M (2025) Mulberry trees: a sustainable solution for urban forestry and improved air quality. In: Sustainable urban environment and waste management: theory and practice. Springer Nature Singapore, Singapore, pp 223–246

Rowe A (2021) Silk production: global scale and animal welfare issues

Sahu R, Niranjan TT (2025) Exploring the psychological antecedents of suppliers' sustainability practices: a qualitative investigation in the Indian textile industry. Corp Soc Responsib Environ Manag 32(1):522–542

Saikia S, Saikia M (2023) Smart Sericulture and IoT: a Review. Biol Forum-An Int J

Samanta P, Konar A, Muhopadhyay A (2023) Colorimetric characterisation and process standardisation for application of natural dyes on textiles: a research review. Adv Colorim

Sannapapamma KJ, Naik SD (2015) Ahimsa silk union fabrics–A novel enterprise for handloom sector

Senapati R, Parapurath F, Pandey3and S, Behera DK (2024) Climate change and its impact on sericulture industry

Shanmugam R, Murugesh KA, Tilak M, Senguttuvan K, Sivakumar K, Radha P, Jothimani P (2024) Boosting the excellence of cocoon production through silkworm excreta

Sharma A, Gupta RK, Sharma P, Qadir J, Bandral RS, Bali K (2022a) Technological innovations in sericulture. Int J Entomol Res 7(1):7–15

Sharma V, Rattan M, Chauhan SK (2022) Potential use of sericultural by products: a review. Pharma Innov 1154–1158

Sharma P (2024) Sustainable development and traditional industries: Embracing epistemic syncretism in silk industry for success. Futur Trends Soc Sci 3:12

Shen X, He J, Zhang N, Li Y, Lei X, Sun C, Muhammad A, Shao Y (2024) Assessing the quality and eco-beneficial microbes in the use of silkworm excrement compost. Waste Manage 183:163–173

Silva AS, Costa EC, Reis S, Spencer C, Calhelha RC, Miguel SP, Ribeiro MP, Barros L, Vaz JA, Coutinho P (2022) Silk sericin: a promising sustainable biomaterial for biomedical and pharmaceutical applications. Polymers 14(22):4931. https://doi.org/10.3390/polym14224931

Solomun JI, Totten JD, Wongpinyochit T, Florence AJ, Seib FP (2020) Manual versus microfluidic-assisted nanoparticle manufacture: impact of silk fibroin stock on nanoparticle characteristics. ACS Biomater Sci Eng 6(5):2796–2804

Ssemugenze B, Esimu J, Nagasha J, Masiga CW (2021) Sericulture: Agro-based industry for sustainable socio-economic development: a review. Int J Sci Res Publ 11(9):474–482

Sudhakar P, Vijaya NB, Kiran Kumar KP (2019) Comparative studies on V1 and G4 high yielding mulberry varieties at nursery level. Int J Curr Res 11(04):2745–2750

Sujatha GS, Kumar GA, Teja KS, Devi DL, Rupali JS, Gautam SK (2024) A comprehensive review of the effect and mitigation of climate change on sericulture. Int J Environ Clim Change 14(7):776–788

Sun Y, Ma L, Wei T, Zheng M, Mao C, Yang M, Shuai Y (2024) Green, low-carbon silk-based materials in water treatment: current state and future trends. Chemsuschem 17(11):e202301549

Susheelamma BN, Jolly MS, Giridhar K, Sengupta K (1990) Evaluation of germplasm genotypes for drought resistance in mulberry. Sericologia (France) 30(3)

Swamy U (2024) From soil to salesfloor: exploring the potential for retailer integration into the fibershed soil-to-soil framework. The University of North Carolina at Greensboro
Upadhyay N, Kamble A (2025) The role of environmental concerns and self-expression in ethical fashion consumption: a mediated model of consumer values. J Glob Mark 1–23
Valiyev S, Rajabov T, Kabulova F, Khujanov A, Urokov S (2024) Changes in the amount of photosynthetic pigments in the native Artemisia diffusa in the semi-desert rangelands of Uzbekistan under the in luence of different sheep grazing intensities and different seasons. J Plant Biota
Vambol V, Kowalczyk-Juśko A, Vambol S, Khan NA, Mazur A, Goroneskul M, Kruzhilko O (2023) Multi criteria analysis of municipal solid waste management and resource recovery in Poland compared to other EU countries. Sci Rep 13(1):22053
Wang CN, Nhieu NL, Chen CM (2024) Charting sustainable logistics on the 21st-Century Maritime Silk Road: a DEA-based approach enhanced by risk considerations through prospect theory. Hum Soc Sci Commun 11(1):1–12
Wang M, Murphy R, Christie I (2025) Bringing sustainable practices, fashion shows, and sociological insights together to reinvigorate sustainable fashion education. Sustainability 17(2):631
Wilhelm MM, Blome C, Bhakoo V, Paulraj A (2016) Sustainability in multi-tier supply chains: understanding the double agency role of the first-tier supplier. J Oper Manag 41:42–60
Wu Z, Xu W, Wu X, Ding X (2025) Life cycle environmental and economic assessment of raw silk production in China. Sustain Product Consump
Wu Z, Xu W, Wu X, Ding X (2025) Life cycle environmental and economic assessment of raw silk production in China. Sustain Product Consumpti
Xu F (2025) Liang Shaoji's silkworm art: untangling multispecies craft. Taylor & Francis
Yadav M, Gupta R, Nair K (2024) Time for sustainable marketing to build a green conscience in consumers: evidence from a hybrid review. J Clean Prod 443:141188
Zeginis D, Kalampokis E, Palma R, Atkinson R, Tarabanis K (2024) A semantic meta-model for data integration and exploitation in precision agriculture and livestock farming. Semantic Web 15(4):1165–1193
Zhang Y (2018) Determination of optimal temperature for production of quality eri silkworm cocoon and seed. Agricult Res Technol Open Access J 17(3):2–9
Zhao M, Duan X (2023) RETRACTED ARTICLE: assessing financial performance through green bond markets and energy reliance in Asian economies. Environ Sci Pollut Res 30(27):70421–70436
Zhou W, Arcot Y, Medina RF, Bernal J, Cisneros-Zevallos L, Akbulut ME (2024) Integrated pest management: an update on the sustainability approach to crop protection. ACS Omega 9(40):41130–41147

# Chapter 13
# From Cocoon to Community: The Social Impact of Silk

**Veena S. More, Aishwarya Lomte, R. Janavi, R. Srinivasan, P. Darshan, and B. Kiran Kumar**

## 13.1 Introduction

Around 2700 BCE, silk—one of the world's oldest and most luxurious fabrics—originated in China. Legend has it that silk was discovered by Empress Leizu when a cocoon dropped into her tea and unraveled into tiny strands. The renowned Silk Road trade routes that linked Asia and Europe were made possible by the Chinese keeping sericulture, or the manufacture of silk, a secret for millennia. The production of silk eventually extended to Persia, India, and other regions. Silk, which is produced in large quantities in countries like China and India, remains a highly prized fabric today due to its silky texture and inherent shine (Gilroy 2023). This highly valued fabric, used for high-end apparel, faces competition from synthetic fibers due to production costs, despite silk being beautiful and long-lasting (Popescu 2020). Although it accounts for only 0.2% of natural fibers worldwide, Asia dominates the production of silk, a common textile derived from silkworms, primarily in developing countries (Popescu et al. 2024). India, the world's second-largest silk producer, produced 34,903 tons in 2021–2022, offering numerous job opportunities and minimal investment (Dasari et al. 2024). People across the world are familiar with the Silk Road's rich history and legacy—a historic trade route that symbolizes friendship, trade, and cultural exchange, supported by various organizations (Winter 2021). The Silk Road, a pivotal trading route, significantly influenced the cultural and economic advancement of civilizations in Asia, the Middle East, and Europe (Ibrahim and Kizito 2021). Emperor Wu established the Silk Road as a network of pathways connecting China and Central Asia, promoting economic expansion and cultural exchange (Xia 2025). China's Silk Road exhibits a distinctive phenomenon—the adaptive evolution of cultural ecosystems, characterized by a changing relationship between environment and culture (Yu et al. 2023). In 2014,

V. S. More (✉) · A. Lomte · R. Janavi · R. Srinivasan · P. Darshan · B. Kiran Kumar
Department of Biotechnology, Sapthagiri College of Engineering, Bangalore, India
e-mail: veenasmore@sapthagiri.edu.in



D. Kumar and M. Shahid (eds.), *Science and Technology of Silk*, Textile Science and Clothing Technology, https://doi.org/10.1007/978-3-032-16230-4_13

China, Kyrgyzstan, and Kazakhstan declared the "Silk Road: The Routes Network of Chang'an-Tianshan Corridor" in Qatar, adding 22 cultural locations to UNESCO's World Heritage List, showcasing regional traditions, religious beliefs, and folk-art cultures (Hussain et al. 2024). This millennium-long trade route connected Europe, Asia, and the Middle East, promoting economic expansion and intercultural interactions through goods like textiles and spices (Muthukumaran 2023). This network of trade routes linking cities and empires, significantly boosted economic growth, facilitated market and commerce growth (Brown and O'Brien 2024). Constantinople, Samarkand, Kashgar, and Baghdad were significant cultural and commercial centers, attracting economic benefits through the construction of financial institutions, roadways, and caravanserais (Dalrymple 2024). Long-distance trade increased the appeal of paper money, credit, and banking, driving global economic expansion and gaining advantages for Byzantine, Persian, Mongol, and Ottoman Empires (Kharé 2024). The Silk Road facilitated intellectual growth, cross-cultural interaction, and religious concepts, with Indian Buddhists influencing Chinese philosophy and art, and Islamic culture advancing Central Asia and Europe (Asadov 2024).

The Renaissance saw the rise of hybrid civilizations, blending regional architectural and artistic influences, with Chinese skills significantly impacting European culture, and intellectual advancements via the Silk Road (Ubaydullaeva and Karimov 2023). In the thirteenth century, the Pax Mongolica provided secure trade routes and fostered economic activity despite facing political instability, severe weather, and banditry (Nyamdaa 2023). New sea routes discovered by European explorers and the growth of maritime commerce in the fifteenth and sixteenth centuries caused the Silk Road's popularity to diminish (Naderi Beni et al. 2024; Chandrasekaran 2025).

Silk, a luxury material, has significantly influenced technological advancements, cultures, civilizations, and economies, dating back to 2700 BCE in China, despite being kept a secret for centuries (Lo 2024). Silk's fineness, beauty, and durability significantly influenced ancient civilizations and cultures, signifying connections and authority (Bonino and Carota 2025). Silk's global economic impact led to the establishment of the Silk Road, connecting East and West, and its high value in Rome, Persia, India, and medieval Europe facilitated significant commerce and economic expansion (Henderson et al. 2024). The expansion of Samarkand, Baghdad, and Constantinople facilitated the silk trade, influenced banking systems, and impacted the economies of Lyon and Venice (Kaplan 2025).

Silk, a symbol of wealth and power, was used by royalty, aristocracy, and religious leaders globally, with the Roman Senate restricting its usage (Mijares 2024). Silk was used as a diplomatic tactic by foreign monarchs to define Byzantine identity and adorn royal clothing, church vestments, and mosques in medieval Europe and Islamic countries (Khalil 2025). Silk revolutionized art and technology, enhancing textile skills, weaving, and dying, leading to intricate needlework designs and influencing European, Ottoman, and Persian textile production (Gansane 2023). Silk, a blend of Eastern and Western motifs, was crucial for ancient Chinese writing before paper, promoting cross-cultural creativity (Spicer 2023). Silk's biocompatibility and molecular composition enhance its use in biomedical engineering, tissue engineering, and medication delivery systems, promoting eco-friendly textiles and synthetic fibers in contemporary commerce and science (Bitar et al. 2024). Silk has significantly influenced various aspects of society, culture, technology, human ingenuity, global history, business, craftsmanship, and intercultural connections since ancient trade (O'Neill 2022).

Figure 13.1 illustrates the framework of cultural textiles, which examines how traditional textile processes, such as silk weaving, contribute to community identity, economic empowerment, and cultural preservation. Analyzing silk-based crafts helps women artisans maintain livelihoods and enhance local heritage, uniting tradition, economy, and community well-being. The Silk Road revolutionized commerce by facilitating the exchange of Chinese silk, Roman glassware, Indian spices, and Persian stones, ultimately benefiting individual traders, towns, and empires (Bellina 2022). The Silk Road's money flow facilitated the growth of trading cities like Samarkand, Kashgar, and Baghdad, enhancing financial systems, banking, credit, and currency swaps for kings and nations (Mir-Makhamad et al. 2021). The Silk Road facilitated cross-cultural exchange and facilitated the spread of Buddhism from India to China, influencing literature, art, and religious customs in China, Central Asia, and Japan (Li 2023). The Silk Road's cultural influence is evident in Dunhuang's Buddhist cave temples, alongside its rich religious diversity, which includes Nestorian Christianity, Islam, and Zoroastrianism (Xiangrong 2022). The Silk Road facilitated easier communication between religious and scientific communities, enabling cultures to exchange ideas and advance discoveries (Wu and Zhang 2023). From East to West, Chinese innovations such as paper and gunpowder influenced military strategies, record-keeping, and education in many locales (Lado 2025). In the thirteenth century, early Silk Road travelers enjoyed some protection under strong empires, such as the Mongols, despite obstacles like inclement weather and bandits (Wang and Tian 2022). The Mongols' establishment of the Pax Mongolica simplified travel and trade throughout Eurasia, thereby supporting the Silk Road's operations by fostering more cross-cultural exchanges and commerce (Avdaliani 2025). The Silk Road, a vital link between civilizations, significantly influenced international trade, culture, and diplomacy, facilitating globalization through its influence on history, innovations, economics, and ideologies (Winter 2022). Sericulture, the traditional art of silk production, has profoundly influenced communities worldwide by fostering social and economic transformation, cultural identity, sustainability, innovation, and new growth opportunities (Somsiriwattana et al. 2025). It empowers rural populations by engaging farmers who cultivate mulberry and artisans who craft intricate textiles (Zhirong 2024). In nations such as China, India, Thailand, and Brazil, silk production contributes significantly to local economies—especially in rural areas of India, where sericulture programs supported by governments and development agencies generate employment opportunities for underprivileged groups (Ssemugenze et al. 2021). Local silk industries reduce dependence on global markets, promote entrepreneurship through high-quality handcrafted textiles, and improve living standards and economic resilience (Dutta 2023). Sericulture also drives social change by empowering women through flexible employment, financial independence, and leadership opportunities. Women gain skills, fair compensation, and recognition through cooperatives and self-help groups, advancing gender equality and social progress (Rani 2024; Javed et al. 2024). Blending ancient craftsmanship with modern technology, silk production preserves cultural identity in regions renowned for their traditional textiles, such as brocades and sarees (Le et al. 2025). Support for local silk enterprises safeguards heritage skills and fosters intercultural exchange in global handmade textile markets (Alfawaz 2024). Government-backed GI designations ensure product authenticity and fair remuneration for artisans, helping silk-producing villages protect their traditions while expanding into international markets (Borsaikia et al. 2023). As a natural, biodegradable material, silk plays a crucial role in sustainable and ethical fashion, with many producers adopting organic and cruelty-free methods, such as Ahimsa or peace silk (Ozkavruk Adanir et al. 2024). Sericulture by-products—mulberry leaves, silkworm pupae, and fibers—are repurposed for use as fertilizers, animal feed, and in biotechnological applications, thereby linking sustainability with innovation (Rawat et al. 2024). Environmentally responsible practices not only protect ecosystems but also open new livelihood opportunities for communities engaged in eco-friendly silk production (Ermini et al. 2022). Silk manufacturing drives inclusive economic growth, social empowerment, cultural preservation, and ecological responsibility, promoting fair trade, ethical fashion, and sustainable development that benefit both people and the planet (Connors and Trites 2025).

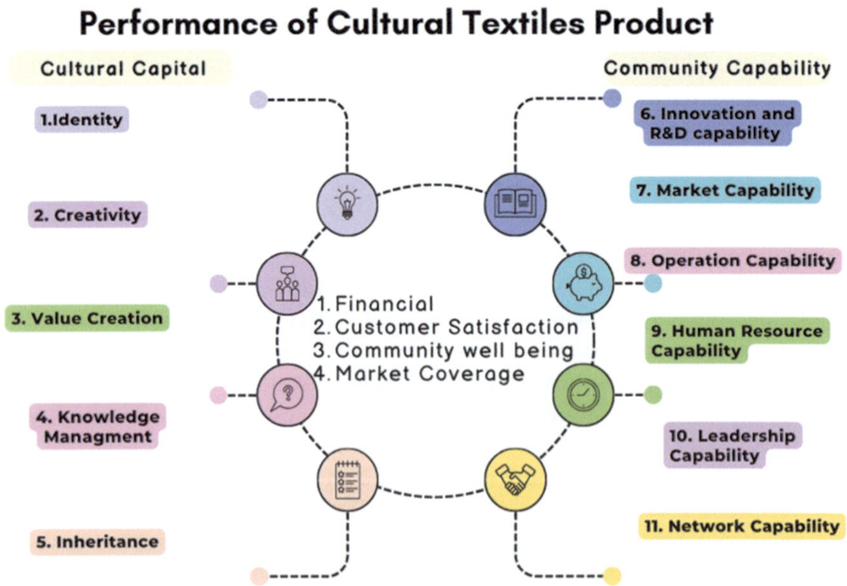


**Fig. 13.1** Performance evaluation for the framework of cultural textiles

## 13.2 Silk as a Status Symbol

Silk, a luxury material, has been associated with prestige, wealth, and power since ancient China. It gained popularity in Rome and medieval Europe, and is still used in haute couture and ceremonial clothing (Kelley 2024).

### *13.2.1 Silk's Role in Differentiating Social Classes and Creating Social Hierarchies*

Silk, a symbol of prestige and luxury, has played a pivotal role in establishing social hierarchies and maintaining social structures throughout history. Due to scarcity and complex manufacturing processes, it was exclusively worn by high-ranking officials in ancient China (Downing et al. 2023). The Chinese government strictly regulated the use of silk, ensuring that its brilliance was only appreciated by the ruling elite, with commoners often forbidden to wear it. Only intellectuals and military officials were permitted to wear silk (Acquaye et al. 2022). The silk trade expanded along the Silk Road, influencing social systems in ancient cultures, such as ancient Rome, where the wealthiest could afford it due to its high price (Marimón 2025). The Byzantine Empire strictly regulated silk manufacturing, allowing only imperial family members and high-ranking officials to wear specific types of clothing. The court established complex customs, with colors and patterns indicating different grades (Winnik 2024).

Silk was used in medieval Europe for ceremonial robes, religious vestments, and royal clothing, symbolizing nobility and divine authority, but access was limited due to sumptuary laws (Butler 2025). Silk production in Europe led to labor divisions and economic systems, with skilled craftspeople controlling output and aristocrats often achieving the same social status (Caracausi 2024). Silk, a symbol of social stratification and exclusivity, is a staple in luxury clothing, symbolizing wealth, prestige, and the connection between textiles, privilege, power, and social divisions (Kılınç and Nasır 2023).

### *13.2.2 The Formation of Silk-Based Cliques: Shared Consumption and Social Identity*

Silk symbolizes material prosperity and has historically shaped social cliques and identity-based societies, allowing individuals and organizations to differentiate themselves through shared consumption and artistic expression (Borges et al. 2024). Silk, a symbol of aristocracy and political power in ancient China, was exclusively worn by royal family members and senior officials, strengthening hierarchical boundaries and establishing social identities (Shea 2021). Silk-wearing groups, recognized by affluent merchants and kings, were used to demonstrate wealth and social control over lesser strata on the Silk Road (Anderson 2024). Ancient Rome's nobility coveted silk, an exotic luxury from the East, establishing restricted social networks focused on silk consumption (Inglis et al. 2021). Roman senators and patricians wore silk clothing at opulent feasts, showcasing cultural sophistication and exacerbating the class divide, thereby transforming silk into a symbol of social identity (Wärnberg 2023).

During the European Renaissance, silk symbolized social identity and class differentiation, particularly among the aristocracy and merchant families. Guilds in Venice and Florence produced top-quality fabrics (Wall 2025). Royal court members and powerful families commissioned luxury silk clothing, tapestries, and accessories to showcase their wealth and social status, preserving cultural authority and improving social ties (Fülemile 2023).

The contemporary silk industry's relationship with social identity has evolved, but it remains firmly rooted in high society and fashion. Luxury fashion houses' silk-based cliques, recognizing haute couture and silk couture creations by designers and celebrities (Pagán et al. 2020). Elites gather at events like the Met Gala and Paris Fashion Week to showcase their wealth and silk-based luxury, wearing custom-made silk dresses and suits from renowned designers (Song 2024). Silk consumption has influenced creative and cultural circles, leading to eco-conscious cliques that prioritize ethically sourced Ahimsa Silk and Fair-Trade Certified silk, thereby promoting social identity and belonging (Nayyar 2024). Silk usage has shifted to artistic and cultural circles, forming eco-conscious groups that prioritize ethical and sustainable silk items like Fair-Trade Certified and Ahimsa Silk, fostering social identity and belonging (Zhong and Wu 2025).

### *13.2.3 Traditional Role of Women*

Women have historically played a vital role in the silk industry, particularly in sericulture, contributing significantly to the production process across nations such as China, India, and Thailand (Madhu et al. 2023). Women in sericulture are responsible for providing silkworms with mulberry leaves, ensuring optimal temperature and humidity, and meticulously extracting cocoons (Liu et al. 2024). Women excel in this step due to their traditional knowledge and abilities, which require patience and attention to detail (Kiplagat 2022). Rural women skillfully remove silk strands from cocoons, spin, and weave intricate patterns, creating a highly prized market textile that reflects their cultural background (Rautio 2024).

Women play a crucial role in the dying and stitching processes of silk fabrics, contributing to their unique colors and intricate decorations. However, they often face challenges like low pay, lack of recognition, and limited access to training and modern technologies (Rahman et al. 2024). Women's participation in the silk industry is crucial for maintaining customs and preserving the industry's cultural heritage, as they provide for their families and contribute to the industry's longevity (Barron 2021).

#### 13.2.3.1 Historical Significance of Women in Sericulture

Women have been central to the development and sustainability of the silk industry for centuries, making significant contributions to its growth and advancement (Islam 2024). In ancient China, Empress Leizu's innovations in silkworm rearing and weaving not only established a highly profitable trade but also relied on women to care for silkworms and harvest delicate threads, preserving this knowledge across generations (Hancock 2021). Across India, Persia, and Europe, women played a significant role in guiding the silk trade along the Silk Road, engaging in all stages, from raising silkworms to spinning and weaving (Jayaram et al. 2024). In Assam, tribal women have maintained traditional Muga silk techniques passed down from mother to daughter, while in Mysore, they play a vital role in operating silk reeling mills and producing premium silk (Khadse et al. 2018). During the Industrial Revolution, women made significant contributions to the silk industry by operating looms and creating intricate patterns, despite working in challenging conditions in automated mills (Macha-Bizoumi et al. 2025). Emerging nations, including China, India, Thailand, and Vietnam, are experiencing a surge in women's participation in sericulture, which is promoting the revival of ancient silk weaving techniques and providing financial support (Escap 2024).

Women's involvement in sericulture has significantly impacted social and cultural aspects, preserving cultural heritage, supporting families, and achieving financial security. Meanwhile, silk, a globally upscale fabric, has gained global recognition for its artistry (Singha and Singha 2020). Women have historically played a crucial role in the silk industry since ancient China, showcasing their talent, inventiveness, and tenacity, which have supported the silk trade and facilitated women's emancipation and regional economic growth (Cheshmehzangi 2024).

#### 13.2.3.2 Women's Knowledge and Skills in Silk Production

Since ancient times, women have been the backbone of the silk industry, contributing to every step of the process. Their sophisticated skills and vast knowledge have been passed down through generations. Women's outstanding workmanship and passion are evident in everything from caring for silkworms to designing intricate patterns on silk textiles (Craig 2021). Sericulture involves women caring for silkworms, providing a healthy habitat with mulberry leaves. The process requires patience and precision, as temperature and humidity can significantly impact the quality of silk fibers. Women's maternal nature and observational abilities help create an ideal environment for silkworms (Rezić 2022). Women play a crucial role in the process of spinning and twisting silk threads after collecting cocoons. Their skill in managing delicate fibers ensures the creation of smooth, long-lasting silk yarns, while also ensuring texture and strength through their spinning and twisting abilities (Duan 2022). Women showcase their creativity in dyeing silk items using natural dyes and traditional methods, resulting in vibrant colors and distinctive designs. Their expertise in fabric treatment and color mixing further enhances their appeal (Ghosh et al. 2024). The success of the silk business is largely due to women's extensive knowledge and expertise, which have improved the quality and cultural relevance of silk goods, and continue to be essential for the global silk industry's expansion and sustainability (Wang and Suwanthada 2025).

## 13.3 Silk and Cultural Exchange

Silk, first produced in ancient China, facilitated the Silk Road, promoting technology, art, religion, and knowledge, and enhancing international understanding and collaboration through global trade (Berthon 2021).

### *13.3.1 The Silk Road: Facilitating Trade, Cultural Exchange, and the Emergence of Cosmopolitan Cliques*

The Silk Road, a trade route connecting East and West, facilitated cultural exchange, facilitated trade, and fostered cosmopolitan communities. Originating in ancient China, it spread across Central Asia, the Middle East, and Europe, significantly influencing advancements in civilization (Kudabay et al. 2021). The Silk Road, facilitated by trade, technology, and Islamic advancements, facilitated economic growth, creativity, and scientific achievement in the Roman Empire and other regions (Hillman 2021). The Silk Road facilitated cultural exchange and the sharing of ideas among diverse peoples, influencing regional customs and spiritual practices through the spread of religions such as Buddhism, Islam, Christianity, and Zoroastrianism (Hai 2024). The Silk Road facilitated cultural exchange, blending philosophical ideas, languages, and creative forms, thereby enhancing communities through Chinese sculptures, paintings, and designs that were influenced by Indian and Persian art (Krotova 2024). The Silk Road fostered cosmopolitan communities in trading centers like Samarkand, Bukhara, and Constantinople, promoting intellectual and cultural exchange, and fostering tolerance and understanding among diverse groups (Mott 2014).

### *13.3.2 Silk's Impact on Fashion and Aesthetics: The Spread of Styles and the Creation of Cultural Trends*

Silk has significantly influenced global fashion and aesthetics through the Silk Road trade, promoting various cultural trends (Hunt 2021). The Silk Road enabled various regions to incorporate silk into traditional clothing, with China's aristocracy and monarchy often using it as emblems of longevity and wealth. During the Renaissance,

European designers embraced silk as a symbol of wealth and power, experimenting with ornate dresses, corsets, and silk capes to establish new fashion norms (Huber and Van Godtsenhoven 2023). Silk's shine and texture influenced tailoring, leading to the development of haute couture and opulent brocades. Florence and Venice's silk weaving hubs produced accessories such as scarves and gloves (Montemezzo 2024). Silk significantly influenced traditional clothing design, particularly in Japan and India, where it inspired the design of high-end garments, including sarees, which later influenced modern fashion designers and their evening dresses and red-carpet gowns (Tarai 2020). Silk, a versatile and sustainable material, is a popular choice in the fashion industry due to its sophistication, versatility, and timeless appeal (Gundev 2023). Silk, a crucial textile in fashion history, has significantly influenced clothing designs, cultural trends, and cross-cultural interactions since its introduction along the Silk Road (Krishnamurthy 2023).

### *13.3.3 The Role of Silk in Diplomatic Relations and the Formation of Political Alliances*

Silk, first used in ancient China, significantly influenced political alliances and diplomatic connections, demonstrating China's craftsmanship and securing peace agreements with foreign rulers (Boon and Duong 2022). The Silk Road was a vital network for political and cultural exchange between East and West, promoting harmony and avoiding hostilities due to its opulence (Vilar 2025). Chinese kings fortified political ties and extended cultural influence to the Middle East and Central Asia, while European kings, such as those of France and Italy, used silk as a mark of distinction to cultivate connections with powerful foreigners (Wilson 2023). Silk had a significant influence on political coalitions and international diplomacy, fostering cultural exchange and confidence among empires. Its historical ability to promote international cooperation and solidarity demonstrates its historical significance (Rong 2022).

### *13.3.4 Empowerment Through Silk Production*

Sericulture, or the production of silk, is a potent instrument for social and economic development, especially in rural areas. It gives farmers, craftspeople, and business owners a steady stream of revenue, opening doors to self-sufficiency and financial freedom. Thousands of people, particularly women and members of marginalized groups, are employed throughout the whole process, which includes several steps from raising silkworms to weaving silk garments. Farmers can reduce their reliance on traditional agriculture, which is often affected by erratic weather and market fluctuations, by diversifying their revenue streams through sericulture. Furthermore, silk farming is a viable option for small-scale farmers and rural communities due to its high returns and minimal investment requirements (Chandler et al. 2021).

Women's contribution to the silk industry is especially important as it allows them to work from home and support themselves. Numerous governmental and non-governmental organizations have recognized the potential of sericulture to empower women and have implemented market connections, financial assistance, and training programs to support them. Women who work in the silk industry not only become financially independent but also acquire entrepreneurial skills that raise their social standing and self-esteem. As a result, community development is strengthened and gender equality is promoted. In addition, small-scale farmers and craftspeople in silk-producing areas can unite through cooperative societies and self-help organizations to enhance production quality, bargain for lower pricing, and get access to wider markets (Ramnath and Kim 2024). In addition to its financial advantages, silk manufacturing is essential for sustainable development and cultural preservation. The increasing demand for silk contributes to the survival of many traditional silk weaving methods in various regions, which have been passed down through the centuries. By utilizing organic methods, silk farming contributes to environmental sustainability through eco-friendliness, reducing the use of synthetic fibers, and promoting biodiversity. In order to increase silk production while guaranteeing its sustainability, governments everywhere—particularly in nations like China and India—have started programs to support sericulture through funding, education, and research (Le et al. 2025).

#### 13.3.4.1 Breaking Down Gender Barriers and Promoting Social Equality

The silk industry has historically promoted social equality and challenged gender conventions. Women have traditionally been responsible for tasks such as spinning, dyeing, weaving, and raising silkworms. However, male-dominated positions have sometimes underappreciated their achievements. Initiatives to empower women and underrepresented groups have helped break down gender barriers and advance social equality (Diwakar and Rehman 2025). Women-led cooperatives and self-help groups in Vietnam, Thailand, and India are providing female artisans with training, fair compensation, and access to international markets. These programs not only close the gender gap but also empower women to lead their companies and improve working conditions. The silk industry supports cultural heritage, lowers inequality, and empowers women to contribute to local economies (Sulolipu et al. 2022).

## 13.4 Silk Production and Economic Development

Sericulture is crucial for supporting rural livelihoods and reducing poverty, particularly in developing countries. It provides steady income and job opportunities for disadvantaged populations, encouraging local ownership, and sustainable development. The expansion of the silk sector often necessitates new infrastructure, such as processing plants and transportation systems, benefiting local communities. Sericulture's expansion beyond its traditional regions reflects its economic potential and attractiveness, offering jobs, revenue, and foreign exchange (Rasuljonovich 2025).

### *13.4.1 Rural Livelihoods and Poverty Reduction*

Sericulture in rural areas fosters social networks and cooperative activities, promoting community development. Cooperatives and community-based groups offer financial support, technical assistance, and training to silk growers, enabling them to access shared resources, exchange information, and enhance their market position. This cooperative model strengthens community bonds and social capital, contributing to the development of regional infrastructure, such as mulberry farms and silk mills. Sericulture projects also foster mutual support and teamwork, enhancing community resilience and achieving common objectives, ultimately improving socioeconomic progress and quality of life in rural communities (Mala et al. 2024).

#### 13.4.1.1 Silk Production as a Source of Income Generation and Economic Independence for Women with Employment in Rural Areas

Women have greatly benefited from the silk industry's ability to generate wealth and achieve financial independence. From caring for silkworms to spinning and creating elaborate 34 textiles, women have always played a major role in the manufacture of silk. In addition to supporting household finances, this engagement has provided women with the opportunity to become financially independent and gain social respect. Women have utilized their skills to establish successful companies and support their families in nations with strong cultural ties to silk production, such as China and India (Yadav et al. 2022). Silk farming and weaving are now significant sources of income for women from low-income families in rural areas, particularly in countries such as Thailand, Vietnam, and India. These women can overcome the constraints of conventional roles that prevent them from being financially independent by engaging in the silk industry. For example, the manufacture of high-quality silk sarees in Tamil Nadu and Karnataka, India, is mostly dependent on women silk farmers and weavers. Women may create sustainable livelihoods by learning how to raise silkworms, dye threads, and weave elaborate designs with the help of government assistance and self-help organizations (Banerjee 2024). Fair-trade programs and cooperative organizations, such as the Thai Silk Association and the Self-Employed Women's Association in India, have significantly enhanced women's economic empowerment. These organizations provide access to markets, resources, and financial assistance, fostering confidence, leadership, equal pay, and improved working conditions. They also enable women to invest in healthcare and education (Malhotra et al. 2024).

The silk industry benefits both economically and socially by preserving traditional handicrafts and cultural heritage. Women artisans pass down their skills, fostering pride and a sense of identity within their communities. Financial independence and cultural preservation empower women to combat gender inequality. Women entrepreneurs are also leading the way in creating organic, ethically sourced silk items, creating new markets (Ghose and Ali 2023). Sericulture is a socially and economically equal industry that benefits society regardless of caste, creed, gender, or religion. Farmers, regardless of wealth, receive the same revenue. Women play a significant role in sericulture, providing opportunities and independence in various aspects. The labor-intensive sector provides jobs and extra revenue for weaker groups. India benefits from mulberry and non-mulberry sericulture, creating jobs and improving socioeconomic status. Sericulture is practiced in over 58 countries worldwide (Sharma and Kapoor 2020). The labor productivity rate has consistently increased over the years, with the highest increases occurring in 2009–10 and 2018–19. However, employment growth in 2020–21 was significantly negative, likely due to the COVID-19 pandemic. Labor productivity has shown a steady increase over time, with notable high points in 2010–11 and 2017–18. The growth rates for both production and employment leveled off between 2021 and 2022, with labor productivity peaking in 2023–2024 at 4.10. The overall trend is modest expansion, with notable declines in 2009–10, 2016–17, and 2020–21. The global pandemic may have contributed to a steep drop in both employment and output growth rates in 2019–20. Despite some oscillations, the overall trend is generally upward, with a peak in 2023–224 at 0.0041. Between 2016 and 2017, there was consistent growth, indicating increases in productivity per worker (Ssemugenze et al. 2021). Silk demand surges domestically and internationally due to its properties. Literate farmers in rural areas are embracing technological advancements, creating jobs in the industry (Kumar 2017).

#### 13.4.1.2 Contribution of the Silk Industry to Local and Regional Economies

The Silk Road Economic Belt has garnered global attention due to its significant economic impact. A study using the GTAP model examines the impact of financial cooperation on international trade, GDP, and overall well-being. The findings suggest that adopting a local currency settlement system could boost GDP in North Africa, Europe, and Central Asia, while optimizing China's international trade structure. However, this shift in international commerce could decrease export and import volumes in sectors not part of the Silk Road, but increase national welfare levels (Zheng and Jia 2017). Asia is the largest contributor to the world's output, including silk, with China being the largest producer. Sericulture is expanding in Africa, South Asia, and Latin America to improve women's quality of life and gender equality. Asia's rising demand for textile and silk goods is driving growth in the global silk market. Advancements in technology in the sericulture industry are expected to help the silk market. The World Bank's 2021 income categorization highlights the importance of sericulture as a marginal and landless subsistence alternative (Mushtaq et al. 2023). A circular economy is a regenerative industrial system that extends product life through robust design, repair alternatives, and sharing options. It involves the upstream value chain, midstream value chain, and downstream value chain. Research suggests that transitioning from linear to circular behaviors with minimal environmental impact can significantly influence the future of the silk industry, promoting a more sustainable and circular approach to managing the silk value chain (Hassan et al. 2023).

#### 13.4.1.3 Development of Silk-Based Industries (Textiles, Handicrafts, Tourism)

Sericulture, an agro-industry that involves mulberry cultivation, silkworm egg production, silkworm rearing, and textile spinning, is gaining popularity due to its immediate market value (Ntaanu 2006). Queen Victoria's 1851 Stuart Ball fancy dress, rooted in mid-nineteenth-century materials and modern British colonial networks, featured an Indian silk underskirt, showcasing the monarchy's past, present, and future. Studying the Banaras brocade used in a unique costume from Varanasi, India, reveals its opulent and richly embellished design, which paved the way for the growth of the silk industry and tourism (Hodge 2025). Handwoven textiles, valued in Bangladesh, Indonesia, and India, are intricately crafted by weavers on traditional looms. These unique, time-consuming items showcase the skill and originality of weavers, reflecting their cultural roots.

Figure 13.2 gives a generalized categorization of handloom products. Handloom fabrics, a sustainable and popular choice for formal events, account for 15% of textiles in India, characterized by unique weaving communities and materials such as chemicals, cotton, wool, jute, dyes, and silk (Sharma 2025). The Indian government estimates that 22.9% of all handloom weavers, including 6.83 lakh families in Varanasi district, weave sarees (Naik et al. 2025).

#### 13.4.1.4 Role of Cooperatives and Self-help Groups in Promoting Sustainable Silk Production

Community-based organizations (CBOs) significantly contribute to rural development, with a substantial number of women involved in the sericulture industry. Their

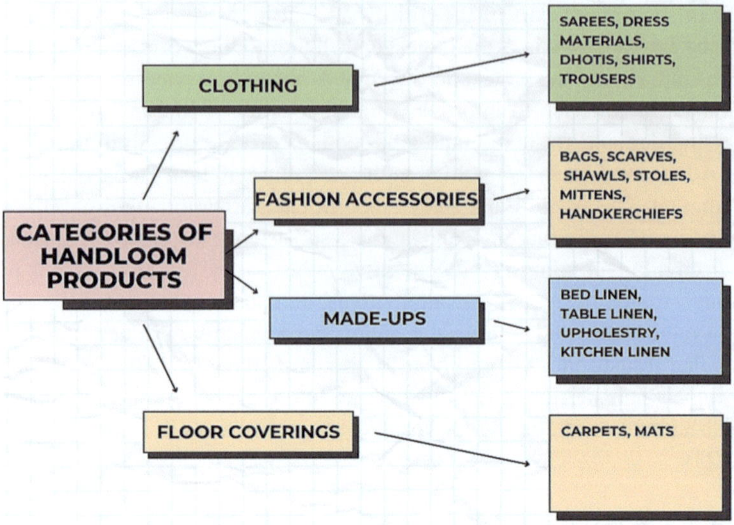


**Fig. 13.2** Categories of handloom products

emancipation can provide valuable insights for government, policy, and organizational development (Eswarappa 2020). Policymakers worldwide are promoting the entrepreneurial ecosystem (EE), a network of interconnected individuals, organizations, and cultural values, to enhance local and regional economies through innovation and revitalization (Pathak and Mukherjee 2021). Sericulture empowers women through self-help organizations, cooperatives, and information exchange, thereby boosting regional economies, creating job opportunities, and promoting community development, which extends beyond individual homes (Hazarika et al.). Government and non-governmental organizations are promoting women's empowerment through programs such as Women's Cooperatives and Women's Self-Help Groups, providing regular income, education, and training (Arora and Chawla 2023). The Central Silk Board, India in collaboration with NGOs like PRADAN and BAIF, has launched a multi-state initiative to enhance the livelihoods of impoverished tribal districts (Venkataronappa et al. 2019). India is leveraging Self-Help Groups (SHGs) to bridge the gap between government institutions and impoverished rural women, promoting collective savings, borrowing, and lending, thereby reducing poverty (Vijayalakshmy et al. 2023).

### 13.4.2 *Impact of Silk Production on the Development of Local Infrastructure (Roads, Markets, Etc.)*

The Belt and Road Initiative (BRI), launched by Chinese President Xi Jinping in 2013, is a comprehensive global connectivity framework that encompasses economic integration, cultural exchange, and infrastructure development. It consists of the Silk Road Economic Belt and the Maritime Silk Road, spanning Asia, Africa, and Europe. The program aims to establish a new paradigm for global collaboration by investing in energy grids, transportation networks, and marine ports. Its ramifications extend beyond economic progress, as it supports China's soft power strategy by exporting its development model and influencing global governance standards (Chiebuka et al.). The Forum for China-Africa Cooperation (FOCAC) has been instrumental in fostering closer ties between China and Cameroon since 2000. Since the 2010s, Sino-Cameroonian collaboration intensified with significant investments in infrastructure, often funded by Chinese loans. Cameroon, a member of the African Union, has aligned with China's foreign policy to capitalize on its economic prospects (Tsafack et al.). The Belt and Road Initiative (BRI) is a strategy by China to advance its geopolitical and economic objectives. After the global crisis, China made significant domestic investments in heavy industries and infrastructure. However, the economy was overheated, leading China to seek external investments. The BRI has contributed to the growth of Central Asian nations, particularly in energy and transportation. The Chinese government plays a crucial role in the development of BRI projects by serving as the primary source of funding for state-owned policy banks. China has also signed trade and economic cooperation agreements with over thirty nations to maintain the BRI (Wang et al. 2024).

## 13.5 Silk Production in Community Development and Community-Based Enterprises

Entrepreneurship makes it possible to start new companies and create jobs, it is increasingly seen as a catalyst for growth and wealth as well as an alternative source of economic expansion. The entrepreneur's personality has been the subject of extensive research in entrepreneurship studies. Other research has examined the process of entrepreneurship and explained it from the viewpoints of new forms of production, cultural values, a strong drive for success, and a means of acknowledging social marginality (Pathak and Mukherjee 2021). The importance of sericulture and the livestock sector in community development is covered in S. R. Charsley's 1982 book "Culture and Sericulture." It serves as both a case study of the South Indian silk industry and an excellent illustration of the contribution of social anthropology to development. Anthropological explanations of comparative development have become increasingly popular when economic ones are deemed inadequate. This research examines a local community in detail, focusing on its industries, organizations, regulations, and implementation. It can provide an anthropological understanding of development and offer suggestions on how anthropology can contribute to development (Kasi 2013). To ensure sustainable silk production, it is also essential to identify the sources and processes of environmental damage, as well as explore potential economic and environmental improvements (Wu et al. 2025). Over 40,000 registered textile businesses use OTOP, but controlling manufacturing costs is challenging due to fluctuating yarn prices. Commercialization could offer more yarn options and imaginative products (Inpakdee and Soodsang 2025).

## *13.5.1 Social Cohesion and Community Building*

Social cohesion is the continuous process of promoting well-being, belonging, and voluntary participation in society, while creating communities that tolerate diverse values and cultures, while granting equal rights and opportunities (Moustakas 2023). Social cohesiveness is a society's social capital, encompassing trust, harmony, and cooperation, despite demographic variations, promoting cooperation and coordination for mutual gain (Jayakody et al. 2022). The Sustainable Development Agenda's 16th goal, "Peace, Justice, and Strong Institutions," advocates for inclusive, accountable, and efficient institutions, community relationships, and policies that recognize the rights of ethnic groups (Khuenkhaew et al. 2025). Understanding the impact of local culture and assessing available resources are crucial for the sustainable growth of the silk industry. This aids in the creation of plans to concurrently accomplish the goals of sustainable development (Siewsamdangdet 2010). Thailand's cultural capital fosters cooperation among ethnic groups, promotes education, and enhances community collaboration. It is integrated into development initiatives for long-term self-reliance (Somsiriwattana et al. 2025). Community leaders play a vital role in sustainable development, promoting growth and empowerment through local knowledge exchange and economic networks, which in turn strengthen national economic growth (Pradhan and Samanta 2025). Figure 13.3 illustrates the key components of the conductive policy framework and proposed interventions to support the silk industry.

### 13.5.1.1 Role of Silk Production in Fostering Community Spirit and Cooperation

Community solidarity promotes gender equality, social norms, and multicultural understanding, uniting individuals and fostering inclusion, collaboration, and trust through social bonds and shared activities, such as social networks (Setiadi et al. 2025). Miyazaki categorizes community challenges into five groups: people, culture, place, industry, and scenery, emphasizing community involvement, preservation of resources, cultural protection, and sustainable development (Fig. 13.4).

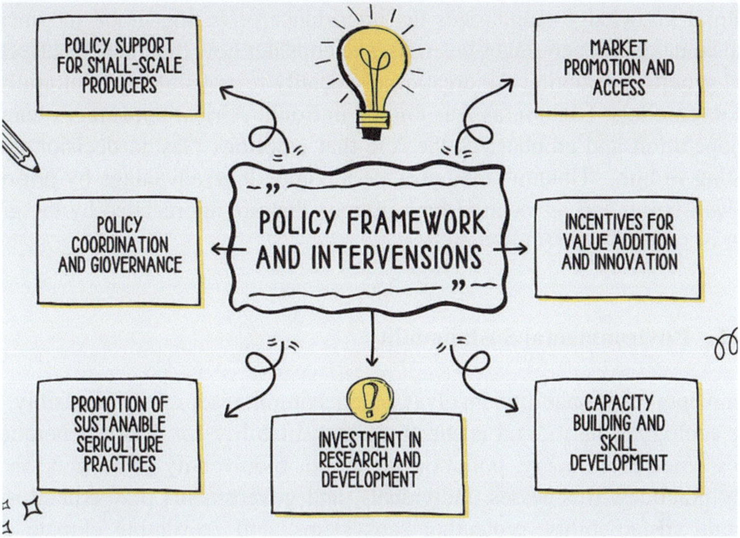


**Fig. 13.3** Key components of the conductive policy framework and proposed interventions to support the silk industry

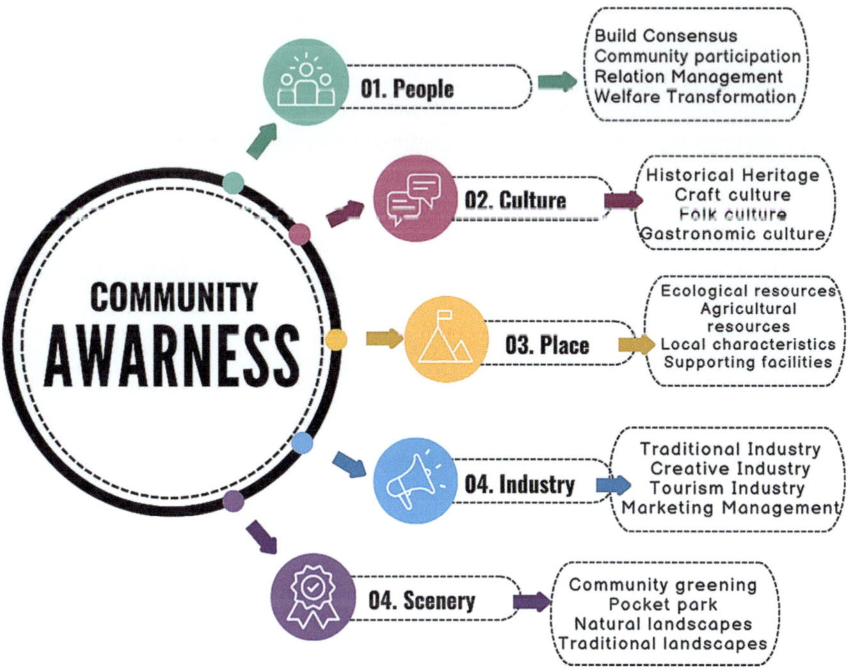


**Fig. 13.4** Five major resources of community construction

Cultural knowledge emphasizes the importance of basing moral judgments on cultural contexts. It encourages businesses to consider how their actions affect other cultural groups, promoting the creation of socially responsible and culturally flexible moral models. Ubuntu, as it is known in South Africa, encourages teamwork and cooperation and emphasizes the role that emotions play in decision-making. According to him, "Ubuntu" may provide a competitive advantage by prioritizing social well-being and advocating for solutions that are appreciated by the greatest number of stakeholders (Goswami et al.).

#### 13.5.1.2 Environmental Sustainability

Environmental sustainability involves using natural resources responsibly, minimizing ecological harm, and ensuring their availability for future generations. It involves renewable energy, pollution reduction, biodiversity protection, and eco-friendly practices. Businesses, individuals, and governments play crucial roles in promoting sustainability, protecting ecosystems, and preventing climate change (Doo et al. 2024). Sericulture promotes sustainable growth by using environmentally friendly techniques, including waste reduction and soil conservation. It can help achieve SDG 13 by reclaiming wastelands and reducing climate change. One example of how conventional practices might support a sustainable society is the Silk Industry (Sharma 2024). To safeguard the planet and advance a sustainable future, the SDGs demand action. To combat climate change and advance sustainability, scientists and engineers play a crucial role. Silk and other bio-based materials have drawn attention within the SDGs. Due to its biocompatibility and biodegradability, silk has supplanted conventional materials in photonic and electrical devices. It has shown promise for various applications, including light-emitting technologies and biomedical sensing. Silk's special qualities, such as its high concentration of antiparallel β-strands, make it a useful component for organic electronics (Baruah and Kalita 2023).

##### 13.5.1.2.1 Eco-friendly Silk Production Practices

Silk, valued for its strength, lightness, and benefits to the wearer, qualifies as an eco-friendly textile when produced and processed with minimal environmental impact, in line with the principle "Think globally, act locally." A material's eco-friendliness depends on factors such as its ecological footprint, renewability, and the amount of chemicals required for production (Suparna and Antony 2016). Green chemistry promotes sustainability, while plant-derived colorants, though less common and low in fastness, offer environmentally safer alternatives. Conventional processing often consumes significant energy, solvents, and time; however, methods such as cationization, bio-polishing, and mercerization enhance fabric absorbability, while innovative techniques like ultrasonic rays accelerate dyeing and extraction (Adeel et al. 2022).

Sustainable fabrics are made with minimal chemicals, sourced from renewable materials, which reduces health risks such as headaches, allergies, and respiratory issues, while maintaining a favorable ecological footprint. Silk processing involves the separation of fibroin and sericin due to their differences in appearance, solubility, and amino acid composition. Scouring and bleaching are then applied to achieve a cleaner, softer, lustrous fabric with a brighter white color (Zhu et al. 2022).

#### 13.5.1.2.2 Conservation of Biodiversity and Traditional Silk-Producing Ecosystems

Biodiversity protection is crucial for sustainable silk production in traditional silk-producing regions, such as China, India, and Thailand. Mulberry plant cultivation and silkworm rearing are essential, while organic farming and avoiding harmful chemicals maintain soil health and local plant and animal health, thereby enhancing silk quality (Fambayun et al. 2022). Wild silk variants, such as Tussar, Eri, and Muga silk, along with mulberry silk, help preserve forest ecosystems by utilizing natural plants and trees. Sustainable harvesting practices and community-led conservation programs enhance the environment and improve the living standards of local communities (Dwivedy 2014). The practice of producing silk, known as sericulture, has undergone tremendous development due to technical breakthroughs. The conventional approaches to sericulture have been completely transformed by these developments, which have improved sustainability, quality, and production (Attria et al. 2024).

## 13.6 Challenges and Opportunities

Reducing rural poverty is a top priority for developing nations. Most of the world's impoverished reside in rural regions. One important tactic is to create jobs in rural areas. Agriculture-related issues impede the establishment of jobs. It is essential to diversify the rural economy. Sericulture helps prevent people from migrating from rural to urban areas and creates employment opportunities. The sericulture sector and its contribution to development are reviewed in this section. Plans are put in place to support the sector and address socioeconomic problems (Gangopadhyay 2008).

### *13.6.1 Challenges Faced in the Silk Industry*

- **The major constraints for silk producers**

Silk producers face significant limitations in productivity and profitability, including high manufacturing costs, climate-dependent pests and diseases, and limited access to modern technology. This leads to lower-quality silk and decreased market competitiveness, as silk production relies heavily on climate change and mulberry farming. Market volatility, a lack of official assistance, insufficient financial support, and challenges in accessing fair-trade markets hinder the expansion potential of small-scale silk growers, necessitating technical breakthroughs, financial support, and the adoption of sustainable practices.

- **The major constraints for processors**

The silk industry faces challenges due to outdated technology, high equipment costs, and uneven raw silk quality. Additionally, the sector faces a scarcity of experienced workers and high labor costs, which negatively impact efficiency and profitability. Sustainability challenges in silk processing include the use of chemicals, water consumption, market fluctuations, competition from synthetic fibers, and limited financial support. Technological advancements, skill development programs, and sustainable practices can improve efficiency and profitability.

- **The major constraint for experts in the public sector**

Public sector experts in the silk industry face challenges such as a lack of funds, budgetary restrictions, inefficient bureaucracy, and a lack of cooperation among government organizations, industry players, and research institutions, which hinder innovation and growth. Insufficient training programs and personnel in public sector organizations hinder the impact of skill development projects. Balancing sustainability and industry development is challenging. To enhance the silk sector's global competitiveness, increased government funding, streamlined regulations, and enhanced collaboration are necessary (Sime and Siraj 2020).

### *13.6.2 Opportunities of the Silk Business*

There are several ways that the silk business can improve its social impact, especially by supporting sustainable development and strengthening the lives of rural people. Economic empowerment through fair-trade policies and cooperative models is a key option. Small-scale farmers and craftspeople, especially women, are the backbone of many silk-producing regions. They profit from established cooperatives and moral trading networks. Silk producers can improve their overall quality of life by gaining access to international markets, earning higher incomes, and securing job stability by strengthening these systems. The shift of disadvantaged workers from subsistence labor to self-sufficient business can also be facilitated by government and non-governmental organization assistance for microfinance and skill development programs (Sharma et al. 2022). Sustainability-driven innovation presents yet another significant potential. Sustainable sericulture techniques, such as organic mulberry farming and cruelty-free silk harvesting techniques like Ahimsa silk, can be adopted by the silk industry as a result of growing worldwide awareness of ethical fashion and environmentally friendly production. These developments not only reduce environmental damage but also attract ethical customers, creating a market for silk that is ethically sourced. Additionally, utilizing contemporary technology in weaving, dyeing, and marketing—such as digital storytelling and e-commerce platforms—can help craftspeople establish direct connections with global consumers, thereby increasing their market reach and financial success. The silk industry may develop into a sector that transforms society and the economy while maintaining traditional craftsmanship by taking advantage of these prospects (Cappellozza et al. 2022; Bhagya et al. 2024).

## 13.7 Conclusion

The chapter examines the intricate history, economics, and societal impact of silk, highlighting its transformative potential. Silk production creates jobs, generates income, and empowers rural communities, particularly women. It enhances living conditions through improved infrastructure, healthcare, and education, ultimately leading to social transformation and improved quality of life. China, India, Thailand, Brazil, and Uzbekistan are major global silk producers, significantly influencing the industry and impacting lives. They must ensure ethical sourcing, eco-friendly practices, and fair treatment of employees to achieve ethical growth in the sector. The silk industry needs a vision that prioritizes sustainability, ethical production, and a fair share of benefits. This requires collaboration between consumers, industry players, and governments. Funding research and development can enhance living standards, mitigate ecological impacts, and promote biodiversity. Promoting traceability and transparency can encourage ethical decision-making. Silk can empower women, support rural communities, and contribute to a more equitable world.

**AI Usage Declaration** AI tools were used solely for language refinement and formatting. The authors have reviewed and verified the final content of the chapter and take full responsibility for its accuracy and integrity.

## References

Acquaye R, Sawyerr NO, Kusi CA (2022) West African prints: fusion of global textile design traditions. In: Textiles and clothing along the silk roads, pp 243–258

Adeel S, Habib N, Batool F, Rahman A, Ahmad T, Amin N (2022) Eco-friendly approach towards isolation of colorant from Esfand for bio-mordanted silk dyeing. Environ Sci Pollut Res 29:13523–13533

Alfawaz MA (2024) Evolution of handicraft industries in the textile sector in Saudi Arabia: a comprehensive analysis of trends and innovations. J Text Sci Technol 10(3):82–100

Anderson JA (2024) The dong world and imperial China's southwest silk road: trade, security, and state formation. University of Washington Press

Arora N, Chawla RA (2023) Empowering women through self-help groups and women cooperatives-a tool for sustainable development. DME J Manag 4(01):25–30

Asadov F (2024) Arab geography as a source of knowledge and study of the history of silk road trade. In: Showcasing the role of the Arabic language along the silk roads, p 193
Attria K, Sujatha GSB, Jekinakatti B, Thrilekha D, Devi DL, Garai I, Satapathy SN, Tripathy A (2024) Advancements in sericulture: innovations and sustainability in silk. Uttar Pradesh J Zool 5(16):305–317
Avdaliani E (2025) Khvarazmians, Mongols, and the decline of the silk roads in the Caucasus. In: The Caucasian silk roads and Eurasian connectivity, 500–1405: trade, culture, and warfare in transit. Springer Nature Switzerland, Cham, pp 93–141
Banerjee A (2024) The handloom industry of Begampur in transition: technology, disjuncture, and development. Rowman & Littlefield
Barron L (2021) The creative influence of history in fashion practice: the legacy of the silk road and Chinese-inspired culture-led design. Fash Pract 13(2):275–295
Baruah MB, Kalita P (2023) Muga silk: sustainable materials for emerging technology. In: Advanced materials from recycled waste. Elsevier, pp 295–316
Bellina B (2022) Southeast Asian evidence for early maritime silk road exchange and trade-related polities. In: The oxford handbook of early Southeast Asia, pp 458–500
Berthon MA (2021) Silk and post-conflict Cambodia: embodied practices and global and local dynamics of heritage and knowledge transference (1991–2018). Royal College of Art (United Kingdom)
Bhagya KPN, Karunaratne PVM, Ranathunga GM, Ranaweera A (2025) Fashion niche market strategies: a systematic literature review. J Fashion Mark Manag 29(1):137–163
Bitar L, Isella B, Bertella F, Vasconcelos CB, Harings J, Kopp A, van der Meer Y, Vaughan TJ, Bortesi L (2024) Sustainable *Bombyx mori's* silk fibroin for biomedical applications as a molecular biotechnology challenge: a review. Int J Biol Macromol 130374
Bonino M, Carota F (2025) New silk road: the architecture of the belt and road initiative. Birkhäuser
Boon K, Duong V (2022) China-Vietnam-France relations on the verge of the collapse of the tributary system in the 19th century. Jebat 49(2):44
Borges A, Calvo MLM, Vaz JA, Calhelha RC (2024) Enhancing wound healing: a comprehensive review of Sericin and *Chelidonium majus* L. as potential dressings. Materials 17(17):4199
Borsaikia T, Kalita A, Dutta M (2023) Scope for branding Dora Baran Gamosa through intellectual property rights and technology. J Heritage Manag 8(2):195–209
Brown MS, O'Brien D (2024) The silk roads and shared heritage in Europe: beyond 'China to Rome'. China Inf 0920203X241281989
Butler W (2025) The land of the Veda. Prabhat Prakashan
Cappellozza S, Casartelli M, Sandrelli F, Saviane A, Tettamanti G (2022) Silkworm and silk: traditional and innovative applications. Insects 13(11):1016
Caracausi A (2024) Between innovation and resistance: the role of guilds in early modern Italian ribbon manufacturing. Technol Cult 65(3):791–817
Chandler DP, Roff WR, Smail JR, Taylor RH, Woodside A, Wyatt DK (2021) In search of Southeast Asia: a modern history (revised edition). University of Hawaii Press
Chandrasekaran A (2025) China-US relations handbook. Educohack Press
Cheshmehzangi A (2024) China under construction: shaping cities through recent urban transformation. Springer Nature
Chiebuka AQ, Janet OO, Oluchi OC. The belt and road initiative: China's vision for global connectivity and soft power influence. Asian J Edu Social Stud 51(1):262–272
Connors SP, Trites RS (2025) Neoliberalism and young adult fiction: exceptionalism, exploitation, and erasure. University Press of Mississippi
Craig C (2021) Middlemen of modernity: local elites and agricultural development in modern Japan. University of Hawaii Press
Dalrymple W (2024) The golden road: how ancient India transformed the world. Bloomsbury Publishing
Dasari J, Venkataramana M, Nandini H (2024) Revealed comparative advantage analysis of trade of Indian silk and silk products. J Crop Weed 20(1):49–55

Diwakar K, Rehman PBDM (2025) Building inclusive and supportive ecosystem for nurturing trans-social entrepreneurship in India. Cuestiones De Fisioterapia 54(2):1169–1192

Doo FX, Vosshenrich J, Cook TS, Moy L, Almeida EP, Woolen SA, Gichoya JW, Heye T, Hanneman K (2024) Environmental sustainability and AI in radiology: a double-edged sword. Radiology 310(2):e232030

Downing BC, Silk MJ, Delahay RJ, Bearhop S, Royle NJ (2023) Culling-induced perturbation of social networks of wild geese reinforces rather than disrupts associations among survivors. J Appl Ecol 60(12):2613–2624

Duan X (2022) An object of seduction: Chinese silk in the early modern transpacific trade, 1500–1700. Rowman & Littlefield

Dutta M (2023) Creative economy and sustainable development: the context of Indian handicrafts. Taylor & Francis

Dwivedy N (2014) Assisting female farmers in exploiting the alternate gainful areas, MPRA Paper 54697, University Library of Munich, Germany

Ermini C, Visintin F, Boffelli A (2024) Understanding supply chain orchestration mechanisms to achieve sustainability-oriented innovation in the textile and fashion industry. Sustain Prod Consumption 49:415–430

Escap U (2024) Course on promoting women's entrepreneurship through digital transformation. https://www.unescap.org/events/2024/promoting-womens-entrepreneurship-through-digital-transformation-policymaker-track

Eswarappa K (2020) Community-based organizations (CBOs) and their role in the development of women: a case from Andhra Pradesh. Contemp Voice Dalit 13(1):34–49

Fambayun RA, Agustarini R, Andadari L (2022) Cultivation and breeding techniques for increase silk productivity in Indonesia. In: IOP conference series: earth and environmental science, vol 995, no 1. IOP Publishing, p 012055

Fülemile Á (2023) Monarchs in national dress-sartorial expressions of national image in the representational practices of 19th-century European courts. Acta Ethnographica Hungarica 66(2):439–495

Gangopadhyay D (2008) Sericulture industry in India–a review. India Science and Technology. SSRN

Gansane AK (2023) The global goods and their impact on society, economy, and environment: analysing the phenomenon from the early modern age to the contemporary age. Ca' Foscari University of Venice

Ghose A, Ali SM (2023) Protection and preservation of traditional cultural expression & traditional knowledge in handicraft industry: advocating the need for a global cultural policy framework. Braz. J. Int'l l. 20:473

Ghosh J, Khan F, Noor T, Rupanty NS, Das SC (2024) Union dyeing of cotton-silk blended woven fabric using tanin bio-mordant and thyme as a colourant. Text Leather Rev 7:303–326

Gilroy CG (2023) The history of silk, cotton, linen, wool, and other fibrous substances: including observations on spinning, dyeing, and weaving. Good Press

Gundev GKMS (2023) Documentation of phulkari motifs for it's revival and sustenance. Doctoral dissertation, Maharaja Sayajirao University of Baroda (India)

Hai DH (2024) Messianism in civilizational history: the transformation of the Buddhist messiah via maitreya. J Daesoon Thought Religions East Asia 3(2):71–92

Hancock JF (2021) Spices, scents and silk: catalysts of world trade. CABI

Hassan R, Terzi S, Rosa P (2023) Silk industry value chain and circular economy. XXVIII Summer School "Francesco Turco"

Hazarika S, Saikia B, Saikia AR, Devi KB. Sericulture: an approach for sustainable livelihood of women. Adv Agric Environ Sustain 74

Henderson J, Morgan SL, Salonia M (2024) 1 Reimagining the silk roads. In: Reimagining the silk roads: interactions and perceptions across Eurasia

Hillman JE (2021) The digital silk road: China's quest to wire the world and win the future. Profile Books

Hodge SJ (2025) Queen Victoria's banarasi brocade: a transcultural approach to royal fashion. Fashion Theor 29(1):51–73
Huber M, Van Godtsenhoven K (2023) Women dressing women: a lineage of female fashion design. Metropolitan Museum of Art
Hunt T (2021) The radical potter: Josiah Wedgwood and the transformation of Britain. Penguin UK
Hussain F, Hussain Z, Khan MI, Imran A (2024) The digital rise and its economic implications for China through the digital silk road under the belt and road initiative. Asian J Comp Polit 9(2):238–253
Ibrahim SG, Kizito A (2021) Developmental historiography of the ancient Silk Road. Afr J Cult Hist Relig Tradit 4:69–80
Inglis D, Almila AM, Ho HK (2021) Is fashion universal? Towards a sociology of proto-fashion. In: Advances in sociology research. Nova Science Publishers, pp 1–46
Inpakdee C, Soodsang N (2025) The design and development of community textile products from banana fiber blends as commercial prototypes. Commun Soc Dev J 26(1):112–125
Islam TU (2024) The historical journey of sericulture: insights into sustainability from past to present. Int J Emerging Knowledge Stud 03(09):719–726
Javed T, Zhao X, Mahmoud AB, Fuxman L, Houjeir R (2024) Promoting social justice and women's rights: mitigating the negative impacts of large hydropower projects on tribal women in the Hindu Kush Himalaya region. Environ Dev 50:101001
Jayakody C, Malalgoda C, Amaratunga D, Haigh R, Liyanage C, Witt E, Hamza M, Fernando N (2022) Approaches to strengthen the social cohesion between displaced and host communities. Sustainability 14(6):3413
Jayaram H, Mahadevegowda LG, Boregowda MH (2024) Seri-entrepreneurship: current status and potential opportunities. Curr Agric Res J 12(1)
Kaplan RD (2025) The loom of time: between empire and anarchy, from the Mediterranean to China. Random House
Kasi E (2013) Role of women in sericulture and community development: a study from a south Indian village. SAGE Open 3(3):2158244013502984
Kelley AC (2024) (Re) using Byzantine textiles: adapting and reinventing material identities through the connected Mediterranean, seventh–twelfth centuries. In: Travelling matters across the Mediterranean
Khadse A, Rosset PM, Morales H, Ferguson BG (2018) Taking agroecology to scale: the zero budget natural farming peasant movement in Karnataka, India. J Peasant Stud 45(1):192–219
Khalil MH (2025) Holy homes, hollow houses: the evolution from sacred dwelling to unsettled domesticity. In: The international conference of utopian and sacred architecture. Springer, Cham, pp 69–88
Kharé V (2024) Questioning Ladakh: towards understanding a region and its people. Notion Press
Khuenkhaew Y, Phromlah W, Chueasakhoo C, Singhapat S (2025) Democracy in the Phutai ethnic group community in Kalasin province, Thailand. Sustainability 17(2):484
Kılınç N, Nasır E (2023) Clothing according to social classes in the Mughal empire: a study based on Akbarnama miniatures. Motif Akademi Halkbilimi Dergisi 16(42):864–880
Kiplagat J (2022) Performance of *Bombyx mori* (L) and *Samia cynthia* ricini silkworms under controlled environmental conditions in Uasin Gishu county. Doctoral Dissertation, University Of Eldoret
Krishnamurthy S (2023) Fashion killa: how hip-hop revolutionized high fashion. Simon and Schuster
Krotova T (2024) Feng Zhang. *АКТУАЛЬНІ ПИТАННЯ ГУМАНІТАРНИХ НАУК*, p 2024167
Kudabay AA, Mukanova GK, Yanchuk VA (2021) The silk road and business communications in the ancient east. Istorie/History 703:7
Kumar DS (2017) Income and employment generation through sericulture in Dharamjaigarh block, Chhattisgarh, India. Int J Acad Res Dev 2(6):1149–1155
Lado MJ (2025) The echo of the past: information networks from stone to silicon and beyond AI. Amazon Digital Services LLC-Kdp

Le THT, Singyabuth S, Nguyen TTT, Tran TTT (2025a) Quang Nam silk, Vietnam: material culture and storytelling in the process of cultural revitalization. Arts Educa 42
Li Y (2023) Networks of faith and profit: monks, merchants, and exchanges between China and Japan, 839–1403 CE, vol 13. Cambridge University Press
Liu Y, Lv Z, Wei J, Liu P, Pan M, Ma H, Lin T (2024) Comparative physiological, proteomic, and metabolomic insights into a promising low-pruning mulberry cultivar for silkworm rearing. Int J Mol Sci 25(24):13483
Lo J (2024) From cocoon to community: the role of silk in intercultural learning, research-creation and art education. Doctoral dissertation, Concordia University
Macha-Bizoumi N, Tranta A, Rigopoulou E (2025) Reflections on women's labour in museum narratives: the case of silk workers. J Textile Eng Fashion Technol 11(1):21–28
Madhu DM, Sinha DK, Singh KM, Kumari T, Singh RP, Kumari S, Ahmad N (2023) Achieving sustainable development goal by empowering women through mulberry silk production in Karnataka, India. Agric Assoc Text Chem Crit Rev J 10:1–6
Mala PH, Thrilekha D, Reddy H, Gulabrao DP, Shree D, Karan S (2024). The socioeconomic impact of sericulture on rural development. Int J Agric Extension Soc Dev 7(8):631–636
Malhotra SK, Mantri S, Gupta N, Bhandari R, Armah RN, Alhassan H, Young S, White H, Puskur R, Waddington HS, Masset E (2024) Value chain interventions for improving women's economic empowerment: a mixed-methods systematic review and meta-analysis. Campbell Syst Rev 20(3):e1428
Marimón AP (2025) Silk consumption and dressing practices in late-medieval Catalonia: the cases of Barcelona and Vic (1400–1460). Revista de Historia Economica J Iberian Latin Am Econ Hist 1–31
Mijares TC (2024) A comprehensive history of American law enforcement. Charles C Thomas Publisher
Mir-Makhamad B, Mirzaakhmedov S, Rahmonov H, Stark S, Omel'chenko A, Spengler RN (2021) Qarakhanids on the edge of the Bukhara oasis: archaeobotany of Medieval Paykend. Econ Bot 1–20
Montemezzo S (2024) Networks in the early history of capitalism: merchant practices in renaissance Venice. Taylor & Francis
Mott CD (2014) The formless empire: the evolution of indigenous Eurasian geopolitics. Doctoral dissertation, University of St Andrews
Moustakas L (2023) Social cohesion: definitions, causes and consequences. Encyclopedia 3(3):1028–1037
Mushtaq R, Qadiri B, Lone FA, Raja TA, Singh H, Ahmed P, Sharma R (2023) Role of sericulture in achieving sustainable development goals. Problemy Ekorozwoju 18(1)
Muthukumaran S (2023) The tropical turn: agricultural innovation in the ancient Middle East and the Mediterranean. Univ of California Press
Naderi Beni A, Lahijani H, Tofighian H, Morshedloo J (2024) Geohistorical and geoarchaeological evidence of seafaring in the Caspian Sea: lessons from the past to draw the future. J Island Coast Archaeol 1–33
Naik MKP, Bhardwaj P, Mishra V (2025) Post-COVID assessment of small business weavers in an Indian handloom industry: identifying and prioritizing key challenges. Res J Text Appar 29(1):40–60
Nayyar J (2024) Environmentally Benign & Banned Dyes. Green Fashion and Sustainability. pp 22–37
Ntaanu PK (2006) Development of silk cottage industry and production of traditional silk "Kente" cloth in Ghana. In: Black, Caspian seas and central Asia silk association (BACSA), p 115. www.bacsa-silk.org
Nyamdaa O (2023) Silk road and trade of the Mongol empire. Mongolian Diaspora J Mongolian Hist Culture 3(1):69–79
O'Neill N (2022) The political economy of taste: the state and the porcelain industry in France, 1682–1815. The University of Chicago

Ozkavruk Adanir E, İleri B, Can F, Ulasli B (2024) An ethical approach to sericulture: production of peace silk in Hatay/Turkey. Textile 22(1):20–30
Pagán EA, Salvatella MDMG, Pitarch MD, Muñoz AL, Toledo MDMM, Ruiz JM, Vitella M, Lo Cicero G, Rottensteiner F, Clermont D, Dorozynski M (2020) From silk to digital technologies: a gateway to new opportunities for creative industries, traditional crafts and designers. The SILKNOW Case. Sustainability 12(19):8279
Pathak S, Mukherjee S (2021) Entrepreneurial ecosystem and social entrepreneurship: case studies of community-based craft from Kutch, India. J Enterprising Commun People Places Glob Econ 15(3):350–374
Popescu A, Şerban V, Ciocan HN (2024) New trends in the global silk production in the period 2011–2022. Sci Pap Ser Manag Econ Eng Agric Rural Dev 24(1)
Popescu A (2020) Silk trade in the European union: trends in the period 2010–2019
Pradhan S, Samanta S (2025) Viability of community-based enterprises in community development: an empirical study. Commun Dev 1–26
Rahman M, Dip TM, Nur MG, Padhye R, Houshyar S (2024) Fabrication of silk fibroin-derived fibrous scaffold for biomedical frontiers. Macromol Mater Eng 309(5):2300422
Ramnath MA, Kim SJ (2024) The characteristics of contemporary Indian fashion designs using traditional handicraft-Focusing on Sabyasachi Mukherjee, Manish Malhotra, and Ritu Kumar. The Research Journal of the Costume Culture 32(2):299–320
Rani GS (2024) Women in sericulture. Discovery Publishing House
Rasuljonovich KR (2025) The development strategies ways of silk industry of the Republic of Uzbekistan. IMRAS 8(2):1–8
Rautio S (2024) Women's work as weavers of tradition. In: The invention of tradition in china: story of a village and a nation remade. Springer Nature Singapore, Singapore, pp 147–175
Rawat A, Verma S, Bhimte A, Narvade VU (eds) (2024) Frontiers in animal science: research and development volume II. Biotechnology 5:326–331
Rezić I (2022) Historical textiles and their characterization. Cambridge Scholars Publishing
Rong X (2022) The silk road and cultural exchanges between east and west, vol 14. Brill
Setiadi H, Ambiapuri E, Parwitasari NA, Wiyono D (2025) The impact of multicultural awareness, social practices, and gender equality on community solidarity through humanistic values in private universities. Ganaya: Jurnal Ilmu Sosial dan Humaniora 8(1):189–205
Sharma K, Kapoor B (2020) Sericulture as a profit-based industry—a review. Indian J Pure Appl Biosci 8(4):550–562
Sharma A, Gupta RK, Sharma P, Qadir J, Bandral RS, Bali K (2022) Technological innovations in sericulture. Int J Entomol Res 7(1):7–15
Sharma P (2024) Sustainable development and traditional industries: embracing epistemic syncretism in silk industry for success. In: Futuristic trends in social sciences volume 3 book, vol 12
Sharma P (2025) Climate change and the creative economy in India: disadvantaged, disturbed and dislocated Indian weavers. Taylor & Francis
Shea E (2021) The spread of gold thread production in the Mongol period: a study of gold textiles in the China national silk museum, Hangzhou. J Song-Yuan Stud 50(1):381–415
Siewsamdangdet P (2010) Sustainable development of silk industry in northeast Thailand. Charles Darwin University (Australia)
Sime D, Siraj Z (2020) Sericulture in Ethiopia: production status, opportunities, challenges and potential areas. A review. J Entomol Zool Stud 8(6):1–10
Singha R, Singha S (2020) Women's empowerment through the silk industry of Assam, India, and its underlying economy. J Int Women's Stud 21(5):9–31
Somsiriwattana U, Hongsiriwat A, Chandrachai A, Pungpapong V (2025) Identifying cultural capital and community capability factors related to the performance of Thai cultural textile products: a preliminary study. Rev Integr Bus Econ Res 14(2):298–316
Song KW (2024) Backyard-degradable interactive electronics. University of California, Berkeley

Spicer T (2023) Man-made: how the bias of the past is being built into the future. Simon and Schuster
Ssemugenze B, Esimu J, Nagasha J, Masiga CW (2021) Sericulture: agro-based industry for sustainable socio-economic development: a review. Int J Sci Res Publ 11(9):474–482
Sulolipu AA, Wahyono H, Soetjipto BE, Haryono A (2022) Silk weaving business sustainability as a cultural heritage of Indonesia: a case study in Wajo regency, South Sulawesi. Industria Textila 73(2):184–190
Suparna MG, Rinsey Antony VA (2016) Eco-friendly textiles. Int J Sci Technol Manag 5(11):67–73
Tarai S (2020) Entrepreneurial development in handloom sector through effective strategy formulation: an empirical study in Odisha, state of India. J Textile Eng Fashion Technol 6(4):133–139
Tsafack O, Mfewou A. China's presence and regional development in Cameroon
Ubaydullaeva S, Karimov N (2023) The role of Samarkand as a center of Islamic scholarship and cultural exchange in medieval Uzbekistan. SGS Eng Sci 2(02)
Venkataronappa MA, Kutala S, Madappa DC (2019) Assessing social and solidarity economy in India's tropical Tasar silk
Vijayalakshmy K, Biswal J, Rahman H, Chakraborty S (2023) Role of Indian women's self-help groups for the development of community resilience in the agricultural and allied sectors. Eur J Dev Stud 3(4):1–5
Vilar P (2025) A history of gold and money: 1450–1920. Verso Books
Wall B (2025) The historical waves of three dimensions of value: 13th to 19th centuries, and still impactful today. Available at SSRN 5129102
Wang XY, Chen B, Hou N (2024) Political relations and trade cooperation of the belt and road initiative: research on trade network based on strategic emerging industries. Int J Emerg Markets
Wang R, Suwanthada P (2025) Dushan Jade in Nanyang, China: literacy education and local wisdom from the perspective of intangible cultural heritage protection. Int J Educ Literacy Stud 13(1):129–137
Wang L, Tian G (2022) Breaking the containment: Horse trade between the Ming empire and its northern neighbors, 1368–1570. J World Hist 33(1):37–71
Wärnberg J (2023) City of echoes: a new history of Rome, its popes, and its people. Simon and Schuster
Wilson TM (2023) European culture and society, and culture and society in Europe: an introduction. In: Europe: an encyclopedia of culture and society (2 volumes)
Winnik A (2024) Dressing the non-elite in the Byzantine Empire. In: Byzantine dress: a guide. Routledge, pp 104–121
Winter T (2021) The geocultural heritage of the Silk Roads. Int J Herit Stud 27(7):700–719
Winter T (2022) The silk road: connecting histories and futures. Oxford University Press
Wu Z, Xu W, Wu X, Ding X (2025) Life cycle environmental and economic assessment of raw silk production in China. Sustain Prod Consumption
Wu Y, Zhang K (2023) From the silk road to the world: historical changes and influence of Guangdong embroidery's export embroideries. Przestrzeń Społeczna (Soc Space) 23(4):303–330
Xia Y (2025) Exploration of historical sites of art and culture exchange on the silk road of pre-Qin and Han dynasties. Culture Text 2(1):63–74
Xiangrong D (2022) Silk road and exchanges among different countries. In: The Routledge handbook of the belt and road. Routledge, pp 83–88
Yadav US, Tripathi R, Tripathi MA (2022) Global handicraft index: a pioneering approach and developing strategies for promotion completion and welfare of artisan in the digital world. Bank Policy 2(1):59–80
Yu J, Safarov B, Yi L, Buzrukova M, Janzakov B (2023) The adaptive evolution of cultural ecosystems along the silk road and cultural tourism heritage: a case study of 22 cultural sites on the Chinese section of the silk road world heritage. Sustainability 15(3):2465
Zheng X, Jia L (2017) An empirical study on the economic effect of financial cooperation among countries of the "silk road economic belt." Eng Econ 28(5):542–551

Zhirong Z (2024) A study of Shang dynasty aesthetic consciousness. Taylor & Francis
Zhong W, Wu W (2025) Vintage cameras: young people's nostalgic practices on social media and the pursuit of identity in contemporary China. J Youth Stud 28(2):374–390
Zhu L, Lin J, Pei L, Luo Y, Li D, Huang Z (2022) Recent advances in environmentally friendly and green degumming processes of silk for textile and non-textile applications. Polymers 14(4):659

MIX
Papier aus verantwortungsvollen Quellen
Paper from responsible sources
FSC® C105338

If you have any concerns about our products,
you can contact us on
**ProductSafety@springernature.com**

In case Publisher is established outside the EU,
the EU authorized representative is:
**Springer Nature Customer Service Center GmbH**
**Europaplatz 3, 69115 Heidelberg, Germany**

Printed by Libri Plureos GmbH
in Hamburg, Germany